I0887735

Collaborative Knowledge in Scientific Research Networks

Paolo Diviacco
Istituto Nazionale di Oceanografia e di Geofisica Sperimentale (OGS), Italy

Peter Fox
Rensselaer Polytechnic Institute, USA

Cyril Pshenichny
Geognosis Project, ITMO University, Russia

Adam Leadbetter
British Oceanographic Data Centre, NERC, UK

A volume in the Advances in Knowledge Acquisition, Transfer, and Management (AKATM) Book Series

Information Science REFERENCE
An Imprint of IGI Global

Managing Director: Lindsay Johnston
Acquisitions Editor: Kayla Wolfe
Production Editor: Christina Henning
Development Editor: Allison McGinniss
Typesetter: Kaitlyn Kulp
Cover Design: Jason Mull

Published in the United States of America by
Information Science Reference (an imprint of IGI Global)
701 E. Chocolate Avenue
Hershey PA, USA 17033
Tel: 717-533-8845
Fax: 717-533-8661
E-mail: cust@igi-global.com
Web site: http://www.igi-global.com

Copyright © 2015 by IGI Global. All rights reserved. No part of this publication may be reproduced, stored or distributed in any form or by any means, electronic or mechanical, including photocopying, without written permission from the publisher. Product or company names used in this set are for identification purposes only. Inclusion of the names of the products or companies does not indicate a claim of ownership by IGI Global of the trademark or registered trademark.

Library of Congress Cataloging-in-Publication Data

Collaborative knowledge in scientific research networks / Paolo Diviacco, Peter Fox, Cyril Pshenichy, and Adam Leadbetter, editors.
pages cm
Includes bibliographical references and index.
ISBN 978-1-4666-6567-5 (hardcover) -- ISBN 978-1-4666-6568-2 (ebook) -- ISBN 978-1-4666-6570-5 (print & perpetual access) 1. Knowledge, Theory of. 2. Research--Methodology. 3. Science--Methodology. I. Diviacco, Paolo.
Q175.32.K45C65 2015
001.4'2--dc23
2014029336

This book is published in the IGI Global book series Advances in Knowledge Acquisition, Transfer, and Management (AKATM) (ISSN: 2326-7607; eISSN: 2326-7615)

British Cataloguing in Publication Data
A Cataloguing in Publication record for this book is available from the British Library.

All work contributed to this book is new, previously-unpublished material. The views expressed in this book are those of the authors, but not necessarily of the publisher.

For electronic access to this publication, please contact: eresources@igi-global.com.

Advances in Knowledge Acquisition, Transfer, and Management (AKATM) Book Series

Murray E. Jennex
San Diego State University, USA

ISSN: 2326-7607
EISSN: 2326-7615

Mission

Organizations and businesses continue to utilize knowledge management practices in order to streamline processes and procedures. The emergence of web technologies has provided new methods of information usage and knowledge sharing.

The **Advances in Knowledge Acquisition, Transfer, and Management (AKATM) Book Series** brings together research on emerging technologies and their effect on information systems as well as the knowledge society.**AKATM** will provide researchers, students, practitioners, and industry leaders with research highlights surrounding the knowledge management discipline, including technology support issues and knowledge representation.

Coverage

- Cognitive Theories
- Cultural Impacts
- Information and Communication Systems
- Knowledge Acquisition and Transfer Processes
- Knowledge Management Strategy
- Knowledge Sharing
- Organizational Learning
- Organizational Memory
- Small and Medium Enterprises
- Virtual Communities

IGI Global is currently accepting manuscripts for publication within this series. To submit a proposal for a volume in this series, please contact our Acquisition Editors at Acquisitions@igi-global.com or visit: http://www.igi-global.com/publish/.

The Advances in Knowledge Acquisition, Transfer, and Management (AKATM) Book Series (ISSN 2326-7607) is published by IGI Global, 701 E. Chocolate Avenue, Hershey, PA 17033-1240, USA, www.igi-global.com. This series is composed of titles available for purchase individually; each title is edited to be contextually exclusive from any other title within the series. For pricing and ordering information please visit http://www.igi-global.com/book-series/advances-knowledge-acquisition-transfer-management/37159. Postmaster: Send all address changes to above address. Copyright © 2015 IGI Global. All rights, including translation in other languages reserved by the publisher. No part of this series may be reproduced or used in any form or by any means – graphics, electronic, or mechanical, including photocopying, recording, taping, or information and retrieval systems – without written permission from the publisher, except for non commercial, educational use, including classroom teaching purposes. The views expressed in this series are those of the authors, but not necessarily of IGI Global.

Titles in this Series

For a list of additional titles in this series, please visit: www.igi-global.com

Enhancing Qualitative and Mixed Methods Research with Technology
Shalin Hai-Jew (Kansas State University, USA)
Information Science Reference • copyright 2015 • 643pp • H/C (ISBN: 9781466664937) • US $225.00 (our price)

Knowledge Management Practice in Organizations The View from Inside
Ulla de Stricker (de Stricker Associates, Canada)
Information Science Reference • copyright 2014 • 318pp • H/C (ISBN: 9781466651869) • US $205.00 (our price)

Knowledge Discovery, Transfer, and Management in the Information Age
Murray E. Jennex (San Diego State University, USA)
Information Science Reference • copyright 2014 • 302pp • H/C (ISBN: 9781466647114) • US $175.00 (our price)

Harnessing Dynamic Knowledge Principles in the Technology-Driven World
Mark Nissen (Royal Oaks, USA)
Information Science Reference • copyright 2014 • 291pp • H/C (ISBN: 9781466647275) • US $175.00 (our price)

Emerging Pedagogies in the Networked Knowledge Society Practices Integrating Social Media and Globalization
Marohang Limbu (Michigan State University, USA) and Binod Gurung (New Mexico State University, USA)
Information Science Reference • copyright 2014 • 352pp • H/C (ISBN: 9781466647572) • US $175.00 (our price)

Knowledge Management and Competitive Advantage: Issues and Potential Solutions
Michael A. Chilton (Kansas State University, USA) and James M. Bloodgood (Kansas State University, USA)
Information Science Reference • copyright 2014 • 387pp • H/C (ISBN: 9781466646797) • US $175.00 (our price)

Ontology-Based Applications for Enterprise Systems and Knowledge Management
Mohammad Nazir Ahmad (Universiti Teknologi Malaysia, Malaysia) Robert M. Colomb (University of Queensland, Australia) and Mohd Syazwan Abdullah (Universiti Utara Malaysia, Malaysia)
Information Science Reference • copyright 2013 • 423pp • H/C (ISBN: 9781466619937) • US $175.00 (our price)

Knowledge Management and Drivers of Innovation in Services Industries
Patricia Ordóñez de Pablos (Universidad de Oviedo, Spain) and Miltiadis D. Lytras (The American College of Greece, Greece)
Information Science Reference • copyright 2012 • 349pp • H/C (ISBN: 9781466609488) • US $175.00 (our price)

IGI GLOBAL
DISSEMINATOR OF KNOWLEDGE
www.igi-global.com

701 E. Chocolate Ave., Hershey, PA 17033
Order online at www.igi-global.com or call 717-533-8845 x100
To place a standing order for titles released in this series, contact: cust@igi-global.com
Mon-Fri 8:00 am - 5:00 pm (est) or fax 24 hours a day 717-533-8661

This book is dedicated to Giovanna Rossa Colussi

Editorial Advisory Board

Robert Arko, *Lamont-Doherty Earth Observatory, USA*
Jon Blower, *University of Reading, UK*
Roberto Carniel, *Università di Udine, Italy*
Simon Cox, *Commonwealth Scientific and Industrial Research Organisation (CSIRO), Australia*
Mustapha Mokrane, *World Data System, Japan*
Andrei Prigov, *LUKOIL Overseas, UK*
Jordi Sorribas, *Unidad de Tecnología Marina (CSIC), Spain*

List of Reviewers

Roberto Biloslavo, *University of Primorska, Slovenia*
Boyan Brodaric, *Geological Survey of Canada, Canada*
Vasiy Bunakov, *Science and Technology Facilities Council, UK*
Gabriella Carrara, *GEOPHI S.R.L., Italy*
Cyndy Chandler, *Woods Hole Oceanographic Institute (WHOI), USA*
Andrew Curtis, *University of Edinburgh, UK*
Rogier De Lange, *Ghent University, Belgium*
Douglas Fils, *Consortium for Ocean Leadership, USA*
Panagiotis Fotaris, *University of West London, UK*
Janet Fredericks, *Woods Hole Oceanographic Institute (WHOI), USA*
Tatyana Gavrilova, *St. Petersburg University, Russia*
Helen Glaves, *British Geological Survey, UK*
Katjuša Gorela, *Regional Development Centre, Slovenia*
Marc Hedley, *Met Office, UK*
Christine Hine, *University of Surrey, UK*
Renzo Kotman, *Max Planck Institute for Marine Microbiology, Germany*
Sergey Kuznetsov, *Higher School of Economics, Russia*
John Laxton, *British Geological Survey, UK*
Victor Loudon, *British Geological Survey, UK*
Kit MacLeod, *The James Hutton Institute, UK*
Giuseppe Manzella, *ETT, Italy*
Stefano Nativi, *National Research Council of Italy (CNR), Italy*

Natasha Noy, *Stanford Center for Biomedical Informatics Research (BMIR), USA*
Carole Lynn Palmer, *University of Illinois – Urbana-Champaign, USA*
Giovanni Paronuzzi, *University of Trieste, Italy*
Monique Petitdidier, *Université Versailles St-Quentin, France & Sorbonne Universités, France*
Debbie Polson, *University of Edinburgh, UK*
Ruth Pordes, *Fermilab, USA*
Roger Proctor, *University of Tasmania, Australia*
Vyacheslav Rogalciuk, *St. Petersburg State Transport University, Russia*
Marc Schaming, *Institut de Physique du Globe (CNRS/UNISTRA), France*
Adam Shepherd, *Woods Hole Oceanographic Institute (WHOI), USA*
Kostantin Solomin, *National Research University of Information Technologies, Mechanics, and Optics (ITMO), Russia*
Milojevic Stasa, *Indiana University, USA*
Gerry Sutton, *Coastal and Marine Research Centre, Ireland*
Sarolta Takacs, *University of California – Irvine, USA*
Sytze Van Heteren, *Geological Survey of The Netherlands (TNO), The Netherlands*
Vera Van Laken, *Royal Belgian Institute of Natural Sciences, Belgium*
Valentina Volpi, *Istituto Nazionale di Oceanografia e di geofisica Sperimentale (OGS), Italy*
Dan Wright, *Centre for Ecology and Hydrology, UK*

Table of Contents

Section 2
Case Studies in Different Domains

Section 3
Formalization of Syntax and Semantics of Static Knowledge

Detailed Table of Contents

Ian McHarg's "The Theory of Creative Fitting" describes the collaborative integration of data into knowledge products for use in ecological architecture developments. In McHarg's system, the provenance of data layers and trust in their creators are vital to the creation of knowledge, ideas recently described as being "urgent and relevant." This chapter begins with these ideas and explores provenance propagation and data publication. Provenance propagation has been a topic under discussion by the World Wide Web Consortium, while data publication has been a focus of the global marine science community. From these projects, the issue of uniquely identifying individuals in a distributed, collaborative network has emerged. Possible implications of this problem are discussed, as are technical solutions to the problem. The principles described within this chapter are illustrated by examples of uncertainty propagation in chained Web Processing Service networks and data publication activities at the British Oceanographic Data Centre.

Philosophical theories of scientific progress are typically disconnected from citation data because a citation to a paper does not necessarily justify the content of the cited paper. Citation data can however be used to test whether scientific contributions coevolve and as such discriminate indirectly between the two main theories of scientific progress: cumulative and non-cumulative progress. This chapter presents this novel approach. First, agent-based models are used to discover essential differences between both patterns of progress. The systematic exploration they allow of their respective entailments reveals four conflicting empirical predictions. These could in principle be tested against citation data, thus operationalizing two important philosophical conceptions of progress. The proposed approach relies heavily on two recent developments, the use of agent-based modeling in philosophy and the availability of vast citation datasets. The research program it suggests offers a unique opportunity to bridge the gap between descriptions of science and explanations of why it is successful.

The inherent uncertainty in information about the Earth's subsurface requires experts to interpret and reach judgements about geological data based on their individual experience and expertise. This is particularly true for the geological storage of CO2 in subsurface saline aquifers where the fate of the injected CO2 needs to be predicted far into the future. In this chapter, linear modelling is used in a structured elicitation exercise to estimate the relative influence of individual experts within a group and to assess whether a group consensus reflects a genuine shared opinion or is biased towards or away from any dominant member or subgroup. The method is applied to a real expert evaluation of the carbon storage potential of a siliciclastic formation. This reveals herding behaviour amongst the experts, and levels of inter-expert influence that are undue given individual experts' levels of expertise, though neither phenomena was apparent during the meeting.

International collaboration and research funding in Sea Level Rise (SLR) research are investigated in this chapter. SLR can be taken as a paradigmatic research area to study the international scientific collaboration and research funding efforts because it is affecting the whole planet and is an interdisciplinary research area involving disciplines belonging to the geosciences but also the life sciences, technology sciences, and social sciences. The aim of the chapter is to identify the main stakeholders in the topic, institutions, and countries; analyze overlapping efforts; identify possible research gaps; and to study the role played by the funding agencies. Bibliometrics and a social network analysis approach are applied. Co-occurrence networks of keywords, affiliations, and funding agencies among scientific papers in Thomson Reuters' Web of Science Core Collection in the SLR topic are analyzed. Conclusions show that international scientific collaboration is common in SLR, but international co-financing is less frequent.

The professional development of a young researcher is mainly influenced by the organizational culture and the relationship with academic colleagues, especially mentors. This chapter presents the results of a qualitative study conducted on a sample of 16 researchers employed in three Slovenian public universities. The study examines the senior-junior researcher relationship, as the relationship between mentor and mentee, the nature of their cooperation, and how new knowledge is generated and transmitted. In-depth semi-structured interviews and content analysis for structuring qualitative data was used. The results indicate that the professional development of junior researchers is based mostly on their independent work where the mentor guidance is only occasional. Despite the development of information and communication technology, and in particular the Social Web tools, the mentorship in the framework of higher education institutions is still based on a face-to-face relationship.

Section 2
Case Studies in Different Domains

This section provides case studies that show how different communities tend to separate creating different and incommensurable cultures. If this is quite easy to understand, considering different research domains; it is also important to recognize this trend within the same research domain when different communities are identified by different schools, paradigms, or traditions.

The research enabled by national and international photon and neutron source facilities makes a key contribution to the modern scientific community supporting thousands of researchers around the world to explore and understand the structure of materials. In this chapter, the authors describe the characteristics of facilities science with a distinct common research lifecycle which requires the provision of facility-based centralized IT infrastructure and data management platforms. The chapter then considers how the nature of facilities science is changing and what opportunities this brings for a more cohesive approach to data management. It then goes on to consider investigation research objects and to formalize the aggregation of related and contextual information which is important for re-use of research outputs by those who were not directly involved in the original experiment. Finally, there is consideration of what infrastructure can support the use of investigation research objects.

This chapter discusses the uncertainties associated with sediment data, which are a commonly neglected obstacle in collaborative seabed-habitat mapping, complicating harmonization. Standardization of these data, although an important prerequisite in the harmonization of surrogate-based habitat maps, is not enough to achieve full map uniformity. Through time and between research institutes, data have been acquired, described, analyzed, and classified using a range of methods and protocols. The effects of differing error margins and indicative meanings remain hidden for most end users in environmental science and policy. The interpolation required to generate full-coverage data products from sedimentological point data creates additional uncertainty. Finally, small-scale habitat heterogeneity not captured by even the densest sampling places limits on the usability of sediment-based surrogates as well. Despite these limitations, there is much room for improvement when it comes to using surrogates in habitat mapping. Multiple visualizations generated from merged digital sediment datasets with flexible querying protocols are especially promising.

The possibility of having free and open access to data, information, books, and conference papers allows each person to improve his/her comprehension and personal beliefs, so that he/she can embark on a social process of interaction with other people and with shared scientific and technological knowledge. This chapter describes some essential components of a knowledge-building environment based on open and free access to data. The internet-based knowledge-building environment also allows access to books, articles, and any publications that transform data into knowledge. The primary objective of the proposed knowledge-building environment is to document, store, and share all data and methods, making them available for review by other scientists or men of learning, who can then verify and reproduce the results.

The chapter analyzes the theme of collaboration in scientific research through Web 2.0 IT tools and a constructivist approach to knowledge, learning, and e-learning. It analyzes, in particular, the case of Sulcis CCS Summer School. Organized by ENEA, Sotacarbo, and the University of Cagliari, the school provided information and materials on technological aspects of Carbon Capture and Storage (CCS). This work might become the model for a new approach to education and research thanks to collaborative learning using the LMS Moodle. In this case study, the construction of knowledge moves from a teacher-centered model to one based on a community of practice. This model is naturally a "prototype," an experiment in progress, that needs to be refined through further experiences.

Catchments are increasingly used to structure collaborative knowledge to advance understanding and management of terrestrial and freshwater natural resources. To produce collaborative knowledge of catchment systems requires greater levels of integrative research based on networks of researchers, policymakers, and practitioners. To enable increased levels of integrative research depends on development and application of digital technologies to improve collection, use, and sharing of data, and devising new knowledge infrastructures. Catchment observatories that integrate existing and novel physical, social, and digital networks of knowledge infrastructures are urgently needed. This chapter presents three leading international examples of collaborative networks of catchment researchers. In particular, the digital infrastructures they have developed to support collaborative knowledge in catchment research networks. This chapter provides future research directions required for the production, sharing, and use of collaborative knowledge in catchment research networks based on catchment systems science.

Section 3
Formalization of Syntax and Semantics of Static Knowledge

After providing theoretical contributions and case studies confronting communities and domains, what follows is related to practical proposals. We recognized that there can be essentially two approaches in this, namely: those that rely on the possibility that knowledge can be made explicit and therefore formalized and those that instead, considering the limitations imposed by tacit knowledge and by the social aspects of scientific research, try to propose solutions where knowledge is left implicit but where its management can anyway be supported by IT systems. If knowledge is eligible for formalization, two paths can be identified, namely: one based on static knowledge formalization, such as semantics, controlled vocabularies, and Web ontologies (this section), and one based on dynamic knowledge formalization (the following section).

This chapter focuses on improved access to marine science data, enabling researchers to generate new information and knowledge products. The history of controlled vocabulary developments in marine sciences, from paper publications to the Semantic Web, is explored in detail. This history is being furthered through the publication of Linked Open Data, meaning: the publication of clearly identifiable entities; a simple, universal mechanism for retrieving resources; a generic graph-based data model; and publishing explicit relationships to other resources. Progress towards Linked Open Data for marine science is reported in this chapter. As shown by the Data-Information-Knowledge ecosystem, the approach of "small pieces of data loosely joined" provides presentation and organisation to data, which creates information. The use of query endpoints to integrate this information from multiple locations into a knowledge base, which required active collaboration between cooperative partners to truly generate new knowledge and to address emerging science questions, is described.

In this chapter, a model demonstrating methods for integration of semantic technologies within observational data services is described. Implementation of the model captures knowledge about data provenance where it is best understood and also enables its persistence across architectural layers through the use of standards-based technologies. Domain experts can build upon the semantic layer to create meaningful ontologies. Brokering services can utilize the ontologies for automated mediation of terms and translation between standards-based technologies. Research communities will be enabled to operate within their own framework, utilizing their familiar, specialized terminology and tools. The role of communities of practice is explored relating to knowledge management across layered architectures. Implementation of semantic technologies within Web-based data and brokering services will minimize the operational barriers to data discovery and access and provide mechanisms that enable the formation of collaborative environments that will facilitate repeatable, well-documented research.

Collaborative research poses important challenges and opportunities to scientific communities, such as the realization of effective interactions between different groups and organizations. Data sharing is growing increasingly important, as repositories are growing in size, number, and variety in the different domains. The underlying challenge is interoperability: between data, services, applications, models, and ultimately people. Abstraction and standardization (e.g. of data and services) have historically played an essential role in enabling interoperability. However, they proved not to be sufficient alone to let large-scale collaborative research thrive. For different spheres (e.g. social/economic realms), intermediation approaches were successfully applied to pursue similar goals in the so-called brokering approach. This chapter argues that brokering is useful to enable collaborative research as well, both by addressing technological interoperability and supporting socio-organizational challenges. A technological brokering framework, implemented to help the earth system science collaborative research, is finally presented, along with success stories.

The chapter describes the research performed within the KOMET (Knowledge and cOntent structuring via METhods of collaborative ontology design) project, which was aimed at developing a new paradigm for knowledge structuring. By knowledge structure, the authors define the main domain concepts and relations between them in a form of graph, map, or diagram. The approach considers the specifics of individual cognitive style. Two stages of research have been completed: research into correlations between the expert's individual cognitive style and the peculiarities of expert's subject domain ontology development; and study of correlations between the expert's individual cognitive style and the group ontology design (including the design performed in groups consisting of experts either of similar or of different cognitive styles). The results of this work can be applied to organizing collaborative ontology design (especially for research and learning purposes), data structuring, and other group analytical work. Implications for practice are briefly delineated.

enhanced by the introduction of semantic rules for formulation of expressions in their places. In such form, they are quite apt to be used as a knowledge representation and visualization method in computer-based tools for support of collaborative research, and their quantification potential (i.e., marking of the net and attributing weights to its arcs) may appear very useful for this task, which is usually accomplished by purely qualitative tools.

Section 5
Knowledge Handling Avoiding Formalization

This section addresses how knowledge can be handled without formalization. As mentioned above, this need comes from the issues related to the difficulties in conceptualizing and communicating knowledge.

Scientific communities tend to speciate in tribes that evolve their culture so that after some time they do not have the same understanding of the same terms or concepts. Tacit knowledge, that is the knowledge that cannot be expressed or formalized, complicates this even more. Since we cannot fully know what a concept refers to, we would have limitations in developing IT systems to support collaborative and distributed cognition activities and knowledge networks trying to map knowledge between paradigms. To address this, the authors propose to switch to another perspective where knowledge is kept implicit and the referential communication function (iconic signification) is exploited instead. In this perspective, they develop a specific virtual research environment named COLLA-ANT, which, albeit being still a prototype and needing more use cases, has proved successful in addressing the above-mentioned issues.

Foreword

The conventional image of scientific research includes two contrasting memes: the dogged/brilliant individual researcher making their breakthroughs in isolation or in a small laboratory and the industrial-sized team working on big problems, which may or may not be relevant to the lives of the masses in the short term. The reality encompasses both of these extremes and many points in between. What is not in dispute is the proposition that scientific progress involves the generation of shared knowledge, systematically expressed, perhaps systematically gathered or generated. What definitely varies is the degree of direct collaboration involved and the way that the networks that support it are constructed.

For some areas of observationally centered research, scale is a determining factor. Ships, satellites, radio telescopes, and particle accelerators require a collaborative approach, at least to the management and funding of the initiatives that use them. But the emergence of ubiquitous communication channels has encouraged an expectation that researchers will collaborate more generally. My own engagement with national and particularly international collaborators was only possible by the emergence of email, the Web, wikis, and other collaborative online workspaces, which support construction of virtual teams that can interact daily or better. A significant impact of this toolkit might be expected for informatics-based projects, but perhaps more impressive examples of large-scale collaborative knowledge development using online tools are found in a variety of natural sciences, most notably biomedicine and biodiversity. Meanwhile, e-research has emerged as something of a discipline in its own right, with an ecosystem of funding calls and conferences in support.

What are the ingredients for successful collaboration networks, and how can we evaluate them? Dipping into my experience, the development of rich data transfer standards through the International Union of Geological Sciences (IUGS) Commission for Geoscience Information (CGI) was grounded in a strong existing consensus on the basic scientific model (historical geology, exemplified in geological maps with their stratigraphic legends) supported by very standardized institutional arrangements (government-backed geological surveys). Each of these is possibly unique in the scientific realm in terms of their longevity and near-ubiquity, so the project was more about *formalizing* than *developing* knowledge. But while the semantic sophistication of the results is clear from a cursory inspection of the elaborate models and encodings, their impact is less clear. The primary delivery portal "OneGeology" now boasts over 100 national members, but fewer than one-fifth are delivering actual data (as opposed to map images), and many of these are limited to what is essentially a cartographic, rather than scientific, product. In response, CGI is now stripping the models back to a more implementable core, meeting the community where it is, rather than where we hoped it was. Another large-scale collaborative knowledge initiative I participate in—technology standards for spatial data infrastructures—has struggled with mandate and conflicting technical demands of different parts of the community. The requirements of experts from the defence/intelligence sector and other industries dominated by large enterprises are largely alien to myriad developers of location-aware applications supported by GPS and Google maps. The latter have never heard of, or see little application for, "coordinate reference systems" or "interior boundaries," for example. In this domain, the European experiment of using legal instruments to enforce compliance with formal standards by public sector agencies is also looking distinctly shaky and has little impact on the mass market anyway, which is often where the most dynamic applications are found.

Now we are developing some experience with self-consciously collaborative knowledge projects, a more systematic and perhaps formal approach and assessment is needed. That is what this book addresses. There are numerous concerns. The issue of frameworks, paradigms, and biases is tackled in the first group of chapters, which considers the degree to which collaboration is reinforced in certain networks and undermined when network participants are talking past each other. Back to the geosciences, the incomprehension, even disdain, between geologists and geophysicists is a conventional trope (storytellers both, but with different weightings to means and end), and I recently encountered a disjuncture between a pair of sub-disciplines whose names are identical, except that the qualifier and noun are flipped (geo-hydrology and hydro-geology). The functioning of collaboration networks in terms of a workflow, and the utility of standardized tools and components emerges from a set of case studies presented in the second group. The addition of a provenance language to the gamut of Semantic Web vocabularies is notable here. The third group considers the formalization of syntax and semantics in a community, approached in a number of ways. Top-down ("the experts") vs. bottom up ("the people") is a common contrast. The optimal point of adoption of standards (which *should* merely formalize community practice) is unclear: while some initiatives attempt to gain efficiencies by pushing responsibility for conformance back to source, this is not always achievable, and intermediaries are likely to have a role. The fact that the consensus may be continuously evolving is tackled in the fourth group, which discusses various approaches to formalizing dynamic knowledge. Finally, a challenge is presented: perhaps knowledge is best expressed informally, thus accommodating conflicts without breaking the system.

This book does not present a uniform theory, but it highlights some key considerations through a fascinating range of case-studies, mostly drawn from natural sciences but leavened with a couple of excursions into the humanities. Sharing of data has an associated requirement for propagation of provenance and uncertainty information. Formalization needs to accommodate distribution requirements and also support change in the consensus. Above all, collaborative knowledge development needs to be sensitive to the issue of reconciliation of different points of view, paradigms, and values. As a scientist whose (second) career has been based on almost daily participation in collaboration networks, I have reflected on why it appeared to work, and then on whether it really had, and what the alternatives are (Cox, Simons, & Walker, 2011). This book presents some frameworks for understanding this better. Any scientist who is not in their own ivory tower and is inclined to think about method will find it a stimulating and worthwhile read.

Simon Cox
CSIRO Land and Water in Melbourne, Australia
24th June 2014

Simon Cox *trained as geophysicist, with a PhD from Columbia University. His work on informatics started in the early 1990s with GIS, moved quickly to the World Wide Web, and he became involved in metadata standards on the Dublin Core Advisory Council. Work on XML for mineral exploration data led on to the GeoSciML project and participation with the Open Geospatial Consortium, where he co-edited the Geography Markup Language standard. He developed Observations and Measurements as an OGC and ISO standard, which now forms the basis for operational systems in diverse fields including air-traffic, water data transfer, and environmental monitoring applications. He spent a year as a senior fellow at the EC Joint Research Centre in Italy working on integration of GEOSS and INSPIRE. He currently has leadership positions in the OGC, ISO/TC 211, and the Research Data Alliance. In 2006, he was awarded OGC's Gardels Medal, and he presented the 2013 Leptoukh Lecture for the American Geophysical Union. Simon is currently based in CSIRO Land and Water in Melbourne, working on a variety of projects across environmental informatics, linked data, and semantics.*

REFERENCES

Cox, S. J. D., Simons, B. A., & Walker, G. (2011). Adoption of earth science informatics standards: Key factors. *Geophysical Research Abstracts*, *13*, 5407.

Preface

Starting a scientific book, one should naturally make it clear for a reader from the very beginning what the book is about. The normal way to do this is to explain its title, word by word. This works fairly well in most cases but here defining each term independently would be rather misleading. Only the title as a whole conveys the full message of the editors and authors.

Knowledge can be associated with the acquisition and processing of information to retain an image or meaning that under certain conditions can be used, expressed, shared, and transmitted. Knowledge is generally associated with science, because, traditionally, knowledge is conceptualized as the ability to provide warranted assertions. The term "warranted" refers to the possibility that a value is provided (true or false) for an assertion. However, in the second half of last century, it became apparent that these categories are not easy to handle and that knowledge or understanding is not always reducible to language. As Michael Polanyi wrote, "we know more than we can tell" (Polanyi, 1966). Thus, while one can know something by oneself, sharing knowledge in a network can be different and provide rich opportunities for shared knowledge.

The word "network" has been used across many sectors at different times: from the 1940s in reference "to broadcast over a (radio) network"; around 1972 in reference to computers; then in the 1980s in reference to groups of people.

Joining the terms in the book title together, however, is harder to trace. While the term "knowledge network" seems to have appeared in the 1980s, anecdotal evidence suggests that the makings of complex networks and knowledge enterprises was injected into corporate organizational structures by Ken Olsen in 1957 (Amidon, 1991). Olson's desire to create an organization that was dynamic and a complex network, rather than managed and hierarchical, reflects some of the early research and directions of organizations from the 1950s.

While the role(s) and value of individuals in collaborative environments (and organizations in general) has been known for some time (McGregor & Cutcher-Gershenfeld, 2005), it is only relatively recently that a diverse range of individuals have themselves become major contributors to information (and even knowledge and data) in complex research networks. In scientific research, traditionally, the principal investigator and collaborators were primarily associated with the generation of knowledge artifacts. Large projects, such as those funded by the National Science Foundation (NSF) in the USA or the European Union (EU) commission in Europe instead foster the collaborative opportunities among institutes and across borders. Today, each stakeholder in a collaborative science network is also a repository of knowledge (via blogs, tweets, and social network postings, as well as more traditional exchanges via email) about her or his domain; albeit incoherent and difficult to find, let alone utilize, in a knowledge network.

Thus, today we see a "de-centering" trend that is absolutely necessary to account for in how knowledge is actually constructed and defined. Traditional models that focused on the cognitive activity of a subject or of a highly hierarchical structure thus have to be revised in order to fit complex, horizontally distributed, and mutually dependent network-based relations.

A fourth word from the title then appears: collaboration. Collaboration in its classical formulation refers to the interaction between humans towards a common aim.

Knowledge produced by a top-down structure can be instilled, which supposedly standardizes meaning. In contrast, knowledge defined collaboratively within a network will be highly conditioned by the set up of the network itself. This means essentially that knowledge is situated. It depends on the context in which it is produced.

In the case of scientific research, this of course contrasts sharply with its depiction as an unconditional process of problem solving where all the available evidence is taken into consideration. Every day we see technical achievements in various domains from computers to new medicines that seem to confirm this latter viewpoint. However, these achievements show rather the façade of scientific research than real scientific life. It is important, instead, to dispel myths and unveil the scientist's complex, and often contradictory, everyday experiences in order to help them in their everyday work in an age of mounting professional and resource pressures.

PHYSIOLOGY OF COLLABORATIVE KNOWLEDGE

To understand how this can be done, we need to unveil a first truth, which is that not all sciences work the same way. In many scientific fields, natural phenomena are based on complex interactions over a range of spatial and temporal scales that could be difficult to break down into elementary processes, as is done, for example, in physics and chemistry. Laws must be inferred from a relatively limited number of cases that generally vary so much in detail that it can be difficult to recognize analogies in manifestations of the same phenomenon (Engelhardt & Zimmermann 1982).

Popper (1963) noted that positive outcomes of experiments cannot confirm a scientific theory but that a single counter-example can demolish it. In physics and chemistry this makes sense, while, in other domains, such as history, medicine, or the geosciences, it is less relevant. What happens in these cases, instead, is that ideas, visions, and traditions tend to fight each other without a winner, and as a consequence, they proceed in parallel.

Choosing among an excessive number of possible causes for a phenomenon is a complex task for researchers, and they tend to focus on what they deem to be the more likely (or simpler) causes and exclude others. Following Woo (1999), conceptual (epistemic) and aleatoric uncertainties may be distinguished. Epistemic uncertainty is rooted in knowledge itself, while aleatoric uncertainty is rooted in the belief of knowledge. Every model has its own intrinsic approximation and associated uncertainty (epistemic), but at the same time, every researcher may trust each model to a greater or lesser degree (aleatoric uncertainty). This degree of trust is called the researcher's (cognitive) bias. Bias may seem to be a negative aspect of research, but it could be thought of as the ability of an expert physician to selectively attend to relevant information and to narrow the set of diagnostic possibilities (Patel, et al., 2004).

Considering the sense of our endeavour, we cannot stop here since researchers confront each other and grow their biases to form the relations among them. Researchers are individuals rooted in different traditions, schools, cultures, and some may say generations (Diviacco, 2012). Following Kuhn (1962), scientists live within paradigms, philosophical or theoretical frameworks, a tradition or school that conditions their way of thinking. Different, concurrent, and incommensurable paradigms exist within any discipline. Following Lakatos (1970), scholars adhering to some theories or paradigms, changing a set of auxiliary and peripheral hypotheses (the protective belt) protect their core from falsification and therefore leave essentially unaltered their tenets.

Paradigm anchoring and such protective belts drive scientists to gather in communities that, after Becher and Trowler (2001), can be thought of as "tribes." These can become so highly specialized to eventually be not able to understand each other. Different tribes can have different languages, myths, rites, territories, tools, or in other words, cultures. It is easy to understand that cultures can clash and prevent collaborative creation (let alone exchange) of knowledge.

Different cultures are reflected in different cognitive models and perceptions (Bruner & Postman, 1949; Warren, 1970) so that people with different backgrounds will shape reality differently. To this, Whitley (2000) adds the role of more "prosaic" circumstances, such as funding, how universities, faculties, and departments are organized, or the control over facilities. Following Latour and Woolgar (1979), "Science is a social construct."

We now explain the title and the message of the book: *to explore the conditions, the processes associated with, and the tools and solutions to handle scientific knowledge, where this latter is intended as a contingent artifact constructed collaboratively over the interaction of a network of individuals and cultures.*

E-RESEARCH

Considering the framing of this volume, scientists essentially need to build knowledge together with other researchers, using Web-based collaborative tools that can access online data held in distributed facilities.

Generally known as collaborative e-Research, this capability could ease the increasing burden of conducting scientific activities and thus boosts the results of their projects, since it would:

- Allow easier and faster access to data, information, and knowledge.
- Speed up the decision-making process.
- Improve the transparency of the decision-making process.
- Improve accessibility of the decision-making process to all participants in a project or programme.
- Allow for easier and greater reuse of knowledge and products of any scientific project.
- Reduce the costs, for example both financial and environmental traveling.

All this of course is very well understood by funding agencies who generally are positive on investing in collaborative e-Research.

While the aforementioned seems quite straightforward, the speed at which scientists actually adopt or adapt to such technologies is not as fast as expected (Wilbanks, 2009; Wolski & Richardson, 2010), which suggests that the available approaches or tools may not fully meet the complex needs of real scientific research. Often what is seen as a straightforward application in e-Research of certain technologies developed for other environments can instead negatively affect the productivity of researchers, and thus a scientists' participation within communities given their noted complex equilibriums.

Issues linked to knowledge-building itself can be very important, too. Different domains can have very different reasoning modes, traditions, and social habits. It is very likely then that no single (technical) solution can encompass all disciplines, fields, or even perhaps cases. At the same time, all of them should have in common the capability to compensate or bridge the gaps among researchers. This capability tension is another very important issue this book will focus on.

What kind of new approach(es) is (are) needed, then? Which new visions should they implement, and how can they be used to build concrete and re-usable knowledge?

TARGET AUDIENCE

This book addresses the issues mentioned above, extending from the conventional paths that tend to separate the domains of science, philosophy, culture, and technology, toward an integrated approach based on the awareness that all of them are increasingly intermingled. This is because human beings do not (and cannot) simply process data to get other data, while instead they live within their ways of thinking, history, and culture. We cannot therefore rely solely on standards but must also understand contexts, and in this sense, the book can be of interest for readers with different backgrounds. Such interests can span from theorists of collaborative science to researchers involved in semantics, from knowledge engineering to software and Web system developers.

TO FORMALIZE OR NOT TO FORMALIZE?

There are essentially two paradigms in knowledge handling: one which deems that knowledge can be formalized; while the second, less optimistically, invokes issues as tacit knowledge (the portion of knowledge that cannot be made explicit because it is obscured in socio-political phenomena or embedded in practices or even tools) or incommensurability to state that knowledge can be formalized only locally.

In the first paradigm, several approaches can be listed, and chapters describing collaborative controlled vocabulary and ontology development are included within this book. Controlled vocabularies were first developed by library and information sciences researchers and provide a mechanism for encapsulating knowledge at a given level of specificity. In their earliest form, controlled vocabularies were used to drive the classification of books in the study of bibliography. These studies have since grown in scope to aid communication in information systems and are a building block in the Semantic Web (2014) concept. The most well used and well respected controlled vocabularies for given domains are those that have been built up over many years by the collaboration of scientific experts in those domains to formalise the granularity of the knowledge being captured. In many cases, groupings of controlled vocabularies have been created in order to formalise the knowledge of a domain into different facets and at different granularities. As communities of researchers collaborate on defining the relationships between these controlled vocabularies, a bottom-up approach to ontology building emerges, where the formal definition of knowledge within a discipline or domain is described by ontology. This view of "ontology" within the information sciences is in stark contrast to the original definition of "ontology" as seen in metaphysics, in which the totality of what exists, or can be said to exist, is encapsulated and classified. The pragmatic view of ontology as seen in the properties and hierarchies emerging from controlled vocabularies is mirrored in the development of the ontology in information sciences, as distinct from the metaphysical ontology, in which the desire of philosophers and computer scientists to describe the world around them has merged and emerged. The second approach to domain ontology building is partially a top-down approach, in which a group of domain experts collaborate to abstract their domain into classes and attributes, which are then captured in the ontological representation. Chapters in Section 3 of this book explore both approaches to formalized knowledge building.

The ontologic vision of the world, at least in its computational sense, however, lacks one feature that becomes crucial in natural sciences, and often in humanities too: an ability to describe the knowledge about the scenarios of evolution of the world and handle situations when relations between the objects change (e.g., man is sleeping - man wakes up). This brings us away from the firm solid ground of the classical true-false way of reasoning that the ontology approach is based on, and makes us "sail in a stormy sea" of dynamic knowledge where the "winds and currents" (i.e., the logical or mathematical formalisms that govern the reasoning) are largely unknown. Still, some methods (continuing the "sea" metaphor), like navigation charts for near-coast voyages, already exist, and their potential is explored in Section 4 (Chapters 16-18) of the book.

The last section of the book proposes methods to handle knowledge that avoid its direct formalization.

ORGANIZATION OF THE BOOK

The organization of the book follows the above-drawn path. It has an introduction section that highlights theoretical issues, a second section oriented to list real examples where the issues highlighted in the first section can be traced. The following sections propose solutions: the third section on the basis of static knowledge formalization, the fourth section on the base of dynamic knowledge formalization, and the fifth section on knowledge handling without formalization. Since the book is on collaboration, we tried to foster as much as possible the collaborative attitude of authors providing, during writing, controlled access to the work of the others. In this, we were aiming at cross-fertilizing contributions and avoiding a mere gathering of separate visions.

Section 1: Introducing Collaborative Networks of Knowledge

Chapter 1 of the introductory section, titled "Reconciling Knowledge and Collaborative E-Research," by Paolo Diviacco, analyses the mechanisms involved in collaboration. The philosophical background of the chapter aims to define simple and useful models that can be used in developing tools to support such activities. The analysis starts from how the researcher understands what is around him, defining concepts such as objectivity, subjectivity, opinions, and facts, in order to overcome some of the myths of science. Different backgrounds, traditions, paradigms, cultures, and therefore, communities co-exist in the same domain. Each of these will have difficulties in understanding the others. To overcome these difficulties, it proposes exploring how knowledge is represented in order to understand cultures and create boundary objects (Star & Griesemer, 1989) to bridge them.

Chapter 2, "Examining Trust in Collaborative Research Networks," by Adam Leadbetter, explores issues such as the provenance of data and trust in the creators of those data in the light of two recent developments in the informatics community: provenance propagation and data publication. His analysis starts from the fact that modern informatics developments are built around use cases, which ask questions of the data and systems in order to produce information and knowledge. Use case driven development in the area of trust has led to research data outputs that are citable, capable of being integrated, or assimilated with other research outputs, and which carry identity or provenance.

Chapter 3, "Collaborative Progress in Citation Networks," by Roger De Langhe, presents a framework to combine descriptive data of how science works with normative insight into why it is so successful. Theories of scientific progress are either cumulative (where theories can only change if our knowledge about the world changes) or non-cumulative (where there is a feedback loop through which theory change affects the values which led to that change). This chapter shows how agent-based simulation can be used to improve our understanding of patterns in citation data and how they can be used to evaluate alternative theories of scientific progress.

Chapter 4, "Assessing Individual Influence on Group Decisions in Geological Carbon Capture and Storage Problems," by Debbie Polson and Andrew Curtis, discusses issues related to the relative influence of individual experts on group decisions and the potential for biased outcomes to occur where relative influence of each expert does not reflect their relative expertise. This chapter provides extensive analysis of the issue as it arises in geological (underground) storage of CO_2 in subsurface saline aquifers. Large uncertainties in subsurface geological information imply that expert judgment is essential for forecasting the long-term fate of stored CO_2. It is therefore important to apply these new techniques for conducting elicitation sessions and analysing their results, in order to minimise phenomena such as herding and undue inter-expert influence.

Chapter 5, "International Funding and Collaboration in Sea Level Rise Research," by Mayte López-Ferrer, studies the panorama of collaboration among universities, institutions, nations using bibliometrics and social network analysis. The study is focused on sea level rise research, which is a global issue and is studied all around the word and therefore can be taken as a paradigmatic research area to study the international scientific collaboration and research funding efforts.

After having explored many issues in knowledge building, Chapter 6, "Relationship between Senior and Junior Researcher: Challenges and Opportunities for Knowledge Creating and Sharing," by Katjuša Gorela and Roberto Biloslavo, concentrates on the transmission of knowledge. The study, exploiting in-depth semi-structured interviews and content analysis, examines the senior-junior researcher relationship as the relationship between mentor and mentee, and the nature of their cooperation. The study highlights that in knowledge transmission the same issues that can be found in other chapters are present here, such as the importance of tacit knowledge and personal relations.

Section 2: Case Studies in Different Domains

Chapter 7, "Towards the Interoperable Data Environment for Facilities Science," by Vasily Bunakov, Catherine Jones, and Brian Matthews, concentrates on the case of scientific collaboration in the field of photon and neutron sources. This domain is particularly interesting since large infrastructure is needed to underpin the research and to provide support for worldwide communities and project teams. In this domain, many skills and cultural backgrounds need to collaborate. To minimize the gaps, the authors propose to integrate various and domain-specific aspects of the life of researchers within an overall common full science lifecycle sequence, from the generation of data to the publication of scientific results.

Chapter 8, "Collaborative Seabed-Habitat Mapping: Uncertainty in Sediment Data as an Obstacle in Harmonization," by Sytze van Heteren and Vera Van Lancker, explores the uncertainty effects related to single parameters on separate research domains, in this case seabed-sediment and habitat mapping. Sediment data from visual observations and laboratory analyses of seabed samples do not only serve as data points for geological mapping but are also widely used surrogates of biological diversity and habitat heterogeneity so that any change in standardization in the former affects the latter. Standards reflect the needs of the communities adopting them and have evolved over time. Therefore, surrogate-based habitat mapping should consider carefully the changing standards adopted in seabed mapping. Using examples from the North Sea area, the chapter shows that such consideration by itself is only one aspect in the interpretation of surrogate-based data. Various sources of uncertainty associated with these kinds of data must also be addressed.

Chapter 9, "Knowledge Building and Computer Tools," by Giuseppe Manzella, draws a perspective that joins today's society and science in an effort to create open spaces for sharing data, information, and knowledge. The collaboration for society+science needs to be based on a common understanding of issues. This requires the adoption of, for example, data and metadata standards or controlled vocabularies and an infrastructure to support the requirements for a dialogue. The chapter proposes solutions and references on how this can be put in place.

Chapter 10, "Beyond Learning: E-Learning Models for Scientific Research," by Andrea Corleto and Veronica Tomassetti, reports on the experience of the Sulcis Carbon Capture and Storage (CCS) Summer School, a joint effort of the Italian National Agency for New Technologies Energy and Sustainable Economic Development (ENEA), the University of Cagliari, and the Sotacarbo company. The collaboration aims to create a permanent venue not only for research and training but also in which to inform local and national administrators and stakeholders to facilitate dialogue, increase confidence in these technologies, and thus contribute to public acceptance of a CCS farm planned in the area of Sulcis (Sardinia). The approach used is that of social constructionism, which emphasizes the active role of the student in the process of knowledge construction, and therefore lets one understand fundamental mechanisms of cooperation such as the negotiation of meaning and the creation of artifacts like symbols or rules. Technologies (mostly Web-based and social media related) are employed and reactions to them by users and communities of practice are detailed.

Chapter 11, "Collaborative Knowledge in Catchment Research Networks: Integrative Research Requirements for Catchment Systems Science," by Kit Macleod is another example of the analysis of what is required to gather different disciplines in a collaborative environment. The chapter focuses on the development of catchment system science with an analysis of three case histories of digital catchment infrastructures. This chapter highlights that greater levels of integrative research requires networks of researchers, policymakers, and practitioners, but also the enabling physical, social, and digital knowledge infrastructures.

Section 3: Formalization of Syntax and Semantics of Static Knowledge

Chapter 12, "Loose Integration of Local Information to Generate Collaborative Marine Science Knowledge," by Adam Leadbetter, Robert Arko, Cynthia Chandler, Adam Shepherd, and Roy Lowry, focuses on the formalization of knowledge in controlled and shared vocabularies, with a specific focus on the marine science domain. From their traditional use for filling metadata fields, controlled vocabularies have evolved to represent concepts and their relations through their publication as Linked Open Data (LOD, 2014). One component of this publication is that they are made available on the Web as Uniform Resource Identifiers with a simple, standard mechanism for retrieval (de-referencing), and their content encoding uses the Resource Description Framework (RDF, 2014). This allows the vocabularies to be integrated from multiple locations to create a knowledge base. The formed knowledge base required active collaboration between cooperative partners to truly generate new knowledge, answer data management questions, and address emerging science questions.

Chapter 13, "Persistence of Knowledge across Layered Architectures," by Janet Fredericks, considers the integration of Semantic Web technologies and observational data services. The capture of provenance close to the source of measurement and processing steps is shown to avoid errors and omissions in the collection process. The chapter highlights the role of communities of practice in defining how knowledge can be captured and translated into controlled and meaningful vocabularies and ontologies, minimizing the operational barriers to data discovery and highlighting the use of brokering technologies. Enabling data quality assessment through communication of provenance is shown to foster engagement in collaborative research and build trust in global data resources.

Chapter 14, "The Brokering Approach for Enabling Collaborative Scientific Research," by Enrico Boldrini, Max Craglia, Paolo Mazzetti, and Stefano Nativi, focuses on the challenges of interoperability between data, services, applications, models, and ultimately, people and organizations contributing to the scientific research. The chapter highlights how standardization alone is not sufficient for large-scale collaborative research and applications to thrive in a sustainable or effective way. The proposed solution is based on the brokering approach, where brokers act as mediating entities standing between the heterogeneous parties. Examples of brokers from different domains are provided in order to present the brokering approach applied to collaborative research, along with real use cases. In this context, brokers serve as mediators between the diverse research groups, organizations, and systems they connect, easing interoperability. Sociological and organizational implications of brokering are explored as well.

Chapter 15, "Building Collaborative Ontologies: A Human Factors Approach," by Tatiana Gavrilova and Irina Leshcheva, focuses on the comparison of individual and group domain ontologies and seeks an understanding of the causes of differentiations between the form and the content of individual ontologies. Studies were conducted in two scientific domains: medicine and computer science. The outcomes were that information on the cognitive personal peculiarities helps to capture the subjectivity of the individually designed ontologies in collaborative ontology development.

Section 4: Formalization of Syntax and Semantics of Dynamic Knowledge

Chapter 16, "Grammar of Dynamic Knowledge for Collaborative Knowledge Engineering and Representation," by Cyril Pshenichny and Dmitry Mouromtsev, introduces a group of chapters dedicated to dynamic knowledge representation where changing environments indicate that the relations between objects cannot be considered fixed. The chapter highlights that very few formal tools are available today to represent such situations. It presents an attempt to formulate a unified grammar to encode the knowledge of changing environments in any field of science. It thus gives a grounding both to existing and new methods of engineering of dynamic knowledge and simultaneously serves as a bridge from these methods to logical and mathematical formalisms.

Chapter 17, "Constructing Historical Knowledge through Graphic Boundary Objects," by Kostantin Solomin and Maria Mandrik, puts in practice what is introduced in Chapter 16. One particular method of dynamic knowledge engineering, the event bush, is for the first time applied in the field of history. History is a good example of a domain characterized by abductive reasoning, so that the use of a dynamic knowledge representation method is very efficient. The authors make use of a practical example considering the period from February to November 1917, analyzing the possible choices and motivations of the Russian army in relation to the decrowning of Russia's last Emperor, Nikolai (Nicholas) the 2nd.

Chapter 18, "Perspectives of Use of Petri Nets in Collaborative Research," by Vyacheslav Rogalchuk and Konstantin Solomin, concentrates on the use of Petri nets as a way of formalizing knowledge in dynamic systems, while at the same time considering collaborative research scenarios.

Section 5: Knowledge Handling Avoiding Formalization

Chapter 19, "Maps, Graphs, and Annotations as Boundary Objects in Knowledge Networks, Distributed Cognition, and Collaborative E-Research," by Paolo Diviacco and Alessandro Busato, closes the path of the book by exploring how knowledge can be represented without formalization. Their proposal is to keep knowledge implicit and exploit instead the referential communication function and the iconic signification mode. This vision has been translated into a real Web-based collaborative environment (COLL-ANT), which is described in detail in the chapter. Such an environment seeks to avoid the clash of different and contrasting cognitive modes.

CONCLUSION

Collectively, the sections and chapters in this book aim at triggering a new approach to issues of collaborative knowledge development and sharing in scientific research that, we claim, should stem from an integrated vision spanning the spectrum from theoretical to technical issues.

We think that this integrated approach is very important since the exciting technological opportunities currently offered entail also the risk that if these developments are driven by technology only the real needs of scientific research will be overlooked. It is, therefore, very important to consider first the science needs and only then devise the technology that could support them (Beyer & Holtzblatt, 1998).

The chapters gathered in the book address themes such as the philosophy of science, sociology, collaborative research, shared knowledge, and learning. Specific cases of scientific collaboration within and across scientific domains, mostly in the geo-sciences but also in the humanities, have been analyzed to understand how different communities can overcome difficulties in communication, sharing concepts and terms, and in ways of thinking. Building on these findings, paths towards possible solutions have been proposed that we estimate can contribute and impact very much the development of new systems for the managing of collaborative knowledge development in scientific research networks.

Paolo Diviacco
Istituto Nazionale di Oceanografia e di Geofisica Sperimentale (OGS), Italy

Peter Fox
Rensselaer Polytechnic Institute, USA

Cyril Pshenicny
Geognosis Project, ITMO University, Russia

Adam Leadbetter
British Oceanographic Data Centre, NERC, UK

REFERENCES

Amidon, D. M. (1991). *Memo to Ken Olson*. Retrieved from http://www.entovation.com/gkp/origins.htm

Becher, T., & Trowler, P. R. (2001). *Academic tribes and territories intellectual enquiry and culture of disciplines*. Buckingham, UK: The Society for Research into Higher Education and Open University Press.

Beyer, H., & Holtzblatt, K. (1998). *Contextual design: Defining customer centered systems*. New York, NY: Academic Press.

Bruner, J. S., & Postman, L. (1949). On the perception of incongruity: A paradigm. *Journal of Personality*, *18*(2), 206–223. doi:10.1111/j.1467-6494.1949.tb01241.x

Diviacco, P. (2012). Addressing conflicting cognitive models in collaborative e-research: A case study in exploration geophysic. In *Collaborative and distributed e-research: Innovations in technologies, strategies and applications*. Hershey, PA: IGI Global Press; doi:10.4018/978-1-4666-0125-3.ch012

Engelhardt, W., & Zimmerman, J. (1982). *Theory of earth science*. Cambridge, UK: Cambridge University Press.

Kuhn, T. S. (1962). *The structure of scientific revolutions*. Chicago, IL: University of Chicago Press.

Lakatos, I. (1970). Falsification and the methodology of scientific research programmes. In *Criticism and the growth of knowledge*. Cambridge, UK: Cambridge University Press. doi:10.1017/CBO9781139171434.009

Latour, B., & Woolgar, S. (1979). *Laboratory life: The construction of scientific facts*. Princeton, NJ: Princeton University Press.

LOD. (2014). Retrieved from http://linkeddata.org

McGregor, D., & Cutcher-Gershenfeld, J. (2005). *The human side of enterprise*. McGraw-Hill (first published in 1960).

Patel, V. L., Arocha, J. F., & Zhang, J. (2004). Thinking and reasoning in medicine. In *Cambridge handbook of thinking and reasoning*. Cambridge, UK: Cambridge University Press.

Polanyi, M. (1966). *The tacit dimension*. New York, NY: Anchor Day Books.

Popper, K. (1963). *Conjectures and refutations*. London, UK: Routledge.

RDF. (2014). Retrieved from http://www.w3.org/RDF/

Semantic Web. (2014). Retrieved from http://www.w3.org/standards/semanticWeb/

Star, S. L., & Griesemer, J. R. (1989). Institutional ecology, translations and boundary objects: Amateurs and professionals in Berkeley s museum of vertebrate zoology. *Social Studies of Science*, *19*(4), 387–420. doi:10.1177/030631289019003001

Warren, R. (1970). Perceptual restoration of missing speech sounds. *Science*, *167*(3917), 392–393. doi:10.1126/science.167.3917.392

Whitley, R. (2000). *The intellectual and social organization of the sciences*. Oxford, UK: Clarendon Press.

Wilbanks, J. (2009). I have seen the paradigm shift, and it is us. In T. Hey, S. Tansley, & K. M. Tolle (Eds.), *The fourth paradigm* (pp. 209–214). Microsoft Research.

Wolski, M. P., & Richardson, J. P. (2010). Moving researchers across the eresearch chasm. *Ariadne*, (65): 1–10. Retrieved from http://www.ariadne.ac.uk/issue65/wolski-richardson/

Woo, G. (1999). *The mathematics of natural catastrophes*. London, UK: Imperial College Press; doi:10.1142/9781860943867

Section 1
Introducing Collaborative Networks of Knowledge

This section delves into fundamental issues related to how knowledge is built when scientists interact with each other, how reasoning can be biased, and how social issues can condition ideas and their transmission.

Chapter 1
Reconciling Knowledge and Collaborative E-Research

Paolo Diviacco
Istituto Nazionale di Oceanografia e di Geofisica Sperimentale (OGS), Italy

ABSTRACT

The aim of this chapter is to analyze the mechanisms and the philosophical background behind collaborative e-research in the perspective of defining simple and useful models that can be used in developing Web-based tools to support such activities. In its simplest form, collaboration refers to the interaction between humans. This is problematic, as the interaction between the subject and the outer world is always uncertain. To address this issue, a framework is proposed here, where metaphysics and epistemology are confronted, clearing up concepts as objectivity, subjectivity, opinions, or facts. The limits of "knowing," of being aware of the knowledge we have, and of the possible transmission of knowledge are highlighted through comparison of science and other domains. A new vision becomes urgent and can be addressed through a "deliteralization" of knowledge grounded on "representation" since this is the projection of knowledge and the source of cultures.

INTRODUCTION

Collaboration is about working with someone else. It is not an easy task to perform since issues can rise from misunderstanding, distance, lack of awareness and many other factors. Different scientific domains can have different needs. Motivations and social relations between partners can be different as well. E-research is a horizon we are aiming to, where collaboration takes place on the web. This adds a further layer of complications, since while linking partners in an easier way that before, at the same time, it bias the traditional way in which people are used to work together. Several examples of implementation of virtual research environments are already available and some have been proposed by us (Diviacco, & Busato, 2015 see chapter 19 this book; Diviacco, 2012). However, we claim that a theoretical approach to e-research is still missing, that could somehow fix the horizon and let us explore it. This chapter (together with Diviacco 2015, see chapter 19 this book) is intended to propose some keys to achieve this aim.

In this we would like to start from the very simple point that if collaboration is about work-

DOI: 10.4018/978-1-4666-6567-5.ch001

Copyright © 2015, IGI Global. Copying or distributing in print or electronic forms without written permission of IGI Global is prohibited.

ing with someone else, one of the first things to consider in the exploration of this domain is to analyze the subject as the necessary actor of collaboration. Once the subject is posed the focus can move to the outer word until it reaches other actors. In this, a huge amount of problems immediately arise and a wide literature advises us on the dangers of this situation. When the subject tries to understand the outer world most of the philosophers in the history of mankind warns us about misunderstanding, incommensurability, bias, subjectivity, up to psychosis. Most of the philosophical traditions are rather pessimistic on the possibility to access the outer world.

Of course here we will not dare to propose solutions for such overwhelming issues, we will only make use of some ideas developed by "giant" thinkers in order to compare domains that are generally considered different and too far from each other, such as Science and the Arts. We claim that this distance is absolutely not so wide as most of the people think, and that many commonalities exist instead. What follows, in fact, will demonstrate that many regularities can be found in the two domains and that this can help us, building common models and paths, in the task of devising tools to support collaborative work and knowledge networks in particular for e-research.

BACKGROUND

The Dichotomy Subjective/Objective

In the path from the subject towards the outside, rises the dichotomy of the subjective/objective, which is one of the hinge around which Science is normally identified with the Truth and other forms of Culture are not.

The issue of subjectivity is such an important matter in recent culture and society, it needs to be explored as much as possible. One of the first things to do in this sense is to try to establish properly what do we mean exactly with the terms subjectivity and objectivity. This is very important since Science is founded on the pretension to be essentially objective.

A first misconception is that subjectivity and objectivity are antinomic concepts, that the term subjective is the complete opposite of the term objective. If something is subjective it is not objective and vice-versa. The term subjective reasonably means "from someone's point of view", while generally the term objective is understood as something related to real facts or objects. This of course is a gross oversimplification.

The term objective can be explained only as "not just from someone's point of view", then in a sense objectivity is a vague concept that demands judgments to be done by a group of sane, rational and sufficiently informed people.

Subjective is private, while Objective is public. It goes without saying that being public does not mean true, while associating objectivity with the point of view of a community, instead, opens a wide range of issues linked to social building of the content of what referring to.

To understand better these issues, it can be interesting to refer to two contrasting visions that marked contemporary philosophical debate, identifying two philosophical communities: the "analytical" and the "continental" ones, a distinction drawn also geographically between mainly Anglophone (Great Britain and United States) thinkers, in the first case, and those residing in the rest of Europe (mainly Germany and France) in the second.

The continental perspective is rooted in the *dualistic* tradition that starting from Descartes separates the mind from an external world. If one can only be certain of the existence of oneself and one's mental content, then the existence of the external word and other minds is unjustified. Descartes relied on God to ensure knowledge of external states of affairs. Those who do not believe can find this insufficient.

Idealism is the viewpoint that precribes that the physical world in some sense is a product of the mind. There are various flavors of idealism. The vision that material things owe their existence to mind is called Metaphysical idealism.

Kant's transcendental idealism, Berkeley's subjective idealism are forms of Metaphysical Idealism. Nietzsche's belief that most of what we consider to be true is entirely unjustified (a myth), the idea to start from subjective experience and the suspension of the common sense realist beliefs as in the Phenomenology, are all in this tradition too, and lead to the skeptical vision of Postmodernists such as Deridda or Foucault, which spans not only philosophy but culture in general, from Architecture to Music to Economics.

On the other side, analytical philosophers such as Moore or Russell reject idealism. John Searle (1995) maintains that idealism is "ridiculous" since no aspect of life, no utterance can be intelligible without the presupposition that there is a real word existing independently of us (external realism).

For the sake of completeness, other philosophical traditions grew in between the Analytical and Continental ones, such as for example Instrumentalism that remained within the scope of anti-realism, and where the emphasis was put on the predictive success of models rather than on their truthful and accurate descriptions, or Pragmatism, that in a tradition from C.S. Peirce to H. Putnam acknowledged an external world which must be dealt with, while accounting for the limitations in accessing it and for the fallibilism of Science in general. However, notwithstanding the trials to bridge the gap between realism and idealism these remained separated.

Accessing What Is Outside

What kind of access do the subject have to the real word?

If we follow continental philosophy we do not go very far from mind. Lets see what is the proposal of analytical thinkers as Searle, Audi or Foss. These add to the dichotomy between subjectivity and objectivity another dichotomy: that between Metaphysics (ontology) and epistemology. Metaphysics is the branch of philosophy concerned with the basic causes and nature of things, while epistemology is the study of nature and grounds of knowledge with reference to its limits and validity. Epistemology is the study of knowledge and justified belief. Epistemology is about what makes a statement true, bearing in mind that the words true and false can be applied only to statements.

The two dichotomies (Subjectivity/Objectivity and Metaphysics/Epistemology) are orthogonal, they do not map on each other, so that four combinations are possible:

- **Metaphysical Objectivity:** Refers to things that do NOT depend on experience for their existence. For example Australia exists even if I never have been there. With some poetry this is often referred to the sound of trees that keep falling in the forest even if I am not aware of them.
- **Metaphysical Subjectivity:** Refers to things that need to be experienced for their existence. For example the taste of an apple. Of course the chemical composition of the apple juice is metaphysically objective, but the brain related process that builds the taste of the apple in my mind is subjective. In fact, an alien (having, very likely, different brain processes than mine) tasting the same apple will experience a different taste.
- **Epistemic Subjectivity:** Refers to claims or judgments which truth value is determined privately, such as for example "Bach is better that Mozart". It is an opinion, a taste and no one can debate on that: "de gustibus non est disputandum" said the Romans.
- **Epistemic Objectivity:** Refers to non-contradictory claims or judgments which truth values are not determined only privately,

they can be determined inter-subjectively by general agreed methods or procedures. For example the statement "Mount Everest is 100 cm tall" is epistemological objective since the value of measurement can be reasonably determined by a measuring device, and people can agree on the procedure and vocabulary to be used. Applying these methods, it should be possible to realize if the statement is true or false. Of course this depends on the premise of having had a convergence of a community.

The two dichotomic axes define a sort of "Cartesian coordinate system" that we can use in navigating through various aspects of our endeavor. This allows us to understand the position of the various aspects of culture we want to analyze so that we can understand the overlap between them and Science.

Facts and Opinions

An interesting case where this approach can be used is to analyze the distinction between facts and opinions.

Facts are by definition metaphysically objective data. For example a fact can be whether this morning you were at the office or conversely if you were ill at home. An issue is an opinion if it is determined epistemically subjectively, for example, again, whether Bach can be considered better than Mozart.

Common sense confronts facts and opinions, meaning that they are (antonymic) extremes of the same path that connects what is allegedly true or false. This vision is widespread at all levels, up to scholars that shelter behind sentences like "mine are facts, yours are opinions" (heard personally many times at congresses).

From an analytic philosopher's point of view facts and opinions are different, because, referring to the above schema, they pertain to different areas. They do not overlap and theoretically, from this point of view, there should be means to understand if an issue is a fact or an opinion.

However, this is not easy to put in practice and the fact that people can disagree both on opinions and matters of facts or conversely that an issue cannot be automatically considered an opinion because people cannot agree on a shared position, can be a demonstration that this approach has severe flaws.

Limiting the point of view to Epistemological processes, on the other hand, facts and opinions cannot be distinguished, since in this area there is no access to metaphysical objective data.

If it is not possible to understand if something is a fact because it is not possible to establish its metaphysically objective condition, in opposition to opinions, we should not talk of facts rather of states of affairs. States of affairs are merely possible situations that can be expressed by any non-contradictory true or false sentence.

Epistemological Objectivity

While epistemological subjectivity is a rather clear and closed issue (non est disputandum), instead, more can be said about epistemological objectivity.

Contemporary philosophy, and philosophy of science in particular, followed essentially three path to try to prove the possibility of an objective epistemology an namely: (1) consensus theories (2) indexical theories (3) correspondence theories.

1. Consensus theories (Gauker 1995) define the epistemic objectivity of a judgement in terms of agreement, in terms of likelihood that it would be agreed upon by rational and sufficiently informed people. This is the approach used above in defining epistemical objectivity as a form of public convergence.
2. Indexical theories (Lycan 1996) rely on the possibility to avoid indexicalities in representations. Indexicalitites refer directly to

the circumstances or context in which an utterance takes place. An indexical expression (such as: that, here, and you) is a word or phrase that is associated with different meanings (or referents) on different occasions, for example the sentence "I am here" can be used by different person at the same time but depending on the context (which of these person) will refer to a different location. Indexical theories postulate that if representations can be devised that do not make use of indexicalities, these will be objective. This is essentially a linguistic approach that holds in linguistic cases.

3. Correspondence theories (Rorty 1989) try a different approach bridging the gap between Epistemology and Metaphysics. The central idea of this class of theories is that epistemically objective beliefs of sentences must in some sense relate to something metaphysically objective. If the process of representation is refused the problems of epistemology are dissolved[1]. Quoting Rorty (1989, p.4):

The word is out there, but descriptions of the word are not.

If epistemology is based on judgments of representations:

Only descriptions of the world can be true or false.

Then epistemology in itself does not have much sense since:

The world on its own unaided by the describing activities of humans cannot. [Be true or false, it is as it is]

In Rorty's vision, problems of Epistemology are dissolved simply because they do not exist. In practice his approach tries to neglect completely Epistemology and focus on Ontology only, which, of course, has been a very controversial point of view.

Summing up, even confronting the analytical and continental points of view, if we want to address the problem of objectivity and subjectivity we need to separate metaphysics and epistemology and handle them separately. The problems of metaphysics, the existence itself of a world outside, is of course relevant, but the limited access to reality makes us move the attention towards Epistemology, which becomes the only possible path to follow.

MAIN FOCUS OF THE CHAPTER

Knowledge

Defining Knowledge is very difficult. The term can have a multifaceted life and strongly depends on the context in which it is used. To overcome these limitations, we can say that it is associated with the acquisition and processing of information to retain an image or meaning that under certain conditions can be used, expressed and transmitted.

This model does not limit Knowledge to any specific domain.

Generally Knowledge is associated with Science, because, traditionally, knowledge is conceptualized as the ability to provide warranted assertions. The term "warranted" refers to the possibility to provide a value (true or false) for an assertion. However it became largely known in the second half of last century that these categories are not easy to handle and that knowledge or understanding is not always reducible to language. As Michael Polanyi said, "we know more than we can tell" (Polanyi, 1966/1983 p.4).

Knowledge, therefore, comes in different forms and through different paths, so that the pressure to systematize Knowledge using language only is only one of the possible means to address it.

The "deliteralization" of Knowledge opens the door for inclusion of multiple forms of human expression that can be gathered around the concept of Culture. Here Science and Technology are only a part of a whole picture that is composed also of values (ideologies), social structures and conventions (norms and taboos), Religion and aesthetics (Art, Music, Literature, Fashion, Architecture, Design).

How information is acquired, processed, retained and transmitted in all these cases is important to understand similarities among all these domains and in particular that under analysis in this chapter which is scientific research. Of course there is not enough space here to analyze all domains. We will try to highlight some cases where the mechanism involved can shed a light on our target field.

Knowledge can be analyzed also in an historical and socio-political perspective, which contrasts the axiom that knowledge can be only true or false.

The idea of Progress is a trend towards better quality of life, economic and cultural development. Criticisms are known and mythologies too.

It is possible to identify two constituents of the path toward social progress: one which pass through matters of political debates, the other which seems to be more intimately intermingled with an ethical-wise vision of Progress.

In Art, for example, the connections between ethics and aesthetics were central to much philosophy from Plato on. At the same time artists often invoke the notion of Progress in support of their choices, in order to see themselves in a perspective that consider their historical precedents they attach to them.

Science has always been rated as a human activity devoted to the acquisition and systematization of knowledge. Progress in Science, focusing only on its ethical aspects, has always been seen straightforwardly as a cumulative process, while, similarly to Art, it is profoundly conditioned by historical and socio-political issues. Observations up to theories can be biased by norms, taboos and the socio-political significations linked to them

Differently from the case of Art these latter aspects of Science have been only very seldom admitted or confessed.

Besides, Knowledge accumulated in Science, and in technology in particular, has been always very influential on Society. Science, technology and society (STS) studies have been the subject of a large corpus of worldwide literature triggered in the '60 by the work of Thomas Kuhn (1962) and that recently has given birth to a whole new sensibility also in international organizations as the UNESCO (UNESCO 2005) that adopted the concept of knowledge society and used it in the plural form 'knowledge societies' to consider it in the perspective of the transformations taking place in the world in which the technological dimension represents the cornerstone of society and in which the new economy and communication networks represent the main features of its general structure.

A very important aspect to study in this endeavor is the transmission of knowledge. The Arts often can convey meaning as "evocation", a form of resonance in someone else that leaves the modes in which this is obtained somehow implicit. Science, traditionally, contrasts this vision and is seen as limited to formalized expressions that can be articulated, codified, stored in certain media, and beyond any doubt readily transmitted to others. Most of the contemporary epistemology demonstrated this is not the case.

To study all these issues the path we will follow, will start from analyzing the subject (the scientist, the author) how he/she relates to the external word, to proceed towards the community he/she needs to relate with.

Representation

We said above that it is possible to speak of the epistemic as judgments and as the things that are similar to judgments such as beliefs, percepts, news reports, books, articles, sentences and pictures, in that they represent or portray things. They are about things. They express intentionality. In other words judgments are connected to representations. Representations can be theories and models, linguistic and mathematical entities, computer simulations, concrete objects and so on and so forth. Language here is only one of the possible ways of representing the epistemic so that this deliteralization allows us to move the focus toward Culture.

Only things that are representational can be epistemically subjective or objective.

Representation is particularly puzzling because it is prone to omissions and commissions (Callender, & Cohen, 2006) having features reality does and does not have. Omission is the act of neglecting some possible causes among those that can explain a phenomenon; commission is the act of deliberately change the network of possible causes and explanations of a phenomenon.

Strictly speaking descriptions of entities and processes afforded by scientific representations are generally false (Chakravartty,1995). Even the most realists among philosophers and scientists will admit that representations cannot be perfectly and comprehensively true. History of Science is full of episodes where this has been demonstrated to be the case, and since the eventual test must be on real observations, contemporary epistemologists are generally rather pessimistic on the possibility to produce an account of the physical word that could be literally true. Rather, the aim of science should be to produce representations that could be empirically adequate (Van Fraassen 1989).

What constitutes the representational relation between a model and the word?

Grice (1989) hold that there are core representational statuses common to every entity: In this perspective we do not need to separate the various forms of representations, so that scientific theories, or linguistic or artistic representations can be analyzed using the same approach.

Goodman (1976), focusing on art but projecting this vision to Science also, maintained that truth can be understood in terms of a ‘matter of fit’ between the representation and the world. Knowledge is built through representation, and knowledge is transmitted through representation as well. This fit should be understood in terms of approximating reality. The problem is that, in addition to proper representations, users of the model should be acculturated with the conventions of the representation. Chakravartty (1995) similarly maintained that representation in Art and Science have many commonalties and suggested to handle them within a common framework.

He calls abstraction (omission in Callender, & Cohen (2006) terminology) a process of representation where only some of the potentially many factors that are relevant to the behavior of a target system are built into the representation. In such a process other parameters are ignored in order to allow the construction of a tractable representation. This is the case of assumptions in the construction of the model.

He calls idealization (commission in Callender, & Cohen (2006) terminology) the process in which at least one of the parameters of the target system is represented in a way that constitutes a distortion of its true nature. They are fictional, as in the case of physical models where the mass is concentrated in one point.

Following Goodman (1976) it should be possible to distinguish realistic and non-realistic representations. Realism stands not in the quantity of information but in the ease to access it. Realistic and non-realistic representations can issue the same information but in the second case conventions need to be internalized by the viewer or user. Of course, in the perspective of collaborative e-science this issue is a (maybe THE) central problem, since to access information within the same non realistic representation users should share the same conventions. This of course opens a wide domain to explore, that will be handled partially, for the connection between representation and referencing in Diviacco (2015, see chapter 19 this book), regarding the transmission of knowledge (learning) in Corleto and Tomassetti (2015, see chapter 10 this book) and Gorela and Biloslavo (2015, see chapter 6 this book) and in several others chapters of the book regarding semantic mediation between communities.

It is to be said that the border between realist and non realist representations is not easy to draw, more likely this can be described as a spectrum of possibilities where realism and non-realism are placed at the extremes and where the user position him/herself in dependence of his cultural status.

The comparison with artwork fruition is particularly suitable. The amount of information that a non-expert viewer can extract from a sculpture or a painting depends on how much the artwork is realistic or non realistic. Basic drawings or pictures, convey information that can be captured by anyone, but when to this basic level an artist would like to impose further coded messages or information, things can become rather tricky. This is the turning point, for example, where contemporary art restricts its public to the more acculturated elites. Goodman (1976) introduces in Art the distinction between depictive and symbolic representation. Non-realist representation is achieved referring to an entity or concept x through the use of a symbol y. Following Goodman (1976) Symbols can be letters, words, texts, pictures, diagrams, maps and more that in Goodman's view are conventions. In Diviacco and Busato (2015, see chapter 19 this book) symbols are analyzed more in detail within a semiotic approach, identifying multiple modes (indexical, iconic and symbolic) of signification.

Following Huges (1997) scientific representation typically has elements of denotation(elements of the model), demonstration (use of the representation to get a result) and interpretation (the results is then interpreted physically).

In scientific representation, abstraction, idealization and the introduction of a semiotic level introduce the need of sharing codes to understand models that are almost always non realistic. The possession or not of those codes induce the distinction in acculturated and naïve users. This eventually pinpoints representation as the trigger for the institution of a community that can "read" and interpret the information contained in a representation and separate it from those that cannot.

Representation is at the source of the "culture" of such communities.

Reasoning in the Sciences

Indeed from the very title of this section of the chapter, the choice of saying Sciences instead of Science reveals the message that different paths of reasoning exist in different scientific domains. We claim in fact that the traditional view of a single Method and Reasoning unifying all the domains of Science is just a myth. While Physics and Chemistry are mostly induction/deduction based disciplines, other domains as Medicine, History, the Human Sciences, Sociology and the Geo-Sciences are strongly dependant on abductive reasoning.

A detailed description of each type of these reasoning modes can be found in Diviacco (2012). It could be, however, of some help here to recall, and tailor for our discussion, several key concepts that are very important in relation to representation.

Induction is a generalization from a series of observations. If I observe that after some ringing of a bell the food comes, I'm induced to think that every time I'll hear the bell, my dinner will be ready, so I can start salivating. Induction produce law-like statements. A form of precarious but seemingly reasonable knowledge.

Deduction is a form of transformation of knowledge in another form of the same knowledge through a law, a process. Deduction does not allow to discover anything new, it is just a reconfiguration of knowledge.

Abduction goes from the resulting state of affairs to the controlling state of affairs. In fact it is often used in causal contexts. Abduction requires an act of creative imagination from the author, where creativity does not even need to be subject to logics. The problem is that Abduction cannot guarantee truth since the same resulting state could be produced also by other premises. Essentially, quoting Peirce (1958, p. 171):

Abduction merely suggests that something may be.

The fundamental issue by which inductive/deductive methods cannot apply in several domains have to do with the possibility to verify the results or predictions of a model. No one was there when the Dinosaurs disappeared. Only with many difficulties we would be able to sample the rocks under a very high mountain range. In the case of a physician, he/she cannot be there to see how the patient got sick and an historian cannot witness the killing of Cesar.

Abduction, instead, fits perfectly with this way of reasoning.

What are the consequences from the point of view of collaborative knowledge?

Essentially that abductive reasoning allows multiple and contrasting explanations of the same phenomenon to coexist. This has a great impact on a scientific community that tends to speciate in tribes, schools and traditions.

Kuhn (1962) postulated that scientists live within paradigms: an environment that condition their way of thinking. Different, concurrent and incommensurable paradigms exist within any discipline.

Lakatos (1970) maintained that theories or paradigms can shield their core from attempts at falsification by changing a set of auxiliary and peripheral hypotheses (the protective belt) while leaving essentially unaltered the kernel of the theory. If this happens in any scientific discipline, the difficulties in handling experiments to prove or falsify theories in abductive domains as those mentioned above makes it very easy to protect core ideas from external attacks, so that in many cases theories can pass through the time with only cosmetic rearrangements.

To further complicate this already intricate situation we have to consider the sociological issues related to scientific debates, which often comes from political interests, positioning in the community and ultimately the access to funding.

Whitley (2000) postulated that external conditions and circumstances, such as how universities, faculties and departments are organized, or the control over facilities, can and do make a difference to disciplinary status and identity.

Latour and Woolgar (1979) analyzed how this can be extended to include the position that the single researcher can maintain in a network of allies, stating eventually that Science is a social construct.

Summing up, a scientific community that needs to access a shared representation can be seen actually as a mosaic of contrasting cognitive models that (consciously or unconsciously) will inevitably projects private perspectives and filters on the representation of the state of affairs. This will consequently shape their view of the information contained in the representation itself. Diviacco (2015, see chapter 19 this book) will stem from this point to further explore the mechanisms of sharing coded representations, and study how coded representations in a distributed cognitive system can be implemented in a web-based collaborative tool devoted to support knowledge networks.

In the continuation of this chapter, instead, in the perspective of distributed cognition, we will delve into the problem of how the subject interact with the designated community and how outcomes are built collaboratively confronting also Science with other form of Culture.

The Role of the Author (a Cult of Personality or a Market Issue)

To understand the path that leads us through representation, from the *inside* of the subject, towards the *outside* of collaboration and networks of knowledge, and since, as we said, Science is based on similar or common processes as those in other domains where representation is evoked; we can explore other forms of culture which embeds many of the characteristics that Scientific research has, such as for example contemporary art.

Regarding the role of the subject, two different sensibilities can be traced, and namely on one hand what we can call the *avant-garde* approach, such as that expressed for instance by Picasso or Schoenberg, and on the other the approach of what is commonly called *alternative culture*, such as that expressed by for example Kaprow (Kaprow 2003) or Cage.

The former has a clear signature in its attempt to fight bourgeois mainstream cultural values on a "shock and estrangement" project largely based on the personality of the author.

In the latter case, artists, in an ethical attempt of recognition of the symbolic universe of the other (Žižek 1991), experience the need to shift away the focus from the author to the context. By context here we mean the contingent set of entities, concepts, situations, people and (material and immaterial) things brought around a subject. Multiple and different context can coexist at the same time. In Diviacco (2015, see chapter 19 this book) the context is further explored as an integral part of the process of knowing.

Both avant-garde and alternative approaches live together so that one can ask what is the reason for maintaining the importance of the author (subject) in the former? Is that related to ethic or esthetic issues, or merely to economic reasons?

Contemporary art market has been described in a masterly manner by Plattener (1998) as one where producers don't make work primarily for sale, where buyers often have no idea of the value of what they buy and where middlemen routinely claim reimbursement for sales of things they've never seen, to buyers they've never dealt with. Such markets are called asymmetrical markets. In these cases there is a disparity in the information available to the dealer-seller and to the buyer, which can lead to unconventional behaviors such as the Veblen paradox (1899), where, raising the prices may stimulate demand. The role of the Author's (subject) reputation is extremely important in this mechanism to estimate a *"price"* for an artwork, while estimating its *"value"* could be almost impossible.

To contrast the Author-centric approach a very different one emerged in the sixties. In Allan Kaprow's happenings there were no distinction or hierarchy between artist and viewer. The aim was to foster the desire to break the boundaries between art and life. In the same direction, the work of John Cage has been particularly important, pioneering and groundbreaking. He was probably the first that could be seen as a "context provider" rather than an Author. As an Example Cage's work 4'33" (http://www.youtube.com/watch?v=JTEFKFiXSx4) was an open system in which the audience or subject participated as interpreter, author and actant in the system. In 4'33" Cage used chance operations to compose a temporal frame which played off the formal conventions of the recital performance. The pianist opening and closing the keyboard framed a silence, which revealed the sounds produced by the audience. Cage repositioned his audience as author and interpreter. He eliminated the distinction between musical and environmental sound, thus achieving a fusion of art and context. Cage intent was political too; he saw the analogy between the conventional process of making music and a political structure that resembled dictatorship[2], suggesting a radical reshaping of political and social structures. These ideas were later picked up by other artists such as for example Christian Wolff whose compositional technique involved the construction of systems resembling social communities, turning the making of music into a collaborative and transforming activity where the performer can be a composer and a listener too.

The world of Cage, Kaprow and the like is exemplary in representing the reflections on the complexity and instability of human being highlighted by Post-structuralist authors as Deridda or Lacan. These speculations led Roland Barthes to think that all western culture is based on mere beliefs of constancy and standards, and that since it is not possible to understand the state of the author's mind there's no way to know his/her real intentions. The possibility of providing the meaning of an artwork is just a myth and when this is postulated it is made so in order to consider it consumable, something that can be used up and replaced in a market. These considerations led Barthes not only to shift the focus away from the Author but even to state the death of the Author. Quoting him, the Author subjective meaning is:

A multidimensional space in which all quotations of a variety of texts, none of them original, blend and clash (Barthes 1977, p.146).

Even more, Foucault (2002, p.117) shifts the focus from the subject as originator to the conditions in which he/she operates.

The modes of circulation, valorization, attribution and appropriation of discourses vary with each culture and are modified within each.

Tagging artworks by author (very likely celebrities) might, often, be needed to support its market position. Can we identify something similar in Science?

Reputation and positioning of artists should immediately evoke in the reader a parallel with scientist rating, bibliometric index, impact factors of journals or research evaluation agencies.

The Academia is renowned to be the putative place for never ending conflicts and diatribes resembling medieval disputations. It is a fertile ground to grow personal identification in specific visions, backgrounds and schools. It fosters in scientists the attitude to focus on personal reputation, strategic positioning in the community and eventually fund-raising. Being on one side or the other, pushing one vision or another is a decision each scientist has to face everyday and has to take very carefully to avoid isolation or budget problems. Personality and charisma of the researcher becomes an important aspect to consider in relation to the scientific research market. Therefore it seems reasonable to answer to the above question, that, yes, there is an economic interest in preserving the role of the Author in Scientific research, and that we have to consider carefully this aspect when devising collaborative networks and the share of knowledge.

In this perspective, another interesting parallel that can be drawn between art and Science is related to the status of the Author and the protection of his/her intellectual property. In the case of figurative Arts, Music or Literature this has always been a concern. Copyright was introduced to regulate printing in the XVIIth century, to be extended lately to cover all forms of creative activities. It is beyond everyone eye what is the current complex situation of intellectual property protection in the era of digital replication of contents.

In the Scientific domain these issues are mirrored in the retentivity of sharing data and observations that researchers can experience while trying to protect their work (Leadbetter 2015, see chapter 2 this book; Diviacco 2012; Diviacco 2005). This is a serious problem that can harm the possibility to collaborate among scientists, since scientific reasoning is based on observations and data. The control over data is an extremely important issue that crosses many scientific domains. Diviacco and Busato (2013), explore several fields and focus on geophysical data and seismic data in particular. These data are very important in oil prospecting and are normally confidential. At the same time, hiding them to the community will prevent any collaboration from taking place. Diviacco and Busato (2013), propose a new approach that leverages a web-based system that allow to work remotely on data without needing a local copy of them. Data, observations, maps and graphs access together with means to include distributed knowledge and reasoning using annotation will be the subject of a specific chapter of this book (Diviacco, & Busato 2015, see chapter 19 this book).

Towards a Simple Model of Collaboration

We saw that the common sense antinomy between facts and opinions is actually an ill posed problem, and that defining subjectivity and objectivity is more complex than it seems. That the metaphysical limitations in accessing the outer world allow meaning to be expressed only through representation. Reasoning and in particular abductive reasoning can speciate communities that can have difficulties in understanding each other. Abstraction, idealization and symbols allow multiple and concurrent representations to exists, so that each community identifies itself with its own set of representation modes. Representations become "coded" so that external users need a key to access their meaning.

Poststructuralist pessimism in the possibility to hold this key drove, in many domains, to the death of the author. These ideas were very important to resize the importance of the single subject/originator and move the focus to the outside world, to the context. The emphasis is no longer on searching the meaning of things, but rather now to search for what they mean in specific situations. Meaning and then knowledge became situated. The term situated has been coined by Donna Haraway to link knowledge to a particular situation or context, seeing it embedded in language, culture, or traditions. It stems from the limitations of the human perception, as well as the limitations of Science

Given that we can rely on the Epistemic only and have no access to facts, given that knowledge is defined by the context, then what happens when the subject interacts with other subjects?

So far we claimed that it is possible to trace a parallel (meaning not touching each other, maybe, but in the same direction and versus) between the Arts and Science. This can be explained in the light of being both epiphenomena of the same socio-political, and therefore by nature mobile, but also cognitive, and therefore not only, but also, innate, mechanisms. We used this parallelism to analyze the issue of representation. Analyzing the case of the Arts we have been enabled to delve easier into the case of Science. We will, here, continue with this approach, in order to devise a simple model of how interactions between subjects can take place.

Collaboration in Art

In the case of the Art it is possible to identify two paths of interaction stemming from the subject, and namely on one side that towards an audience, and on the other side that towards other Artists with whom the subject is performing or collaborating.

In the first case, as mentioned above, this can take place as an unidirectional movement from the Artist to the audience only, or as a bidirectional flow between the Artist and the audience, where, following the trend in shifting the focus away from the Author, the audience is involved in the creation of the artwork as well.

In both cases, the representational code is an important factor, since, as already mentioned, the public can be not fully aware of what it is experiencing.

In case of interaction among Artists, during a performance for example, the situation can be a bit more complex. There are several ways in which this can take place, but a distinction is often made between *collaborative art*, which implies the involvement of a number of artists that seek some kind of consensus around an artwork and *collective art*, which has as a result an art piece with a distinct and cohesive art language. In the first case the aesthetics is generally defined by an initiator or leader, while in the second case it is common, shared and determined by the community as a whole. Collective works, eliminates personal aesthetics over collective directions. Being by nature a multiple authors work, all authors are entitled for equal intellectual rights[3]. Collaborative works are often the case of a collaboration of a minor subjects with a major artist with an established reputation.

Artist collectives work in the direction of hiding member's identity and breaking up their narcissistic egos. Examples of this approach are common from the antiquity such as for example the ancient sculpture workshops in Greece, to modern times such as for example the COBRA group in the last century, from Gilbert and George who refer to themselves not as two artists but as just one, to the Ant Farm, to The Residents, to many more.

If this mental attitude and status can be met, collective decisions are naturally taken.

Collaboration in Science

In the case of Science, we can distinguish between the interaction of the researchers with the society and that with other scientists.

In the first case, the most common perspective has been that of an unidirectional flux of information from a researcher/project/Institute towards a community. This can be explained in the light of the already mentioned political and economical role of Science in our Society. Aside from this, a new paradigm is emerging where a bidirectionl flux can be established as well.

More in details, the unidirectional flux can have two aspects, and namely, on one hand, that used to push information to public communities "sensu stricto", such as for example natural hazard alert systems, while, on the other, that devised to help people understanding (decoding) scientific visions, such as for examples: "open days", lectures at schools, or social media presence.

In the case of a mutual exchange of information between the scientist and an external "audience", this latter becomes an active part of the process of knowing, with the limitation that this generally takes place at the level of observation not at that of theorization, which is, instead, within the scope of work of the scientist. An example of this case is the extremely interesting new paradigm that is *crowdsource observations*. Gathering observations from the laymen has, of course, been always used in the history of science, but the power of the web allows us to reach, now, results that could not have been imagined just a few years ago. A very impressive case of this approach is, for example, the use of Twitter (https://twitter.com/) to detect earthquakes events (Allen 2012). In this project, the United States Geological Survey (USGS) monitors Twitter messages containing the word "earthquake". This allow to detect earthquakes that are broadly felt, while, so far, the method cannot provide accurate location or magnitude information. The idea, of course, can be projected in other domains such as meteorological studies, pollution and many more and we are pretty sure that, very soon, this will become a common practice.

The de-coding skills needed by contributors can vary largely from project to project. In the example above, no scientific knowledge is needed since everyone knows what an earthquake is. Other initiatives could require higher levels of acculturation; this can be the case for example for the skills in reading an instrument or identifying instances of a class or category, such as, for example, animal species. When this happens the set of contributors is generally restricted[4].

If we move to the case of scientists working together, as in the case of artists, things gets more complicated and exploring this issue we have to refer to the work of Karin Knorr Cetina. In her seminal book *Epistemic Cultures* (Knorr Cetina 1999), maintaining that different scientific fields exhibit distinct "epistemic cultures". She compares the field of High Energy Physiscs (HEP), and in particular the case of the Large Electon Positron Collider (LHC) coupled with a large detector called Atlas, with the field of Microbiology.

In her vision, HEP research have a "post-traditional communitarian structure" where authority is distributed because no one person can know everything that must be known to make the experiment work. Along with authority goes also responsibility. This must also be distributed, which, fosters a positive mental habit towards trust and cooperation. As a demonstration, rewards tend to be shared so that the common practice and tradition dictates that papers published on scientific journals in this community normally have a long list of authors, listed alphabetically, which means that no prominent actor or principal investigator is identified. Several researchers in this community argued that authorships in this case becomes not useful, and that in the perspective of researchers reputation, the focus should be moved on letters of recommendation, informal records of experience within the experiment, and the number of conference presentations given on behalf of the entire collaboration project. This argument has given rise to a debate that is described in greater detail in Birnholtz (2006).

Turning to Microbiology, Knorr Cetina highlights instead the importance of each researcher working individually that together with the attitude of Microbiologists to aim at turning biological processes into industrial "output" creates a completely different scenario. Here decisions are more hierarchical, with lab leaders managing lab products and expertise.

In the case of HEP she invokes the Durkheimian notion of "collective consciousness": an erasure of the individual as an epistemic subject, to support the idea that we should consider the post-traditional communitarism as a form of collective cognition. Within HEP projects, one only with difficulty can isolate any person, or even a small group of individuals producing the resulting knowledge. Some authors, in other fields such as biomedicine, (Olson et al. 2008) insist on the importance of the dimension of the initiative and of the facilities (which would explain the HEP case). Others in the field of humanities (Bowman, Crowther, Pybus, & Kirkham, 2008), for example, highlight the need of a balance between the role of facilities and team-work and reasoning. Giere (2002) extends the perspective of collective cognition to include also interaction with artifacts to obtain what he calls *Distributed Cognition*. In the parlance we used so far in this chapter, and that will be used also in Diviacco (2015, see chapter 19 this book), the scope of this extension that includes then people, their way of reasoning, their socio-political presence, their backgrounds and the artifacts connected to them, is essentially the context in which cognition takes place.

In the case of Microbiology, researchers work together towards the construction of a common product but under the steering force of a principal investigator, while in the case of HEP, the process, the products, the responsibility and rewards are "open".

Judgment Aggregation

True collective decisions should be based on the will to avoid individual interests. Unfortunately this is not enough since group decisions are anyhow influenced and limited by "connaturated" mechanisms of opinion and judgement building (see also Polson, & Curtis, 2015, chapter 4 of this book).

Condorcet's theorem states that a majority is more likely to be correct on a judgement than each individual juror, and that the probability of a correct judgment by a majority increases as the jury size increases. Following List (2005) it is necessary to distinguish two challenges in collective judgment and namely: (1) the correspondence challenge, in the sense of its correspondence to facts, and (2) the coherence challenge, which is the ability to organize judgment in a coherent, logical manner. For a collective to be wise it is needed that its judgments meet both challenges. A judgment is made on a set of propositions, technically named the *agenda*, on the base of each member judgment, and through an aggregation procedure. A simple example of aggregation procedure is majority voting. The work of List and Pettit (2002) showed that while often meeting the expectation for the correspondence criterion, majority voting often does not meet the coherence criterium. To rewind to our case, mapping between representations, basing the efforts on consensus gained through majority voting, can lead easily to inconsistencies. Dietrich and List (2007) further extends these conclusions to any procedure of aggregation that is not dictatorship[5], unless strategies are applied to the procedure itself in order to reduce risks.

One possible strategy to guarantee consistency is to increase the majority threshold. This is used for examples in international initiatives as UN Security Council that can request two-third or even unanimity for a resolution. This unfortunately expose decisions to judgment suspensions and vetoes.

A different approach is to assign to some propositions of the agenda the value of premises and devise a deductive path towards some other propositions of the agenda that can be considered then conclusions. This, although not always possible, focuses the conflicts on the premises, lowering the load of the discussion. Additionally, if a division of the cognitive work can be established so that subgroups can elaborate on the premises, better judgment aggregations can be met if the epistemic gain obtained from specialization is not hampered by the epistemic loss due to the lower number of participants.

A Simple Model

Summing up, from the comparison between the cases of Art and Science we do see some regularities.

In both cases we can identify two paths, and namely: (1) one towards an external audience and (2) one towards other co-actors.

In the first case the interaction (1a) can be passive or (1b) somehow also active. The distinction made between a passive or active interaction does not depend directly on coding, while, mostly in Science, it is the level to which an active contribution can aspire that depends on the level of acculturation. Path 1b is a first evidence of a trend to shift the focus outside the subject to include the context.

Turning to the interaction between co-actors, which is of course our main target, the comparison between Arts and Science allows us to identify two cases and namely: (2a) proper collaboration: where the work is done independently and supervised by a leader, such as for example the case of a classical music orchestra and its director, or the case of microbiology in Science. (2b) Collective work: where there is a tendency to minimize the individual as an epistemic subject and conversely to emphasize distributed cognition in a post-traditional communitarian structure, such as for example in artistic collectives or in the case of the HEP. A very interesting case we can associate with class 2b is that of the Communities of practices/interests (CoP/CoI). These are informal groups largely kept together by the social presence of members and where the aim in itself is sharing knowledge (Wenger, 1998).

We would like to stress as much as possible the importance of collective work, since we claim that, in the field of Science, only very seldom its importance has been considered.

On contrary, in other fields such as for example music, a lot is available since the beginning of last decade (see for example Jordà, Kaltenbrunner, Geiger, &Alonso, 2006). Many experiences are available, from the Princeton Laptop Orchestra (http://plork.cs.princeton.edu/) to the vast number of improvvisative experimental music made with Max/MSP (http://cycling74.com/products/max/).

In the field of e-research nothing like this has been devised so far and we claim that this has to be related mainly to the misconception of the role of Science and to its mythology that obscured how research actually takes place in particular in abductive sciences.

The Epistemic Subject

Throughout this chapter the role of the subject is the criterion to identify various concepts and categories. While in the case of collaborative[6] work it seems pretty clear who the epistemic subject is: the author, the researcher; in the case of collective work, instead, this identification seems more complex.

Knorr Cetina maintains that if knowledge is being produced, there must be anyway an epistemic subject, so that at this point a very important issue emerges. Since, as we said above, in the case of distributed cognition no proper and unique "human", "single", "isolated" subject can be identified, the epistemic subject must be identified with the project itself and with the community and the artifacts and tools related to it.

CONCLUSION

The Reconciliation

What described above is a very important point for our discussion. It means that any tool we aim to develop to support collaborative e-research in the case of a post traditional communitarian structure, is an integral part and condition of the process of cognition.

It does not only aim at knowledge, it is knowledge.

We can therefore reverse reasoning and say that, especially in the case of abductive sciences or where traditional and hierarchic collaborative work is not the right framework for scientific research, we need to devise tools that not only support an activity but that can be the activity itself: that not only store information but that can be the information itself, that not only support reasoning but that can be part of the process of reasoning itself.

How can this be put in practice? We cannot be exhaustive here, but can suggest that if the system has to be the activity, it needs to embed the work, actors presence, their cognition and representation.

It needs therefore to allow access to observations and information.

It needs to offer data processing and representation.

It needs to expose workflows, concept or mind maps where ideas can be expressed and shared. (please see the last section of the book on knowledge engineering)

It needs to offer tools to map different projections of knowledge between communities, such as ontologies or controlled vocabularies (please see the many chapters of the book where this argument is handled in detail).

It needs to promote mutual awareness through communication systems.

This last point is of particular interest. Communication can be synchronous such as face-to-face videoconferencing, or asynchronous, such as the kind of interaction that can be found in forums or some already existing CoP, and that showed to be very effective especially in learning and in overcoming the problems of tacit knowledge.

FUTURE RESEARCH DIRECTIONS

Our intentions throughout the chapter was mainly to identify a trend to shift the focus from the subject working in isolation to a post traditional structure, where the processes of cognition could be distributed among the members of a community. We identified and motivated this perspective confronting Science and other domains of culture. Many aspects can be recognized as similar, so that a simple model can be drawn where collective work has a clear position and where its characteristics can be easily identified.

In the last section we proposed some ideas that can be practically implemented in IT systems designed to support e-research and knowledge networks. We cannot consider these exhaustive, and in fact this chapter is intended not as a conclusion, rather a starting point.

From here many directions of study can stem. We will here report on the directions currently under analysis in our research group, while to delve into other directions we send the reader to the other chapters of the book.

A very important issue in our endeavour, that emerged in many point of this discussion, and that is the main object of our future research is the problem of coding.

If we expect researchers to perform distributed cognition how can we manage to overcome possible conflicting coding. A method should be introduced that is able to cross paradigms, backgrounds, schools and traditions. The solution we will explore is that of developing Boundary Objects, meaning artifacts that can be used by different communities to bridge their differences while at the same time preserving their identities. In Diviacco and Busato (2015, see chapter 19 this book), this is applied to observations, measurements, information that can be handled within an iconic signification mode. Here the perspective of the deliteralization is very important since it let cultural issues emerge not framing them within a rigid case. In this perspective they can be used as Boundary objects that allow to bridge different contexts. We expect that this will be beneficial in enhancing both information sharing and also awareness in all cases of knowledge networks and collaborative e-research

ACKNOWLEDGMENT

Živa Pahor, Jaren Diviacco, Ilija Diviacco.

REFERENCES

Allen, R. (2012). Transforming Earthquake Detection? *Science*, *335*(6066), 297–298. doi:10.1126/science.1214650 PMID:22267802

Barthes, R. (1977). *Image, Music, Text*. New York: Hill and Wang.

Birnholtz, J. (2006). What does it mean to be an author? The intersection of credit, contribution, and collaboration in science. *Journal of the American Society for Information Science and Technology*, *57*(13), 1758–1770. doi:10.1002/asi.20380

Bowman, A., Crowther, C., Pybus, J., & Kirkham, R. (2008). *Virtual Research Environment for the Study of Documents and Manuscripts*. Oxford eResearch Conference 2008. Retrieved from http://bvreh.humanities.ox.ac.uk/files/Oxford-eResearch2008-VRE-SDM.pdf

Callender, C., & Cohen, J. (2005). There Is No Special Problem About Scientific Representation. *Theoria*, *55*, 67–81.

Callender, C., & Cohen, J. (2006). There is No Special Problem About Scientific Representation. *Theoria*, *21*(1), 67–85.

Carnap, R. (1942). *Introduction to semantics*. Cambridge, MA: Harvard University Press.

Carnap, R. (1947). Meaning and Necessity. Chicago, IL: The University of Chicago Press.

Chakravartty, A. (1995). Truth and Representation in Science: Two Inspirations from Art. In R. Frigg & M. Hunter (Eds.), *Beyond Mimesis and Convention: Representation in Art and Science*, (pp. 33-50). Berlin: Springer.

Corleto, A., & Tomassetti, V. (2015). Beyond Learning: E-learning Models for Scientific Research. In P. Diviacco, P. Fox, C. Pshenichny, & A. Leadbetter (Eds.), *Collaborative Knowledge in Scientific Research Networks*. Hershey, PA: IGI Global.

Dietrich, F., & List, C. (2007). Judgment aggregation under constraints. In T. Boylan & R. Gekker (Eds.), *Economics, Rational Choice and Normative Philosophy*. London: Routledge.

Diviacco, P. (2005). An open source, web based, simple solution for seismic data dissemination and collaborative research. *Computers & Geosciences, 31*(5), 599–605. doi:10.1016/j.cageo.2004.11.008

Diviacco, P. (2012). Addressing Conflicting Cognitive Models in Collaborative E-Research: A Case Study in Exploration Geophysics. In *Collaborative and Distributed E-Research: Innovations in Technologies, Strategies and Applications*. Hershey, PA: IGI Global Press. doi:10.4018/978-1-4666-0125-3.ch012

Diviacco, P., Busato, A. (2013). The Geo-Seas Seismic data viewer: a tool to facilitate control of data access. *Bollettino di Geofisica Teorica ed Applicata, 54* (3), 257-270.

Diviacco, P., & Busato, A. (2015). Maps, graphs and annotations as Boundary Objects in Knowledge Networks, Distributed Cognition and Collaborative e-Research. In P. Diviacco, P. Fox, C. Pshenichny, & A. Leadbetter (Eds.), *Collaborative Knowledge in Scientific Research Networks*. Hershey, PA: IGI Global.

Eco, U. (1976). *A Theory of Semiotics*. London: Macmillan.

Engelhardt, W., & Zimmerman, J. (1982). Theory of Earth Science. New York: Cambridge.

Foucault, M. (2002). Archaeology of Knowledge. London: Routledge.

Gauker, C. (1995). *Thinking Out Loud*. Princeton, NJ: Princeton University Press.

Giere, R. (2002). Distributed Cognition in Epistemic Cultures. *Philosophy of Science, 69*(4), 637–644. doi:10.1086/344627

Goodman, N. (1976). *Language of Art: An Approach to a Theory of Symbols*. Indianapolis, IN: Hackett.

Gorela, K., & Biloslavo, R. (2015). Relationship between Senior and Junior Researcher: Challenges & Opportunities for Knowledge Creating and Sharing. In P. Diviacco, P. Fox, C. Pshenichny, & A. Leadbetter (Eds.), *Collaborative Knowledge in Scientific Research Networks*. Hershey, PA: IGI Global.

Grice, H. P. (1989). *Studies in the way of words*. Cambridge, MA: Harvard University press.

Haraway, D. (1988). Situated Knowledges: The Science Question in Feminism and the Privilege of Partial Perspective. *Feminist Studies, 14*(3), 575–599. doi:10.2307/3178066

Hughes, R. I. G. (1997). Models and representation. *Philosophy of Science Supplement: Proceedings of the 1996 Biennial Meeting, 64*, 325-336.

Jordà, S., Kaltenbrunner, M., Geiger, G., & Alonso, M. (2006). *The reacTable: a tangible tabletop musical instrument and collaborative workbench*. ACM SIGGRAPH. doi:10.1145/1179849.1179963

Kaprow, A. (2003). *Essays on the Blurring of Art and Life*. University of California Press.

Knorr-Cetina, K. (1999). *Epistemic Cultures: How the Sciences Make Knowledge*. Harvard University Press.

Kuhn, T. S. (1962). *The structure of scientific revolutions*. Chicago, IL: University of Chicago Press.

Lakatos, I. (1970). Falsification and the methodology of scientific research programmes. In Criticism and the Growth of Knowledge. Cambridge, UK: Cambridge University Press. doi:10.1017/CBO9781139171434.009

Latour, B., & Woolgar, S. (1979). *Laboratory life: The construction of scientific facts*. Princeton, NJ: Princeton University Press.

Leadbetter, A. (2015). Examining trust in collaborative research networks. In P. Diviacco, P. Fox, C. Pshenichny, & A. Leadbetter (Eds.), *Collaborative Knowledge in Scientific Research Networks*. Hershey, PA: IGI Global.

List, C. (2005). The Probability of Inconsistencies in Complex Collective Decisions. *Social Choice and Welfare*, *24*(1), 3–32. http://personal.lse.ac.uk/list/PDF-files/inconsistencies.pdf doi:10.1007/s00355-003-0253-7

List, C., & Pettit, P. (2002). Aggregating Sets of Judgments: An Impossibility Result. *Economics and Philosophy*, *18*, 89–110.

Lycan, W. (1996). *Consciousness and Experience*. Cambridge, MA: MIT Press.

Olson, J., Ellisman, M., James, M., Grethe, J., & Puetz, M. (2008). Biomedical informatics research network (BIRN). In G. Olson, A. Zimmerman, & N. Bos (Eds.), *Scientific Collaboration on the Internet*. Cambridge, MA: MIT Press. doi:10.7551/mitpress/9780262151207.003.0013

Peirce, C. S. (1958). *The Collected Works of Charles Sanders Peirce*. Harvard University Press.

Plattener, S. (1998). A Most Ingenious Paradox: The Market for Contemporary Fine Art. *American Anthropologist*, *100*(2), 482–493. doi:10.1525/aa.1998.100.2.482

Polanyi, M. (1983). *The tacit dimension*. Gloucester, MA: Peter Smith Publisher. (Original work published 1966)

Polson, D., & Curtis, A. (2015). Assessing individual influence on group decisions in geological carbon capture and storage problems. In P. Diviacco, P. Fox, C. Pshenichny, & A. Leadbetter (Eds.), *Collaborative Knowledge in Scientific Research Networks*. Hershey, PA: IGI Global.

Popper, K. (1934). *The logic of scientific discovery*. New York, NY: Basic Books.

Popper, K. (2004). Conjectures and refutations: the growth of scientific knowledge. (Reprinted. ed.). London: Routledge.

Putnam, H. (1975). Mind, Language and Reality. In Philosophical Papers, (vol. 2). Cambridge, UK: Cambridge University Press. doi:10.1017/CBO9780511625251

Rorty, R. (1989). *Contingency, Irony and Solidarity*. Cambridge, UK: Cambridge University Press. doi:10.1017/CBO9780511804397

Searle, J. (1995). *The Construction of Social Reality*. New York: The Free Press.

Van Fraassen. (1989). *Laws and Symmetry*. Oxford, UK: Oxford University Press.

Veblen, T. B. (1899). *The Theory of the Leisure Class. An Economic Study of Institutions*. London: Macmillan Publishers.

Wenger, E. (1998). *Communities of Practice: Learning, Meaning and Identity*. Cambridge, UK: Cambridge University Press. doi:10.1017/CBO9780511803932

Whitley, R. (2000). *The intellectual and social organization of the sciences*. Oxford, UK: Clarendon Press.

Žižek, S. (1991). *Looking Awry. An introduction to Jacques Lacan through Popular Culture*. Cambridge, MA: MIT Press.

ADDITIONAL READING

Sheriff, R. E. (1973). *Encyclopedic dictionary of exploration geophysics*. Tulsa, OK: The Society of Exploration Geophysics.

KEY TERMS AND DEFINITIONS

Boundary Object: An artifact that allow two different and incommensurable communities to nevertheless collaborate. They are weakly structured in common use, and become strongly structured in individual use.

Context: The set of circumstances to which something is bound. In the chapter it is used to highlight the possible co-existence of multiple contexts linked to the same entity, object or a concept. This co-existence can trigger the issue of having possibly contrasting interpretations or visions of the same concept or object can coexist considering different circumstances.

Data: Is raw observations or measurements. It is not organized and does not present any meaning. Adding meaning to data we get Information.

Knowledge: Is what we know. It is the output of the process of reasoning which is based on Information and it is conditioned by the context in which the researcher lives. Knowledge mirrors back in all human activities and decisions so that its shadow is projected over any scientific experiment, data or information.

Tacit Knowledge: Knowledge that cannot be made explicit because obscured in socio-political phenomena or embedded in practices or even tools.

ENDNOTES

1 Dissolved not resolved.

2 See section of this chapter on judgment aggregation.

3 We have here a connection with information sharing.

4 Such as in the case of non realistic representation and of some form of "elite" art.

5 Dictatorship, of course, is no form of collaborative or collective knowledge building or judgment aggregation, and therefore is outside the scope of this work.

6 Here the term collaborative refers to the distinction made between collaborative and collective mode of interaction with a community and a context. The term here means a non distributed cognition process.

Chapter 2
Examining Trust in Collaborative Research Networks

Adam Leadbetter
British Oceanographic Data Centre, NERC, UK

ABSTRACT

Ian McHarg's "The Theory of Creative Fitting" describes the collaborative integration of data into knowledge products for use in ecological architecture developments. In McHarg's system, the provenance of data layers and trust in their creators are vital to the creation of knowledge, ideas recently described as being "urgent and relevant." This chapter begins with these ideas and explores provenance propagation and data publication. Provenance propagation has been a topic under discussion by the World Wide Web Consortium, while data publication has been a focus of the global marine science community. From these projects, the issue of uniquely identifying individuals in a distributed, collaborative network has emerged. Possible implications of this problem are discussed, as are technical solutions to the problem. The principles described within this chapter are illustrated by examples of uncertainty propagation in chained Web Processing Service networks and data publication activities at the British Oceanographic Data Centre.

INTRODUCTION

In his 1976 lecture "The Theory of Creative Fitting" (Margulis, Corner, & Hawthorne, 2007), Ian McHarg lays forth the principles of collaborative Geographical Information Systems (GIS). In general terms, McHarg's theory comprises four main themes beginning with the acceptance that creativity is employment of energy and matter to raise matter and energy to higher levels of order. Thus, creativity naturally finds an antithesis in reduction. The second element of this theory is the existence of criteria by which directionality and attributes of the creative process may be identified. By this definition, creativity always shows the tendency to move from greater to lesser randomness, from simplicity to complexity or from instability to dynamic equilibrium. McHarg identifies the verb to fit as being the vital criteria here as there is a requirement for any system, be

DOI: 10.4018/978-1-4666-6567-5.ch002

Copyright © 2015, IGI Global. Copying or distributing in print or electronic forms without written permission of IGI Global is prohibited.

it organic or artificial, to find the most fit of all environments, and to adapt both the environment and the system itself. This leads into the third pillar of McHarg's theory, the definition of creativity as this act of fitting. Finally McHarg looks at analysis of a creatively fitted system, using the idea of the health of the system, which is the ability of a system to recover from a negative perturbation, or the ability of the system not only to solve problems, but to seek them out.

In applying this general theory to his field of expertise, environmental architecture, McHarg introduces the concept of the "overlay map", that is a map with layers provided by various experts familiar to GIS users today. McHarg's overlay maps make a fundamental use of the collaboration of geologists, climatologists, hydrologists, limnologists and biologists to build a holistic ecological model for the purposes of designing architectural projects in harmony with the surrounding environment. In his discussion of McHarg's lecture, James Corner (Margulis et al., 2007) describes this collaborative network as innovative and urgent. Throughout the collaborative network he describes, McHarg was keen that trust is in built and that each data point used to create the information layers and to abstract the knowledge had provenance: identified by the name of the person who prepared the analysis; should be the best available data; and should be substantive.

Establishing trust in knowledge sharing networks has been well documented in the business domain for a decade (Abrams, Cross, Letter & Levine, 2003). In the quoted study, two main themes emerged in the development of trust between participants in a network: the competence of an individual and the belief that an individual is working towards the improvement of the collaborative network ("their benevolence"). These factors are deemed to be difficult to influence for the coordinators or managers of a network. There are clearly identified lines of action, activity and behaviour which can build trust between individual participants within a collaborative network and this chapter will explore how these can be implemented within software tooling for distributed collaborations. However, this does raise the level of information tooling required in building the collaboration software (Cheshire & Cook, 2004).

Cheshire and Cook (2004) identify the need to exchange reputational information across networks as key to establishing and building trust. The propagation of trust is an important aspect of the scientific knowledge building network, not just on the social level - for example where an instrument make or model is known to produce results with less accuracy than another. At the social level, this could be seen as one researcher's methods being less trusted by the network than another's. Ziegler and Lausen (2005) introduced a trust model to the Semantic Web space, covering the domain of internet connected networks. More recently, the World Wide Web Consortium (W3C) has produced the PROV model (Groth & Moreau, 2013). Within the PROV model, provenance is defined as information about entities, activities, and people involved in producing a piece of data or thing, which can be used to form assessments about its quality, reliability or (and most importantly in the context of this chapter) trustworthiness. The goal of PROV is to enable the wide publication and interchange of provenance on the Web and other information systems. PROV enables one to represent and interchange provenance information using widely available formats such as the Resource Description Framework (RDF, Manola & Miller, 2004) and the eXstensible Markup Language (XML, Quin, 2013).

This chapter takes up McHarg's themes and examines the philosophical issues of building trust between participants in a collaborative research network, and explores the technical issues of propagating and encoding trust.

Figure 1. The data-information-knowledge ecosystem (Shedroff, 1994) illustrates McHarg's statement that "information fragmented is of no use to anybody"

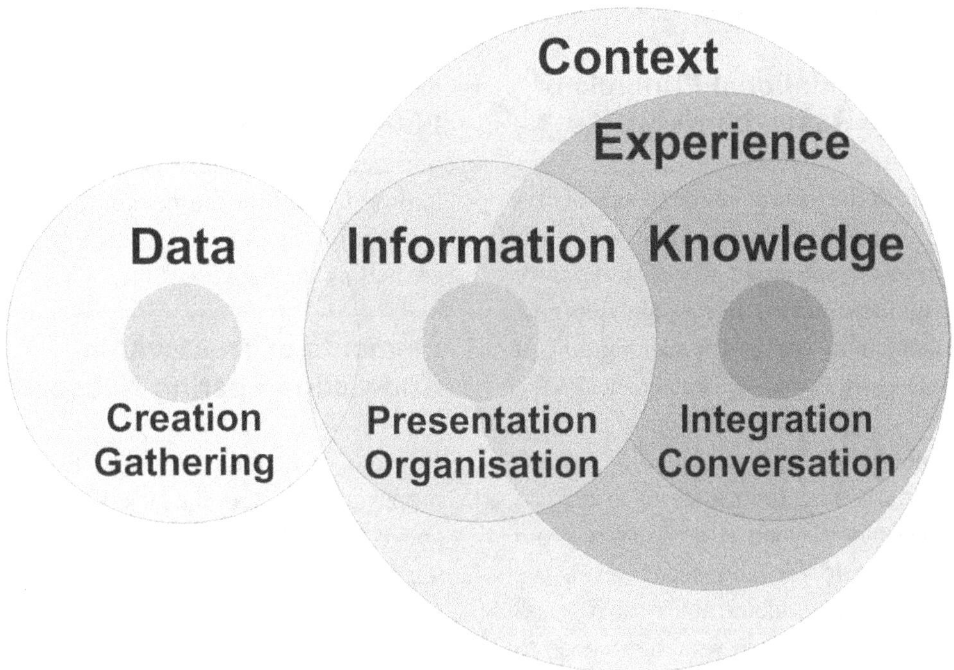

BACKGROUND

Ian McHarg's 1976 lecture "The Theory of Creative Fitting" (Margulis et al., 2007) outlines his vision for a collaborative knowledge building network. In his words, "information fragmented is of no use to anyone". This statement fits well with our current understanding of the data-information-knowledge ecosystem illustrated in Figure 1 (Shedroff, 1994).

McHarg continues in his lecture to describe his solution to the problem of fragmentation of information with respect to architectural and civil engineering planning in a sustainable manner, and presents the foundations of what we know today as Geographical Information Systems. In McHarg's system all the available geospatial data are available as overlay layers on top of a simple map of the region of interest. However, he also believed that the most difficult part of the system was to present the "value system" in which the data and information must be interested in order to create knowledge. Due to the perceived conflicting "value systems" of his data contributors, at least if these were not formally stated, McHarg's solution was to propose that "values systems" became explicit in his information presentations. He believed that every data point supplied to the presentation system should be identified with the name and the institution of the person who had prepared the analysis. In more recent discussions, the "value system" of an individual has been shown to be a motivator for becoming involved in knowledge sharing networks, if those networks share an interest and are meaningful to the individual (Gangné, 2009). McHarg was also aware that some analyses carry more weight, or trust, within a community stating that "one should generally be able to get agreement among scientist in [a given] discipline that this statement by that scientist represents the most current, best statement of the phenomena and processes within the discipline". It is in light of this analysis that this chapter continues its consideration of trust within knowledge sharing networks.

Trust as a Foundational Principle of Collaborative Research Networks

In his analysis of the lecture on the theory of creative fitting, James Corner states that McHarg's message is relevant and urgent to a contemporary audience as his ideas allow us new insights and new knowledge for a more complete understanding of the world (Margulis, Corner, & Hawthorne, 2007). As stated by Fox (2013a) modern informatics, in contrast to nearly all-existing technology-based approaches, starts with the use case. Instead of building successive layers of technology from the data level up to the knowledge level, the practice of informatics determines what questions people want to ask, and what tasks need to be performed. Modern informatics works from the upper levels down to meet (and often build or adapt) the underlying and evolving technical infrastructure. Fox goes on to describe the use case of a researcher wishing to explore, confirm or deny their results; or to discuss and collaborate in the creation of knowledge. In order for this to take place, research outputs must be citable, capable of being integrated or assimilated with other research outputs, and carry identity, or provenance. These criteria are fulfilled through the documentation of the processes used to create the research outputs with reference to explanation, justification, verification, proof and trust in the process used to create the research output and the information presented. This gives rise to accountability, transparency and translucency in research outputs, which has become increasingly important in recent years, particularly in the Earth sciences following on from the media hype surrounding the so-called "Climategate" incident (Nature, 2010). Leiserowitz, Maibach, Roser-Renouf, Smith, and Dawson (2013) report that this incident caused a measurable decline in the belief in the general public that climate change is happening, and may have caused erosion in public trust in research scientists as authoritative sources of information on global warming. It can therefore be seen that the importance McHarg places on trust in collaborative research networks not only affects the presentation of the results and the generation of knowledge within the network, but also the interpretation by outside parties of the knowledge created by the network.

Elements of Trust within Knowledge Sharing Networks

In order to better understand the mechanisms propagating trust, or the lack of trust as described above, both within and outside a knowledge sharing network, it is important to understand what factors make up "trust". In their study of trust in knowledge sharing networks, Abrams et al. (2003) identified two forms of interpersonal trust which enable effective knowledge creation and sharing. These are trust in a person's competence and trust in a person's benevolence. Abrams et al. (2003) define competence as "you have relevant experience and can be depended upon to know what you are talking about" and benevolence as "you care about me and take an interest in my well-being and goals". This has echoes of the value system of experts proposed by McHarg and presented earlier. A third element of trust in knowledge sharing networks was defined earlier by Mayer, Davis and Schoorman (1995) as integrity, defined as consistently adhering to a set of principles. This model of trust is visualized in Figure 2.

Despite their differing starting points of environmental architecture and business management, this fits neatly with McHarg's analysis that explicit value systems are required to build trust in collaboratively built knowledge.

Competence is viewed intuitively in a hierarchically restrictive schema (Kim, Ferrin, Cooper, & Dirks, 2003). This is manifested by a general belief that those who have a high degree

Figure 2. The model of trust used within this chapter. The critique of Abrams et al. (2003) that integrity is not a vital component of trust in collaborative knowledge sharing networks is reflected by the smaller set for that element.

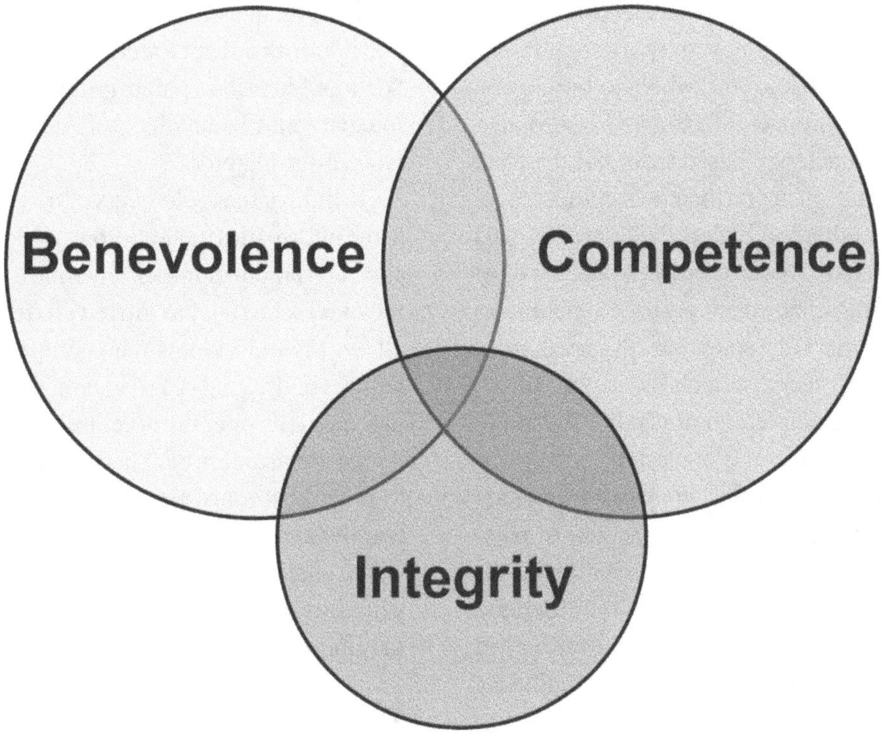

of competency can perform at a wide range of levels, whilst those who are perceived as of low competence can only perform at or below their perceived low competency level. This means that a person's competence tends to be inferred from the outcome of a single task: performed well it is used to infer competence; performed poorly and incompetence is inferred. In some ways this harks back to McHarg's idea that the recognized leaders in a field should be given the task of generating the knowledge for his data layers and needs to be overcome for a truly productive collaborative research network to succeed. Within the digital economy there is increasing use of "digital badges" to acknowledge and accredit competence (Adams & Demaiter, 2010). A digital badge is a validated indicator of competence that can be earned in many and various online environments, such as StackExchange (http://stackexchange.com/), Mozilla Open Badges (http://openbadges.org/) or the Society for Science and the Public (https://student.societyforscience.org/ssp-intel-digital-badging). Open badging makes it easy for anyone to issue, earn, and display badges across the web—through a shared infrastructure that's free and open to all.

Sharrat and Usoro (2003) acknowledge that one of the main barriers to knowledge sharing is the fear of "losing face" through adding an incorrect or misleading contribution to a discussion or network. This may also be coupled to a belief, however false, that one's contribution to a network may not be deemed sufficiently important or relevant. Belief that a network is benevolent will help with overcoming the first of these issues, as an increased belief that a network is benevolent lowers the barrier to contribution. Therefore, Sharrat and Usoro state that "the greater the perceived benevolence in a community the greater one's participation in knowledge sharing". Thus, engendering benevolence based trust in a collaborative research network makes knowledge sharing more likely amongst the participants and allows the knowledge of the whole to both grow and have greater impact. Sharrat and Usoro also considered competency-based trust, but the trust that an individual places on their own competence, and concluded that the higher the degree of this form of competency based trust, the lower the likelihood of one engaging positively in a collaborative research network. They state that "the greater the trust in the competence of one's own community, the less one's participation knowledge-sharing."

Integrity, which is the consistent adherence to a set of principles which the consumer finds trustworthy, may be seen as a key point of trust in a knowledge sharing network. Abrams et al. (2003) critique this standpoint as they believe that the consumer of knowledge or information is unlikely to find an issue with information or knowledge which is created under a different value system to that held by the consumer. However, this is slightly at odds with McHarg's original position as knowledge creation may be informed by the value systems at play at that time. As we also see from the data-information-knowledge ecosystem, the presentation and context of data and information play key roles in the creation of knowledge. It is therefore important that the consumer can make a judgment based on the value system of the data, information or knowledge creator as to the integrity of what they are being presented with in a collaborative network. In academic research networks, the new knowledge generated may carry much of the information required to make these value system judgments if it is published in an academic journal and has been subject to the peer-review process. For researchers in the sciences, the scientific method and the perceived reproducibility of the results of a knowledge building exercise will also be part of the process of identifying the integrity of a network. As noted above, there has been work undertaken in the software engineering realm to accredit competence and to propagate this accreditation, but similar examples for benevolence and integrity are not found within the literature. A "data publication" paradigm which encapsulates integrity and benevolence is described in detail later in the chapter.

In light of these elements of trust, we can return to the informatics methodology (Fox, 2013a) of use case driven development in small, expert teams. It can be seen that having a well defined use case allows the individuals who comprise the team to believe in the goals of the team, as they are transparent, increasing the perception of benevolence. By selecting a team of experts in the required fields to reach the goals, there will be a perception of competence within the team. Therefore, it can be stated that the informatics methodology is well suited to building teams inside which there is a likelihood of a high degree of trust.

Trust in a Distributed Collaborative Network

More and more often, collaboration in knowledge building is facilitated by World Wide Web tools to create virtual, distributed networks of researchers. It therefore follows that the above elements of trust need to be placed into the context of a distributed network, mediated through the Web. As reported by Hsu, Ju, Yen, and Chang (2007), many of the issues described above are also of relevance if the collaborative network is such a distributed, virtual community. Within their study they show that the belief of individuals to complete tasks and reach goals (self-efficacy) plays a crucial role in the engagement with a distributed community. Bearing this in mind, the technical infrastructure should be hidden from the users as much as possible to promote engagement, and this is a topic addressed below. Their findings also show that members of distributed collaborative networks become more likely to share knowledge if they believe it is for the greater good of the network (analogous to the discussion of benevolence above) and Hsu et al. (2007) suggest that credit mechanisms should be identified for promoting this benevolence, and approaching this from the "data publication" perspective is discussed later in this chapter. Another of their findings is that information based trust must be established and then identity based trust must follow. The next section of this chapter discusses provenance, which allows for the distribution of information based trust, and a sub-issue, that of persistently and uniquely identifying individuals within a distributed, Web based framework, is also considered. It is worth noting that there is evidence, as reported by Panteli and Sockalingam (2005), that the information and identity based trust form much stronger bonds within the collaborative network than the calculus based trust of a reward centred network.

PROVENANCE

The society of my new school, like that of provincial England as a whole, was so alarmingly codified and stratified that I couldn't conceive of anyone whose provenance and emotional valency weren't absolutely fixed. (Self, 1993)

The above quote, from Will Self's 1993 novel "My Idea of Fun", illustrates that by having strict values and practices, provenance can be fixed. In research environments, provenance is vital to be able to properly cite and give credit to the creator or creators of an output. Research outputs are generally held to be of good provenance when they are documented in detail sufficient to allow reproducibility (Altinas et al., 2004). Scientific workflows assist scientists and programmers with tracking their data through all transformations, analyses, and interpretations. Datasets are reliable when the process used to create them is reproducible and analyzable for defects (Boose et al., 2007). Current initiatives to effectively manage, share, and reuse ecological data are indicative of the increasing importance of data provenance. Examples of these initiatives are the United States' National Science Foundation Datanet projects, DataONE (https://www.dataone.org/) and Data Conservancy (http://dataconservancy.org/).

As described above, it is desirable for research knowledge to be citable in order to progress the discussion around research outputs and to enable accountability and transparency in the scientific or other research method. With reference to provenance, this means that research outputs must be fully documented such that a collaborator or consumer can use the data without further recourse to the originator. Renzl (2008) notes that the major stumbling block to achieving this goal is the fear of an individual researcher, or group of researchers, losing their unique value through creating and publishing the required documentation. In Renzl's analysis this may be mitigated through a belief that the "benevolence" of managers and peers will provide credit for the research and that this belief has a positive impact on the documentation of knowledge, and in turn the documentation of knowledge has a positive impact on the sharing of that knowledge. Within the domain of the Earth sciences, the desire to give quantifiable credit for the proper documentation of knowledge has seen a recent push towards the "Data Publication" paradigm.

"Data Publication"

It is possible to "publish" data relatively easily: at its most basic all a researcher has to do is to put the files on a website somewhere. This makes the data accessible, but without any form of long-term commitment. There are no guarantees that the data will still be in the same location in six months, or that the files have not become corrupted. Furthermore, it is possible that a scientist who is not the data creator will not be able understand the contents or even open the files. Even if the dataset is readable and has sufficient metadata, there is no information about the scientific quality of the dataset, other than that associated with the creator's reputation.

By contrast, a formal "publishing" process adds value to the dataset for the data originator and for future users of the data. Publishing may provide an indication of the scientific quality and importance of the dataset (as measured through a process of peer review), and by ensuring that the dataset is complete, frozen and has enough supporting metadata and other information to allow it to be used by others in the years to come. "Publishing" also implies a commitment to persistence of the data and allows data producers to obtain academic credit for their work in creating the datasets.

There have been many discussions held about closed versus open data, and there will be many more in the future. What is generally well agreed is that it is no longer appropriate to keep significant datasets stored on a single hard drive, or several CDs in a drawer in an office somewhere.

Previously, there was little benefit to a scientist in making their dataset available as a free download from a webpage, unless they worked in certain areas of science where this is expected (e.g., for genetic sequence data from GenBank, http://www.ncbi.nlm.nih.gov/genbank/). In fact, prior to this, the reputational risk of doing so (others might find errors, or worse, take advantage of the dataset to earn new research funding) and the extra work involved in doing so, might mean that the scientist would prefer to store the data on a closed server. However, if the dataset author could receive full citation credit for their data collection effort, thus contributing to measurable performance metrics, motivation for data publication would be greatly increased. Additionally, funding agencies are requiring that data gathered through funded research be made accessible. These funding agencies are requesting a Data Management Plan to be submitted in grant proposals. Thus, data centres are working with scientists to bring data from the closed servers and CDs into archives where they can be properly curated, with the eventual aim of publication and the dataset author receiving full academic credit for their efforts.

The advent of funding agency mandates for open data, such as the National Science Foundation requirement that a data management plan be included in proposals and the European Commission's recent recommendation for open access to scientific data, is expected to provide incentive for authors to make data available. The Scholarly Publishing & Academic Resources Coalition has recently published guidelines on implementing an Open Data Policy (Tananbaum, 2013). The ability to cite one's data accurately makes openness advantageous to research scientists.

The assignment of persistent identifiers, specifically Digital Object Identifiers (DOIs, Bunakov, Jones, and Matthews, 2015, see chapter 7 this book), enables accurate data citation. Data publication that enables data citation can certainly be an incentive to make data accessible.

There are many examples of data publication from throughout the sciences: from data journals in Facilities Science (Bunakov, Jones, and Matthews, 2015, see chapter 7 this book), chemistry and physics (Journal of Chemical and Engineering Data, Journal of Physical and Chemical Research Data) to the Earth Observing System Data and Information System providing DOIs to data from NASA. An example from the oceanographic domain shows that marine science librarians, such as those at the Marine Biological Laboratory of Woods Hole Oceanographic Institution, are increasingly becoming an integral partner in this data publication process. Through their knowledge of cataloguing and metadata creation, and their role in maintaining institutional repositories, they play an important part in making data accessible to all those interested in scientific information and data: science colleagues, policy makers, academics and the interested public. The professional librarian has long-standing knowledge of the publishing process, the new online scholarly information cycle, and the standards necessary to link data to publications.

There has been much discussion around the metaphor of "publication" and whether this is the most suitable term or not. Many of the salient arguments and alternative metaphors are discussed by Parsons and Fox (2013). An alternative to the "publication" metaphor is "data release", analogous to the product release pattern used by the software industry (Humble & Farley, 2010). Relevant to the discussion here are alpha, beta and production, or live, releases. Alpha release in terms of software is the first phase to begin software testing. In this phase, developers generally test the software using white box techniques. Additional validation is then performed using black box or gray box techniques, by another testing team. Software in the beta phase will generally have many more bugs in it than completed software, as well as speed/performance issues and may still cause crashes or data loss. The focus of beta testing is reducing impacts to users, often incorporating usability testing. The process of delivering a beta version to the users is called beta release and this is typically the first time that the software is available outside of the organization that developed it. Production release is a term used when a software product is ready to be delivered or provided to the customer. This build may be digitally signed, allowing the end user to verify the integrity and authenticity of the software purchase. Carbotte et al. (2013) present a workflow which lends itself to this second analogy, publishing data collected by the United States' academic research vessel fleet at various stages of their lifecycle by the Rolling Deck to Repository (R2R) programme. The original workflow is presented here, annotated with analogous software release cycle points.

R2R provides a uniform approach to the stewardship of routinely collected environmental sensor data collected by the 26 active service vessels in the United States' academic research fleet. It is a multidisciplinary collaborative programme, with participants ranging from technicians to schedulers, data curators, research scientists and software developers and covering fields of oceanography, geophysics and meteorology. R2R follows the DataCite (http://www.datacite.org/) policy for issuing DOIs for datasets, using an international standard metadata kernel populated using controlled vocabularies and conforming to best practises. R2R first publishes data with a DOI for the original field data from a ship's environmental sensors. These data have not been processed in any way, and is the first phase of the data release, analogous to an alpha software release. There may be errors in the data caused by any number of reasons, including sensor errors or documentation issues. R2R then also publishes DOIs for datasets which have been used to create the standard data products for the programme. These data products include quality controlled navigation data and geophysical profiles. The lineage of the raw environmental sensor data used to create these products is preserved and explicitly stated through the DOI of the raw dataset. As these data products have a greater usability for general consumers of the data and may therefore be thought of as analogous to a beta software product. Finally, a domain specific repository (such as the Biological and Chemical Oceanography Data Management Office (http://www.bco-dmo.org/) at the Woods Hole Oceanographic Institution) publishes fully processed, calibrated and amalgamated datasets with DOIs, maintaining the reference to the raw data streams and intermediate data products used to create the integrations through their DOIs. This final data version may be used to support journal papers and may be considered analogous to the publication release in the software lifecycle.

In a move to give credit to researchers who publish their data in this manner, in what may be seen as fulfilling Renzl's noted requirement that the sharing of documentation is promoted by giving credit, a Data Citation Index has been started by Thomson-Reuters (http://wokinfo.com/products_tools/multidisciplinary/dci/). The Data

Citation Index will allow researchers to receive credit for the creation of scholarly datasets and to track the use and importance of their datasets as they are used to generate new knowledge across multiple domains. Datasets catalogued within the Data Citation Index will be displayed in context along with all the academic articles which cite that dataset, and therefore the research which that dataset has influenced.

Whichever metaphor is chosen for this activity, and no matter how the impact is measured or the credit is given to the originator, sight must not be lost of the overarching goal: to facilitate reproducibility of the research.

The Semantic Web and Linked Data

The solution to the issue of provenance given above is based on viewing the World Wide Web (WWW) as a vast library, storing and documents, images and other media in formats accessible to a human audience through a browser. However, the content of the WWW is largely unstructured. More advanced viewpoints exist, and provide their own mechanisms for attributing and propagating provenance. The Semantic Web is an extension of these fundamental ideas of the WWW into a web of data. This is achieved through the annotation of documents with hyperlinks to machine-readable documents. These machine-readable documents describe the meaning of given fields or keywords (Berners-Lee, Hendler, & Lassila, 2001; Shadbolt & Berners-Lee, 2008).

The Semantic Web relies on machine readable resources for the annotation of data and metadata elements. Within these resources there exists a broad spectrum in the complexity and the level of detail encoded. The points of this spectrum, often called the "ontology spectrum" or "semantic spectrum", are defined by McGuinness (2003). In overview, the simplest point on the spectrum is the catalogue (or glossary, as in Navigli & Velardi, 2008) that simply provides a list of terms used to annotate data and their definitions. A catalogue is not necessarily published to either the WWW or Semantic Web. Nor does a catalogue require a group of experts to provide governance over its content. Complexity is increased in the controlled vocabularies (Warner, 2002) and data dictionaries (Kumpati, 1985, IBM, 1993) which add, respectively, naming conventions and publication on top of the catalogue. A data model (van Renssen, 2005) then adds some definition of the relationship between the entities of a data dictionary. A taxonomy further adds a tree hierarchy with the concepts at the bottom of the tree inheriting the properties of those concepts at the top of the tree (Navigli & Velardi, 2008). A concept can be viewed as the definition of an idea or notion; a unit of thought (Miles & Bechhofer, 2009).

A more complex point of the semantic spectrum is the ontology. An ontology describes all the concepts in a given domain and the relationships between them, and following best practice (Cox, 2012), is conceptualised using Uniform Resource Identifiers (URIs) for each of the data elements, relationships and property types. An ontology can be parsed by a reasoning engine and then used to infer knowledge about the domain.

Berners-Lee (2006) uses these fundamental design tenets to formulate the idea of a Web of Data., The solution he proposed here is commonly known as Linked Data. Linked Data is built on four pillars. Firstly, URIs are used to name data or metadata resources published on the World Wide Web. Second is that users should be able to look up the URIs using a web browser (that is, the URIs use the HyperText Transfer Protocol, HTTP, so that they appear as web addresses). Third, the document to which the URI points should be encoded in a standard, as laid out by the World Wide Web Consortium. Finally, there should be links from one Linked Data document to the URIs of other Linked Data documents, so that users can discover more data or metadata (data about data) resources.

PROV Ontology

A group of documents describing a formal model of provenance is defined by Groth and Moreau (2013). They take as their starting point the view that "provenance is information about entities, activities, and people involved in producing a piece of data or thing, which can be used to form assessments about its quality, reliability or trustworthiness". The documents which they present are known as The PROV Family. PROV defines a model to enable the interchange of provenance information in heterogeneous environments such as the World Wide Web. The PROV data model has three simple structures, or classes, at its heart: the entity; the activity; and the agent. The entity is an object of which the provenance is being described, such as the data layers in McHarg's presentations. The activity is the generation, processing, modification or use of an entity, and could describe the process gone through in creating a data layer. Finally an agent is someone or something responsible for making an action that has bearing on an entity, for instance the expert who collected and processed the data to create a data layer. The PROV model and its associated documentation is an important project as it is defined and standardised by the World Wide Web Consortium (W3C), the body who sets the standards on which the Web is built.

In addition to the data model itself, a range of representation formats for the data model are also offered by the PROV documents. The PROV ontology represents the PROV data model in the Web Ontology Language, a W3C standard for encoding in the Semantic Web. The PROV ontology provides a set of classes, properties, and restrictions that can be used to represent and interchange provenance information generated in different systems and under different contexts. It can also be specialized to create new classes and properties to model provenance information for different applications and domains.

Ontology Extension

Developing any level of semantic resource, from a data dictionary to a formal ontology (for example the PROV ontology), is a valuable activity as it provides a new level of interoperability between the metadata and data annotated with terms defined by that resource. However, a much richer semantic resource may be developed by looking for commonality between concepts defined in different resources. The relationships between these common concepts can then be described and encoded. This is the process known as ontology extension. Ontology extension ensures that a resource specific to a domain is compatible with more general, high level semantic resources or ontologies. This can be seen to promote interoperability between domain specific semantic resources. Ontology extension may also be of particular use when an authoritative or legally required resource exists and it is wished to create links to the concepts published by that resource. Examples of these resources include: the World Register of Marine Species (WoRMS, http://www.marinespecies.org/) which aims to publish the most authoritative list of names of marine species; the European Research Vessel Infobase (http://www.rvinfobase.eurocean.org/) which maintains a searchable database of the European research vessels including vessel specifications and contact information; and European Commission INSPIRE directive code lists. The ontology extension approach also allows the reuse of classes defined by other ontologies or to define the class to which the concepts defined in a data model or taxonomy belong.

The ontology extension approach has been taken in the Deep Carbon Observatory project (Fox, 2013b). The starting point was the PROV ontology described above, and terms were integrated from the VIVO (http://www.vivoweb.org/), Friend of a Friend (FOAF, http://www.foaf-project.org/), the BIBO bibliographic ontology (http://bibliontology.com/), Observations and Measurements, Virtual Solar-Terrestrial Observatory (Fox et al., 2009) and Ocean Data Ontology (http://ocean-data.org/odoBrowse/odoBrowse.html) resources. The information models of these ontologies provide domain level views of the data and cover a wide knowledge base encoded in Semantic Web standards. This approach has integrated, at the application level, many of the key research functions required by the Deep Carbon Observatory community in a single, collaborative environment hosted on the World Wide Web. Fox (2013b) reports that most importantly, the technical details have been hidden from the users: they see the benefits but do not need to be informatics experts in order to use the system. Continuing with this approach, an overview example of mixing the PROV ontology with classes from the Ocean Data Ontology and the Observations and Measurements standard is shown in Figure 3.

Unique and Persistent Identifiers for People

As we have seen, the Semantic Web and Linked Data rely on the identification of objects within the World Wide Web, normally through the use of URIs. An interesting issue within this arena was identified during the preparation of the Ocean Data Publication Cookbook (Leadbetter et al., 2013) namely the unique and persistent identification of individual people within a system. The importance of identifying individuals is highlighted in the book *Neosentience*, where Rösseler and Seaman (2011) claim that "identity is the greatest miracle". In countries where data protection acts exist (for example in the United Kingdom, OPSI, 1988) there is a legal obligation to obtain permission of individuals before storing and disseminating information that makes them identifiable. It is therefore difficult under these circumstances to

Figure 3. A practical example of ontology extension. The base ontologies here are indicated by: a) the prov: prefix for the PROV ontology; b) the foaf: prefix for the Friend-of-a-Friend ontology; c) the ocean-data: prefix for the Ocean Data Ontology; and d) the o&m: prefix for the Observations and Measurements ontology.

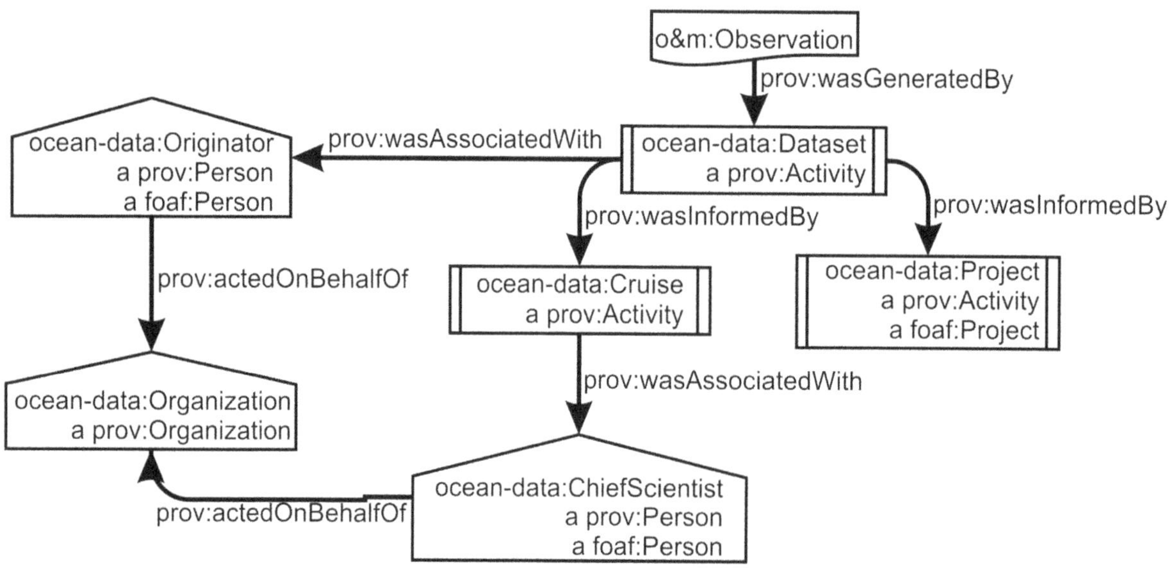

create an exhaustive database of all researchers in a given field, let alone all researchers who may contribute to a cross-disciplinary collaborative project. Indeed this may also be seen as a fruitless exercise as the database created would require a significant maintenance overhead, and would probably be little more than a fire fighting exercise.

Therefore two options exist. The first is to identify users uniquely within a given, smaller, context (for example a research project or programme), and then to provide relationships using RDF semantics to identify instances from different contexts as synonymous. In systems adopting this approach, best practise is to use existing Semantic Web resources such as the FoaF vocabulary such that machine-to-machine interactions are more easily facilitated. The second is to make use of other services which researchers may have subscribed to in order to provide the unique persistent identifier. One such system is the ORCID Registry (http://orcid.org/) which provides both the unique identifier to an individual and an Application Programming Interface allowing the development of tools and systems against the unique identifiers it serves. An alternative system to ORCID is VIVO, which assigns persistent identifiers to people, organisations and data entities. The VIVO ontology has also been extended to represent other areas of science research networks such as datasets, instruments and field sites (Fox, 2013b). This use of VIVO is perhaps closer to McHarg's original vision than any of the other approaches discussed here.

There are positive and negative aspects to both of these solutions, both of which derive from the desire to create identifiers which are both "persistent" and "unique". Where a system is used over which the collaborative network has full control, such as the Deep Carbon Observatory identifier (Fox, 2013b), the persistence of the identifiers is almost guaranteed as, even if a project is no longer funded, research groups and institutes will ordinarily find ways of keeping systems alive. However, there is no guarantee that all the RDF relationships to other instances of a researcher in different systems have been accounted for and so the uniqueness of the identifier is at risk. In contrast, a system over which the collaborative network which desires to use persistent identifiers has no control (e.g. leveraging ORCID identifiers) there is an issue of persistence. While the chances are low, there is a risk that the centralised database may become unusable for a time, and so the opaque identifiers cannot be resolved to give meaningful information, or the central database may cease to exist. This may be mitigated by caching the information from the central database, but there are considerations to be made regarding access, data duplication and relevancy and copyright if this path is pursued. The plus side of adopting this approach is that the identifier to the researcher will be almost certainly unique.

Practical Implementations of the Ideas in this Chapter

In this section, two projects which have begun to implement the ideas are discussed. The first is a practical implementation of a mechanism for propagating the "value system" of a set of measurements through a chain of processing services which may be accessed through an online interface. The second is a reference implementation of the data publication paradigm, which will be shown to positively influence the perception of integrity and benevolence in the research outputs of knowledge building networks.

The NETMAR project developed a pilot European Marine Information System (EUMIS) for searching, downloading and integrating satellite, in situ and model data from ocean and coastal areas. It is a user-configurable system offering flexible service discovery, access and chaining facilities using World Wide Web and Open Geospatial Consortium standard technologies. It uses a semantic framework coupled with ontologies for identifying and accessing distributed data, such as near-real time, forecast and historical data.

EUMIS also enables further processing of such data to generate composite products and statistics suitable for decision-making in diverse marine application domains. One of the main developments within EUMIS was a service chaining system allowing data and processing tools to be created, stored and deployed in a distributed manner but providing a collaborative Web based interface to their use and execution. Due to the nature of the chained data and services (for example, estimating phytoplankton concentrations from ocean colour values monitored by satellites) the generation of errors or uncertainties in their outputs is an inherent risk. These errors or uncertainties may be seen to be the "value system" of the measurements, and therefore, in light of McHarg's theory of creative fitting, identifying and codifying them is important for full understanding of the system. The project therefore also investigated the propagation of these uncertainties through a data and service chain (Lassoued, Leadbetter, Clements, & Grant, 2011).

The methods introduced by NETMAR provide the generic mathematical framework for estimating uncertainty propagation in the service chains. Note that in order to estimate the uncertainty of the output data of an operation; it is necessary to know what the operation does exactly. Therefore, computing the output uncertainty must be the responsibility of the service performing the operation. The general uncertainty propagation formulae suppose that the operation or function being dealt with is continuous and derivable. In reality this cannot be guaranteed, as functions may be non-derivable or even discrete. For instance, an operation that applies a filter to drop those pixels whose values are below or above a specified threshold is not continuous or derivable. In these cases it is up to the service operation developer to define the appropriate uncertainty propagation method, which must be part of the service itself or, more specifically, the operation in question. It is important to note that uncertainty has a scope, that is it must always be related to a parameter or attribute, such as position (e.g. positional accuracy), time (e.g. temporal accuracy) or temperature. With reference to the Semantic Web methods described above, NETMAR collaborated with the UncertWeb project in the semantics used to describe the uncertainty. The UncertWeb project has its own semantic resources for describing statistics and distributions within UncertML. The UncertWeb project has both content and technical governance responsibility for this resource and progress has been made to making the UncertML definitions available in a Semantic Web standard form.

A second example is driven by a use case arising from the fact that there is an increasing pressure from the scientific community for an intimate linkage between papers published in the scientific literature and the data upon which they are based and for a mechanism to reward data collection through citation. As identified above, the data publication paradigm can be seen as a mechanism to increase the perceived benevolence of an individual or a group. However, these requirements expose a philosophical difference between the data serving practices of many data centres and the publication model. The publication process exposes a fixed copy of an object and then manages that copy in such a way that it may be located and referred to over an indefinite period of time. There is a fundamental assumption that the copy of the object will be exactly the same each time it is referenced. Many environmental data centres, including the British Oceanographic Data Centre up until 2011, consider data to be a dynamic entity that improves with time and that users should be served with the best currently available version of the data of interest. Past versions of data objects are not retained, particularly those resulting from relatively trivial changes such as improved wording of parameter definitions. Consequently, were a data object to be cited as published there is no guarantee that it would be the same each time it was accessed which may necessitate a perceived reduction in integrity, particularly in reference to the rigorous scientific method which demands exact reproducibility. To address this BODC has developed the 'Published Data Library' (PDL, Urban et al., 2013). The PDL enables specific datasets to be cited in journal papers through the assignment of a DOI in collaboration with the British Library and the global DataCite initiative. There is a fundamental assumption that the copy of the dataset will be exactly the same each time it is referenced.

It should be noted that the assignment of a DOI is not a substitute for long-term data management and curation by a suitable environmental data centres, which enables users to construct their own datasets from all data holdings. The PDL is designed for base datasets suitable for future re-use in other applications, rather than data reworked specifically for a single research publication — the latter sometimes termed 'data behind the graph'. It is based on a model in which multiple copies of datasets are stored indefinitely and in full. Data storage, however, is not an infinite resource. Therefore, the system has obvious limits to the size and number of datasets that it can handle. Procedures are in place to decide which datasets should be included. As the number of fully published datasets grow and their citations in the literature increase, there will become opportunities to use the positive impact os systems like this to gain funding to increase the storage base and so begin to relax some of these restrictions.

PDL datasets are of two types. The first is datasets that have not yet been ingested into the BODC system, but are destined for future ingestion. Candidate datasets are identified through negotiation between data originators and BODC. The technical quality of these datasets (including metadata) to an acceptable standard is the responsibility of the data originator. BODC judges the acceptability of candidate datasets in terms of their completeness, but not in terms of their scientific quality or value. The second PDL data type is datasets that have been ingested into the BODC system and subsequently exported. Candidate datasets of this type are identified through discussion between the scientists who supplied the data and BODC. The technical quality of these datasets is BODC's responsibility.

In response to the ideas of the Semantic Web and Linked Data, outlined above, the web pages to which a DOI assigned to PDL resource resolves are published in both human and machine readable formats allowing for both human and computer networks to assess the provenance of a dataset. Ongoing work is using the PROV ontology to serialise the authors of datasets and the actions of the data publishing centre in minting the DOI in a way which is compatible with these methodologies.

FUTURE RESEARCH DIRECTIONS

Future research directions should: emphasize the adoption of mechanisms for accrediting competence recognition to individuals, for instance through the use of digital badges, and how these are adopted by research communities; drive the harmonization of unique and persistent identifiers for individual researchers; and designing software tools beyond those described here to attribute levels of benevolence, and where appropriate integrity, and propagate these through the collaborative research space. Within the data publication paradigm expanded above, there is work to be carried out to identify how the proposed data citation indices will carry weight within the community, and how this will translate into increased trust within knowledge creation networks, and within the users of the research outputs of these networks.

CONCLUSION

In the forty years since McHarg proposed his "Theory of Creative Fitting", one of his central tenets – trust in the collaborative research network

– has remained of paramount importance, to the extent that even now the theory can be described as “relevant and urgent”. In examining trust in collaborative research networks, it has been seen that there are mechanisms emerging which can promote trust both within and without the network. Three main behaviours emerge which should be encouraged within collaborative networks: competency, benevolence, and integrity. Research from the business community has shown that trust is enhanced through the sharing of documentation, but that that sharing of documentation is itself promoted when credit may be received for the action. Cutting-edge technologies deployed within the framework of the World Wide Web and Linked Data have given methods for the identity of data creators to be preserved as their datasets are passed along research chains, and for the provenance of datasets to be explicitly stated in standardized ways. Activities from the sciences, highlighted in this chapter by projects in the marine environmental sciences domain, have shown that mechanisms are in place for sharing of data and its descriptions. The academic publishing industry is beginning to put weight behind these activities and to supply mechanisms for the attribution of credit to the researchers who participate in them. For the continued productivity of collaborative research networks, and for trust to be engendered in their results beyond the borders of the networks, all of these techniques must be combined to create a lasting legacy.

REFERENCES

Abrams, L., Cross, R., Lesser, E., & Levin, D. (2003). Nurturing interpersonal trust in knowledge-sharing networks. *The Academy of Management Executive*, *17*(4), 64–77. doi:10.5465/AME.2003.11851845

Adams, T. L., & Demaiter, E. I. (2007). *Knowledge workers in the New Economy: Skill, flexibility and credentials.* American Sociological Association. Retrieved from http://citation.allacademic.com/meta/p178393_index.html

Altintas, I., Berkley, C., Jaeger, E., Jones, M., Ludäscher, B., & Mock, S. (2004). Kepler: an extensible system for design and execution of scientific workflows. In *Proceedings of 16th International Conference on Scientific and Statistical Database Management*, (pp. 423–424). doi:10.1109/SSDM.2004.1311241

Berners-Lee, T. (2006). *Linked Data – Design Issues*. Retrieved from http://www.w3.org/DesignIssues/LinkedData.html

Berners-Lee, T., Hendler, J., & Lassila, O. (2001). The Semantic Web. *Scientific American*, *284*(5), 34–43. doi:10.1038/scientificamerican0501-34 PMID:11396337

Boose, E., Ellison, A., Osterweil, L., Clarke, L., Podorozhny, R., & Hadley, J. et al. (2007). Ensuring reliable datasets for environmental models and forecasts. *Ecological Informatics*, *2*(3), 237–247. doi:10.1016/j.ecoinf.2007.07.006

Bunakov, V., Jones, C., & Matthews, B. (2015). Towards the interoperable data environment for Facilities Science. In P. Diviacco, P. Fox, C. Pshenichny, & A. Leadbetter (Eds.), *Collaborative Knowledge in Scientific Research Networks*. Hershey, PA: IGI Global.

Carbotte, S., Arko, R., Ferrini, V., Stocks, K., Miller, S., Maffei, A., Chandler, C., Smith, S., & Bourassa, M. (2013). Rolling Deck to Repository (R2R): Working with international partners to improve data discovery and access. *Bollettino di Geofisica Teorica ed Applicata, 54*(Supplement 1), 38-39.

Cheshire, C., & Cook, K. (2004). The emergence of trust networks under uncertainty - implications for internet interactions. *Analyse & Kritik*, *26*, 220–240.

Cox, S. (2012). *Best practice in formalizing a SKOS vocabulary*. Solid Earth and Environment GRID, Commonwealth Scientific and Industrial Organisation (CSIRO). Retrieved from https://www.seegrid.csiro.au/wiki/Siss/VocabularyFormalizationInSKOS

Fox, P. (2013a). *How environmental informatics is preparing us for the era of Big Data*. Paper presented at 2013 Fall Meeting, San Francisco, CA.

Fox, P. (2013b). *Progress in open-world, integrative, collaborative science data platforms*. Paper presented at 2013 Fall Meeting. San Francisco, CA.

Fox, P., McGuinness, D., Cinquini, L., West, P., Garcia, J., Benedict, J., & Middleton, D. (2009). Ontology-supported scientific data frameworks: The Virtual Solar-Terrestrial Observatory experience. *Computers & Geosciences*, *35*(4), 724–738. doi:10.1016/j.cageo.2007.12.019

Gangné, M. (2009). A model of knowledge-sharing motivation. *Human Resource Management*, *48*(4), 571–589. doi:10.1002/hrm.20298

Groth, P., & Moreau, L. (2013). *An overview of the PROV family of documents*. World Wide Web Consortium. Retrieved from http://www.w3.org/TR/prov-overview/

Hsu, M.-H., Ju, T., Yen, C.-H., & Chang, C.-M. (2007). Knowledge sharing behaviour in virtual communities: The relationship between trust, self-efficacy, and outcome expectations. *International Journal of Human-Computer Studies*, *65*(2), 153–169. doi:10.1016/j.ijhcs.2006.09.003

Humble, J., & Farley, D. (2010). *Continuous Delivery: Reliable Software Releases Through Build, Test, and Deployment Automation*. Addison Wesley.

IBM. (1993). *IBM Dictionary of Computing*. New York: McGraw-Hill.

Kim, P., Ferrin, D., Cooper, C., & Dirks, K. (2003). Removing the shadow of suspicion: The effects of apology versus denial for repairing competence-versus integrity-based trust violations. *The Journal of Applied Psychology*, *89*(1), 104–118. doi:10.1037/0021-9010.89.1.104 PMID:14769123

Kumpati, M. (1985). *US Patent 4774661*. Washington, DC: US Patent Office.

Lassoued, Y., Leadbetter, A., Clements, O., Grant, M. (2011). *NETMAR semantic framework specification – Version 1*. NETMAR: Open Service Network for Marine Environmental Data. European Commission within the Seventh Framework Programme (2007-2013) Theme ICT 2009.6.4 Project 249024. Deliverable 4.3.1.

Leadbetter, A., Raymond, L., Chandler, C., Pikula, L., Pissiersens, P., & Urban, E. (2013). *Ocean Data Publication Cookbook*. Paris: UNESCO.

Leiserowitz, A., Maibach, E., Roser-Renouf, C., Smith, N., & Dawson, E. (2013). Climategate, public opinion, and the loss of trust. *The American Behavioral Scientist*, *57*(6), 818–837. doi:10.1177/0002764212458272

Manola, F., & Miller, E. (2004). *RDF Primer*. World Wide Web Consortium. Retrieved from http://www.w3.org/TR/rdf-primer/

Margulis, L., Corner, J., & Hawthorne, B. (Eds.). (2007). *Ian McHarg: Conversations with students: Dwelling with nature*. Princeton Architectural Press.

Mayer, R., Davis, J., & Schoorman, F. D. (1995). An integrative model of organizational trust. *Academy of Management Review*, *20*(3), 709–734. doi:10.5465/AMR.1995.9508080335

McGuinness, D. (2003). Ontologies come of age. In D. Fensel, J. Hendler, H. Lieberman, & W. Wahlster (Eds.), *Spinning the Semantic Web: bringing the World Wide Web to its full potential* (pp. 171–196). Cambridge, MA: MIT.

Miles, A., & Bechhofer, S. (2009). *SKOS simple knowledge organization system reference*. World Wide Web Consortium. Retrieved from http://www.w3.org/TR/2009/REC-skos-reference-20090818/

Nature. (2010). Editorial: Closing the climategate. *Nature, 468*, 345.

Navigli, R., & Velardi, P. (2008). From Glossaries to ontologies: extracting semantic structure from textual definitions. In P. Buitelaar & P. Cimiano (Eds.), *Ontology learning and population: bridging the gap between text and knowledge* (pp. 71–87). Amsterdam: IOS Press.

OPSI. (1988). *Data Protection Act 1998*. London: Office of Public Sector Information. Retrieved from http://www.legislation.gov.uk/ukpga/1998/29/contents

Panteli, N., & Sockalingam, S. (2005). Trust and conflict within virtual inter-organizational alliances: A framework for facilitating knowledge sharing. *Decision Support Systems*, *39*(4), 599–617. doi:10.1016/j.dss.2004.03.003

Parsons, M., & Fox, P. (2013). Is data publication the right metaphor? *Data Science Journal*, *12*(0), 32–46. doi:10.2481/dsj.WDS-042

Quin, L. (2013). *Extensible markup language (XML)*. World Wide Web Consortium. Retrieved from http://www.w3.org/XML/

Renzl, B. (2008). Trust in management and knowledge sharing: The mediating effects of fear and knowledge documentation. *Omega*, *26*(2), 206–220. doi:10.1016/j.omega.2006.06.005

Rösseler, O., & Seaman, B. (2011). *Neosentience: The benevolence engine*. Intellect Ltd.

Self, W. (1993). *My Idea of Fun*. London: Bloomsbury Publishing.

Shadbolt, N., & Berners-Lee, T. (2008). Web science emerges. *Scientific American*, *299*(4), 76–81. doi:10.1038/scientificamerican1008-76 PMID:18847088

Sharratt, M., & Usoro, A. (2003). Understanding knowledge-sharing in online communities of practice. *Electronic Journal of Knowledge Management*, *1*(2), 187–196.

Shedroff, N. (1994). *Information interaction design: A unified field theory of design*. San Francisco: Vivid Studios.

Tananbaum, G. (2013). *Implementing an open data policy: A primer for research funders*. Scholarly Publishing and Academic Resources Coalition. Retrieved from http://www.sparc.arl.org/sites/default/files/sparc-open-data-primer-final.pdf

Urban, E., Leadbetter, A., Moncoiffe, G., Pissiersens, P., Raymond, L., & Pikula, L. (2012). Pilot projects for publishing and citing ocean data. *Eos, Transactions, American Geophysical Union*, *93*(43), 425–426. doi:10.1029/2012EO430001

van Renssen, A. (2005). *Gellish: A generic extensible ontological language*. (Doctoral dissertation). Delft University of Technology, Delft, The Netherlands.

Warner, A. (2002). A taxonomy primer. Austin, TX: School of Information, University of Texas at Austin. Retrieved from http://www.ischool.utexas.edu/~i385e/readings/Warner-aTaxonomyPrimer.html

Ziegler, C.-N., & Lausen, G. (2005). Propagation models for trust and distrust in social networks. *Information Systems Frontiers*, *7*(4/5), 337–358. doi:10.1007/s10796-005-4807-3

ADDITIONAL READING

Bachmann, R., & Zaheer, A. (Eds.). (2006). *Handbook of Trust Research*. Edward Elgar. doi:10.4337/9781847202819

Herrington, S. (2010). The Nature of Ian McHarg's Science. *Landscape Journal*, *29*(1), 1–20. doi:10.3368/lj.29.1.1

KEY TERMS AND DEFINITIONS

Benevolence: An element within the overall concept of trust, which is defined as care about the well-being of other individuals and the goals of a community.

Competence: An element within the overall concept of trust, which is defined as the relevant experience and knowledge of an individual.

Creative Fitting: Ian McHarg's scientific theory explaining and validating designing with nature, using ecological data and producing outcomes matching the processes of adaptation and evolution.

Integrity: A minor element within the overall concept of trust, which is defined as an individual's consistent adherence to a set of principles which they find trustworthy.

Linked Data: Linked Data is a technique which uses the World Wide Web to connect related data that was not previously linked, or using the Web to lower the barriers to linking data currently linked using other methods. In practice, Linked Data often uses the Resource Description Framework model and Web addresses (as Uniform Resource Locators, or URLs) to achieve these linkages.

Provenance: Provenance is information about entities, activities, and people involved in producing a piece of data or thing, which can be used to form assessments about its quality, reliability or trustworthiness.

Resource Description Framework (RDF): RDF is a standard model for data interchange on the World Wide Web. The RDF data model is based upon the idea of making statements about resources in the form of subject-predicate-object expressions, known as triples. A classical example of the subject-predicate-object triple is *the sky-has colour-blue*.

Semantic Web: The Semantic Web is the World Wide Web of data built on the Resource Description Framework model as a foundation for publishing and linking data.

Trust: The reliance of an individual on another person or entity. Trust can be attributed to relationships between people. Trust is also attributable to relationships within and between social groups (families, friends, communities, organisations, companies, nations etc.).

Value System: A set of consistent personal and cultural values used for the purpose of ethical or ideological integrity.

Chapter 3
Collaborative Progress in Citation Networks

Rogier De Langhe
Ghent University, Belgium

ABSTRACT

Philosophical theories of scientific progress are typically disconnected from citation data because a citation to a paper does not necessarily justify the content of the cited paper. Citation data can however be used to test whether scientific contributions coevolve and as such discriminate indirectly between the two main theories of scientific progress: cumulative and non-cumulative progress. This chapter presents this novel approach. First, agent-based models are used to discover essential differences between both patterns of progress. The systematic exploration they allow of their respective entailments reveals four conflicting empirical predictions. These could in principle be tested against citation data, thus operationalizing two important philosophical conceptions of progress. The proposed approach relies heavily on two recent developments, the use of agent-based modeling in philosophy and the availability of vast citation datasets. The research program it suggests offers a unique opportunity to bridge the gap between descriptions of science and explanations of why it is successful.

INTRODUCTION

Agent-based models allow to study the patterns resulting from the interactions of agents with pre-specified behavior. Agent-based models are instrumental in isolating and describing essential characteristics of processes. Although they are only theoretical tools that can never by themselves identify any causal mechanism, their usefulness is in gaining a deeper understanding of their consequences and their robustness. This paper exemplifies the possibility offered by agent-based models to explore in further detail the consequences of philosophical intuitions as fundamental as the question whether or not scientists have direct access to the world. This leads to novel hypotheses that could in principle be tested against vast and relatively new scientometric datasets.

Accounts of processes of scientific change roughly fall into two categories, one- and two-process change (Godfrey-Smith, 2003; see also chapter 1.4 in this book). One-process views hold that scientific change is a single process of gradual cumulation toward an independent

DOI: 10.4018/978-1-4666-6567-5.ch003

Copyright © 2015, IGI Global. Copying or distributing in print or electronic forms without written permission of IGI Global is prohibited.

point of reference, for example Popper's cycle of conjecture and refutation (Popper, 1959). One of the most influential criticisms is by Thomas Kuhn (1962), who held that standards for science are not independent but coevolve with scientific activity itself. Science then does not evolve toward a goal but away from goals previously set, with no fixed point left to compare its progress against. Without such a stable ''Archimedean platform", there is nothing against which science can meaningfully be said to cumulate. Moreover, changes of standard caused by scientific change can cause cascades of further scientific changes. For both accounts an agent-based model is constructed that explains how possibly cumulative, linear and non-cumulative, nonlinear progress can be generated from the local interactions between individual agents.

Agent-based models are corroborated to the extent that they reproduce the target pattern and useful to the extent that they suggest novel empirical hypotheses. Therefore the model is first validated by showing its capacity to reproduce these respective patterns. Then four conflicting hypotheses are deduced about the expected statistical properties of the resulting system of science.

1. ONE-PROCESS AND TWO-PROCESS CHANGE

Theory change is often conceived in analogy to biological evolution. . One of the more striking ones is the analogy between one- and two-process scientific change and adaptationist and coevolutionary views of change in evolutionary biology. Popper (1959) held that falsification of theories is analogous to biological evolution: random conjectures and selective refutation, with the standard for refutation itself being objective and independently testable. By abandoning unfit theories, science is a gradual and cumulative process of adaptation to that standard. The evolutionary analogue of the one-process view of scientific change is the "adaptationist" program in biology which explains organisms' adaptations by reference to the stable environment they inhabit. Just as organisms adapt to an exogenous environment, so do theories adapt to the world which exists independently of our theories about it. The world is, as it were, lying there waiting to be discovered. Analogously, in biology "if evolution is described as the process of adaptation of organisms to niches, then the niches must exist before the species that are to fit them"(Lewontin, 1978, p. 159). In line with this analogy Bird argues that scientific change is like the evolution of a single species to a given environment rather than coevolution of two species because "the results of experimental tests do not change. A good experiment is one that is replicable; it gives the same results whenever performed" (Bird, 2000, p. 212). There is only one process of change: theories can only change if our knowledge about the world changes.[1] Progress then resembles the discovery of a fixed landscape. Bird (2000) would agree with the assumption of a *fixed* landscape because it "captures the idea that in science our theories may change but the features of the world that they respond to are what they are independently of our theories, and are by and large constant over time" (Bird, 2000, p. 213).

By contrast on a two-process view such as Kuhn's there are two processes. Theories not only adapt to standards, but standards can also adapt to theories. For example in the second half of the 18th century the requirement to explain qualities such as color and texture was a standard in chemistry, and Lavoisier's theory did not meet this standard. With time it was not Lavoisier's theory that was rejected, but the standard that changed.[2] There is "a feedback loop through which theory change affects the values which led to that change" (Kuhn, 1977, p. 336). By way of this feedback loop, standards *coevolve* with the very theories they regulate. So what is particular to this view is not just that standards change, but that they can change *because* of a change in what they regulate: "historically, value change is ordinarily a belated and largely unconscious concomitant of theory choice" (Kuhn, 1977, p. 335). Standards are thus made endogenous to scientific change. They not only regulate the process of change but are themselves also a function of it. As a consequence, "there is no neutral algorithm for theory choice" (Kuhn, 1977, p. 330). Because standards evolve as a function of our scientific knowledge, they will only be fixed after all knowledge has been acquired: "though the experience of scientists provides no philosophical justification for the values they deploy (such justification would solve the problem of induction) those values are in part learned from that experience, and they *evolve with it*" (Kuhn, 1977, p. 335 my emphasis).

Although Kuhn agrees the world exists independently of our theories, for something to be a "fact" involves not only a recognition *that* something is, but also a theoretical understanding of *what* it is.[3] As a consequence, for Kuhn scientists have no access to the world in itself (see chapter 1.4 in this book). There is no stable or pre-existent environment against which our theories could be said to adapt. Rather that environment is constituted in part by those very theories: "theories do not evolve piecemeal to fit facts that were there all the time. Rather, they emerge together with the facts they fit" (Kuhn, 1970, p. 141). There is no "truth" lying there waiting to be discovered. The world does not contain a pre-existent set of puzzles. While the world can discriminate between rival solutions to a puzzle, it is silent as to what the puzzle should be. This is itself a function of scientific knowledge and as such subject to change.

The puzzles of contemporary normal science did not exist until after the most recent scientific revolution. They cannot be traced back to the historic beginning of the science. Earlier generations pursued their own problems with their own instruments and their own canons of solution. Nor is it just the problems that have changed. Rather, the whole network of fact and theory that the textbook paradigms fits to nature has shifted" (Kuhn, 1970, pp. 140-1).

Also for this program there is an analogue to be found in biology. Biologists such as Eldredge and Gould (1972) and Lewontin (1978) have claimed that the local environment to which organisms adapt coevolves with the creatures that inhabit it. Similarly, for Kuhn the standards to which our theories adapt change as a result of those very theories. Scientific successes in one area raise the bar for successes in other areas. Whether or not an approach deserves pursuit is in part affected by the evolution of other approaches. In biology this process of endogenous change is called "coevolution" and the resulting pattern of change called "punctuated equilibrium" is not gradual but largely stationary with violent bursts of extinction, analogous to Kuhn's description of scientific change as "a succession of tradition-bound periods punctuated by non-cumulative breaks" (Kuhn, 1970, p. 208).

In summary, then, both accounts agree that standards regulate science but the essential difference between one- and two-process views is whether these standards are exogenous or endogenous to science.

2. CHANGE OF AND ON EPISTEMIC LANDSCAPES

In as far as agent-based models aim to isolate the essential mechanism behind an aggregate pattern, viz. the mechanism that is sufficient to produce a pattern and without which it would not be generated, they must attempt to replicate the aggregate pattern with as little means as possible (but no less).[4] In the previous section it was argued that the possibility of endogenous standard change is the essential difference between one- and two-process views of change. In this section the epistemic landscapes framework by Weisberg and Muldoon (2009) is extended to construct a very simple agent-based model of theory change that captures this essential difference. The model of change is validated by showing that it can possibly produce both accounts of scientific progress.

Since the seminal work by Sewall Wright evolutionary change in biology is represented as trajectories of organisms on fitness landscapes. Fitness landscapes typically represent the relation between genotypes and fitness as a landscape. The coodinates of the landscape represent different genotypes and the topography represents their associated fitness. Recently this framework was adapted to philosophy of science by Michael Weisberg and Ryan Muldoon to represent different approaches and their associated significance. *Approaches* specify what are the relevant questions, what counts as solutions and how they can be obtained. The height or topography of the landscape corresponds to the *significance* of the results yielded by scientists adopting an approach. Weisberg and Muldoon describe the trajectories of different types of agents on a fixed landscape and argue that mixed strategies foster scientific progress. Progress on their account is defined as the percentage of the landscape explored, or as the speed with which peaks in the landscape are discovered. This is a simple but powerful framework for the study of one-process change. The fixed topography of the landscape elegantly captures the one-process assumption of an exogenous standard. Extending this framework with the possibility of endogenous change means that the significance of approaches can change as a result of the exploration of other approaches. Weisberg and Muldoon do not specify how the landscape can change and what happens when it does. But fitness landscapes in biology have already been developed to the point where they can represent this coevolutionary view. One of the simplest such models is the Bak-Sneppen model (Bak & Sneppen (1993), Sneppen, Bak, Flyvbjerg, & Jensen (1995)). This is the framework within which one- and two-process scientific change will now be modeled. To reduce the complexity of changing landscapes, the Bak-Sneppen model only represents the fitness of the N approaches that are currently being exploited. Relations of geographical proximity between approaches can be preserved by introducing L links. The same is now done for approaches on an epistemic landscape. For the purpose of this paper, let $N = 100$ and $L = 2$.[5] If each approach has a unique combination of 2 links, the resulting network topology is a *ring network*. Initially the significance S of approaches is drawn from a uniform random distribution between 0 and 1. The dynamics of the model is specified by defining two steps: 1) when scientists move on the landscape and 2) what happens when they do. For Popper (1959) the problem of induction means that scientists cannot choose the best theories, but they can abandon the worst, viz. those that are falsified. Similarly for Kuhn (1962) scientists move away from those approaches that show unexpected results rather than toward more promising approaches. So both can agree that the approach that is most likely abandoned in order to explore a new approach is the *least significant* approach. For both the problem of induction means that the significance of the new approach is assigned randomly because it is unknown before it has been explored.

Step 1a: Each step replace the least significant approach with a new approach of randomly assigned significance between 0 and 1.

But both accounts disagree about the consequence of abandoning an approach. On a one-process account nothing happens because the significance of approaches is independent of other approaches and can here be considered an exogenous property of an approach. On the two-process account the origin of significance lies with the other approaches. As a consequence, the abandonment of an approach will result in a re-evaluation of the significance of a number of neighboring approaches. The significance of approaches can *change*, and it can change for better or for worse. For example the discrepancy between Newtonian mechanics and the perihelion of Mercury had been a long-standing puzzle, but it only *became* an anomaly for Newtonian mechanics when Einstein's general theory of relativity succeeded in explaining it. Conversely the inability of Newtonian mechanics to explain the motion of the Moon[6] had been a significant problem in the 18th century until the puzzle was resolved in 1750. The Kuhnian model of landscape exploration can therefore be characterized by the following feedback loop designed to make significance of approaches endogenous:

Step 1b: If an approach is abandoned, replace the significance of linked approaches to a randomly assigned significance between 0 and 1.

In this section I have introduced a Popperian and a Kuhnian model of landscape exploration. Both accounts agree that scientific change is regulated by standards, but disagree about whether standards can be regulated by theory change. The difference between the two is the presence of the feedback loop in step 1b.

3. MODEL VALIDATION

To validate the very simple model for one- and two-process change this section shows that it is sufficient to generate the essential properties and differences of their associated views on scientific progress. One-process change is associated with cumulative, gradual progress. Two-process change is associated with cyclical, nonlinear progress.

3.1 Cumulative vs. Cyclical

On the one-process model the significance of approaches is exogenous and as such provides a stable point against which progress can meaningfully be measured. Figure 1 shows progress as the evolution of average significance. The abandonment of least significant approaches results in a decreasing but cumulative approximation of maximal significance. In the Kuhnian model the notion of average significance used to measure progress in the Popperian model is meaningless because the significance of each approach is relative to the other approaches. And because these approaches change every turn as a result of the abandonment

Figure 1. Progress toward maximal significance. Evolution of average significance, 10000 steps 100-step moving average, Popperian model.

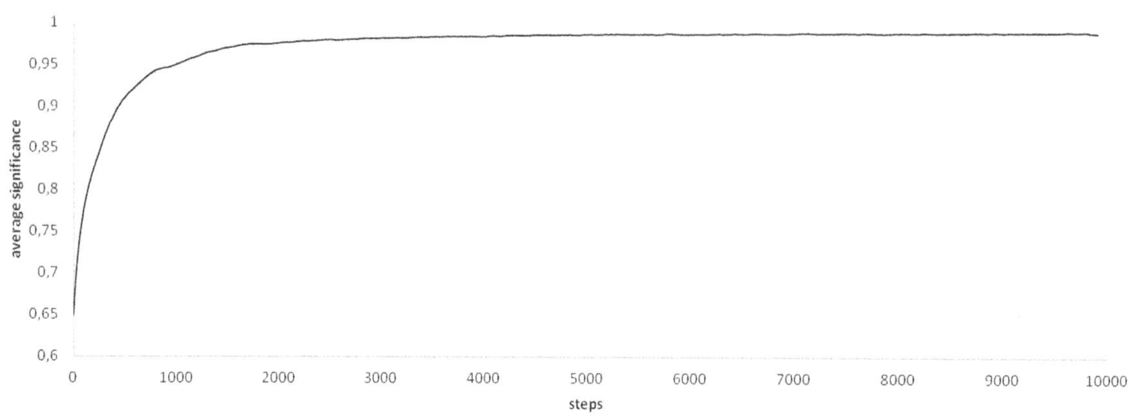

of the least significant, the standard against which progress is measured changes with every step of the model and progress is non-cumulative. As a result of the feedback loop introduced by step 1b, the abandonment of the least significant approach can prompt further replacements. This results in a never-ending process of replacement that does not converge toward maximal significance. However, Bak and Sneppen have found that this circularity does not result in complete arbitrariness. They report the surprising result that there is a threshold above which approaches are immune to abandonment. The abandoned approach with the highest significance is around

$$s = 0.66702 \pm 0.00010$$

(Grassberger, 1995, p. 279) and robust against variations of N (save for finite size effects for very small N). The system reaches a statistically stationary state, a dynamic equilibrium, in which the significance distribution of approaches evolves from a uniform one in the interval $[0,1]$ to a uniform one in the interval $[s,1]$.

Approaches with a significance higher than around 0.667 can be considered stable. This emergent threshold provides a goal for the process of abandonment: approaches are abandoned until one is found that is stable above the threshold. If the goal is to bring all approaches above this threshold, progress can be measured as the average distance of the approaches below the threshold to that threshold. As Figure 2 shows, the resulting pattern of progress is then not cumulative but cyclical because the very activity of replacing approaches causes a re-evaluation of the other approaches, prompting additional replacements. Paradoxically the very quest for stability guarantees that the system as a whole is forever unstable. All approaches have a significance $> s$ save for the unavoidable few that are part of a cascade. Indeed, if all approaches were to succeed in crossing the threshold simultaneously, the threshold would increase. But it does not. By selecting away the least significant approach, the model is attracted toward the equilibrium where all approaches are near or above the threshold. But this state in which all approaches are at least s is also the critical point c of the model. At c a change in the network can result in a cascade of correlated changes of all sizes (see Figure 4). Because $s = c$, the model organizes its own criticality. This also explains why s is robust against variations of N, although it has also been shown that no random approach with significance $s' > s$ will ever be abandoned in the thermodynamic limit $N -> \infty$ (Paczuski, Maslov & Bak, 1996, p. 417).

Figure 2. Progress away from equilibrium. EVolution of the average distance of the approaches below the threshold to that threshold, 10000 steps 100-step moving average, Kuhnian model.

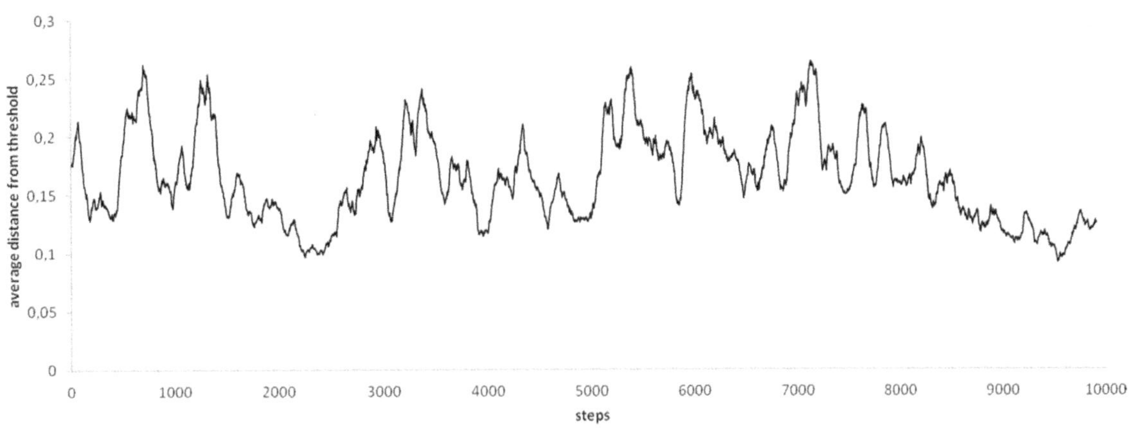

3.2 Linear vs. Nonlinear

On the one-process view of change theories can only change as a result of changes in our knowledge of the world. Theory change must therefore be proportional to the change in our knowledge of the world. And if our knowledge of the world accumulates gradually then theory must change gradually (in this model at every turn one approach is changed, which does not result in any further changes since approaches are causally independent of each other; see Figure 3). A change in theory cannot cause other theory changes, because the world is independent of our theories. Endogenous theory change is however possible on a two-process view of science. Theory change causes a change of standard that can in turn cause further theory change. I will call such a causally connected sequence of theory change a "cascade". The size of these cascades need not be proportional to the size of the initial theory change that triggered it (nonlinearity). With the system tending toward a state in which all approaches have a significance $> s$, the size of endogenous cascades can be measured by monitoring how many changes it takes for the system to restore this temporary equilibrium when disturbed. Bak and Sneppen found that a single such exogenous change of equal size can cause cascades of (endogenous) changes of all sizes (nonlinearity). More specifically the distribution of cascade sizes is a power law with an exponent 1.073 ± 0.003 (Grassberger, 1995, p. 281). Figure 4 shows the distribution of cascade sizes for the current model.

4. HYPOTHESES

Two alternative mechanisms of progressive scientific change have been constructed and validated. The mechanism allowing only exogenous change has been shown to result in the traditional cumulative, linear picture of progress. The mechanism allowing also endogenous change was shown to result in a different understanding of the dynamics of scientific change, namely cyclical and nonlinear. These models allow to make some novel predictions about what science ought to be like with and without endogenous change. This section compares four expected statistical properties of the respective resulting systems.

Figure 3. Cumulated approach change for 10000 steps, Popperian model

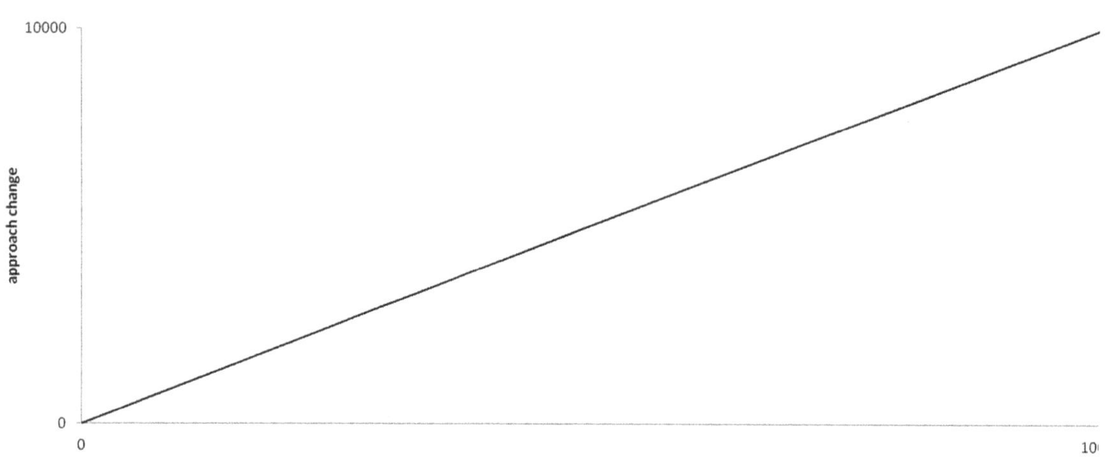

Figure 4. Size-frequency distribution of cascades of approach changes for 10000 steps during stationary state reached after 10000 steps

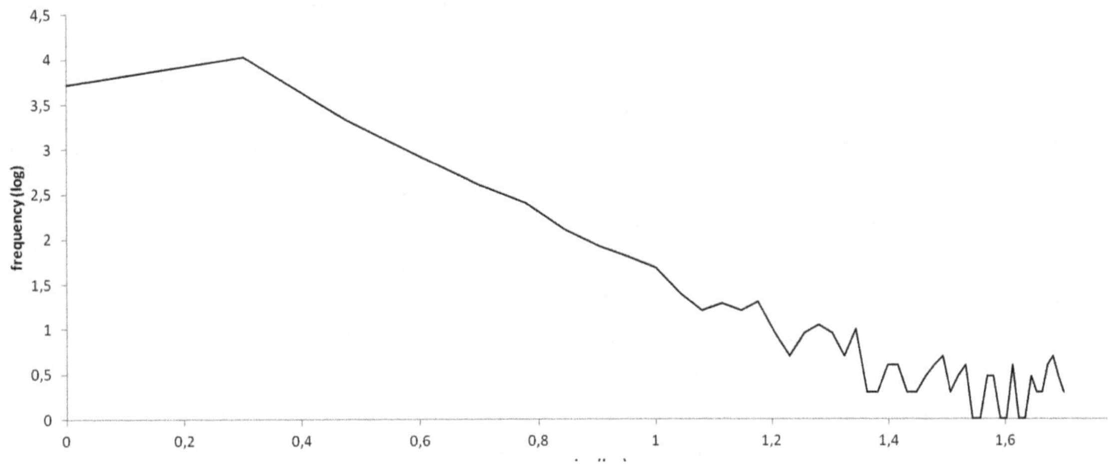

4.1 Correlation between Age and Probability of Abandonment

If approaches are adapting *toward* a given standard, then for the Popperian model older approaches should be more robust to abandonment than younger ones because older approaches have a proven ability to be better adapted. This can indeed be observed for the Popperian model. The concave shape of the trendline in Figure 5 indicates that younger approaches have a higher probability of abandonment.

By contrast if approaches are evolving *away from* previous standards, the standard against which approaches are adapting at every given time is always changing and past performance is not

Figure 5. Age of abandonment of approaches as a function of frequency for 10000 steps, Popperian model

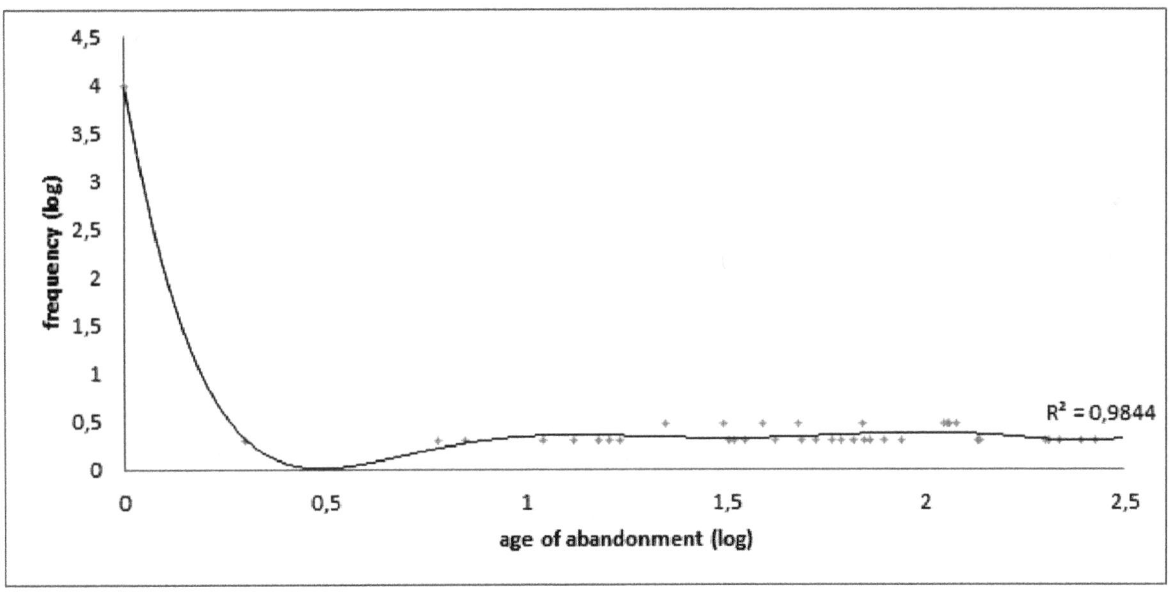

a guarantee of future results. Figure 6 shows that in the Kuhnian model the probability of abandonment is for the most part constant, represented by the straight line.

4.2 Correlation between the Stability of Neighboring Approaches

For the Popperian model the assumption of independence guarantees that there is no relation between the significance of neighbors. Significance is a function of the world only and as such it is independent of other approaches that happened to develop. As a result, seen in Figure 7, there is no correlation between the age of an approach and the age of its $L = 2$ neighbors in the ring network.

On the Kuhnian model, the higher significance of an approach protects its neighbors from changes in significance and vice versa. As such there is a correlation between the stability of an approach and the stability of its neighbors. A characteristic aggregate pattern is that stable approaches will tend to cluster together into what could be called "disciplines", networks of stable approaches with varying borders that merge and split up through time. Figure 8 shows a correlation between the age of an approach and the age of its neighbors, with clusters of approaches of roughly equal age.

4.3 Size Distribution of Approaches

A feature of the original epistemic landscapes model of Weisberg and Muldoon is that scientists occupying approaches produce *publications.* If approaches are assumed to produce a constant number of publications every step, then the distribution of the age of abandonment of approaches corresponds to the distribution of approach sizes, viz. the number of publications produced in an approach. Figure 5 and Figure 6 can thus be reinterpreted to show the expected distribution of

Figure 6. Age of abandonment of approaches as a function of frequency for 10000 steps during stationary state reached after 10000 steps, Kuhnian model

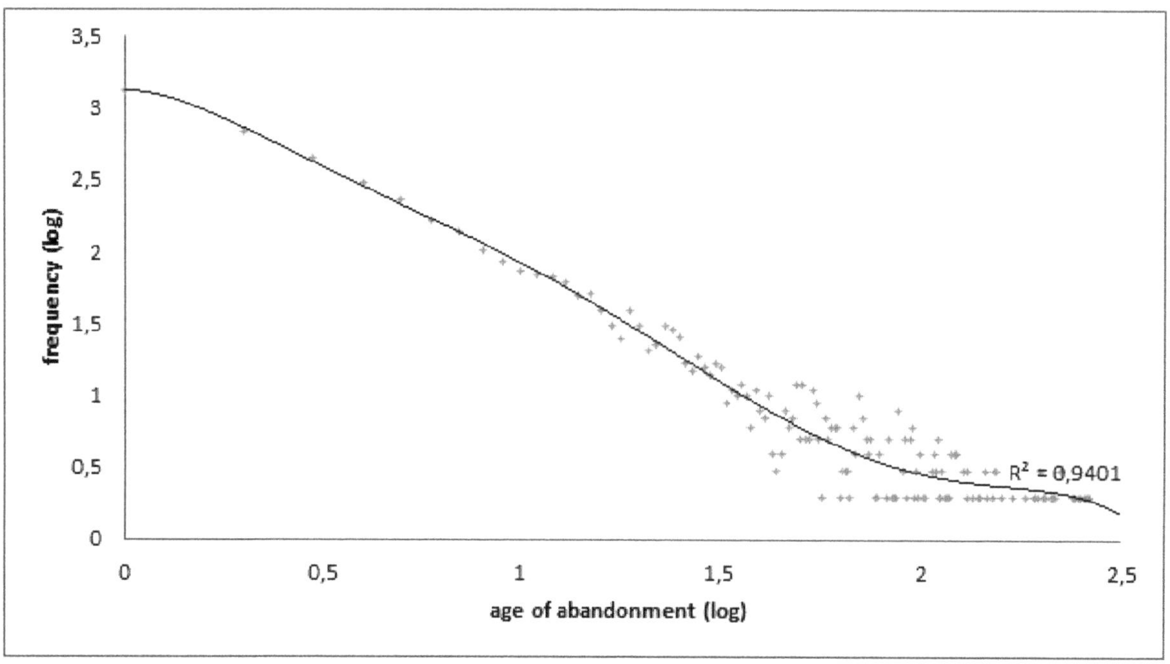

Figure 7. Correlation between age of approach and age of neighbors for 10000 steps, Popperian model

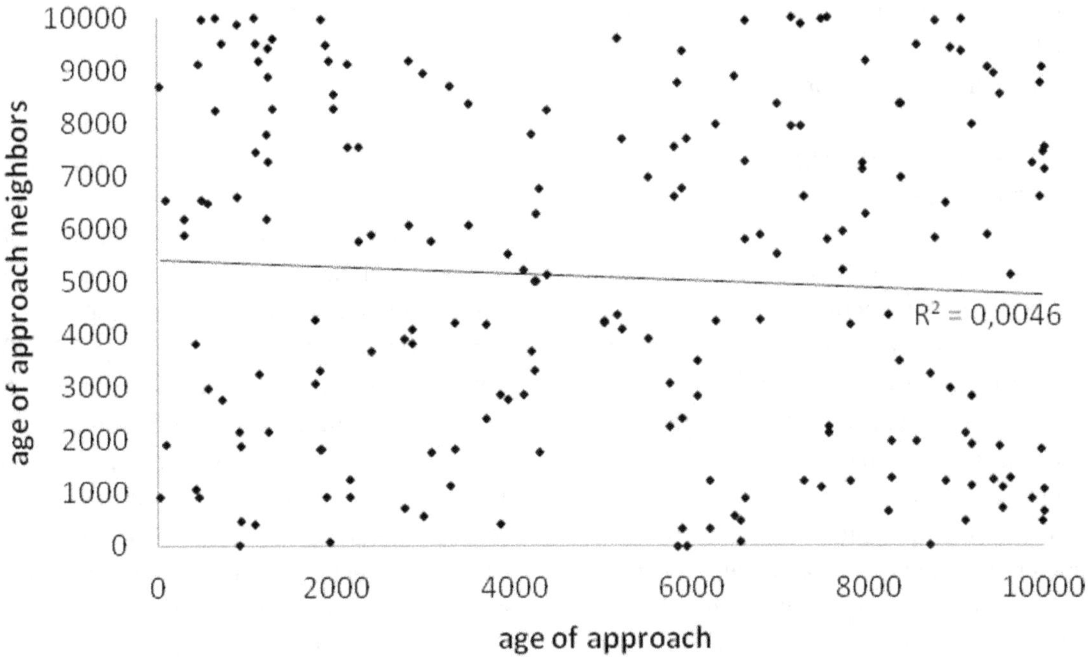

Figure 8. Correlation between age of approach and age of neighbors for 10000 steps during stationary state reached after 10000 steps, Kuhnian model

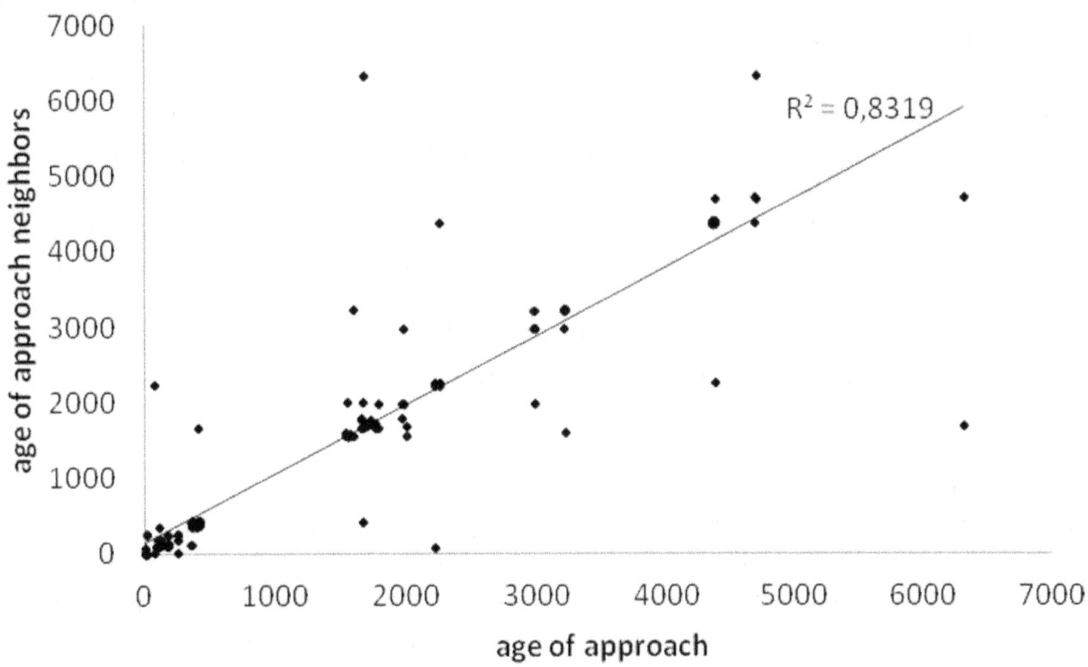

approach sizes. The Kuhnian model predicts approaches of all sizes. The Popperian model predicts a large frequency of very small approaches, with the other approaches of more or less equal size.

4.4 Citation Distribution of Approaches

If publications are made to approaches, a natural way to interpret the function of *citations* is to see them as a means of marking the approach of the paper by relating it to other papers in the same approach. Citations to papers outside of the approach can then be assumed to be randomly distributed. Under these assumptions the number of citations papers receive will be proportional to the size of their approach and Figure 5 and Figure 6 can also be interpreted as the expected citation distribution. Under these assumptions the Kuhnian model predicts the well-known power-law distribution of citations. The Popperian model predicts a random distribution.

5. A BUILDING OR A SANDPILE?

Is scientific progress gradual or revolutionary? Popper answers the former. The associated metaphor is well-known: that of a building to which bricks are added, one on top of the other. Kuhn's answer is less straightforward to envision. His answer is: both. There is a gradual increase of solutions, but this is accompanied by correlated cascades of small changes in problems and standards of all sizes. Why focus on the dynamics of the problems and standards, rather than on that of solutions? Solutions are not a good measure of scientific progress because they are relative to (changing) problems and standards. Rather what improves across revolutions is the problems and standards themselves. Although there is nothing left to progress against, their progress can be measured against the system's own previous states which they move away from based on what according to those very states has become unsatisfactory; "from primitive beginnings but toward no goal" (Kuhn, 1970, p. 172). True novelty, viz. novelty independent of a certain set of problems and standards, then consists not in presenting a better solution to an existing problem, but in an improvement of the problem itself. For example Kuhn mentions "Newton's ... small but revolutionary reformulation in the questions that scientists asked about motion as well as in the answers they felt able to accept" (Kuhn, 1970, pp. 139-40). A normal contribution is a contribution to a given problem, a revolutionary contribution is a solution to a problem nobody knew they had. For example Henry Ford once said that if he had asked people what they wanted him to make, they would have asked for faster horses. This suggests an interesting change of perspective: an account of progress in terms of problems instead of solutions. Puzzles and standards are not instrumental in producing ever-better solutions, rather the solutions are instrumental in producing ever-better puzzles and standards. From the perspective of scientific change, solutions to problems, although of course of practical value, are nothing but an intermediate step in the process of producing better problems. On a global level, then, scientific communities are not concerned with solving problems but with the question what the problem should be.

Such an account is not without consequences for science policy, which aims at designing the institutions of science in such a way as to further the advancement of science. The publication pressure resulting from the recent managerial academic trends based on bibliometric evaluation constitutes a policy aimed exclusively at maximizing solutions. But from the point of view of progress in terms of both solutions and their problems, such a policy is suboptimal because it disincentivize the search for new problems if only one makes the reasonable assumption that only papers sharing the same problems and standards cite each other. In terms of the fitness landscape, such policy invites scientists to stick to a local peak instead of searching for other, potentially higher peaks. Ideally scientists should both solve existing problems and find new ones. This conflict between exploitation and exploration has been thematized by Kuhn as the "Essential Tension" (Kuhn, 1977). New developments in probability theory, such as research on multi-armed bandit problems, take up this problem and are a promising new avenue for policy-relevant philosophical research, see for example Mayo-Wilson, Zollman and Danks (2013).

With a given set of problems and standards, progress can meaningfully progress toward a certain endpoint. However, if problems and standards are themselves constructed from previous solutions, then there is an infinity of ways in which the world can be broken down into puzzles. This is analogous to a biological world consisting of niches where "there is an infinity of ways the world can be broken up into arbitrary niches" (Lewontin, 1978, p. 159). If there is a given set of problems, then it is natural to assume that the set of solutions is finite. But if new problems are created as a result of previous solutions, then there is no in principle limit on the number of possible problems. Scientific progress is potentially infinite.

These individual consequences can be brought together in a new metaphor for scientific progress. While Kuhn merely criticized the gradualist analogy of scientific progress as the addition of bricks to a building, the introduced mechanism now suggests an alternative metaphor. In this picture there is no building, no pre-existent structure to decide what the problems should be and what counts as a solution. Indeed, where would such a thing come from? In this picture problems and standards emerge from the very contributions they regulate. Contributions make their own rules as they go along. They self-organize into a dynamic structure, like dropped grains of sand self-organize into a sandpile. A continuous addition of grains leads to discontinuous "avalanches" in the sandpile, just as on Kuhn's view a continuous increase in "the number of solved scientific problems and the precision of individual problem solutions" (Kuhn, 1977, p. 320) causes discontinuous paradigm change. An area on the sandpile is mostly stable, but every once in a while one newly added grain of sand causes an avalanche which disrupts the structure of the sandpile until a new stable configuration is found. The grains remain the same but their position is rearranged. There is no restriction on how big the sandpile can become. Its potential size is infinite. Therefore its progress can not be measured with respect to a fixed endpoint in its evolution but only how far it has evolved from its base.

In a system where evaluation is relative to the value of other practices, each practice's attempts to avoid losing value relative to the other practices aggregates into a system where each practice needs to evolve just to hold on to its previous value. Stagnation means regress. This is the "Red Queen" effect, so called after the character in Through the Looking Glass who had to keep running just to stay in the same place. This effect has been used to describe coevolutionary progress. In the case of paradigms this means that a paradigm must produce a steady stream of interesting new puzzles in order to remain fruitful. Contributions are made to practices in order to maintain their fruitfulness, but this in turn raises the barrier for satisfactory practices. This "feedback loop through which theory change affects the values which led to that change" (Kuhn, 1977, p. 336) forces scientists away from local peaks. What follows is a cascade of correlated changes to both the shape and size of the landscape until new stable networks of interlocking practices emerge. As such the mechanism proposed is a powerful engine of innovation which can account for Kuhn's claim that "History even suggests that the scientific enterprise has developed a uniquely powerful technique for producing surprises of this sort" (Kuhn, 1970, p. 52). Thus the mechanism introduced is not circular in any damaging way, rather it constitutes a powerful driver of innovation.

6. CONCLUSION

I have no knowledge of previous attempts to empirically establish the distribution of the length of the lifetimes of scientific theories. Nevertheless, for every separate prediction alternative generative mechanisms can be devised: myriad of generative mechanisms exist to explain powerlaw distributions (Newman, 2005). By establishing the link between several separate predictions, agent-based models can give multiple separate observations additional collective weight. The Kuhnian model predicts a system of science in which there is no correlation between the age of approaches and their probability of abandonment, but different approaches of varying size do cluster together in larger and relatively stable networks that merge and split up through time with shifting and unstable borders. The Popperian model predicts non-clustered approaches of roughly equal size in which there is a correlation between age and probability of abandonment. An attempt was even made to extend the observable consequences to distributions of paper citations, one of the most robust and well-established empirical facts about science. This paper is just one step in a larger project of operationalizing philosophical concepts using newly available data and methods (De Langhe, 2012). I can only hope for methodologically sound empirical work to take its lead.

REFERENCES

Bak, P., & Sneppen, K. (1993). Punctuated equilibrium and criticality in a simple model of evolution. *Physical Review Letters*, *71*(24), 4083–4086. doi:10.1103/PhysRevLett.71.4083 PMID:10055149

Bird, A. (2000). *Thomas Kuhn*. Bloomington, IN: Acumen Publishing.

De Langhe, R. (2012). The Kuhnian paradigm. *Topoi*, *32*(1), 65–73. doi:10.1007/s11245-013-9153-3

Eldredge, N., & Gould, S. J. (1972). Punctuated equilibria: an alternative to phyletic gradualism. In *Models in paleobiology* (pp. 82–115). San Francisco: Freeman Cooper.

Godfrey-Smith, P. (2003). *Theory and reality*. Chicago: Chicago University Press. doi:10.7208/chicago/9780226300610.001.0001

Grassberger, P. (1995). The Bak-Sneppen model for punctuated equilibrium. *Physics Letters. [Part A]*, *200*(11), 277–282. doi:10.1016/0375-9601(95)00179-7

Kitcher, P. (1978). *The nature of mathematical knowledge*. Oxford, UK: Oxford University Press.

Kuhn, T. (1962). *The structure of scientific revolutions*. Chicago: Chicago University Press.

Kuhn, T. (1970). *The structure of scientific revolutions* (2nd ed.). Chicago: Chicago University Press.

Kuhn, T. (1977). *The essential tension*. Chicago: Chicago University Press.

Kuhn, T. (2000). *The road since structure: Philosophical essays 1970-1993*. Chicago: Chicago University Press.

Lewontin, R. (1978). Adaptation. *Scientific American*, *239*(3), 155–169. doi:10.1038/scientificamerican0978-212 PMID:705323

Mayo-Wilson, C., Zollman, K., & Danks, D. (2013). Wisdom of the crowds vs. groupthink: Learning in groups and isolation. *International Journal of Game Theory*, *42*(3), 695–723. doi:10.1007/s00182-012-0329-7

Newman, M. (2005). Power laws, Pareto distributions and Zipf's law. *Contemporary Physics*, *46*(5), 323–351. doi:10.1080/00107510500052444

Paczuski, M., Maslov, S., & Bak, P. (1996). Avalanche dynamics in evolution, growth, and depinning models. *Physical Review E: Statistical Physics, Plasmas, Fluids, and Related Interdisciplinary Topics*, *53*(1), 414–443. doi:10.1103/PhysRevE.53.414 PMID:9964272

Popper, K. (1959). The logic of scientific discovery. New York: Basic Books.

Sneppen, K., Bak, P., Flyvbjerg, H., & Jensen, M. (1995). Evolution as a self-organized critical phenomenon. *Proceedings of the National Academy of Sciences of the United States of America*, *92*(11), 5209–5213. doi:10.1073/pnas.92.11.5209 PMID:7761475

Weisberg, M., & Muldoon, R. (2009). Epistemic landscapes and the division of cognitive labor. *Philosophy of Science*, *76*(2), 225–252. doi:10.1086/644786

KEY TERMS AND DEFINITIONS

Agent-Based Model: A model allowing to study the patterns resulting from the interactions of agents with pre-specified behavior.

Epistemic Landscape: Representation of a network of scientific approaches as a function of the epistemic significance of their results.

Fitness Landscape: Representation of a network of niches as a function of the fitness they convey to the organisms that populate it.

One-Process Change: The evolution of science conceived as a process driven by a single driver such as verisimilitude or problem-solving capacity.

Paradigm: A network of scientific contributions characterized by similar problems, standards and aspirations.

Scientific Revolution: A change in the problems, standards and aspirations that characterize a paradigm.

Self-Organized Criticality: Phenomenon by which a system's attractor and tipping point coincide, resulting in a system attracted to its tipping point.

Two-Process Change: The evolution of science conceived as a process resulting from the interaction between two separative processes, for example the evolution of questions and the evolution of answers to those questions.

ENDNOTES

1 According to Kitcher (1978, p. 151) this was the view held by many logical empiricists: "Without new observations, science would be static. I do not know whether anyone has held exactly this picture of scientific change, but something very close to it seems to be implicit in the writings of many logical empiricist philosophers of science."

2 "One of the objections to Lavoisier's new chemistry was the roadblocks with which it confronted the achievement of what had previously been one of chemistry's traditional goals: the explanation of qualities, such as color and texture, as well as of their change. With the acceptance of Lavoisier's theory, such explanations ceased for some time to be a value for chemists; the ability to explain qualitative variation was no longer a criterion relevant to the evaluation of chemical theory"(Kuhn, 1977, p. 335).

3 Kuhn's most elaborate example is the discovery of oxygen, which according to Kuhn was not the instantaneous realization of the existence of a gas called oxygen that can easily be ascribed to a single person, but rather a process of theoretical assimilation of a novel fact that takes time and involves multiple persons (Kuhn, 1962, p. 53-6).

4 Kuhn for one declared at the end of his life that "many of the most central conclusions

we drew from the historical record can be derived instead from *first principles* [... that] are necessary characteristics of any developmental or evolutionary process."(Kuhn, 2000, p. 112-119) For recent work on finding these first principles, see De Langhe (2012).

5 The goal of this paper is a qualitative comparison between a model with and without a feedback loop to neighboring approaches. Differences in approaches and numbers of neighbors will result in quantitative differences, but do not affect the qualitative differences reported in this paper. There exists a large literature on the Bak-Sneppen model and different versions along a variety of dimensions can be found in the literature.

6 "The predicted motion of the Moon's perigee was only half of that observed" (Kuhn, 1962, p. 81).

Chapter 4
Assessing Individual Influence on Group Decisions in Geological Carbon Capture and Storage Problems

Debbie Polson
University of Edinburgh, UK

Andrew Curtis
University of Edinburgh, UK

ABSTRACT

The inherent uncertainty in information about the Earth's subsurface requires experts to interpret and reach judgements about geological data based on their individual experience and expertise. This is particularly true for the geological storage of CO_2 in subsurface saline aquifers where the fate of the injected CO_2 needs to be predicted far into the future. In this chapter, linear modelling is used in a structured elicitation exercise to estimate the relative influence of individual experts within a group and to assess whether a group consensus reflects a genuine shared opinion or is biased towards or away from any dominant member or subgroup. The method is applied to a real expert evaluation of the carbon storage potential of a siliciclastic formation. This reveals herding behaviour amongst the experts, and levels of inter-expert influence that are undue given individual experts' levels of expertise, though neither phenomena was apparent during the meeting.

INTRODUCTION

Subsurface carbon capture and storage requires CO_2 to be stored indefinitely in the intended geological storage reservoir. However, the uncertainty associated with information about the deep subsurface makes prediction of the long term fate of the injected CO_2 difficult. The geological information on which such predictions are based is inherently uncertain, and requires individuals or groups of experts to make interpretations and judgements about the likely properties and characteristics of the reservoir, caprock and overburden layers of rock. A critical component of site evaluation

DOI: 10.4018/978-1-4666-6567-5.ch004

Copyright © 2015, IGI Global. Copying or distributing in print or electronic forms without written permission of IGI Global is prohibited.

is the analysis of risk where risk is defined as a combination of the likelihood and the severity of impact of any event (Chadwick et al., 2008; Det Norske Veritas, 2010, Smith et al., 2010, Polson et al., 2012). As part of this process, experts are asked to assess the risk of CO_2 storage in a given site using available data, simulation results, and their experience and expertise. Uncertainty can lead to different experts forming different opinions based on the same information (Baddeley et al., 2004; Bond et al., 2007; Lowe & Lorenzoni, 2007; Polson & Curtis, 2010).

Group interactions between experts can improve the quality of decisions made as it allows knowledge and experience to be shared between experts (e.g. Phillips, 1999). All individual experts are subject to cognitive biases when making judgements in situations of uncertainty (e.g. Kahneman et al. 1982, Anderson, 1998, O'Hagan et al, 2006, Bond et al. 2012, Curtis, 2012), and evidence suggests that group interaction may help to mitigate the effects of individual bias (Sniezek, 1992, Diviacco, 2015, see chapter 1 of this book). However group interaction may lead to the introduction of other, group-related biases (Sniezek, 1992), and it is towards these that we focus attention in this chapter.

In situations where a group consensus must be reached, individual experts will exert different levels of influence on the group. In an ideal world the influence of each expert will reflect his/her relative expertise, and group performance will improve with more effective use of each additional expert (Hackman & Morris, 1975; McGrath, 1984). However other factors such as individual force of personality and confidence may bias the group decision. Also the relative influence of each expert may not be apparent from the meeting itself: while particularly dominant characters may stand out, their actual influence on the group may be lower than their dominance of the conversation would suggest. Furthermore even where no expert appears to dominate the discussions or decisions, individual experts may still have a disproportionate influence on group decisions, and on other experts.

By using a novel structured elicitation process we are able to identify when such group biases occur, for example in the assessment of geological information (Polson & Curtis, 2010). Expert elicitation theory and practice have been investigated and used in numerous studies in the earth and environmental sciences such as nuclear waster disposal (Bonano & Apostolakis, 1991), interpretation of geological data (Curtis & Wood, 2004a; Bond et al., 2007), risk and impacts of climate change (Morgan et al., 2001; Arnell et al., 2005; Lowe & Lorenzoni, 2007), hydrological modelling (Ye et al., 2008), and risk assessment connected with volcanic eruptions (Aspinall, 2006) and earthquakes (Bommer et al., 2005; Runge et al., 2013). Using a well-structured elicitation method it is possible to track the opinions and judgements of individual experts in a group environment where a consensus opinion or view is required. Polson and Curtis (2010) used this to track the influence of group discussions on the opinions of individual experts, and to compare the judgements of individual experts with the group consensus. This revealed that a variety of identifiable biases had influenced views in a real geological evaluation.

In principle such a process can also be applied to understand the relative influence of individual experts on the group consensus, and on each other. By examining in detail the judgements made at each stage of the process we may be able to identify a pattern of influence. However this analysis is complex and the results are often ambiguous. To fully understand the relative influence of each expert we need to consider each judgement made by each expert at each stage of the process within the wider context of all judgements made during the elicitation session. Even with only three experts in a group this can be a complicated and time consuming task. As the number of experts increases so does the complexity of the analysis.

Numerous studies have been conducted on influence within groups (see e.g. Kirchler & Davis, 1986; Hinsz, 1990; Laughlin, 1988, 1992; Kerr, 1992; Davis et al., 1996a, 1997; Ohtsubo et al., 2002) and on modelling the consensus forming behaviour of groups (see e.g. Lehrer & Wagner, 1981; Davis 1982, 1992, Davis et. al. 1996b). By tracking the opinions of individual members of the group it is possible to identify which experts have the most influence on the group and develop expert-weighted models (Bonner 2000, 2004).

In this chapter we use a simple linear model to characterise the relative influence of individual members of a group in a well-structured group elicitation session. This method allows us to extract the overall trends of influence quickly and without the need for the detailed individual analysis discussed above. Fitting the initial assessments of each expert at the start of the session with the final group consensus allows us to identify which expert appears to have the most influence overall on the consensus judgment. Similarly it is possible to determine the apparent relative influence of individual experts on each other by fitting the initial assessment of each individual expert with the set of individual assessments from the end of the session.

This process was applied to two group elicitation sessions to assess risks involved in a subsurface CO_2 storage project for climate change mitigation, where experts were asked to assess individually, and reach a group consensus about, the likelihood and severity of a set of Features, Event or Processes (FEPs) that might affect the project (see e.g., Maul et al., 2005; Wildenborg et al., 2004, 2005; Chadwick et al., 2008; Polson et al., 2009; Det Norske Veritas, 2010). The elicitation focused on a real onshore saline aquifer formation which was evaluated as an analogue for geological CO_2 storage in formations offshore around the UK. The results reveal which member of the group exerted the most influence, and identifies where a consensus reflected the true shared opinions of the group, and where this was not the case. In the latter situation the results should be treated with more caution than in the former.

GEOLOGICAL CONTEXT

As part of the CO_2 Aquifer Storage Site Evaluation and Monitoring (CASSEM) project (Smith et al., 2010), the region around the Ferrybridge power station in eastern England was investigated to identify a saline aquifer formation suitable for hypothetical CO_2 storage. Using available images constructed from 2D and 3D seismic data acquired for oilfield evaluation purposes in the previous decades (see Smith et al. (2010) for an example cross-section through one image that was used), and existing borehole and well data, a geological model of the subsurface was constructed for a geological area of 945 km^2, extending from Lincoln in the west to the coast in the east. A set of selection criteria were used to identify the most suitable reservoir rock formation, including but not limited to depths >800m, salinity >100 gl^{-1}, porosity >20% and permeability >500mD. The bedrock succession considered ranged in age from Carboniferous to Cretaceous. Within the Triassic strata, the Sherwood Sandstone Group was identified as the primary target aquifer with corresponding Mercia Mudstone Group caprock.

The geological context of the reservoir and caprock is described in Ford et al. (2009). The Sherwood Sandstone Group consists mainly of fluvial sandstones with occasional thin and laterally discontinuous marl seams. The Mercia Mudstone Group is predominantly red mudstones with subordinate siltstones and sandstones. The aquifer formation extends offshore from the coast at Saltfleetby to Lincoln in the west. Generally the caprock layer should provide a physical barrier to the upward migration of CO_2 to the surface. However in this case the formation is shallowly dipping, and in fact both the Sherwood Sandstone Group and Mercia Mudstone Group outcrop at the surface, far up-dip to the west of the hypothetical injection site. Thus there is a potential leakage pathway for CO_2 that may allow it to migrate to the surface. However, over long timescales (hundreds of years), CO_2 will dissolve in the brine. The density of the resulting fluid is greater than the native brine and hence the CO_2 becomes 'dynamically trapped'. Therefore the site is not necessarily ruled out for the storage of CO_2 (although to the authors knowledge there are no current plans to do so) despite the lack of complete structural closure, as the likelihood of leakage of CO_2 may still be low if the CO_2 is able to dissolve before it has time to migrate to the surface. Detailed site assessment would be required to determine the risk of leakage. Furthermore, where the aquifer is shallower in the west, it is also a source of potable water and there is the potential for the contamination by either the injected CO_2 or the saline formation fluids.

METHOD

Expert Elicitation

A project-specific risk register based on the Quintessa CO_2 FEP register (Savage et al., 2004) was constructed to assess risk for the Lincolnshire exemplar storage site (Polson et al., 2012). The register includes a hopefully exhaustive list of FEPs that define relevant scenarios and behaviour of CO_2 in the storage system which may pose risk to the project. These are assessed by experts for their likelihood (K) of impacting the project, and for the severity (S) of this impact. The full register may contain many tens or hundreds of FEPs, covering a wide range of areas of expertise. All must be assessed at multiple stages of a project by a number of experts to track any changes. In practice it is therefore only possible to complete the full risk register by using a number of experts who assess the FEPs independently, and the results of this analysis are described in Polson et al. (2012). However, it is also important to understand how the judgements of experts working in isolation might differ from those that evolve in a group environment where knowledge can be shared.

In each structured group elicitation session, three experts were asked to quantify the likelihood and severity for a subset of FEPs. The subsets of FEPs, which were ranked as moderate or high risk by the individual experts, were selected based on the similar range of expertise required to assess them. The likelihoods K are scored on a scale of 1-5 where 1 is low and 5 is high. Two separate scales for the severity S were used for each FEP: the first was for the impact on health and safety (HS), and the second for the impact on the project's financial viability, the environment, research, and industrial viability (FERI). The experts were asked to assess the lower-bound, best-guess and upper-bound for both the likelihood and severity, resulting in nine quantitative judgements per FEP (three for likelihood, and three from each of two scales for severity).

At the start of the session the experts were asked to assess each FEP independently of the other experts. These individual assessments were then shared with the group, and each expert was asked to explain their reasoning. Through subsequent discussion, the group was required to produce a group consensus assessment for each FEP which all experts could agree to. Immediately thereafter they were again required to complete individual assessments independently at the end of the session. See Polson and Curtis (2010) for more details of the elicitation methodology, and for another application in a geological asset team environment.

Two group sessions were carried out using different experts and assessing different FEPs related to a CO_2 storage project. In the first session four FEPs were assessed, resulting in a total of 36 separate judgements made by the experts at each stage of the elicitation process. This set of FEPs related to the industrial viability of CCS and included:

- (A) Construction and Site Logistics,
- (B) Construction of Pipeline,
- (C) Leakage from Pipeline,
- (D) Public Perception and Security.

In assessing the risk from these FEPs, the experts must consider the characteristics of the intended storage formation and estimate, given available information, the likelihood and severity of a negative impact related to each FEP occurring. For example, in considering public perception, the issue of the incomplete geometrical closure of the aquifer by the caprock may be an important issue; in considering the construction of pipeline, previous experience from existing CO_2 and oil and gas pipelines may contribute significantly.

In the second session three FEPs were assessed related to the transport and dynamics of CO_2 in a porous medium (the subsurface reservoir), resulting in a total of 27 separate judgments per expert at each stage of the process. The FEPs assessed were:

- (E) Advection of free CO_2,
- (F) Buoyancy-driven flow,
- (G) Water mediated transport.

For these FEPs the experts must consider the properties of the aquifer and caprock and how these will influence the transport of CO_2 in the subsurface. For example given available information about the caprock and the associated uncertainties, is CO_2 likely to migrate upwards through the subsurface to the near-surface environment?

More detailed descriptions of each FEP were given to the experts to stimulate ideas and focus attention on potential risk areas (see Appendix).

Fitting Procedure

A simple linear model is used to represent the influence of each expert on the group consensus

$$\underline{C} = \sum_{i=1}^{n} \beta_i \underline{I}_i \tag{1}$$

where $\underline{C}$ is a rank-m vector containing the consensus values (in this case it consists of blocks of three judgements $[L,B,U]$ for each FEP, where L is the lower bound score, B is the best guess score and U is the upper bound score), m is the total number of judgements made at the consensus stage, $\underline{I}_i$ is the corresponding rank-m vector of the initial or prior values for expert i, β_i is the coefficient representing the influence of expert i on the consensus, and there are n experts. Using least squares fitting to determine the values for coefficients $\underline{\beta}$, the relative influence of each expert's prior position on the final consensus can be quantified with larger values of β_i representing greater influence of expert i.

By replacing the consensus values in Equation (1) with the values from the final individual assessment of an expert, the relative influence of all experts on that individual expert can be estimated, including the influence of their own initial assessment on their own final judgments:

$$\underline{F}_i = \sum_{j=1}^{n} \alpha_{ij} \underline{I}_j \tag{2}$$

Here $\underline{F}_i$ is a rank-m vector of the final values of expert i, and coefficients α_{ij} represents the influence of the initial views of expert j on the final view of expert i.

Similarly by using just one expert's own initial assessment and the consensus assessment to fit that expert's end of session assessment it is possible to determine whether the consensus (or emergent group behaviour) has more or less influence on their final assessments than their own prior position:

$$\underline{F}_i = \gamma_i^I \underline{I}_i + \gamma^c \underline{C} \tag{3}$$

Here $\underline{F}_i$ is a rank-m vector of the final values of expert i, and γ_i^I and γ^c are the coefficients corresponding to the influence of the expert's initial values and the consensus values, respectively.

More generally it is possible to find the optimal coefficients, δ, of the equation

$$\underline{\underline{D}} = \underline{\underline{\delta}}^T \underline{\underline{D}} \tag{4}$$

where $\underline{\underline{D}}$ is the matrix containing all FEP assessments at all stages in the process, each row taking the form [L,B,U] for any one FEP, and $\underline{\underline{\delta}}$ being a matrix of mixture coefficients used to weight the various results to produce all other results. Eq. (4) is solved by $\underline{\underline{\delta}} = \underline{\underline{I}}$, the identify matrix. However this is the solution assuming that no group interaction takes place, as $\underline{\underline{\delta}} = \underline{\underline{I}}$ implies that each estimate of [L,B,U] for each FEP is influenced by nothing other than that estimate. While that may be true for the prior pdfs, it is unlikely to be true for most others. Taking the point of view that expert prior positions impact on group interaction, implies that the matrix $\underline{\underline{\delta}}$ may have certain structures imposed a priori. In this work we will use the structure described by Eqs. (1) to (3) (constructed from Eq. (4) by equating different elements of $\underline{\underline{\delta}}$ to each other or to zero), but it may be useful to note that other structures are possible.

To ensure that any distinction in the pattern of influence of individual experts was statistically significant on the consensus assessments (equation (1)), the student t-values for each coefficient were tested assuming a null hypothesis that all terms exert equal influence (i.e., that the t-values of all coefficients are equal):

$$t = \frac{\hat{\delta} - \delta_0}{SE_{\hat{\delta}}} \tag{5}$$

where t is the student t-value, $\hat{\delta}$ is the least squares estimate for a coefficient, δ_o is the specified value for δ assuming the null hypothesis is true, and SE_{δ} is the standard error for the coefficient $\hat{\delta}$. If the t-values of all coefficients do not exceed some statistically significant level (corresponding p-value < 0.05), then the linear model does not distinguish between the influence of the individual terms, and hence does not distinguish between experts. For the models described by equations (2) and (3) we test the null hypothesis that the expert is only influenced by their own original assessment.

RESULTS

Session One

Figures 1-4 show the individual and consensus scores for session 1 for FEPs A, B, C, and D respectively. Analysis and interpretation of the expert interactions exhibited within these plots is complex, requiring each to be examined individually. Figures 1-4 show no immediately obvious pattern as to which expert has the most influence on the consensus or on the other experts. Similarly it is not clear which experts tend to agree the most with the consensus. For example, initial analysis shows that the number of occasions where the consensus is closest to the individual start of session opinion of each expert is almost exactly equal for all three experts. This would suggest that they have similar levels of influence on the

Figure 1. Session 1 scores for FEP A for individual start and end of session (grey) and consensus (black) assessments for (a) likelihood, (b) health and safety (HS) severity and (c) financial, environmental, research and industrial viability (FERI) severity. Solid circles show best guesses, while horizontal lines run from lower-bound to upper-bound estimates.

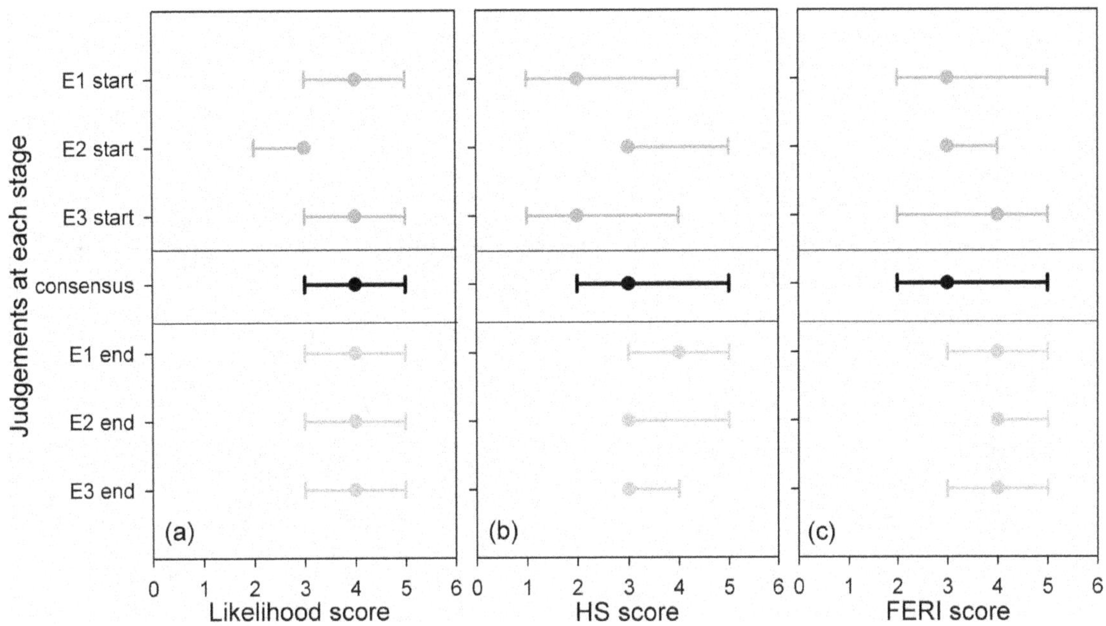

Figure 2. Session 1 scores for FEP B for individual start and end of session (grey) and consensus (black) assessments for (a) likelihood, (b) health and safety (HS) severity and (c) financial, environmental, research and industrial viability (FERI) severity. Key as in Figure 1.

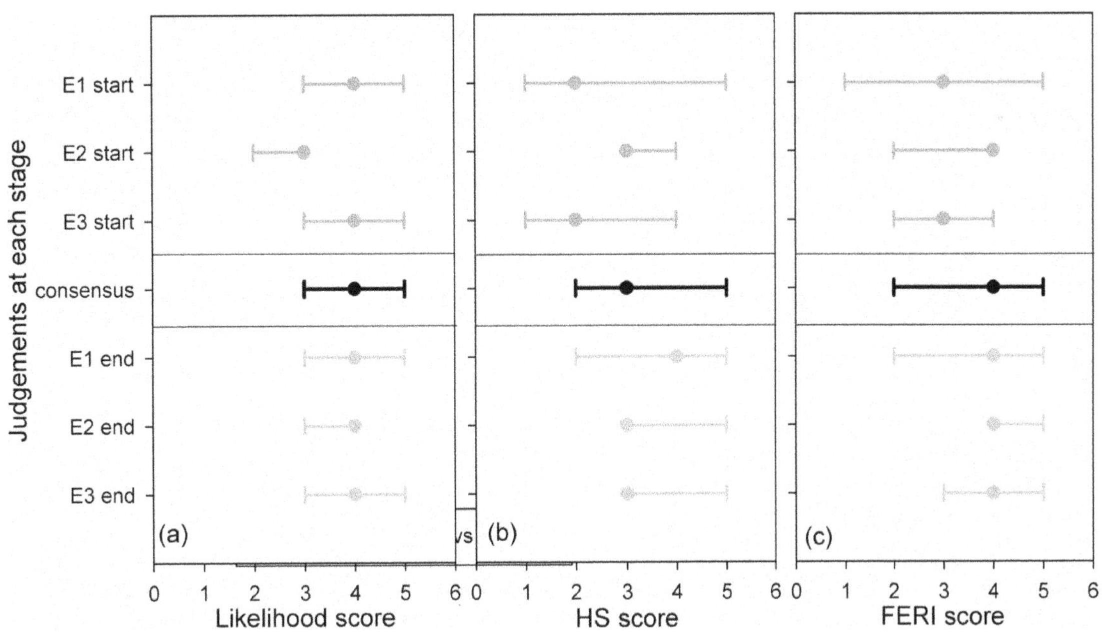

Figure 3. Session 1 scores for FEP C for individual start and end of session (grey) and consensus (black) assessments for (a) likelihood, (b) health and safety (HS) severity and (c) financial, environmental, research and industrial viability (FERI) severity. Key as in Figure 1.

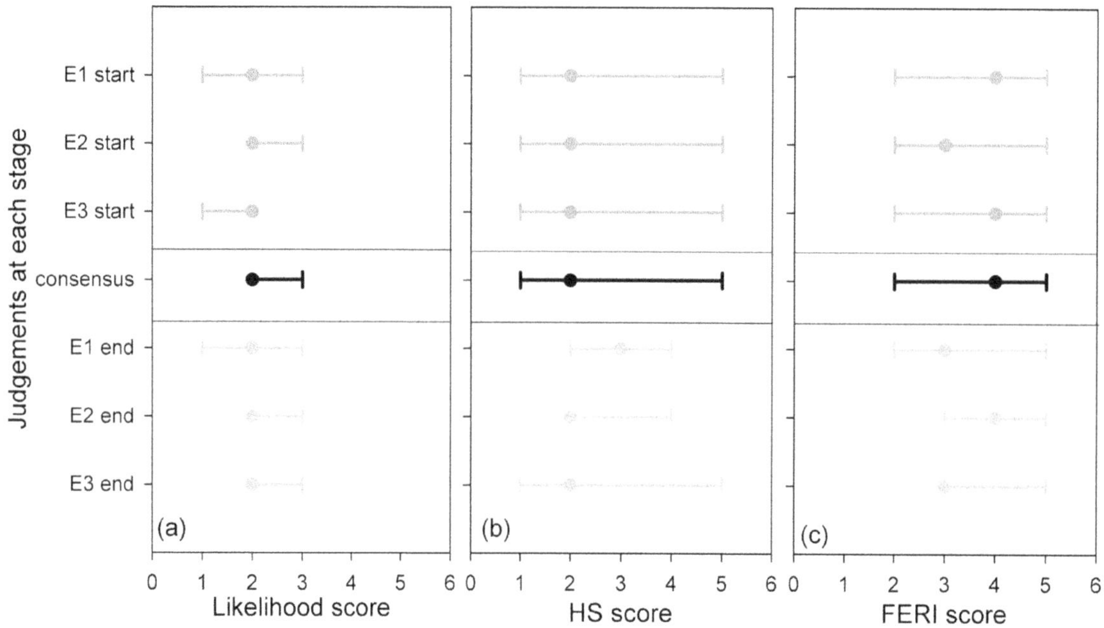

Figure 4. Session 1 scores for FEP D for individual start and end of session (grey) and consensus (black) assessments for (a) likelihood, (b) health and safety (HS) severity and (c) financial, environmental, research and industrial viability (FERI) severity. Key as in Figure 1.

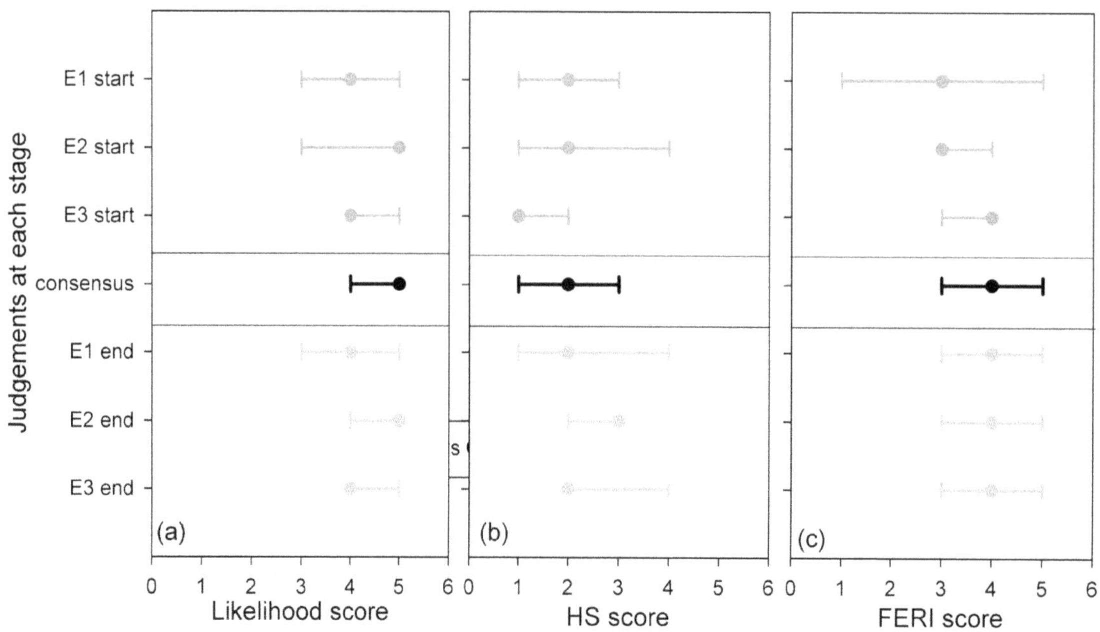

group consensus. However to fully understand the patterns of influence we must consider not only which expert's individual scores tend to be closest to the consensus, but also the overall quantitative degree of similarity between each expert's individual scores and the consensus. We also need to be able to distinguish between experts who have produced exactly the same scores for a particular FEP. These figures demonstrate the complexity of the interpretation of results even with only three experts if the linear fitting technique was not used. As the number of experts grows, so the task of interpreting the results becomes increasingly difficult. Nevertheless, as we show below, the linear fitting technique offers a simple method to identify overall patterns of influence.

The coefficients for the linear models for the individual expert assessments are shown in Table 1. Here the coefficient t-values disprove the null hypothesis that each expert is only influenced by their own original opinion with p-value < 0.05. The coefficients α_{ij} suggests that expert E2 has the most influence on the other experts: the final assessment of expert E1 seems to be more strongly influenced by the initial assessment of expert E2 than by their own initial assessment (i.e., $\alpha_{21}>\alpha_{11}$). Likewise for expert E3, the initial assessment of expert E2 has the most influence on their final assessment ($\alpha_{23}>\alpha_{33}$). The final assessment of expert E2 is most influenced by their own initial assessment, with very little influence at all from expert E1.

Table 2 shows the coefficients produced by using just the expert's own initial assessment and the group consensus to fit each expert's individual assessment at the end of the session. Experts E1 and E3 are both strongly influenced by the group consensus ($\gamma^c \approx 1$), and are not strongly influenced by their own initial assessment ($\gamma_1^I \approx \gamma_3^I \approx 0$). Expert E2 is less strongly influenced by the consensus ($\gamma^c = 0.63$), however in all cases the coefficient t-values disprove the null hypothesis that the experts are only influenced by their own original assessment with $p < 0.05$, demonstrating that some interaction has taken place to significantly alter all of their positions.

The fitting procedure was also applied to the start of session assessments and the consensus assessments producing the model shown in Table 3. The coefficients β_i for the individual start of session assessments reflect the relative influence of each expert on the consensus. The coefficients are similar for all three experts implying they have similar influence as suggested above by direct examination of Figures 1, 2, 3, and 4. In fact t-values of each coefficient from equation (5) have $p > 0.05$ and hence fail to disprove the null hypothesis that all terms are equally influential.

The root mean square error (RMSE) of the experts' end of session assessments with respect

Table 1. Fitted model coefficients, α, for the individual expert (1, 2 and 3) assessments from session 1. Each coefficient α_{ij} shows the influence of each expert i on the individual end of session assessment of expert j. The coefficient showing the influence of the expert's own initial assessment on their own final judgment is in bold.

	Coefficients		
	α_{1j}	α_{2j}	α_{3j}
Expert E1	**0.39**	0.58	0.14
Expert E2	0.08	**0.69**	0.36
Expert E3	0.18	0.61	**0.33**

Table 2. Fitted model coefficients, γ, for the final individual expert (1, 2 and 3) from session 1. Each coefficient γ_i^I shows the influence of the initial assessment of expert i on their own final judgment while γ^c shows the influence of the consensus.

	Coefficients	
	γ_i^I	γ^c
Expert 1	-0.01	1.01
Expert 2	0.45	0.63
Expert 3	-0.03	1.04

Table 3. Fitted model coefficients β for the final individual expert (1, 2 and 3) assessments from the start of the session and the consensus from session 1. Each coefficient, β_i shows the influence of expert i on the consensus.

	Coefficients		
	β_1	β_2	β_3
Start	0.34	0.40	0.37

to the consensus provide a measure of the degree to which the experts tend to disagree with the consensus at the end of the meeting. For experts E1, E2 and E3 the RMSE values are 0.60, 0.69 and 0.53 respectively. This shows that expert E3 tends to agree with the consensus the most and expert E2 agrees with the consensus the least.

Figure 2(c) shows an example of individual start and end of session assessments, and the consensus assessment, which exhibit the effects identified by the linear fitting algorithm. Here we can see that at the start of the session the scores of expert E2 are closest to the consensus scores that the group reached with the same best-guess and lower bound. By the end of the session the assessments of experts E1 and E3 both more closely match the consensus assessment than the assessment of expert E2. Expert E1 gives the same scores as the consensus while expert E3 has the same best guess and upper bound with a slightly higher lower bound. Expert E2 has not changed his/her assessment of the best-guess, however he/she has increased his/her upper bound but also increased his/her lower bound so that he/she is now less in agreement with the consensus of the three experts.

In an earlier assessment, the experts were asked to rate their expertise on a scale of 1 to 5 for each FEP where 1 = novice and 5 = expert. Expert E1 consistently ranked his/her expertise above the other 2 experts as shown in Table 4. However expert E1 has similar levels of influence overall on the consensus as experts E2 and E3, and has the least influence on the final assessments of the other two experts. Expert E2 on the other hand has the most influence on the final assessments of the other two experts even though on one occasion he/she went as far as saying that for one FEP he/she was not particularly knowledgeable in that area. Despite the difference in expertise, expert E2 shows the most influence on the group while being the least influenced by it.

Session Two

A second elicitation session was carried out for three different experts and a different set of FEPs. Figures 5-7 show the individual assessments of each expert at the start and end of the session and the consensus assessment for FEPs E, F, and G respectively. As with Figure 1-4 it is not immediately obvious which expert tends to exert the most influence overall; detailed analysis of each plot would be required to extract this information. One pattern that does emerge is that expert E4 tends not to change his/her views from the start to the end of the session, reproducing exactly the same scores in 7 out of 9 occasions. This suggests that the other two experts in the group have little influence on this expert. In order to better understand the overall trends of influence we use the same linear fitting technique as above.

The coefficients for the linear models produced for each individual expert are shown in Table 5. In this case all three experts are most influenced by their own original assessment from the start

Table 4. Self rated expertise of each expert for each FEP from session 1. Expertise scale 1-5 where 1 = novice and 5 = expert.

FEP	Expert 1	Expert 2	Expert 3
A	4	3	2
B	4	2	2
C	4	2	2
D	3	3	2

Figure 5. Session 2 scores for FEP E for individual start and end of session (grey) and consensus (black) assessments for (a) likelihood, (b) health and safety (HS) severity and (c) financial, environmental, research and industrial viability (FERI) severity. Key as in Figure 1.

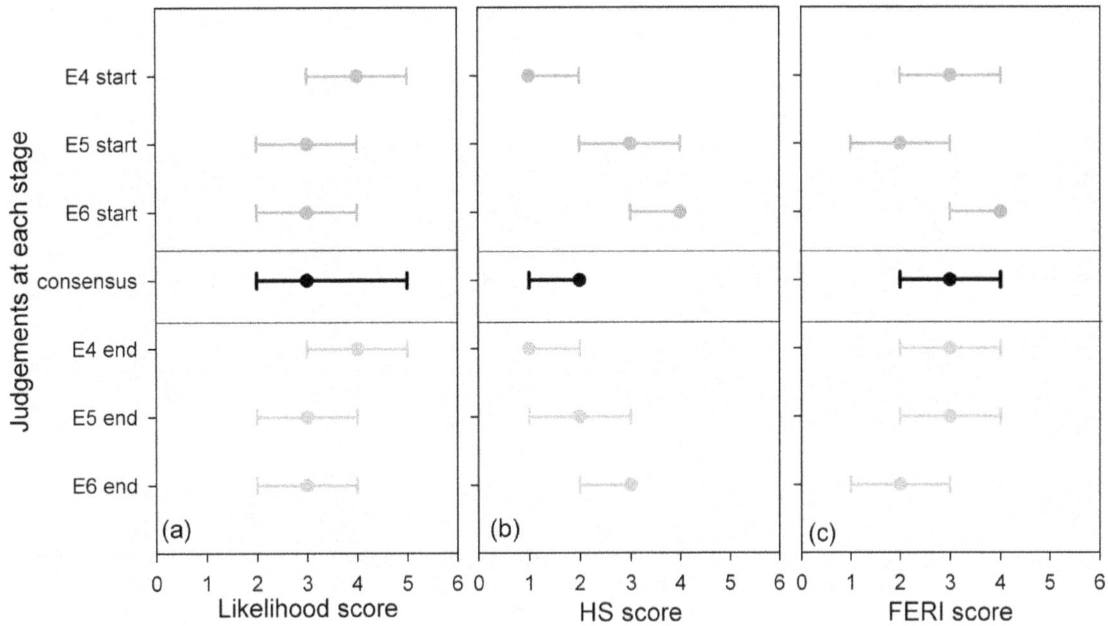

Figure 6. Session 2 scores for FEP F for individual start and end of session (grey) and consensus (black) assessments for (a) likelihood, (b) health and safety (HS) severity and (c) financial, environmental, research and industrial viability (FERI) severity. Key as in Figure 1.

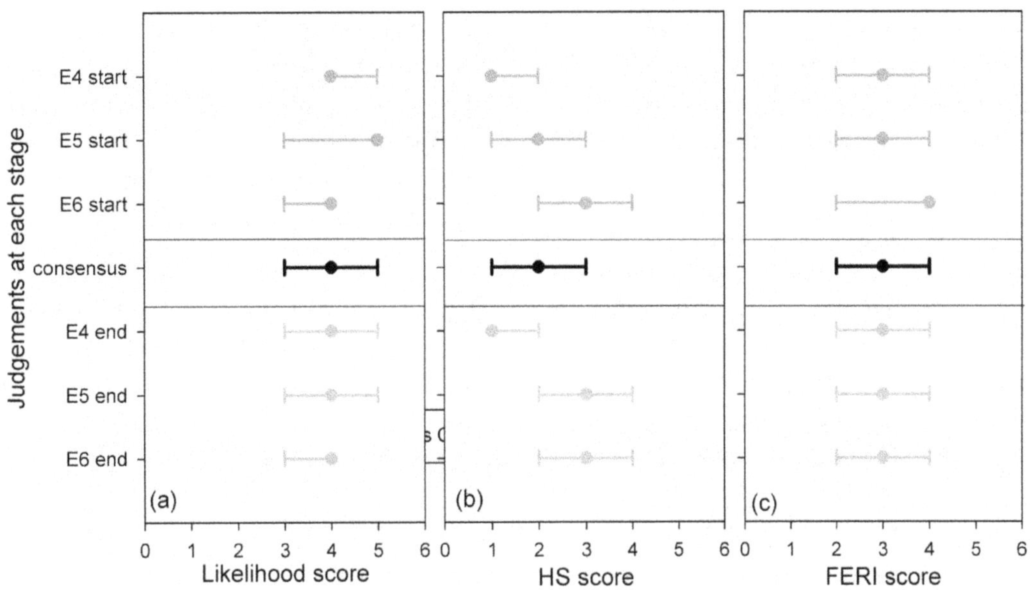

Figure 7. Session 2 scores for FEP G for individual start and end of session (grey) and consensus (black) assessments for (a) likelihood, (b) health and safety (HS) severity and (c) financial, environmental, research and industrial viability (FERI) severity. Key as in Figure 1.

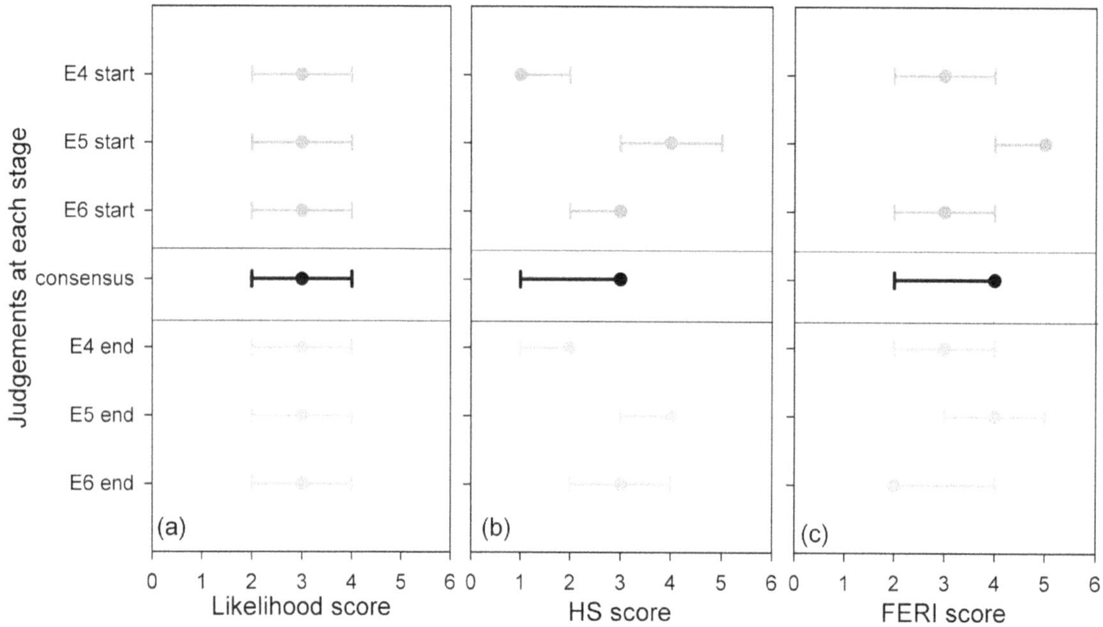

of the session. As identified in the examination of Figures 5, 6, and 7, expert E4 in particular exhibits little influence from the other two experts with $\alpha_{44} \approx 1$ and $\alpha_{54} \approx \alpha_{64} \approx 0$. Experts E5 and E6 are more influenced by the other experts, but the influence of their own prior opinion still dominates with corresponding coefficients of ~ 0.5. Here the t-values for the model for expert E4 fail to disprove the null hypothesis that the expert is only influenced by their own original opinion. For experts E5 and E6 the t-values are statistically significant and hence show that these two experts have been influenced to some degree by the interaction with the other experts, even though they are most influenced by their own opinion.

Table 6 shows coefficients for the linear models produced for fitting the individual end of session opinions to their own start of session assessments

Table 5. Fitted model coefficients, α, for the individual expert (4, 5 and 6) assessments from session 2. Each coefficient α_{ij} shows the influence of each expert i on the individual end of session assessment of expert j. The coefficient showing the influence of the expert's own initial assessment on their own final judgement is in bold.

	Coefficients		
	α_{4j}	α_{5j}	α_{6j}
Expert 4	**0.95**	0.02	0.02
Expert 5	0.20	**0.52**	0.29
Expert 6	0.08	0.33	**0.50**

Table 6. Fitted model coefficients, $\boldsymbol{\gamma}$, for the final individual expert (4, 5 and 6) from session 2. Each coefficient γ_i^I shows the influence of the initial assessment of expert i on their own final judgment and the $\boldsymbol{\gamma}^c$ shows the influence of the consensus.

	Coefficients	
	γ_i^I	γ^c
Expert E1	0.87	0.11
Expert E2	0.48	0.55
Expert E3	0.58	0.35

and to the consensus assessments. Of the three experts, expert E4 is the least influenced by the consensus assessment with their start of session assessment coefficient $\gamma_4^I = 0.87$ compared to the consensus assessment coefficient $\gamma^c = 0.11$, though the t-value for γ_4^I is statistically significant with $p < 0.05$ and hence in this case shows that expert E4 is not solely influenced by their own original opinion. Both experts E5 and E6 are more strongly influenced by the consensus with $\gamma_5^I = 0.48$ and $\gamma^c = 0.55$ for expert E5 and $\gamma_6^I = 0.58$ and $\gamma^c = 0.35$ for expert E6, where the t-values are statistically significant. This again suggests that expert E4 is very confident in his/her original assessment and is the least swayed by the other experts in the group.

Table 7 shows the coefficients β_i for the linear model between the consensus and start of session individual assessments. The coefficient t-values in this case disprove the null hypothesis that all terms are equally influential and hence we can distinguish a pattern of influence. Expert E4 exerts the most influence on the consensus with $\beta_4 = 0.58$ compared to $\beta_5 = 0.22$ for expert E5 and $\beta_6 = 0.18$ for expert E6. Clearly in this instance expert E4 has significantly more influence on the group consensus than the other two experts whose influence seems to be approximately equal. The RMSE values for the experts' end of session assessments and the consensus are 0.54, 0.69 and 0.79 for experts E4, E5 and E6 respectively. This suggests that at the end of the elicitation session, the assessment of expert E4 is closest to the consensus while the assessment of expert E6 is furthest from the consensus.

An example of this is shown in Figure 5(b) where the individual scores from the start of the session show that the consensus is very similar to assessment of expert E4. At the end of the session expert E4 does not change his/her assessment, while experts E5 and E6 do change theirs towards the consensus but are still partly anchored to their assessments from the start of the session. Indeed, expert E6 does not agree with the consensus at all.

The self-assessed expertise of the expert for each FEP are shown in Table 8. From this we can see that experts E4 and E6 consider themselves to have equal expertise for all FEPs while expert E5 considers his/her expertise to be lower. As with the previous session, this was not reflected in the relative influence that each expert had on the group, with expert E4 exerting considerably more influence on the consensus than expert E6. The result is that the consensus closely reflects the start of session assessment of expert E4. This expert is influenced little by the other experts with the result that they change their start of session assessment little when asked to reassess the likelihood and severity at the end of the session. The other two experts have agreed to the consensus of expert E4 but they are less influenced by this expert in their individual assessments. Despite agreeing to a consensus that strongly reflects the start of session assessment of expert E4, personally they still tend to anchor most strongly to their own initial assessments at the end of the session.

Table 7. Fitted model coefficients, β, for the final individual expert (4, 5 and 6) assessments from the start of the session and the consensus from session 2. Each coefficient, β_i shows the influence of expert i on the consensus.

	Coefficients		
	β_4	β_5	β_6
Start	0.58	0.22	0.18

Table 8. Self rated expertise of each expert for each FEP from session 2. Expertise scale 1-5 where 1 = novice and 5 = expert.

FEP	Expert 4	Expert 5	Expert 6
E	4	2	4
F	4	2	4
G	4	1	4

CONCLUSION

From the range of opinions presented in the elicitation sessions, it is clear that experts can reach differing and even contradictory conclusions from the same geological data. To reconcile these differences experts are often asked to reach some consensus position requiring detailed group discussions. However, as this work and other studies have shown, this group consensus may not only reflect the range of opinions of all experts present in the meeting but can also be influenced by the social dynamics in the meeting.

Using simple linear models to represent the behaviour of the group we can begin to understand the relative influence of individual members and the dynamics of the group interactions without the need for complex recording and analysis of group behaviour, nor the complicated detailed analysis of the opinions and judgements of individual experts at each stage of the elicitation process. However, caution must be shown in interpreting these results as this does not conclusively show which expert exerts the most influence. For example, a high value coefficient may simply reflect that the initial assessment of a particular expert is coincidently closer to the final assessment of the group. In addition, as a least squares (L2 norm) fit is used, more emphasis will be placed on choosing coefficients that reduce large differences in individual scores rather than smaller differences. For example, an expert who disagrees with the lower bound of the consensus by two units will tend to be found to be less influential than an expert who disagrees with the consensus lower bound by one unit and the best-guess by one unit. This could of course be altered by using a different norm for the measurement of misfit, such as an L1 norm instead of L2. Repeating the analysis using an L1 norm produces largely similar results. The only notable change of influence is for expert E5, who in Table 5 is identified as most influenced by their own initial opinion using the L2 norm. When we use the L1 norm this expert is equally influenced by their own initial opinion and the opinion of expert E6 and is not longer influenced at all by expert E4; in all other cases, the identity of the most influential expert is unchanged.

In the first session, the start of session opinions of expert E2 are consistently the most influential in the linear models of the individual experts' opinions at the end of the session. This suggests that this expert did indeed have the most influence overall in the group discussions with the other two experts changing their initial assessments to more closely reflect the original assessment of expert E2. In addition, expert E2 is also the least influenced by the group consensus when making his/her individual assessment at the end of the session. Experts E1 and E3 are both primarily influenced by the group consensus suggesting they generally agree with the group judgement. Expert E2 is influenced by the group consensus but less than experts E1 and E3. This suggests that at the end of the session, expert E2 tended to be least swayed by the group. Somewhat paradoxically, it was found that post-consensus expert E2 tended to be furthest from the group consensus position despite apparently having the most influence in the group.

The paradox can be resolved by reasoning that despite his strong influence on the other experts, and his relative intransigence with respect to his personal opinion during the discussion, he was nevertheless willing to sign up to a consensus position that was farthest from his own opinion. This is a complex pattern of behaviour which could have other causative interpretations: for example, it could be that while he presented persuasive evidence, his manner dissuaded the others from agreeing publicly. Either way, we might be reasonably confident that in this case the consensus is relatively conservative in the sense that it has not obviously been dominated by the most influential individual.

For session 2 a different picture emerges. In this session expert E4 seems to exert the most influence on the group with the consensus assessment tending to most closely match the initial assessment of this expert. However, unlike in session 1 where one expert appeared to exert greatest influence over other members of the group, in session two, the experts all seemed to be more influenced by their own initial assessments than by other members of the group. They also tended to be less influenced by the group consensus when making their end of session assessments than the experts in session 1. For example in Figure 7(b) we see that expert E5 agreed to a consensus that is contradictory to (does not agree with) either their initial or final opinions.

From the RMSE of the end of session assessments and the consensus, the expert who appears to most closely agree with the consensus is expert E4. Here the expert who exerted the most influence on the consensus tended to agree with it most strongly. This implies that he/she was more successful at persuading the other experts in the group to agree with his/her opinion during the consensus-reaching part of the discussion. It appears that experts E5 and E6 exhibit herding behaviour: they converge around expert E4 during discussions and diverge immediately thereafter. The consensus should therefore not be regarded as the group's shared opinion in session 2.

Comparing the coefficients for the start of session assessment with the RMSE at the end of the session for session 1, we see that the expert who exerted the greatest influence on the group was the least likely to agree with the group consensus at the end of the session. However in session 2 we see the expert who has exerted the greatest influence on the consensus also agrees with it the most. Thus, by careful elicitation and simple analysis of the results of the various linear models, we can begin to distinguish the different individual and group dynamics that have produced different kinds of results for the two sessions.

The relative influence of individual experts on the group might be expected to correlate with experience or expertise, with the most expert member exerting the most influence on the group and being the least influenced by other members of the group. However, the experts ranked their expertise for each FEP, and this ranking was shown not to be correlated with relative influence in either session. This either indicates that expertise does not correlate with influence, or that experts are poor at assessing their own expertise.

It was not apparent during the group discussions for either sessions 1 or 2 that any one expert was dominating the group, or that any expert was tending to be overruled. Nevertheless the results suggest that in both cases, a single expert has exerted significantly more influence than the other experts despite having less or equivalent relevant expertise to other experts in the group. This was also not obvious from the results by direct inspection. By using a simple linear model we are able to better understand the behaviour of individual experts and the group dynamics, helping to identify where a single expert may be dominant or where bias such as herding and divergence occurs. In this case we are able to identify that in one of the two sessions the consensus position should *not* be used as a fair representation of the group's combined uncertainty or opinions, and either the post-session individual opinions might need to be somehow combined to synthesise a fair post-session group position (see O'Hagan (2006) for a discussion of the difficulties of carrying this out), or another elicitation session, perhaps using a different mode of communication between experts, might be necessary. This method of structured elicitation and linear modelling can be applied to any group decisions where experts are asked to reach a quantitative consensus judgement of some uncertain property (e.g. in an asset team in an oil company). While herding is always likely to be a problem in such situations, this method makes it possible to identify some situations in which the consensus decision is likely to be biased and as such provides a test of the reliability of the final group decision. When bias is identified, it gives the decision makers the opportunity to reassess the outcome of group discussions. Alternatively, the results may provide reassurance that a consensus judgement fairly reflects the opinions of the individual experts. Appropriately communicating the results of this type of analysis may provide greater confidence in expert decisions and help with public acceptance of new technologies like carbon capture and storage.

The results of such risk assessments are critical to evaluating the suitability of a potential subsurface CO_2 storage site. They inform decision makers of areas of high risk, help to prioritise and plan future activities and allocation of resources, and ultimately may lead to rejection or acceptance of a site. It is therefore critical that methods are found to reduce the bias in such assessments and improve the decisions made by groups. The method described in this chapter combined with the elicitation procedure of Polson and Curtis (2010) provides an efficient and easily applied means to understand the dynamics within a group and identify potential bias within group decisions. Crucially, by comparing the influence of individual experts with their level of expertise, we can assess whether the group is making the optimum use of the available personnel's knowledge and experience when making such critical decisions. In this study it was shown that influence does not correlate with expertise, and that herding of opinion towards a consensus position followed immediately by divergence of opinion occurred. For the robust use of group opinions in geological problems it will therefore be important that future research identifies robust strategies for remedial action in cases where biases such as herding and undue influence are detected.

ACKNOWLEDGMENT

The authors wish to acknowledge the support of, the UK Technology Strategy Board (TSB) and the Engineering and Physical Sciences Research Council (EPSRC), and of the project industry partners AMEC, Marathon, Schlumberger, Scottish Power, and Scottish and Southern Energy and the academic partners; the British Geological Survey, Heriot-Watt University, University of Edinburgh, and of the University of Manchester. The research related to this chapter has been carried out within the CO_2 Aquifer Storage Site Evaluation and Monitoring (CASSEM) project, which is a project supported by the UK TSB. We also thank the anonymous reviewers for their useful contributions.

REFERENCES

Anderson, J. L. (1998). Embracing uncertainty: the interface of Bayesian statistics and cognitive psychology. *Conservation Ecology*, *2*, 2. Retrieved from http://www.consecol.org/vol2/iss1/art2

Arnell, N. W., Tompkins, E. L., & Adger, W. N. (2005). Eliciting information from experts on the likelihood of rapid climate change. *Risk Analysis*, *25*(6), 1419–1431. doi:10.1111/j.1539-6924.2005.00689.x PMID:16506972

Aspinall, W. P. (2006). Structured elicitation of expert judgment for probabilistic hazard and risk assessment in volcanic eruptions. In H.M. Mader, S.G. Coles, C.B. Connor, & L.J. Connor (Eds.), Statistics in Volcanology. Geological Society of London.

Baddeley, M. A., Curtis, A., & Wood, R. A. (2004). An introduction to prior information derived form probabilistic judgements: elicitation of knowledge, cognitive bias and herding. In A. Curtis & R. A. Wood (Eds.), Geological Society, London Special Publications: Geological Prior Information: Informing Science and Engineering (vol. 239, pp. 15–27). Academic Press.

Bommer, J. J., Scherbaum, F., Bungum, H., Cotton, F., Sabetta, F., & Abrahamson, N. A. (2005). On the use of logic trees for ground-motion prediction equations in seismic-hazard analysis. *Bulletin of the Seismological Society of America, 95*(2), 377–389. doi:10.1785/0120040073

Bonano, E. J., & Apostolakis, G. E. (1991). Theoretical foundation and practical issues for using expert judgements in uncertainty analysis of high-level radioactive waste disposal. *Radioactive Waste Management and the Nuclear Fuel Cycle, 16*(2), 137–159.

Bond, C. E., Gibbs, A., Shipton, Z. K., & Jones, S. (2007). What do you think this is? 'Conceptual uncertainty' in geoscience interpretation. *GSA Today, 17*(11), 4–10. doi:10.1130/GSAT01711A.1

Bond, C. E., Lunn, R. J., Shipton, Z. K., & Lunn, A. D. (2012). What makes an expert effective at interpreting seismic images? *Geology, 40*(1), 75–78. doi:10.1130/G32375.1

Bonner, B. L. (2000). The Effects of Extroversion on Influence in Ambiguous Group Tasks. *Small Group Research, 31*(2), 225–244. doi:10.1177/104649640003100205

Bonner, B. L. (2004). Expertise in Group Problem Solving: Recognition, Social Combination, and Performance. *Group Dynamics, 8*(4), 277–290. doi:10.1037/1089-2699.8.4.277

Chadwick, A., Arts, R., Bernstone, C., May, F., Thibeau, S., & Zweigel, P. (2008). *Best Practice for the Storage of CO2 in Saline Aquifers. In Observations and guidelines from the SACS and CO2STORE projects.* Amersham: British Geological Survey by Halstan & Co. Ltd.

Cohen, S. G., & Bailey, D. R. (1997). What makes teams work: Group effectiveness research from the shop floor to the executive suite. *Journal of Management, 23*(3), 239–290. doi:10.1177/014920639702300303

Curtis, A. (2012). The science of subjectivity. *Geology, 40*(1), 95–96. doi:10.1130/focus012012.1

Curtis, A., & Wood, R. (2004a). *Optimal elicitation of probabilistic information from experts.* doi:10.1144/GSL.SP.2004.239.01.09

Curtis, A., & Wood, R. (2004b). Geological Society of London Special Publication: Geological Prior Information: Informing Science and Engineering (vol. 239). Academic Press.

Davis, J. H. (1982). Social interaction as a combinatorial process in group decision. In H. Brandstatter, J. H. Davis, & G. Stocker-Kreichgauer (Eds.), *Group decision making*. London: Academic Press.

Davis, J. H. (1992). Some compelling intuitions about group consensus decisions and internal aggregation phenomena: Selected examples, 1950–1990. *Organizational Behavior and Human Decision Processes. Special Issue: Group Decision Making, 52*, 3–38.

Davis, J. H. (1996b). Group decision making and quantitative judgements: A consensus model. In E. Witte & J. H. Davis (Eds.), *Understanding group behavior: Consensual action by small groups* (Vol. 1). Hillsdale, NJ: Erlbaum.

Davis, J. H., Au, W., Hulbert, T. L., Chen, X. P., & Zarnoth, P. (1997). The effects of group size and procedural influence on consensual judgements of quantity: The example of damage awards and mock civil juries. *Journal of Personality and Social Psychology, 73*(4), 703–718. doi:10.1037/0022-3514.73.4.703

Davis, J. H., Hulbert, L., & Au, W. T. (1996a). Procedural influence on group decision making: The case of straw polls—Observation and simulation. In R. Y. Hirokawa & M. S. Poole (Eds.), *Communication and group decision making*. Beverly Hills, CA: Sage Publications.

Det Norske Veritas. (2010). *CO2QUALSTORE Guideline for Selection and Qualification of Sites and Projects for Geological Storage of CO2*. Author.

Diviacco, P. (2015). Reconciling Knowledge and Collaborative e-Research. In *Collaborative Knowledge in Scientific Research Networks*. IGI Global Press.

Ford, J.R., Pharaoh, T.C., Hulbert, A.G., Voncent, C.J. & Cooper, A.H. (2009). CASSEM Work Package One – British East Midlands and Lincolnshire Geological modelling. *British Geological Survey Commissioned Report*, CR/08/152. Available on request.

Hackman, J. R., & Morris, C. G. (1975). In L. Berkowitz (Ed.), *Group tasks, group interaction process, and group performance effectiveness: A review and proposed integration*. In Advances in experimental social psychology (Vol. 8, pp. 45–99). New York: Academic Press.

Hinsz, V. B. (1990). Cognitive and consensus processes in group recognition memory performance. *Journal of Personality and Social Psychology*, *59*(4), 705–718. doi:10.1037/0022-3514.59.4.705

Kahneman, D., Slovic, P., & Tversky, A. (Eds.). (1982). *Judgement under Uncertainty: Heuristics and Biases*. Cambridge, UK: Cambridge University Press. doi:10.1017/CBO9780511809477

Kerr, N. L. (1992). Issue importance and group decision making. In S. Worchel, W. Wood, & J. A. Simpson (Eds.), *Group process and productivity* (pp. 68–88). Newbury Park, CA: Sage.

Kirchler, E., & Davis, J. H. (1986). The influence of member status differences and task type on group consensus and member position change. *Journal of Personality and Social Psychology*, *51*(1), 83–91. doi:10.1037/0022-3514.51.1.83

Laughlin, P. R. (1988). Collective induction: Group performance, social combination processes, and mutual majority and minority influence. *Journal of Personality and Social Psychology*, *54*(2), 254–267. doi:10.1037/0022-3514.54.2.254

Laughlin, P. R. (1992). Influence and Performance in Simultaneous Collective and Individual Induction. *Organizational Behavior and Human Decision Processes*, *51*(3), 447–470. doi:10.1016/0749-5978(92)90021-X

Lehrer, K., & Wagner, C. (1981). *Rational Consensus in Science and Society: A Philosophical and Mathematical Study*. Dordrecht: Reidel. doi:10.1007/978-94-009-8520-9

Lowe, T. D., & Lorenzoni, I. (2007). Danger is all around: Eliciting expert perceptions for managing climate change through a mental models approach. *Global Environmental Change*, *17*(1), 131–146. doi:10.1016/j.gloenvcha.2006.05.001

Maul, P., Savage, D., Benbow, S., Walke, R. R., & Bruin, R. (2005). Development of a FEP database for the geological storage of carbon dioxide. Greenhouse Gas Control Technologies, 1, 701-709.

McGrath, J. E. (1984). *Groups: Interaction and performance*. Englewood Cliffs, NJ: Prentice Hall.

Morgan, M. G., Pitelka, L. F., & Shevliakova, E. (2001). Elicitation of expert judgements of climate change impacts on forest ecosystem. *Climatic Change*, *49*(3), 279–307. doi:10.1023/A:1010651300697

O'Hagan, A., Buck, C. E., Daneshkhah, A., Eiser, J. R., Garthwaite, P. H., & Jenkinson, D. J. et al. (2006). *Uncertain Judgements: Eliciting Expert's Probabilities*. Wiley.

Ohtsubo, Y., Masuchi, A., & Nakanishi, D. (2002). Scheme Model in a Group Polarization Context. Majority Influence Process in Group judgement: Test of the Social judgement. *Group Processes & Intergroup Relations*, *5*, 249–261. doi:10.1177/1368430202005003005

Phillips, L. D. (1999). Group elicitation of probability distributions: are many heads better than one? In J. Shantaeu, B. Mellors, & D. Schum (Eds.), *Decision Science and Technology: Reflections on the contributions of Ward Edwards* (pp. 313–330). Norwell, MA: Kluwer Academic Publishers. doi:10.1007/978-1-4615-5089-1_17

Polson, D., & Curtis, A. (2010). Dynamics of uncertainty in geological interpretation. *Journal of the Geological Society*, *167*(1), 5–10. doi:10.1144/0016-76492009-055

Polson, D., Curtis, A., & Vivalda, C. (2012). The evolving perception of risk during reservoir evaluation projects for geological storage of CO2. *International Journal of Greenhouse Gas Control*, *9*, 10–23. doi:10.1016/j.ijggc.2012.02.010

Polson, D., Curtis, A., Vivalda, C., & Saunier, S. (2009). Process to Track the Evolving Perception of Risk During CO2 Storage Projects. In *Proceedings of Offshore Europe 2009*. SPE 124703.

Runge, A. K., Scherbaum, F., Curtis, A., & Roggelsen, C. (2013). An Interactive Tool for the Elicitation of Subjective Probabilities in Probabilistic Seismic-Hazard Analysis. *Bulletin of the Seismological Society of America*, *103*(5), 2862–2874. doi:10.1785/0120130026

Savage, D., Maul, P., Benbow, S., & Walke, R. (2004). *A generic FEP database for the assessment of long-term performance and safety of the geological storage of CO2*. Retrieved from http://www.co2captureandstorage.info/docs/QuintessaReportIEA.pdf

Smith, M., Campbell, D., Polson, D., & Mackay, E. (2010). *CO_2 Aquifer Storage Site Evaluation and Monitoring*. Edinburgh, UK: Heriot Watt University.

Sniezek, J. A. (1992). Groups under uncertainty: An examination of confidence in group decision making. *Organizational Behavior and Human Decision Processes*, *52*(1), 124–155. doi:10.1016/0749-5978(92)90048-C

Wildenborg, T., Leijnse, A., Kreft, E., Nepveu, M., Obdam, A., & Orlic, B. (2005). Risk assessment methodology for CO2 storage: the scenario approach. In *Carbon dioxide capture for storage in deep geologic formations*. Amsterdam: Elsevier.

Wildenborg, T., Leijnse, T., Kreft, E., Nepveu, M., & Obdam, A. (2004). *Longterm safety assessment of CO2 storage: the scenario approach*. Paper presented at the Seventh international conference on greenhouse gas control technologies. Vancouver, Canada.

Wildenborg, T., Leijnse, T., Kreft, E., Nepveu, M., & Obdam, A. (2004). *Longterm safety assessment of CO2 storage: the scenario approach*. Paper presented at the Seventh international conference on greenhouse gas control technologies. Vancouver, Canada.

Ye, M., Pohlmann, K. F., & Chapman, J. B. (2008). Expert elicitation of recharge model probabilities for the Death Valley regional flow system. *Journal of Hydrology (Amsterdam)*, *354*(1-4), 102–115. doi:10.1016/j.jhydrol.2008.03.001

APPENDIX: DETAILED DESCRIPTION OF EACH FEP GIVEN TO THE EXPERTS IN EACH SESSION

The detailed description of each FEP given to the experts in session 1 are shown below (these are from the Quintessa CO_2 FEP register [Savage et al., 2004]).

Construction and Site Logistics (Non-Pipeline): Construction and site logistics issues can affect the timescale, financial, safety and viability of the project. There are issues surrounding the ability to build on existing sites and the resources available to build carbon capture facilities.
Failures and delays in construction and logistical aspects of the project may have significant impact on viability of the project through impacts on timescale, finance or safety. Site logistics may require compromises in construction and/or retrofitting of carbon capture facilities to existing sites. If many sites decide to introduce carbon capture at the same time, this may lead to a shortage of skills, labour and possibly materials which will likely increase the cost of construction. As well as restricted supplies, more inflexible time slots amongst suppliers may impact the project.

Construction of Pipeline: The construction of the pipeline will be affected by a number of factors. Many of these will not be unique to CO_2, however the pipeline route will be affected by regulations particular to CO_2 such as distance from populated and ecologically sensitive areas. The composition of the CO_2 will affect corrosion rates and therefore impact on the design of the pipeline as will the need for isolation points. In addition the pipeline design and construction will also be affected by the injection pressure which will determine whether it is more cost efficient to have onshore or offshore compression.

Planning and regulations will affect the route pipeline can take with minimum distances from sensitive areas; this affects the length and therefore the cost of the construction. Planning regulations may cause particular difficulties near to the plant or coast where potential routes are restricted. In addition isolation points will be required more frequently near populated areas where the health effects of potential leaks need to be considered. Also the injection pressure required for the aquifer will affect whether it is more cost efficient to compress the CO_2 onshore and transport the CO_2 at higher pressures or to transport the CO_2 at lower pressures and use offshore compression near the aquifer.

Leakage from Pipeline: Onshore, accidental damage to the pipe by works unrelated to the project could potentially produce a CO_2 release resulting concentrations in nearby areas (over 10%) that could have major health implications due to the asphyxiation effects of CO_2. There is also the potential for small undetected leaks from the pipe that could result in the accumulation of CO_2 in low lying areas (e.g. basements). Small undetected leaks could also potentially impact the local ecology both through increased CO_2 concentrations and potential acidification of ground water and through release of contaminants (e.g. heavy metals). Offshore the potential impacts to human health are reduced however small undetected leaks could impact the marine ecology due to acidification or release of contaminants. In areas where trawler fishing or the navy are active, damage to the pipeline due to collisions may result in large scale release. In addition to potential health or ecological impacts, leakage may also result in negative public perceptions which could have a negative impact on future projects.

Leakage from the pipeline provides a potentially important pathway for CO2 to enter the environment with impacts on human health, the environment, public perception and cost.

Public Perception and Security: Public perception could have negative or positive impacts on planning applications. In addition negative public perception may lead to increased security risks with onsite protests and potentially sabotage.

Public perception and security could impact the project in cost, time delays and where public opposition is extreme, potentially hinder or prevent the completion of the project.

The description of the FEPs given to the experts session 2 are shown below (these are from the Quintessa CO_2 FEP register (Savage et al., 2004)).

Advection of Free CO_2: Advection of free CO_2 occurs in response to differences in pressure. The pressure difference may be due to differences in the pressure of injected CO_2 and formation pressures. The rate and direction of advection is affected by the physical properties of the rock, such as porosity and permeability. Advection may also occur though fractures. Fracture flow will be episodic with high transport efficiencies. Resealing of fractures (for example by cementation) will reduce and ultimately block fluid flow.

This includes Fault Valving a process resulting from gradual build up of pore pressure due to fluid generation, causing the subsequent opening of a fault along with fluid escape towards surface. This mechanism has been recognised as causing earthquakes in many parts of the world, as a result of hydrocarbon generation or infiltration of other fluids.

Advective flow is a key transport process for migrating CO2, and associated contaminants, in the geosphere (reservoir, surrounding and overlying rock), near-surface and surface environments. Large releases of pore fluids may occur during fault valving episodes.

Buoyancy-Driven Flow: Different relative densities of fluids in a geological system will result in buoyancy-driven flows as less dense fluids will have a tendency to flow upwards. The density of fluids will depend on their temperature and pressure.

Carbon dioxide can be less dense than water, which may cause injected CO_2 to flow upwards and accumulate above the water phase below the cap rock of a reservoir. Water with dissolved CO_2 is more dense than water, which can result in stratification of water bodies into which CO_2 may leak, if conditions are suitable.

Water Mediated Transport: Processes related to transport of CO_2, and associated contaminants, in ground and surface water, include advection, dispersion and molecular diffusion. Advection is the process by which CO_2, and associated contaminants, are transported by the bulk movement of water in which they are dissolved. Advective groundwater flow can occur along connected porous regions, such as fractures and faults. Processes that affect the movement of groundwater, such as fault valving, will also affect the migration of dissolved CO2, and associated contaminants. Dispersion is the collective name for the consequences of a number of processes that cause 'spreading-out' of CO_2, and associated contaminants, dissolved in water in all directions, superimposed on the bulk movement predicted by a simple advection model. It results in a spatially distributed contaminant plume. Spreading of the solute plume can occur in the direction of advection, in which case it is known as longitudinal dispersion, or it can occur perpendicular to the direction of advection, in which case it is known as transverse dispersion. Diffusion is the process whereby chemical species move under the influence of a chemical potential gradient (usually a concentration gradient). In the storage domain, diffusion of CO_2 might be significant in the cap rock and low permeability sedimentary host rock environments where advective transport is absent or limited, and diffusion of other chemical species may give rise to chemical regimes in parts of the system that inhibit or enhance the transport of CO_2.

The transport of CO_2, and associated contaminants, within groundwater is likely to be a key migration process and therefore an important consideration in determining performance and safety.

Chapter 5
International Funding and Collaboration in Sea Level Rise Research

Mayte López-Ferrer
Universitat Politècnica de València, Spain

ABSTRACT

International collaboration and research funding in Sea Level Rise (SLR) research are investigated in this chapter. SLR can be taken as a paradigmatic research area to study the international scientific collaboration and research funding efforts because it is affecting the whole planet and is an interdisciplinary research area involving disciplines belonging to the geosciences but also the life sciences, technology sciences, and social sciences. The aim of the chapter is to identify the main stakeholders in the topic, institutions, and countries; analyze overlapping efforts; identify possible research gaps; and to study the role played by the funding agencies. Bibliometrics and a social network analysis approach are applied. Co-occurrence networks of keywords, affiliations, and funding agencies among scientific papers in Thomson Reuters' Web of Science Core Collection in the SLR topic are analyzed. Conclusions show that international scientific collaboration is common in SLR, but international co-financing is less frequent.

INTRODUCTION

Globalization, in its broadest sense, "is a process that encompasses the causes, course, and consequences of transnational and transcultural integration of human and non-human activities" (Al-Rodhan & Stoudmann, 2006). It has an undeniable impact. Its geopolitical implications include global economic integration, international politics, stability and durability of cultures and also the transmission of knowledge and science. Science undergoes a process of globalization, which compels to scientists from different geographical and cognitive communities to understand each other. But due to the fact that science is strongly conditioned by social aspects (traditions, schools, institutions, economic interests, and reputation systems), this understanding is not always so evident. And, also, the other way round. Scientific research affect society and human life, as well,

DOI: 10.4018/978-1-4666-6567-5.ch005

Copyright © 2015, IGI Global. Copying or distributing in print or electronic forms without written permission of IGI Global is prohibited.

since their results are at the base of technological developments, that produce new tools that in turn transform the society way of life (cars, freezers, computers are examples).

Interdisciplinary and multidisciplinary science is another remarkable feature of research. Here the research problem is approached by two or more scientists from different disciplines. In words of Boldrini, Craglia, Mazzetti and Nativi (2015, see chapter 14 this book), multidisciplinary research is academic but loosely integrated, i.e. consisting of scientists from distinct disciplines researching the same topic but with multiple disciplinary goals. In the case of interdisciplinarity, research is academic and tightly integrated, and the scientists from unrelated disciplines try to create and share knowledge to pursue a common goal.

The interdisciplinary approach is based on a conceptual model that involves the integration of theoretical frameworks, research designs and methodologies, and requires the use of different perspectives and skills throughout the different phases of the research (Aboelela et al., 2007).

In both multidisciplinary and interdisciplinary research cooperation is always key. Such cooperation can have different levels of intensity, ranging from simple communication of ideas to the mutual integration of conceptual and theoretical frameworks. In addition, this cooperation may result in the solution to a specific problem, in a few co-authored scientific papers or, perhaps, in a new emerging scientific discipline. Both globalization and multidisciplinary or interdisciplinary science are strongly related to scientific collaboration.

BACKGROUND

Data sharing is first step in scientific collaboration. This is particularly true of research carried out on large-scale facilities and when research questions demand not just single experiments but rather sequences of them, which is very common in the geosciences. The experiments undertaken in these facilities are of growing complexity and are expected to soon produce a "data avalanche" (Bunakov, Jones, & Matthews, 2015, see chapter 7 this book).

In the geosciences there are many initiatives in order to share data, e.g.: PANdata (http://www.pan-data.eu) (Bunakov, Jones & Matthews, 2015, see chapter 7 this book); Tatiana (Dyke, Lund, & Girardot, 2009; Dyke, Lund, & Girardot, 2010); Mulce (http://mulce.org)(Reffay, Chanier, & Betbeder, 2012) and (Reffay, Dyke, & Betbeder, 2012); datapublication (http://datapublication.tge-adonis.fr); Dataverse project (King, 2007); and ESSD journal, all in.

Open access and open data movements have encouraged the scientific community to pay more attention to their data by considering the societal impact brought about by making such data available to everyone. But to make these changes sustainable, shaping policies, funding specific research programs, organizing networks, providing infrastructure, promoting standards, etc. are necessary steps. In addition, because researchers' time is limited and publication pressure is high and in many disciplines, collaboration is necessary. However, the quality of collaboration and data publication is not given sufficient importance to make it a priority for researchers. So, innovation is needed to find easier processes for researchers to make their datasets reusable and to render their research results more visible.

Raw research data are the cornerstone with which researchers build knowledge; their findings supports the idea that any paper presenting results from an empirical study should be linked to the data that has been used for it. As such, the aim of data availability is to facilitate validation, since interpretations of data are easier to do with the data alongside it; replication, since opens the door to further analyses on the same data to verify the conclusions; and meta-analysis, datasets may be compared or combined to build higher order analyses.

There have been efforts related to developing ontologies and controlled vocabularies which allow connecting different repositories of data, often from different countries. Semantic mediation tools are necessary in any information system, for both sides of the exchange process: document description and also information retrieval.

There are many examples of knowledge organization systems in the geosciences. For example, those developed by the Open Geospatial Consortium. These ontologies can be utilized in collaborative environments where terms with the same or similar meanings may have different tags but must have established correspondences. Through inclusion of links to encoded terms, these values can be mapped to have the same meaning when data is integrated. As Fredericks (2015, see chapter 13 this book) conclude, even though the use of standards in geospatial data access is important, integration is nearly impossible without the inclusion of references to registered terms in a semantic framework.

Another example are the controlled vocabularies from the British Oceanographic Data Centre used in SeaDataNet in Europe, the Rolling Deck to Repository (R2R) programme and the Biological and Chemical Oceanography Data Management Office (Leadbetter, Arko, Chandler, Shepherd, & Lowry, 2015, see chapter 12 this book), which are also an example of collaboration-driven "bottom-up", as opposed to "top-down", creation of standards.

Once can also note the efforts in achieving interoperability of systems in many multi-disciplinary spatial data infrastructures, such as the Global Earth Observation System of Systems (GEOSS), the INSPIRE Directive in Europe, European Marine Observation and Data Network (EMODnet), and the Earth Cube initiative in the USA (Boldrini, Craglia, Mazzetti, & Nativi, 2015, see chapter 14 this book). Here, the standardization of data and metadata are keys questions, as is the adoption of agreed technologies, and prominently as Boldrini, Craglia, Mazzetti & Nativi (2015, see chapter 14 this book) illustrates from his Brokering Approach an organizational and social perspective which include data policy; and legal, cultural, and people harmonization. These facilitate multidisciplinary and participatory research in a global environment because it respects organizations' autonomy, conciliates infrastructure mandates and governance arrangements, and lowers social and cultural entry-barriers.

In order to establish ontologies for global information retrieval systems in the geosciences for a better data sharing of practices, but also for further communication and comprehension between different scientific communities, the concept of boundary objects, proposed by Diviacco (2015, see chapter 19 this book) is a concept that can be developed and applied en different environments. Knowledge is not always reducible to language (Polanyi, 1966/1983) in Diviacco (2015, see chapter 1 this book). Sometimes one can have different forms such as evocative objects that are able to resonate for another person, allowing him or her to grasp the question (Diviacco, 2015, see chapter 1 this book). Boundary objects are conceptual artifacts that are weakly structured in common use but strongly structured in individual use, and which contain enough detail to be understood by all actors, even though each individual does not fully understand the context in which the other partners use it. There are many of them in science: diagrams, workflows, mind maps, networks, which have been demonstrated to effectively support collaborative research in many scientific domains (Diviacco, & Busato, 2015, see chapter 19 this book). In the geosciences, maps are an excellent example of boundary objects.

Scientific collaboration could be defined as researchers working together to achieve the common goal of producing new scientific knowledge (Katz, & Martin, 1997). This vague definition, in the words of its author, can range from the recognition of someone who contributes only a small part of the research to, at the other extreme, when all scientists contribute directly to all major research tasks throughout the project. Therefore, the conclusion is that scientific collaboration is somewhere on this continuum. Leaving aside the conceptual problem, collaboration is conventionally measured through co-authored papers. Although co-authorship and collaboration are not synonyms, co-authorship is used as an indicator of collaboration.

There are many reasons for scientific collaboration. Lassi and Sonnenwald, (2010) refer to several benefits of scientific collaboration including: works with several authors are more often cited and are cited over longer periods of time; research funding agencies encourage collaborative research; collaboration increases efficiency through better use of existing resources; larger research groups can take on larger projects; collaborative work enables transfer of learning; and being a member of a research community brings prestige. Smykla and Zippel, (2010) add that building an international reputation can play a major role in being promoted, hence the growing importance of international mobility for advancing research careers. Therefore, scientific collaboration is seen as a virtue, and various public policies actively promote scientific collaboration among individuals and institutions. Some studies suggest that the leaders of potential research teams, as well as policy makers, should take into account scientific studies that show a relation between collaboration and productivity when deciding about the composition of teams, and should exploit the qualities that will bring benefits to the whole team and the project (Barjak, 2006).

A wide range of studies tracks the incidence of co-authored articles and uses this as a proxy for collaboration (Birnholtz, 2007). Using the social networks built from co-authorship in scientific publications, we can take clusters of authors as bibliometric footprints to measure the collective production of knowledge (Velden, Haque, & Lagoze, 2010). The research group is the unit of knowledge production, and co-authorship networks enable them to be studied, analyzed and visualized in terms of scientific collaboration (Lee, Kwon, & Kim, 2011; Sorensen, Seary, & Riopelle, 2010).

Geosciences embrace all sciences related to the planet. It includes the study of the atmosphere, hydrosphere, oceans and biosphere, as well as the cryosphere. Typically, geoscientists will use tools from physics, chemistry, biology, engineering, telecommunications, etc. to build an understanding of how Earth evolved to its current state and how it works, as well to predict, when possible, future changes.

Main geoscience research topics can be considered global and interdisciplinary. Its areas of study deal with issues and phenomena pertaining to the entire planet, e.g. plate tectonics, or elements and phenomena that although occur at one point affects the entire planet, e.g. poles melting.

Some bibliometric analyses have been found in the previous scientific literature in subjects related to geosciences such as biogeochemistry and disturbances in coastal ecosystems (Gattuso, Dawson, Duarte, & Middelburg, 2005); coastal and marine environments (Soomai, MacDonald, & Wells, 2013); freshwater marshes loss (Chen et al., 2013); global climate change (Li, Wang, & Ho, 2011; Weichselgartner, & Kasperson, 2010); marine biotechnology (Pudovkin, 1993; Sears, 1993); ocean circulation (Zhang, Qian, & Ho, 2009); ocean engineering (Martin Cantos, 2011), and others.

Within the area of geosciences, the Sea Level Rise (SLR) as an effect of climate change is one of the issues with some of the most important research fronts identified from their respective core papers, according to "Essential Science Indicators" [1].

Many factors produce short-term changes in sea level, such as atmospheric pressure, El Niño

(ENSO), Rotational variations (Chandler wobble), astronomical tides, seasonal water balance among oceans, water density changes, earthquakes, seiches, tsunamis, etc. But most climate scientists admit that it is very likely that global warming has been human-induced, i.e. anthropogenic. Specifically, global warming is the cause of most current SLR, through two main factors: the thermal expansion of ocean water and the melting of the poles and of other major areas of land ice, like glaciers and ice sheets.(IPCC, 2012)

This rise in sea levels around the world potentially affects human populations in coastal and island regions, as well as natural environments like marine ecosystems. And, even worse, this phenomenon is expected to continue for centuries. Also, over longer geological timescales, changes in the shape of oceanic basins affects sea level, but this is not related to anthropogenic effects. (IPCC, 2012)

Most of human population lives on the coast and it this proportion is increasing. The average density of people in coastal regions is three times higher than the global average density. Even though the climate projections that scientific community produces are not predictions, and have been criticized as very conservative, the economic, social and human implications are enormous. This scientific question could be described as one of the most threatening challenges that humankind faces nowadays.

The main international organizations managing initiatives affecting Sea Level Rise as a part of Climate Change challenge are United Nations, UNESCO and Intergovernmental Panel on Climate Change (IPCC).

United Nations has created and supported the United Nations Conference on the Human Environment (Stockholm, 1972); United Nations Conference on Environment and Development, UNCED (Rio de Janeiro, 1992); the World Summit on Sustainable Development, WSSD, (Johannesburg, 2002) and the United Nations Conference on Sustainable Development, Rio+20, (Rio de Janeiro, 2012). These conferences have led to agreements and international treaties such as the Stockholm Declaration, the United Nations Framework Convention on Climate Change (UNFCCC), the Kyoto Protocol, Agenda 21, etc. Specifically related to funding initiatives to prevent and cope with consequences or Climate Change and Sea Level Rise, the Global Environment Facility (GEF) must be mentioned. Because the contribution of countries to climate change mainly through CO2 emissions, vary enormously, The UNFCCC and the Kyoto Protocol foresee financial assistance from countries with more resources to those less endowed and more vulnerable.

The main initiatives of UNESCO Climate Change Platform related to rising sea levels are the Global Ocean Observing System (GOOS) and the Global Sea Level Observing System (GLOSS), as well as the multi-organizational Global Forum on Oceans, Coasts and Islands.

Closer to scientific research environment works the Intergovernmental Panel on Climate Change. It was created in 1988 by the World Meteorological Organization (WMO) and the United Nations Environment Program (UNEP) to prepare, based on available scientific information, assessments on all aspects of climate change and its impacts, with a view of formulating realistic response strategies. Because of its scientific and intergovernmental nature, the IPCC embodies a unique opportunity to provide rigorous and balanced scientific information to decision makers. Since its foundation has been publishing the IPCC First Assessment Report, FAR, (1990) and Supplementary Reports (1992); the IPCC Second Assessment Report: Climate Change, SAR, (1995); the IPCC Third Assessment Report: Climate Change, TAR, (2001); the IPCC Fourth Assessment Report: Climate Change, AR4 (2007); the IPCC Fifth Assessment Report: Climate Change, AR5, (2013).

The authors believe that SLR may be a paradigmatic research area to study international scientific collaboration, since this issue affects countries in both hemispheres, from all continents, with different levels of development, and, therefore, different scientific environments. It is also an interdisciplinary research area involving disciplines such as geology, geophysics, and oceanography belonging to the geosciences, but also life sciences like biology, environmental science and ecology, and technology sciences such engineering and telecommunications, and, finally, social science disciplines such as demography or urban planning.

It would seem reasonable that, facing this global threat, scientific collaboration would be abundant and effective, and scientific efforts were taken to pursue the resolution of the global problem. However, although there is much talk of collaboration, there is another powerful force that organizes science, which is competition.

In our work, we analyze the international scientific collaboration on research relating to SLR from climate change. We are trying to answer many questions, including: Who are the main stakeholders, in terms of institutions and countries? Do they collaborate? Are they organized to avoid overlapping efforts and to fill all possible research gaps? What role do the funding agencies play?; Are the agencies, with their scientific policies, setting the direction of the overall objectives?

To do so we will use a bibliometric and social network analysis approach. We will study the co-occurrence networks of keywords, affiliations, and funding agencies fields among scientific papers in the SLR topic.

Even though that is a new approach to the subject, bibliometrics and Social Network Analysis are widely used methodologies for mapping of science using thematic networks typically built from citations (Boyack, Klavans, & Borner, 2005; Leydesdorff, 2004; Leydesdorff & Rafols, 2009; Moya-Anegon et al., 2004), and studies of scientific collaboration, based on personal networks built from the co-authorship of papers (Bozeman & Corley, 2004).

MAIN FOCUS OF THE CHAPTER

Methods and Materials

The main source of information for this study is Thomson Reuters' Web of Science Core Collection[2].

This source was chosen for its multidisciplinarity and broad international coverage, and mainly because its FU field, which contains the Funding Agencies and Grant Number for each research paper.

We searched the text strings used on the database to describe SLR research in topic fields, i.e. Title, Abstract, Author's Keyword and KeyWords Plus. None period restriction was applied. Search strategy was: TS=("sea level rise" OR "sea level change" OR "sea-level rise" OR "sea-level change"). Record downloading was carried out on 28 January 2014.

A total of 9,364 records were recovered, covering a period lasting from 1929 until now, summarizing international scientific production in the sea level research topic at Thomson Reuters' Web of Science Core Collection.

The software used to process the matrices and graphs is UCINET and Netdraw.

Social network analysis software is a tool created in order to facilitate network analysis. Visual representation of networks also facilitates qualitative interpretation of network data. SNA software often contains visualization capabilities.

UCINET 6 for Windows is a software package for the analysis of social network data (Borgatti, Everett, & Freeman, 2002). It comes with the NetDraw network visualization tool (Borgatti, 2002).

UCINET is used, first of all, to manage matrices containing relational data but also for calculating centrality measures. These measures are important to determine the important elements in a network. These scores are numeric values associated with every node, in general are an indicator of how network structure and position contributes to the relevance of every node.

There are several centrality measures, each one capturing different aspect. Degree tests direct influence of each node and quantitatively the connection; closeness measures how each node is close to all other nodes, it is related to how long whatever flows along the network takes to arrive to it; high scores in Eigenvector centrality mean that a node is connected to the best connected nodes in the network, it is also an index of power; and betweenness quantifies exactly how often a node lies along the shortest path between two other nodes, so that is interpreted as indicator of control, power and access to flows, potential for synthesizing, gatekeeping, brokering, controlling the flow of information, and also the power to leave apart and separate and disconnected parts of the network.

Research Topics

Knowledge, rather than the juxtaposition of fields, is a continuous space in which different scientific disciplines are scattered; some are closely related and these relationships can be represented in a network.

Network in Figure 1 represents the co-occurrence of KeyWords Plus terms in Web of Science retrieved through ID field. KeyWords Plus are index terms created by Thomson Reuters from significant, frequently occurring words in the titles of article.

In order to provide clear representation some of the most common terms have been eliminated from the graphic representation. Those whom appear so frequently that hardly provide any specific information to the subject of the paper, such us Model, Record, Climate, Variability, Ocean, Temperature, Dynamics, Coast, Impact, Growth, System, Water, etc. Nevertheless all place-names (in red) remain, since one of our research questions was if different geographical areas had any influence on research topics and, if so which where the common places of interest. Node size is equivalent to the betweenness centrality score of each term.

Betweenness centrality measures information control (Freeman, 1979). In our analysis, it represents terms that connect different research topics, leaving both sides of them research topics in which the term in question is involved. Figure 1 presents evidence that different zones of the planet articulate the research carried out. So USA or North-America is closely related to Canada and British Columbia. This central cluster is connected on one hand to the Gulf of Mexico and Lousiana, and on the other hand to the North-Atlantic or North-Sea and England, Scotland, and British Isles in general. Disconnected at some level is the cluster including the Indian Ocean and Australia. Other very frequent subject terms in pairs are mass extinction and glaciation, constraints and glacial isostatic-adjustment, and continental-margin and morphology.

2-Mode Network: Funding Agencies-Research Institutions

In order to measure the contribution of the several funding agencies we observe, we take the occurrences in downloaded records, where each mention implies that the funding agency was found in Thomson Reuters' Web of Science FU field. This is not related to the amount of the contribution from the agency to the research paper, which would be the most accurate indicator but is not available to the authors.

In absolute terms, the main funding agencies at international level are the National Science Foundation, NSF and National Aeronautics and Space Administration, NASA from USA; the Natural Environment Research Council, NERC from UK and the European Union.

Every country has its own main agency which concentrates the most of the resources that institutions rely on. The main ones are the Natural

Figure 1. Co-word network in SLR

Sciences and Engineering Research Council (Canada), Deutsche Forschungsgemeinschaft (Germany), Netherlands Organisation for Scientific Research (Netherlands), Australian Research Council (Australia), National Natural Science Foundation (China), Japan Society for the Promotion of Science (Japan), and Spanish Government (Spain), and Centre National d'Etudes Spatiales (France).

International research in SLR has its prominent research centers in the Australian National University; the University of Durham, and the British Antarctic Survey in England; The University of Colorado, CALTECH, Louisiana State University, the National Center for Atmospheric Research in USA; and the University of Toronto in Canada.

At a lower level, each country has its own research centers, such as the Potsdam Institute for Climate Impact Research and the Alfred Wegener Institute for Polar and Marine Research, both in Germany, the East China Normal University and the Chinese Academy Of Sciences, the Utrecht University in the Netherlands and Azti Tecnalia in Spain.

Figure 2 is a 2-Mode Network representing the relations established between Funding Agencies (rounded squares) and Research Institutions (circles), through co-occurrence in papers. Colors

Figure 2. Funding agencies and research institutions in SLR institutions network

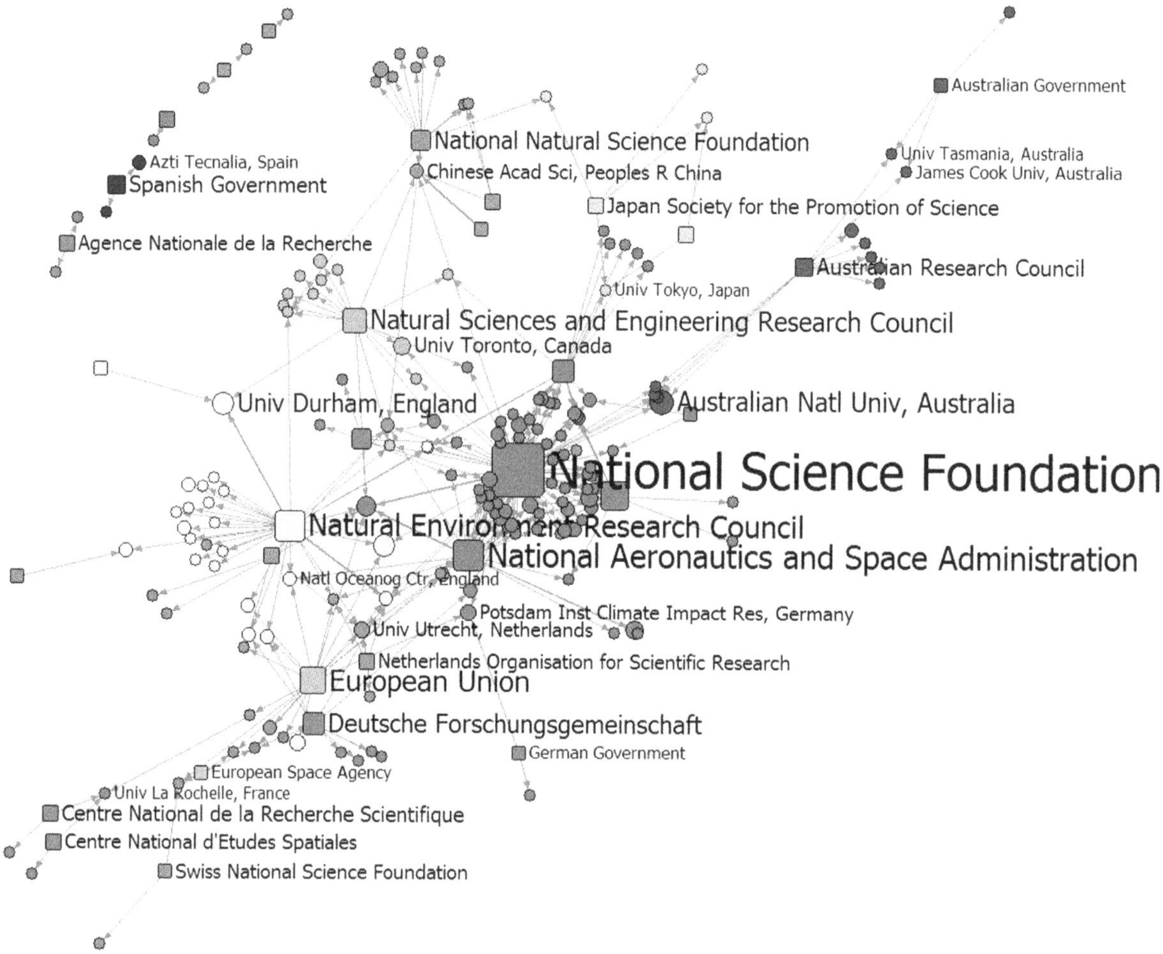

represent the main nationalities in both cases: USA, UK, Europe, Australia, Japan, China, Canada, Germany, France, and Spain.

The arrow's direction shows the agencies funding the research institutions. The most common scheme is national agencies funding their own national institutions.

The authors have detected very little co-financing of agencies from different countries aimed at research institutes. That is because two Japanese research institutions should be highlighted: the AIST and University of Tokyo, the former receiving funding from China and Japan, and the latter receiving funding from the USA.

The USA funds projects at Australia's national university, and at the Geological Survey in Canada.

In Europe, the most successful institutions when it comes to raising funds to finance research in different countries are: the Utrecht University in Holland, the Potsdam Institute for Climate Impact Research in Germany and the National Oceanography Centre in England.

FUTURE RESEARCH DIRECTIONS

Our findings in this study cannot be considered more than exploratory. It would have been desirable to include an analysis at a micro level that would allow us delve into research topics and to analyze each research institution.

The social network analysis applied to scientific collaboration observed from SLR papers is revealing, but it would require an interpretation of different communities involved both from scientific disciplines and countries. However this analysis exceeds the capabilities of our project so it remains as a potential future research line.

This work has provided us with new questions to answer, in addition to those set out in the proposal. Despite this, our findings allow us to point out some tentative conclusions.

CONCLUSION

International scientific collaboration is a common circumstance in SLR, this is explained by the size of many of the projects, the use of large-scale scientific facilities and the need for comparisons. International co-financing is less frequent. The same geographical and linguistic patterns are observed more than in other research areas; the global risk posed by SLR seems not to imply any change in the concept of collaboration and research funding.

Some countries emerge as central to research in SLR, while others are totally absent. Their differences are not dependent on their susceptibility to SLR and its impacts, but rather with who funds and conducts research, so we can say that the effects and implications of SLR in large areas of the planet are being poorly funded.

SLR is a central topic in the geosciences, however, despite the economic and social implications, we find that from demographic, urban planning, economics and, in general, social sciences perspective, it is not treated as an important issue.

Finally, considering the SLR as an example of the risks that humanity as a whole faces, and analyzing the organizative efforts made by science and science policy to face it, we must conclude that political and economic boundaries outweigh scientific evidence.

REFERENCES

Aboelela, S. W., Larson, E., Bakken, S., Carrasquillo, O., Formicola, A., Glied, S. A., ... Gebbie, K. M. (2007). Defining interdisciplinary research: Conclusions from a critical review of the literature. *Health Services Research, 42*(1p1), 329–346.

Al-Rodhan, N. R., & Stoudmann, A. G. (2006). Definitions of Globalization: A Comprehensive Overview and a Proposed Definition. Geneva: Centre for Security Policy. Retrieved from http://gcsp.ch/content/download/1267/9834/file/Definitions%20of%20Globalization%20-%20A%20Comprehensive%20Overview%20and%20a%20Proposed%20Definition.pdf

Barjak, F. (2006). *Team diversity and research collaboration in life sciences teams: Does a combination of research cultures pay off?* University of Applied Sciences Northwestern Switzerland. Retrieved from http://netreact-eu.org/documents/DPW2006-02_Team%20Diversity%20_Barjak_Franz.pdf

Birnholtz, J. P. (2007). When do researchers collaborate? Toward a model of collaboration propensity. *Journal of the American Society for Information Science and Technology, 58*(14), 2226–2239. doi:10.1002/asi.20684

Boldrini, E., Craglia, M., Mazzetti, P., & Nativi, S. (2015). The Brokering Approach for enabling collaborative scientific research. In *Collaborative Knowledge in Scientific Research Networks*. IGI Global Press.

Borgatti, S. P. (2002). *NetDraw Software for Network Visualization*. Analytic Technologies.

Borgatti, S. P., Everett, M. G., & Freeman, L. C. (2002). *UCINET for Windows: Software for social network analysis. Harvard Analytic Technologies.*

Boyack, K. W., Klavans, R., & Borner, K. (2005). Mapping the backbone of science. *Scientometrics*, *64*(3), 351–374. doi:10.1007/s11192-005-0255-6

Bozeman, B., & Corley, E. (2004). Scientists' collaboration strategies: Implications for scientific and technical human capital. *Research Policy*, *33*(4), 599–616. doi:10.1016/j.respol.2004.01.008

Bunakov, V., Jones, C., & Matthews, B. (2015). Towards the Interoperable Data Environment for Facilities Science. In *Collaborative Knowledge in Scientific Research Networks*. IGI Global Press.

Chen, X., Mark, B., Patrick, S., Robertson, M., Jon, K., & Wang, Z. (2013). Bibliometrical study method on tidal freshwater marshes loss and research orientation. *Acta Ecologica Sinica*, *33*(4), 237–240. doi:10.1016/j.chnaes.2013.05.001

Diviacco, P. (2015). Maps, graphs and annotations as Boundary Objects in Knowledge Networks, Distributed Cognition and Collaborative e-Research. In *Collaborative Knowledge in Scientific Research Networks*. IGI Global Press.

Diviacco, P. (2015). Reconciling Knowledge and Collaborative e-Research. In *Collaborative Knowledge in Scientific Research Networks*. IGI Global Press.

Essential Science Indicators. (n.d.). Retrieved September 28, 2013, from http://www.accesowok.fecyt.es/

Fredericks, J. (2015). Persistence of Knowledge Across Layered Architectures. In *Collaborative Knowledge in Scientific Research Networks*. IGI Global Press.

Freeman, L. C. (1979). Centrality in Social Networks Conceptual Clarification. *Social Networks*, *1*(3), 215–239. doi:10.1016/0378-8733(78)90021-7

Gattuso, J. P., Dawson, N. A., Duarte, C. M., & Middelburg, J. J. (2005). Patterns of publication effort in coastal biogeochemistry: A bibliometric survey (1971 to 2003). *Marine Ecology Progress Series*, *294*, 9–22. doi:10.3354/meps294009

IPCC. (2012). *Cambio climático 2007: Informe de síntesis*. IPCC.

Katz, J. S., & Martin, B. R. (1997). What is research collaboration? *Research Policy*, *26*(1), 1–18. doi:10.1016/S0048-7333(96)00917-1

Lassi, M., & Sonnenwald, D. H. (2010). *Identifying factors that may impact the adoption and use of a social science collaboratory: A synthesis of previous research*. Academic Press.

Leadbetter, A., Arko, R., Chandler, C., Shepherd, A., & Lowry, R. (2015). Loose integration of local information to generate collaborative marine science knowledge. In *Collaborative Knowledge in Scientific Research Networks*. IGI Global Press.

Lee, B., Kwon, O., & Kim, H. (2011). Identification of dependency patterns in research collaboration environments through cluster analysis. *Journal of Information Science*, *37*(1), 67–85. doi:10.1177/0165551510392147

Leydesdorff, L. (2004). Clusters and maps of science journals based on bi-connected graphs in Journal Citation Reports. *The Journal of Documentation*, *60*(4), 371–427. doi:10.1108/00220410410548144

Leydesdorff, L., & Rafols, I. (2009). A Global Map of Science Based on the ISI Subject Categories. *Journal of the American Society for Information Science and Technology*, *60*(2), 348–362. doi:10.1002/asi.20967

Li, J., Wang, M.-H., & Ho, Y.-S. (2011). Trends in research on global climate change: A Science Citation Index Expanded-based analysis. *Global and Planetary Change*, *77*(1), 13–20. doi:10.1016/j.gloplacha.2011.02.005

Martin Cantos, A. (2011). Journal and country rankings in instrumentation and ocean engineering. *Martech*. Retrieved from http://upcommons.upc.edu/e-prints/handle/2117/13401

Moya-Anegon, F., Vargas-Quesada, B., Herrero-Solana, V., Chinchilla-Rodriguez, Z., Corera-Alvarez, E., & Munoz-Fernandez, F. J. (2004). A new technique for building maps of large scientific domains based on the cocitation of classes and categories. *Scientometrics*, *61*(1), 129–145. doi:10.1023/B:SCIE.0000037368.31217.34

Pudovkin, A. (1993). Citation Relationships Among Marine Biology Journals and Those in Related Fields. *Marine Ecology Progress Series*, *100*(1-2), 207–209. doi:10.3354/meps100207

Sears, J. (1993). Poster Session - a Bibliometric Analysis of the Marine Biotechnology Literature - a 90s Assessment. In *Aquatic Information Resources: Tools of Our Trade: Proceedings of the 18th Annual Iamslic Conference*, (pp. 147–147). Academic Press.

Smykla, E., & Zippel, K. (2010). Literature Review: Gender and International Research Collaboration. *Report Prepared with Funding from NSF, OISE, 936970.*

Soomai, S. S., MacDonald, B. H., & Wells, P. G. (2013). Communicating environmental information to the stakeholders in coastal and marine policy-making: Case studies from Nova Scotia and the Gulf of Maine/Bay of Fundy region. *Marine Policy*, *40*, 176–186. doi:10.1016/j.marpol.2013.01.017

Sorensen, A. A., Seary, A., & Riopelle, K. (2010). Alzheimer's Disease Research: A COIN Study Using Co-authorship Network Analytics. *1st Collaborative Innovation Networks Conference - Coins2009, 2*(4), 6582–6586.

Velden, T., Haque, A., & Lagoze, C. (2010). A new approach to analyzing patterns of collaboration in co-authorship networks: Mesoscopic analysis and interpretation. *Scientometrics*, *85*(1), 219–242. doi:10.1007/s11192-010-0224-6

Weichselgartner, J., & Kasperson, R. (2010). Barriers in the science-policy-practice interface: Toward a knowledge-action-system in global environmental change research. *Global Environmental Change*, *20*(2), 266–277. doi:10.1016/j.gloenvcha.2009.11.006

Zhang, W., Qian, W., & Ho, Y.-S. (2009). A bibliometric analysis of research related to ocean circulation. *Scientometrics*, *80*(2), 305–316. doi:10.1007/s11192-007-1863-0

ADDITIONAL READING

Borgman, C. L. (1990). *Scholarly communication and bibliometrics*. Sage Publications.

Carrington, P. J., Scott, J., & Wasserman, S. (2005). *Models and Methods in Social Network Analysis*. Cambridge University Press. doi:10.1017/CBO9780511811395

Freeman, L. C. (1979). Centrality in Social Networks Conceptual Clarification. *Social Networks*, *1*(3), 215–239. doi:10.1016/0378-8733(78)90021-7

Hanneman, R. A., & Riddle, M. (2005). *Introduction to social network methods*. University of California Riverside.

Hicks, C. C., Fitzsimmons, C., & Polunin, N. V. (2010). Interdisciplinarity in the environmental sciences: Barriers and frontiers. *Environmental Conservation*, *37*(4), 464–477. doi:10.1017/S0376892910000822

Kadushin, C. (2012). *Understanding Social Networks: Theories, Concepts, and Findings*. Oxford University Press.

Lee, B., Kwon, O., & Kim, H. (2011). Identification of dependency patterns in research collaboration environments through cluster analysis. *Journal of Information Science*, *37*(1), 67–85. doi:10.1177/0165551510392147

Li, J., Wang, M.-H., & Ho, Y.-S. (2011). Trends in research on global climate change: A Science Citation Index Expanded-based analysis. *Global and Planetary Change*, *77*(1), 13–20. doi:10.1016/j.gloplacha.2011.02.005

Moed, H. F., Glänzel, W., & Schmoch, U. (2006). *Handbook of Quantitative Science and Technology Research: The Use of Publication and Patent Statistics in Studies of S&T Systems*. Springer.

Small, H., & Information, I. for S. (1990). Bibliometrics of basic research. National Technical Information Service.

Wasserman, S., & Faust, K. (1999). *Social Network Analysis: Methods and Applications*. Cambridge University Press.

KEY TERMS AND DEFINITIONS

Betweenness: It is a centrality measure of a node in a social network, it quantifies the number of times a node acts as a bridge along the shortest path between two other nodes. It was first proposed by Linton Freeman. It shows to what extent an actor is required to enable the connection between others, so it can be a 'broker' or 'gatekeeper', with the corresponding power and potential to control other points. The relevance of this type of centrality is that the actor is 'in the middle' can control interactions between other actors in the network and therefore has power over certain paths of interaction.

Bibliometrics: A set of methods to analyze quantitative academic literature, bibliometric methods are most often used in the field of library and information science, but have wide applications in other areas. It relies on the scientific documents as a proxy of science in general.

Co-Authorship: It is the fact in which two or many authors share the authorship of a paper, it means they are collaborating or they are a joint author.

Funding Agencies: Organizations that provide funds for the funding of science, the term often connotes funding obtained through a competitive process, in which potential research projects are evaluated and only the most promising receive funding, such processes, usually are run by governments, corporations or foundations.

Scientific Collaboration: Process in which researchers work together to achieve the common goal of producing new scientific knowledge.

Sea Level Rise: Caused by climate change and global warming, consists in the rise of sea level which is expected to cause a great impact on the long-term coastal morphology, and an increased flooding.

Social Network Analysis: Social network analysis views social relationships in terms of network theory, consisting of nodes and ties, representing entities and its relationships, respectively. These networks are often depicted in a social network diagram, where nodes are represented as points and ties are represented as lines.

ENDNOTES

1 Essential Science Indicators is a documentary research source compiled by Thomson Reuters, producer of the Web of Science

databases, which identifies important trends in the world of science. They are made from the most cited papers of each discipline during the recent 10 years, and is constantly updated.

2 Including: Science Citation Index Expanded (SCI-EXPANDED), Social Sciences Citation Index (SSCI), Arts & Humanities Citation Index (A&HCI), Conference Proceedings Citation Index- Science (CPCI-S) and Conference Proceedings Citation Index- Social Science & Humanities (CPCI-SSH).

Chapter 6
Relationship between Senior and Junior Researcher:
Challenges and Opportunities for Knowledge Creating and Sharing

Katjuša Gorela
Regional Development Centre Koper, Slovenia

Roberto Biloslavo
University of Primorska, Slovenia

ABSTRACT

The professional development of a young researcher is mainly influenced by the organizational culture and the relationship with academic colleagues, especially mentors. This chapter presents the results of a qualitative study conducted on a sample of 16 researchers employed in three Slovenian public universities. The study examines the senior-junior researcher relationship, as the relationship between mentor and mentee, the nature of their cooperation, and how new knowledge is generated and transmitted. In-depth semi-structured interviews and content analysis for structuring qualitative data was used. The results indicate that the professional development of junior researchers is based mostly on their independent work where the mentor guidance is only occasional. Despite the development of information and communication technology, and in particular the Social Web tools, the mentorship in the framework of higher education institutions is still based on a face-to-face relationship.

INTRODUCTION

Organizations that wish to remain successful in the market in this age of globalization, marked by the financial and economic crisis, must be able to adapt to its rapid changes and to fulfil every new demand. Universities are also adapting to this trend, as they fight for their survival with the help of public funds or of those acquired with marketing activities. Universities represent a work environment where an individual and the knowledge he possesses are an important source of competitive advantage, which is why universities, in order to avoid losing competitive position,

DOI: 10.4018/978-1-4666-6567-5.ch006

Copyright © 2015, IGI Global. Copying or distributing in print or electronic forms without written permission of IGI Global is prohibited.

are encouraging experts to pass their knowledge onto their younger colleagues. The most common form of such transfer of knowledge takes place during the process of participation of experienced academics and those who are just starting their academic career. The future of universities and higher education institutions depends to a large extent on the quality of cooperation between older, more experienced colleagues and younger ones and on the care for career development of junior faculty (Whitcomb, 2004).

Two of the main forms of cooperation between senior and junior faculty are formal and informal mentoring (Berk, Berg, Mortimer, Walton-Moss, and Yeo, 2005). These two forms of cooperation between already established experts and novices in the field of research differ mainly in the way of the relationship is established and its duration. Formal mentoring is formed systematically, with the involvement of the organization and it mostly lasts a short period of time. It does not necessarily require a personal relationship or understanding between mentor and mentee. On the other hand an informal mentoring relationship is formed spontaneously and without intermediaries (Douglas, 1997), which brings us to the conclusion that an informal mentoring relationship is a relationship between a mentor and a mentee, who have a special affinity for each other and both derive pleasure from the cooperation. While a formal mentoring relationship can be entered into by two individuals from different departments (Douglas, 1997; Murray, 1991) informal mentoring relationships are mostly formed by senior and junior researchers working in the same field of science and who have similar goals and interests (Ragins and Cotton, 1999). However, Boice (Boice, 1992) notes that informal mentoring is very rarely seen in the educational system and that only a third of new teachers are in such a relationship.

The time-limited nature of formal mentoring results in a lower level of trust and emotional closeness between experts and novices. In the case of a prolonged relationship, which is typical for informal mentoring, the mentor and the mentee have more time at their disposal to develop psychosocial and career development functions, which is why the results of their participation may only become apparent after a longer period of time (Ragins and Cotton, 1999).

Because of the benefits of mentoring programs, some organizations, including universities, began organizing "Network Mentoring" programs in the field of pedagogical work, in which large numbers of people from different faculties or departments can participate. This form of mentoring requires a relationship in which all the participants are equal and where mentors and mentees can swap roles and thus actively contribute to the relationship (Haring, 1999). In their study performed at the Purdue University, Wasburn and LaLopa (2003) found that this form of mentoring satisfied the needs of both mentors and mentees. The former because the informal nature of the program and an opportunity to evaluate their teaching style and cooperate with teachers/researchers from other schools. The latter because the program allows them to exchange views on teaching, acquire new ideas on teaching techniques and gain an insight into the teaching style, which is used in various scientific disciplines. Network Mentoring programs have proven to be an opportunity for the mentee to obtain the desired information and reduce the feeling of isolation. At the same time, these programs contribute to the improvement of pedagogical work, for the mentors and the mentees (Wasburn and LaLopa, 2003). A review of the available literature showed no such initiatives in the field of research.

Most of the literature on the subject of mentoring in the field of scientific research focuses mainly on the relationship between mentor and doctoral or post-doctoral students (Olson and Connelly, 1995; Whitley and Oddi, 1998), while the area of cooperation at the individual level between senior - junior researcher remains relatively poorly researched. For those individuals who are just starting their academic career, the right mentor can be a great source of help and support, which is why it is not surprising that in recent years, higher education organizations have opened up to various forms of mentoring, such as mentoring circles or Network Mentoring programs like the one described above (Cawyer, Simons, and Davis, 2002; Darwin and Palmer, 2009; Pierce, 1998; Wasburn and LaLopa, 2003). The proposed chapter focuses on the relationship between senior and junior researchers and on the impact of this relationship on the mutual transfer of knowledge and the creation of new knowledge. The purpose of this study is thus to provide answers to the following research questions:

- How do senior and junior researchers experience their roles as mentors or mentees and how do they see each other's roles?
- How do different perceptions of their roles affect the transfer of knowledge between them and the quality of their cooperation?
- In their opinion, what role does the use of IT technology play in their cooperation and in building their relationship as mentor and mentee and how important is it?

The study was designed as a qualitative study using semi-structured interviews, which enables a greater insight into the opinions, expectations and experiences of those who are directly involved in process of knowledge transfer at the mentor - mentee level. The survey was conducted in Slovenia, a European country, where the financial and economic crisis affected the academic and research community, as a consequence of financial cuts in the public education sector. The study included senior and junior researchers of natural sciences, social sciences and humanities at three Slovenian public universities. This triangulation of sources within the research sample contributes to a higher validity and reliability of the results obtained. Despite different forms of mentoring, for instance peer group mentoring, and network mentoring, the purpose of the study is to discover characteristics of the dyadic relationship, which at higher education institutions is formed between the mentor (senior researcher) and the mentee (junior researcher). The reason for this is that junior researchers in particular at the beginning of their career are to a large extent dependent on their mentors, as the role of the latter is to support and motivate younger and less experienced colleagues, to give them advice and lead them through their development (Lumpkin, 2011).

BACKGROUND

Knowledge

Knowledge, as a complex phenomenon, is constantly changing and upgrading through an individual's own experience and reflection. The literature on the subject provides different definitions and categorizations of knowledge. For example, Zack (1998) distinguishes between five types of knowledge - declarative (know about), procedural (know how), causal (know why), conditional (know when) and relational knowledge (know with), on the other hand, Lundvall and Johnson (1994) state that an individual's knowledge falls under one of four categories - know-what (knowledge of facts), know-why (knowledge of scientific facts), know-who (knowledge that refers to social skills) and know-how (knowledge that refers to skills).

In continuing our study, we adopt the definition of knowledge, as provided by Nonaka (1994), who defined two dimensions of knowledge – the explicit and tacit dimensions of knowledge based on work of Polanyi (1966). Division of knowledge in two types explicit and implicit is largely used in the organizational literature (e.g. Dyck, Starke, Mischke, and Mauws, 2005; Ipe, 2003; Nonaka, and von Krogh, 2009; Ribeiro, and Collins, 2007). Explicit knowledge can be encoded (e.g. words, pictures, mathematical formulas), stored and transferred from person to person; this is the knowledge that is generated through logical thinking and education. On the other hand tacit knowledge is the kind of knowledge that is difficult to encode and transfer directly to another person, as it requires cooperation, understanding and trust between the person giving the knowledge and the one receiving it; such knowledge includes know-how, which are the individual's technical skills, mind-sets, beliefs and positions, which can be obtained through observation, imitation and direct experience (Nonaka, 1994). Through experience and self-reflection, the individual obtains tacit knowledge which he internalizes. The transfer of such knowledge to other person is possible through direct interaction, apprenticeship, networking and action learning, face-to-face interaction and practical experience (Haldin-Herrgard, 2000). The tacit knowledge is one of the elements that the experts lean on, as this is the type of knowledge which to a large extent influences the achievement of excellence in a job and major time saving (Lawson and Lorenz, 1999). Experts use it for making proper decisions and achievement of agreed performance (Uguid, 1998).

At the organizational level, there are two categories of knowledge, namely individual knowledge - the knowledge that each employee carries within himself and collective knowledge, which represents the total organizational knowledge contained in the in rules, procedures, routines and shared norms (Lam, 2000). The first is the so-called human capital of the organization, which can bring many benefits to the organization, but this knowledge is highly fluid and dynamic, as it is lost once the individual is no longer part of the organization. Collective knowledge belongs to the so-called structural capital of the organization and is not affected by the departure or arrival of an individual. Between the two, there is a feedback loop in which the collective knowledge has an impact on which individual knowledge will be considered especially important by the organization and to what degree, as well as how this knowledge will eventually become part of the collective knowledge, shaping it in the process.

Knowledge Creation and Transfer

New knowledge within an organization is created on the basis of individual or team work and out of the cooperation between experts in fields that are more or less related. Regular contact between employees enables mutual observation, creative communication and the exchange of experience. This is the basis of mutual trust and the beginning of the process of creating new knowledge (Nonaka, 1994). When new knowledge is created, there is an ongoing interaction between the individual's explicit and tacit spheres of knowledge (Nonaka and von Krogh, 2009), which means that a person develops new knowledge through the transition between these two dimensions.

The newly created knowledge has to be properly preserved (in printed or electronic form) and, if necessary, passed on to the employees. At this point, we once again observe cooperation between an individual who already possesses a certain piece of knowledge and an individual who does not and who is on the receiving end of said knowledge (Mayor and Cordey-Hayes, 2000). However, the knowledge of an individual cannot be transferred to another person in its entirety, because there is an active process of creating new knowledge on the part of the recipient (Duffy and Cunningham, 1996; Oliver, 2001). In addition, we must take into account the information asymmetries that exist between the person who possesses the knowledge and the one receiving it during the transfer of knowledge (Wilkesmann and Wilkesmann, 2011).

The process of knowledge transfer is affected by the type of knowledge that one wishes to transfer (Inkpen and Dinur, 1998), which can be transmitted directly from one person to another, or indirectly through media (Sveiby, 1997). Explicit knowledge, which can be encoded, is easier to transfer from one person to another, while tacit knowledge is more difficult to transfer and requires cooperation between employees, a sort of shared understanding and a certain degree of mutual trust (Lam, 2000). Cooperation between various persons is thus successful, if it is based on commitment and trust, rather than on power and the ability to condition others (Morgan and Hunt, 1994). Trust among colleagues, therefore believing that the other person is honest and sincere (Claro, de Oliveira Claro, and Hagelaar, 2006), develops over time and all involved benefit from a relationship based on trust. In a relationship based on trust, there is no need to use power (Inkpen and Currall, 2004), as it is easier to define common goals and solve any problems that might arise between the parties involved (Sahay, 2003). In order to establish such relationship, the partners must communicate with each other and the greater the trust, the better the communication (Morgan and Hunt, 1994). Good communication between colleagues contributes to the mutual exchange of knowledge (Hendriks, 1999). Inadequate communication, on the other hand, can lead to a failure in the transfer of knowledge (Gilbert and Cordey-Hayes, 1996; Treleaven, 2004). Trust affects a person's willingness to transfer knowledge (Morgan and Hunt, 1994; Davenport and Prusak, 1998, Dyer and Nobeoka, 2000; Soekijad and Andriessen, 2003; Moller and Svahn, 2004), while a lack of trust between the parties involved can result in a lower accuracy of the information exchanged or of the knowledge transferred (Currall and Judge, 1995).

While being transferred, knowledge does not suffer any damage or loss in value. On the contrary, knowledge increases with use (Sveiby, 2001). In spite of this, some people do not want to share their knowledge. This can be due to a lack of awareness of the benefits of knowledge transfer, being overworked, lack of time, not being proficient in the use of ICT, which significantly facilitates the transfer of knowledge, or to a desire to maintain power and a competitive advantage over peers (KPMG, 2000). The factors that positively affect an individual's willingness to share knowledge are a sense of fulfilment and a desire to increase one's own reputation or a sense of public recognition (Cabrera and Cabrera, 2002). These factors are directly linked to values, rules and norms which represent an organization's culture (DeTienne, Dyer, Hoopes, and Harris, 2004).

Information and Communication Technology

Modern society and modern organizations cannot operate efficiently and effectively without the use of information and communication technology (ICT). Modern and high-quality ICT enables organizations to achieve and maintain a competitive advantage and to respond to the demands of their environment in a rapid manner. ICT plays an important role in the process of transmission and generation of new knowledge and its preservation; modern technology is particularly suitable for the transfer of explicit knowledge, while personal contacts play a crucial role in the transfer of tacit knowledge. Wilkesmann, Fischer, and Wilkesmann (2009) found that the use of the Internet and formal meetings help to ensure knowledge within the organization, but informal contacts between employees are still the most effective channels for knowledge transfer (e.g., coffee breaks).

Some authors (Flanagin, 2002; Johannessen, Olaisen, and Olsen, 2001; Hislop, 2002) claim that ICT enables only limited transfer of tacit knowledge from one person to the other, while others are of different opinion (Harris and Lecturer, 2009; Hildrum, 2009; Falconer, 2006; Murray and Peyrefitte, 2007). In general, it holds true that even the most sophisticated ICT cannot replace face-to-face interaction. However, numerous research confirm that modern social web tools can facilitate the transfer of tacit knowledge and obtaining of experiential knowledge (Hsia, Lin, Wu, and Tsai, 2006; Abidi, Hussini, Sriraj, Thienthong, and Finley, 2009; Osimo, 2008; Steininger, Ruckel, Dannerer, and Roithmayr, 2010).

Social Web

In the opinion of Alavi and Leidner (2001), there are ICT tools which facilitate the knowledge creation and sharing process, as they enable access to information, dialogue among experts, better understanding of some phenomena and achievement of new conclusions. In addition, its users can freely express their opinions, views and arguments. Stenmark (2000) also sees in the modern ICT a tool which supports the process of tacit knowledge transfer, in particular when the environment for networking and communication among experts is created. Among the most frequently used ICT tools for transfer of tacit knowledge are communication tools (e.g. instant messaging and forums), online communities, Web 2.0 tools (e.g. blogs, wikis, social networks), multimedia sharing tools and video conferencing (Song, 2009; Lai, 2005; Harris and Lecturer, 2009; Murray and Peyrefitte, 2007; Khan and Jones, 2011). Modern ICT tools support natural language, immediate feedback, social cues, and social presence (Chua, 2001), and additionally they can support real-time synchronous communications, e.g. catting, commenting and video- and text-based conferencing (Marwick, 2001). All of that contributes positively to the transfer of tacit knowledge. It seems that the new internet-based applications, such as social networks, community of practice, and e-learning, can to a certain extent outweight the negative consequences of physical distance between the sender and the recipient of tacit knowledge (Yi, 2006). However, they cannot entirely replace emotional feelings, eye contact or body language (Hislop, 2002; Hooff and Weenen, 2004). ICT thus remains primarily an excellent supplementary tool to traditional methods for transfer of tacit knowledge (Panahi, Watson and Partride, 2012), in particular when the tacit nature of knowledge is low or medium high.

Social web, such as blogs, wikis, social networking sites etc., can despite mentioned limitations significantly contribute to the transfer of tacit knowledge via ICT. Their characteristics are user-friendliness, they enable multiple-way communication, user-generated content, possibility for global networking, multimedia-oriented (Panahi, Watson, and Partridge, 2012), their users are active, as they are characterized by desire to communicate, collaborate, participate in interest groups and share knowledge (Lindmark, 2009). Online communities enable their members regular communication which leads to a higher level of mutual trust (Raisanen and Oinas-Kukkonen, 2008). The latter then influences the readiness of conveying own knowledge. The advantage of social web is the possibility of anonymous transfer of tacit knowledge. This is particularly important when we're dealing with transfer of delicate knowledge, or when credit card holders lack self-esteem (Raisanen and Oinas-Kukkonen, 2008; Yi, 2006). Among the advantages of social web, Gordeyeva (2010) counts also the fact that an individual's knowledge is available to a wide range of persons and requires less effort and shortens the time for its transfer. Among the most recognized social web tools for knowledge transfer is wiki where people can post information and edit each other's entries. In this way, knowledge is disseminated and expanded through a collaborative effort (Mayfield, 2010). Wiki supports a common understanding of complex and ambiguous concepts, which can act as anchors around which new knowledge is created.

Mentoring

Organizations operating in an increasingly unstable and ever-changing environment, entering the global market with their services and products, must make sure that they preserve the knowledge of their experts even after they leave the organization. One of the possible modes of transmission and preservation of knowledge within an organization include mentoring, which Healy and Welchert (1990, p. 17) defined as "[a] dynamic, reciprocal relationship in a work environment between an advanced career incumbent (mentor) and a beginner (protégé) aimed at promoting the career development of both (…)." Therefore, what happens in a mentoring relationship, is a transfer of knowledge from one person (usually older, with more expertise and experience) to another (usually a younger person with less experience in a particular field of work). In this case, the role of the mentor is very important, because their knowledge is transmitted to the newcomer, making him or her qualified to take on specific responsibilities and functions (Kram and Hall, 1996).

The mentoring includes different stages – an initiation stage, when the mentorship relationship is only being formed, a mentee stage and a break-up stage and, in case of successful relationship between a mentor and mentee, a lasting friendship stage. The transition from one stage to the other depends on the success of the preceding stage (Chong, 2009). The mentoring relationship terminates expectedly when a mentor reaches a desired level of maturity and independence, or unexpectedly in case of illness or death of one of the players (Bhagia and Tinsley, 2000). Upon expected termination of cooperation, usually a collegial relationship or friendship between a former mentor and its mentee develops (Chong, 2009).

Kram (1985), who was one of the first researchers to study mentoring in the workplace, divided mentoring tasks among two groups: career development functions and psychosocial functions. The first group of tasks is to help the mentee's professional advancement. Within this group, Kram distinguishes between five functions of a mentor - sponsoring promotions, coaching, protection, challenging assignments, increasing the mentee's exposure and visibility. Career developmental functions are largely dependent on the status and position held by a mentor within the organization. The function assumed by the mentor also influences the mentee's perception of the organization he is employed with – in case of sponsorship, exposure and visibility, the mentee has a feeling that the organization cares about him, while in case of mentorship as coaching and challenging assignments he doesn't have such feeling (Baranik, Roling and Eby, 2010). The psychosocial functions of a mentor focus on the personal growth and professional development of the mentee. This category includes helping the protégé to develop a sense of professional self, providing problem-solving and a sounding board, friendship and role modelling. Psychosocial functions are dependent on the quality and emotional ties that the relationship between mentor and mentee is based on. A mentor can assume some or all of these roles in his or her relationship with a mentee (Ragins and Cotton, 1999).

The relationship mentor – mentee requires dedication from both of them, as it requires time, energy, and intellectual, physical and emotional resources. The reasons such as lack of time for formation of a relationship, from one or both sides, insufficient communication, different areas of interest or also different personalities, can negatively affect the productivity and success of the relationship mentor – mentee. Since learning is one of the key purposes of mentorship, the mentors have a task of transferring to their younger colleagues the competencies they have developed in their area of work. However, when the mentor is overly focused on the development of his own career, he can neglect the relationship with the mentee and the latter's career path. Successful mentorship requires time, the mentor's openness in the relationship with the mentee and his questions and the mentor's ability to understand phenomena from different perspectives as well as allowing new options (Holmes, Hodgson, Simari, and Nishimura, 2010). The time devoted by the mentor and the mentee to their personal interaction, influences the level of transfer of tacit knowledge, as the latter is transferred through interaction with other people and through experience, exercise and reflection (Haldin-Herrgard, 2000). Through personal contact an unintentional transfer of tacit knowledge can also occur. This can also occur when the holder of knowledge is not aware of the knowledge he has at his disposal (see Polanyi, 1958, 1966). This also is one of the reasons why personal contact between the mentor and the mentee is so important. The transfer of tacit knowledge which can only be subconsciously understood, applied, and developed from direct experience (Zack, 1999), is also influenced by the quality of communication, the level of goodwill and the level of cognitive-style congruency between the owner and the recipient of knowledge, since – as Lin Kao and Chang (2010, p. 290) conclude - "cognitive congruency is particularly critical for tacit knowledge transfer in mentoring systems". The transfer of tacit knowledge between the experts and novices can be hindered due to a gap in their knowledge (more knowledge and experience also means higher tacit nature of knowledge, which additionally aggravates its articulation) and differences in language. The latter means that the experts and novices are forced to find some common language, which would at least partially mitigate the transfer of tacit knowledge (Haldin-Herrgard, 2000).

The mentoring is just seemingly a one way transfer of knowledge, since - as Holmes et al. (2010, p. 337) claim – during the mentoring both participating parties are enriching their knowledge: "[M]entoring is a 2-way street with transmission of experience from mentor to mentee, as well as explanation of new technology by mentees to mentors." The latter applies in particular when between the mentor and mentee is a greater age difference or a difference in terms of cultural environments they both come from. In such case the mentee has significantly different thinking models than the mentor, a fact that can further support a two-way exchange of knowledge.

Higher Education

In a time of financial and economic crisis, when cuts are being made to funding in education and research, the academic world should pay more attention to alternative sources of financing. Higher education institutions and their employees should invest more time and expertise in research and project work, in writing scientific articles and in gaining research credits that allow them to compete for tenders and additional funding sources, either public or private. In this kind of working environment, knowledge becomes an important competitive advantage for an organization and even more for an individual. Based on that it is not surprising that some individuals do not wish to pass on their knowledge to their colleagues who might become one day their competitors in the academic job market or even within their own organization. Thus, even in higher education, the main purpose of which is the creation and transfer of knowledge, you can find individuals who perceive knowledge as their personal property and prefer not to share it with others (Wind and Main, 1999).

In this context, organizational culture which prevails in a specific organization plays an important role, since – as Gupta, Iyer, and Aronson (2000, p.18) claim - "[o]ften, organizational culture itself prevents people from sharing and disseminating their know-how in an effort to hold onto their individual powerbase and validity." O'Dell and Grayson share a similar opinion, as they understand the organizational culture as "a key driver and inhibitor of knowledge sharing" (in Rollett, 2003, p. 24).

Organizational culture of universities can be defined as professional bureaucracies (Mitzberg, 1983) characterized by clear division of work (even within higher education institutions there is accurately defined division of work and tasks among different departments and research areas), clearly defined authority and responsibility (in higher education, autonomy depends on title), a strong emphasis on formal expertise (in higher education institutions there is a big gap among the statuses of a full professor, associate professor, assistant professor and assistant). Professional bureaucracies tipically have high level of expertise and expert autonomy as well as standard, but complex procedures (Martin and Marion, 2005). Srikanthan and Dalrymple (2002, p. 215) also claim that the culture in higher education is "bureaucratic in nature, more prone to conflict than collaboration". Among the reasons for such culture are the need for high level of autonomy, internal competition, different comprehension and understanding of knowledge, negative aspects of expert status and posturing, increased level of non-personal communication (increased use of technology) (Tippins, 2003). As found in some research of the culture in higher education institutions, the cultures differ among institutions - well established institutions with many years of tradition have more hierarchical culture while institutions, which were formed only some decade ago or less are much more market oriented (e.g. Biloslavo and Prevodnik, 2010). However, even those researches confirm that in higher education institutions bureaucratic and hierarchical cultures prevail with formal and rigid structures and procedures. Such organizational culture aggravates the quality of relationship and knowledge transfer between the senior and junior researcher. Therefore, higher education institutions should stimulate the organizational culture based on cooperation and trust, as the research shows that mutual trust is one of the key conditions for successful knowledge transfer (Andrews and Delahaye, 2000; Hislop, 2005; Newell and Swan, 2000; Roberts, 2000). Higher education institutions should change their attitude toward mentoring and should attribute to it greater importance. Thus Ramani, Gruppen, and Kachur (2006, p. 407) claim "Institutions should change their culture to overtly value and reward mentoring so that mentoring does not remain an invisible and only implicitly valuable aspect of their educational programs."

Creation and Transfer of Knowledge in Higher Education

Society holds many different expectations from higher education institutions. For example, we expect them to be stable and reliable (Farjoun, 2010), while at the same time being "innovative, flexible and responsive to change" (Steinkellner and Czerny, 2010). The basic purpose of higher education remains the creation of new knowledge and its translation into a narrower and wider social context and to contribute to the advancement of society as a whole. The knowledge of higher education institutions can be divided primarily into scholarly knowledge and operational knowledge, i.e. the knowledge of administrative staff (Geng, Townley, Huang, and Zhang, 2005). Academics generate new knowledge through studying professional literature, individual work and teamwork, cooperation in research projects, formal and informal meetings, through the exchange of views, critical and analytical thinking as well as personal reflection. The resulting new knowledge is often preserved in the form of scientific articles, monographs, reports and other publications. The knowledge created by the professional staff in these organizations is usually for internal use only and is preserved in the form of manuals, rulebooks, reports, etc.

The primary purpose of higher education institutions is the creation and transfer of knowledge, but it often happens that teachers/researchers do not want to share their knowledge with other colleagues. The reasons behind this behavior, among other things, are a lack of trust, a lack of time, the perception of knowledge as a means to obtain a certain status, the belief that knowledge should not be made available to everyone, intolerance of errors or unwillingness to provide assistance, the recipient's inability to absorb knowledge and differences in the cultural, linguistic and scientific bases of all the parties involved (Davenport and Prusak, 2000).

Modern ICT also contributes to the successful transfer and creation of new knowledge in the academic environment and is becoming increasingly important, especially for the entry, storage, retrieval and distribution of its knowledge (Gill, 2009). ICT makes the pedagogical and research process easier to implement and more efficient, as it grants access to global databases, e- libraries e -learning, sharing common documents, electronic meetings, and much more. Tacit knowledge can also be transferred by using videos, which demonstrate procedures or via storytelling and sharing day-to-day experiences (Hildrum, 2009).

Mentoring in Higher Education

At the beginning of their careers, junior researchers rely on established rules and theoretical knowledge, while senior researchers rely on intuitive thinking developed through experience (Wilkesmann and Wilkesmann, 2011). A senior researcher as an expert in a particular field therefore possesses a lot of knowledge to pass on to the younger and less experienced colleague. Assuming that the knowledge of the employees is "property" of the organization and that the most efficient way to transfer organizational knowledge is to "hire smart people and let them talk to one another" (Davenport and Prusak, 2000), we can conclude that the relationship of trust between senior and junior researchers is the foundation for the creation and transfer of knowledge in the field of science and research.

The relationship between senior and junior researcher can thus be seen as a relationship between mentor and mentee, that is, between an older and more experienced person and a younger, less experienced person, in which the first helps and guides the second (Ensher, Thomas, and Murphy, 2001). Thus, mentoring is a relationship in which a person who has more experience and expertise teaches and advises someone who is a novice in a particular area; the goal of their cooperation is aimed primarily at the professional development of the novice. This form of cooperation or mentoring typically lasts a longer period of time and is characterized by an emotional commitment of both the mentor and the mentee (Stewart and Krueger, 1996; Wocial 1995; Yoder, 1990). In a mentor-mentee relationship, both parties have defined roles and responsibilities arising from them, meaning that they both contribute to the success or failure of their cooperation (Allen, 2007).

Since mentoring is a multi-faceted relationship, a senior researcher (as a mentor) has the task of introducing an inexperienced colleague to the research environment, help him or her to became acquainted with its norms and values (Morrison, 1993) and guide him or her towards new knowledge (Swap, Leonard, Shields, and Abrams, 2001). In this kind of relationship, mutual learning and knowledge transfers are inevitable. An interpersonal knowledge transfer is not an easy process, because the person who has the knowledge and is passing the knowledge onto another, cannot influence the final result and the manner in which the recipient will receive and understand the knowledge. The recipient, on the other hand, mostly has limited opportunities to influence how knowledge is handed to him or her. In general, this is a process in which the recipient integrates the new information into his or her existing knowledge and adapts it to the existing mental image of the world, based on which the recipient acquires new knowledge (Wilkesmann and Wilkesmann, 2011).

A junior researcher can view the mentor as a positive or negative example, from whom to learn the methods and approaches to scientific and research work, as well as planning, critical thinking, introspection and self-evaluation. A successful cooperation between senior and junior researcher not only has positive consequences for the novice, but also for the mentor. The mutual benefits of their relationship are mainly reflected in professional advancement (Burke and McKeen, 1997; Higgins, 2000, 2001), personal growth and satisfaction (Ehrich et al., 2004) as well as increased self-esteem for both researchers (De Vries, 2005). The quality of their cooperation also affects increased research activity and consequently, a higher number of research projects and scientific papers, which also benefits the research institution for which they work (Gardiner, 2005).

Mentoring has the following advantages for the mentee - more promotions (Dreher and Ash, 1990; Scandura, 1992), higher income (Chao, Walz, and Gardner, 1992; Whitely, Dougherty and Dreher, 1991), more mobility (Scandura, 1992; Ragins and Cotton, 1999) and greater career satisfaction (Fagenson, 1989). In a study of mentoring in academia, Pierce (1998) notes that senior researchers who acted as mentors experienced a revived interest in teaching and made stronger connections with their academic colleagues. Organization in which there is cooperation between mentors and mentees reported reduced staff turnover (Viator and Scandura, 1991) and greater job satisfaction (Koberg, Boss, Chappell, and Ringer, 1994), and finally, the effects are also visible on organizational socialization (Ostroff and Kozlowski, 1993).

The factors that could complicate the cooperation between a mentor and a mentee are primarily personality differences, the purpose of the mentoring being defined too loosely and the mentee having the feeling that the mentor is not complying with his or her objectives (Johnsrud, 1990). On the other hand, disappointment and frustration are mainly due to not having enough time to establish the relationship between the mentor and the mentee (Wasburn and LaLopa, 2003). Ehrich, Hansford and Tennent (2004) highlight at least three disciplines - education, business and medicine - where there is a lack of time for meetings between mentor and mentee, bearing in mind that successful mentoring depends largely on the mentor's accessibility (Cawyer et al., 2002).

Cooperation between senior and junior researchers may be organized and determined in advance; in this case, we are dealing with formal mentoring, whereas if it develops spontaneously and without obligations, we are dealing with informal mentoring. The latter is supposed to be the more effective of the two forms, because of its more relaxed nature (Ragins and Cotton, 1999). In a formal relationship between mentor and mentee, the objectives of their cooperation is predetermined and coordinated by the organization where the mentoring takes place, while the goals of an informal mentoring arrangement are always changing, updating and adapting to the professional needs of both the mentor and the mentee (Ragins and Cotton, 1999). In the field of scientific research, both forms of cooperation are applied, but informal, spontaneous mentoring relationships are becoming increasingly rare. The main reason behind the decrease in the number of informal mentorships is the fact that senior researchers are too busy with research and with their main tasks, as well as with publishing scientific papers, which can make it difficult for younger researchers to gain access to their more experienced colleagues (Pololi and Knight, 2005). A more competitive environment within higher education and a wish to protect own power and position expressed by some senior academics is also supporting a knowledge hoarding That is why some higher education institutions began introducing various forms of formal faculty mentoring programs (Meyers and Smith, 1999; Pierce, 1998). Some researchers are even of the opinion that institutionalization of mentorship is the first condition of durable knowledge exchange processes as interaction process between organization's members cannot just be a result of the contingent factors but needs to be a part of existing structure of values and norms that form lasting conditions for actions and mutual transactions (Mesner-Andolšek, 2011).

The relationship between the mentor and mentee can also be influenced by the stage of the mentor's career. A mentor who is climbing the career ladder will be more focused on its own promotion and personal achievements and therefore less ready for mentoring a younger colleague (Chong, 2009). To achieve the highest academic title requires a lot of time and effort; an individual's promotion is linked to conduction of teaching and research work, which *inter alia* encompasses participation in scientific-research projects, writing and publishing of scientific articles and participation on scientific conferences, all of which contributes points for promotion. Independent research work brings more points and thus faster climbing of the career ladder, which can also influence the decision of a senior researcher for neglecting cooperation with the younger colleague.

Based on their research, Sands, Parson and Duane (1991, p. 185) identified the four ideal types of mentoring at the level of faculty-faculty:

- **Friend:** Mentoring as a friendship-based relationship that includes emotional support, counselling and assistance in resolving personal conflicts;
- **Career Guide:** A form of mentoring, based on cooperation in projects and writing scientific articles together, promoting the assistant and introducing him or her to influential persons, giving advice on grants and other funding sources;
- **Information Source:** The mentor provides the assistant with information about the working environment, the formal and informal rules of the organization;
- **Intellectual Guide:** This form of mentoring is mainly characterized by cooperation, equality in the relationship, intellectual guidance, constructive criticism and reviewing scientific articles.

Mentors can combine two or more styles of mentoring, depending on the mentor, his or her relationship with the mentee and other factors (Sands et al., 1991).

EMPIRICAL RESEARCH

Research Method

Data on mentoring at the senior-junior researcher level were collected on the basis of in-depth, semi-structured interviews. A qualitative approach was chosen because of the nature of the research questions. As defined by Kaplan and Maxwell (2005, p. 30) "The goal of qualitative research is understanding issues or particular situations by investigating the perspectives and behavior of the people in these situations and the context within which they act.". The interviewees were asked questions concerning the relationship between the mentor and the mentee, about the nature and characteristics of their cooperation, about the process of transfer and creation of new knowledge and about the role of ICT in the relationship.

The participants were selected based on personal acquaintances and random contacts. Due to the small size of the research sample, the results obtained cannot be generalized to the total population of senior and junior researchers in the Slovenian academic area, which represents a limitation of this study. What gives a greater degree of validity to the results is the fact that the participants in the study are employees of three different Slovenian public universities and operate in three different scientific disciplines (i.e. triangulation of data sources). The three universities are very different in size (the biggest has around 60,000 students, the smallest around 8,000) and age (the oldest is almost 100 years old, the youngest is 10 years old). According to Kitzinger (1995, p. 300), it is "advantageous to bring together a diverse group (for example a range of professions) to maximize exploration of different perspectives within a group setting."

The subjects who were invited to participate in the study were informed that the interview would last from 30 to 90 minutes and that the conversations would be recorded and later transcribed.

They were also informed that the survey would be anonymous and that the data obtained would be used solely for research purposes. All 16 participants agreed to participate in the survey and to the conversation being recorded. The interviews were conducted in a relaxed atmosphere and were trust-based. The interviews were conducted in January of 2014, lasted from 20 to 90 minutes, were recorded using a digital voice recorder and transcribed the same day. To ensure that the data were interpreted as accurately as possible, the interviews were listened to once again and read at the same time, after the transcription was made (DiCicco-Bloom and CrabtreeW, 2006). After the transcription of the interviews, individual segments of the text were coded, related parts of the interviews were classified into separate categories and the broader themes were determined. In order to prevent the misuse of the recordings, after the analysis of the data we will follow the recommendations of DiCicco-Bloom and CrabtreeW (2006, p. 318), who state that "Recorded data should be carefully guarded and generally destroyed after transcription or once analysis is complete."

Most of the study participants work in the field of social sciences, although the electronic invitation to participate in the survey was sent to researchers from various scientific disciplines. One of the possible reasons for this is the nature of the study, which researchers in the field of social sciences may feel closer to than those in natural sciences, because of the qualitative approach used. The invitation to participate in the survey was forwarded to 114 senior and junior researchers at three Slovenian public universities, of which 16 responded - 10 senior and 6 junior researchers, two junior researcher have the title of assistant professor, but for personal reasons they chose to answer for the role of assistants. A demographic profile of the participants is provided in Table 1.

FINDINGS

Mentoring as Guidance and Leadership

The answers of the participants revealed that, in most cases, the concept of mentoring is understood as *guidance and leadership*. Five of the senior researchers put particular emphasis on the role of the mentor in making work easer, that is, "opening doors" for their younger colleagues, while three senior researchers point out that the assistant must surpass his or her mentor in the professional sphere.

All the researchers involved in the study understand the concept of mentoring as providing guidance to the employees in their professional careers. In line with this view, senior researcher A understands mentoring as "/ ... / guiding a colleague on his way towards professional development." Senior researcher D agrees with this statement and says "/ ... / helping a younger colleague, who could someday become established in the same profile or in the same profession as the person mentoring him." Similarly, junior researcher L understands mentoring as "someone guiding you on your career path, introducing you structurally into the system and helping you with the content of your work." Junior researcher N agrees with this view and adds that mentoring is "Having someone to teach you and lead you in your work, the work process itself." Assistant P believes that the relationship between senior and junior researcher is primarily about guidance and support, stressing the importance of having a common goal. "I see it mainly as a mentor providing support to a mentee and a kind of guiding of a process in search of a common goal." Lumpkin (2011) also believes that mentoring is primarily about guiding and directing the development of a younger colleague. He says:

Table 1. Demographic profile of participant pool (N = 16)

Participant	Gender	University	Field of Expertise	Academic Title	Age
A	Male	University of Maribor	Social Sciences	Full Professor	45 or more
B	Female	University of Primorska	Humanities	Associate Professor	45 or more
C	Male	University of Primorska	Social Sciences	Assistant Professor	45 or more
D	Male	University of Primorska	Social Sciences	Associate Professor	45 or more
E	Female	University of Primorska	Humanities	Associate Professor	45 or more
F	Female	University of Primorska	Social Sciences	Assistant Professor	45 or more
G	Female	University of Primorska	Social Sciences	Assistant Professor	45 or more
H	Male	University of Ljubljana	Natural Sciences	Assistant Professor	35-45
I	Male	University of Primorska	Natural Sciences	Assistant Professor	35-45
J	Male	University of Primorska	Natural Sciences	Assistant Professor	35-45
K	Female	University of Ljubljana	Social Sciences	Assistant Professor	35-45
L	Male	University of Primorska	Humanities	Assistant Professor	35-45
M	Female	Univeristy of Primorska	Social Sciences	Assistant	25-35
N	Female	University of Primorska	Social Sciences	Assistant	25-35
O	Female	University of Primorska	Humanities	Assistant	25-35
P	Male	University of Primorska	Natural Sciences	Assistant	25-35

"[M]entors guide, advise, or coach colleagues in their development."

Junior researcher K, who says "To me, mentoring is a great responsibility / ... / "sees the function of the mentor in helping the assistant both in the research field and in teaching." A mentor can be of great help simply by providing advice, experience and with a little guidance / ... / giving tips on where to publish, how to start, how tackle a particular study, how to use the study to produce as many articles as possible, where to go for help, what to check out, all of these things, in addition to the pedagogical work / ... /," which is consistent with Borisoff (1998), who states that mentors and novices in higher education are active mainly in the field of teaching, research and service. A similar opinion is given by senior researcher B, who says that mentoring is about cooperation and coordination at various levels "Cooperation, support in terms of guidance on several levels. In the research community, it is about guiding the student in his work with a specific problem, work technique, methodology, and approaches and, on a higher level, also connecting with colleagues who are dealing with similar problems and are working in the same fields." Here, she adds that sometimes the role of a mentor is also to *open doors* for younger colleagues. "/ ... / Research in the field of history, archaeology and heritage requires a lot of contact with some institutions, where more time is needed for certain procedures, etc. and this is where the role of a mentor is very important and is sometimes referred to as "opening doors". Preceding research (Ramanan, Phillips, Davis, Silen, and Reede, 2002; Jackson, 2003) also shows that that the mentor can be the one who "opens the doors" for the mentee and exposes him to new opportunities.

Senior researcher F shares this opinion and says "For example, the task of the mentor is also to pave the way for a younger colleague and enable him to publish in other journals. This is the norm in other countries. As a senior researcher, you use your influence to support your younger colleague." Senior researcher G also perceives mentoring as a form of assistance for younger colleagues and as a mission at the same time. "/ ... / I see it as a kind of mission. / ... / and as making work easier for the younger generations. In the sense that you can prevent the younger researcher from wasting time with certain dead ends that they can avoid with the help of a mentor." Senior researcher D has a very similar opinion, which he puts into words: "/ ... / we share our experiences with them and help them avoid taking the same wrong paths and perhaps making the same mistakes that we made / ... /."

Four researchers believe that the main purpose of mentoring is to enable their assistants to develop professionally and even surpass them in their respective fields. Thus, senior researcher D says "/ ... / It is in our interest to find colleagues who can become involved in our work, help us and even become creatively superior to us, so that our field can continue to develop." Senior researcher C agrees with this statement and emphasizes that a mentee should not be a copy of the mentor. "/ ... / to guide a young candidate, a colleague, not encouraging him to become a copy of us, but merely to introduce him to this world / ... /". He adds that a mentee must outgrow his mentor and that "A good teacher is one that the students eventually outgrow and you have to outgrow me, not just reproducing my work / ... /." Senior researcher F also understands mentoring in a higher education environment as "Leading, guiding, providing assistance with academic and research work, giving advice," while emphasizing the importance of the assistant developing independently from the mentor and not merely reproducing the mentor's knowledge. "/ ... / of course, one must allow the mentee enough freedom to develop his own knowledge. Therefore, the purpose of mentoring is not to reproduce what I know, but for a person to achieve the most that they can in a certain field." Holmes et. al. (2010) also claim "(M)entees must not become mere extensions of their mentors' careers", as by doing so they could represent a competition when obtaining acknowledgments, awards and other advantages.

Although senior researchers see the role of a mentor primarily as guiding and leading younger researchers, the latter often find themselves alone during discovering of the academic world and directing of their career. A contact between the mentor and mentee seems marginal and on the surface, to which rare personal contacts contribute. The frequency of contact between a mentor and mentee is evidently displayed by senior researcher J "In the past it was like this; it was said once per week, then we met once every two weeks, once per month." Such relationship between the mentor and mentee is against Cox (1997), who says that meetings between the mentor and mentee should be organized and regular, as long as this becomes a habit. The fact that junior researchers are often left on their own even in periods of personal distress is reflected in the statement of a senior researcher F who says "I am not interested in his personal affairs (...) I am not a psychiatrist nor a psychoanalyst, not a social worker nor a confidant, or a shoulder for crying. None of this." It seems that mentors and mentees understand quite similarly the purpose and role of mentoring. Mentoring is viewed primarily as a rational relationship where actors evaluate exchange outcomes of the relationship and only exceptionally express altruistic behavior.

The statements made by junior researchers also show the attitude of the mentors towards them. The person K says the following about her experience as a junior researcher: "I had a feeling as if I was not welcome when I came. It turned out that the vacancy position for assistant was issued against his will. Therefore, the beginning itself was very negative." and she adds "When I was selected for the assistant position, I had to beg if we could meet in order to even see him, in order to even tell me (…) what he expected from me etc." The person L also thinks that the mentors don't have time for their younger colleagues. She says: "I think the mentor doesn't always have time to ask his assistant 'Tell me how are you doing, how are you progressing?'" and emphasizes the junior researcher's anxiety when the mentor gives him a task he cannot cope with due to lack of experience or lack of help "(…) during research (…) we, the young ones always fell into the situation where we actually had to write different projects (…). This was sometimes very stressful for us, the younger colleagues as we had to write about things we didn't have a clue idea about (…) I know a lot of colleagues that left this profession for this reason."

Mentoring as a Knowledge Creation Process

All participants of the research agree that junior researchers must be mainly independent in their work, even when it comes to creation of their know-how.

The senior researcher B during creation of new knowledge emphasizes primarily independent work of the junior researcher "/…/ to direct and on the other hand to leave enough freedom to the mentee, so that he somehow feels also competent enough to find something himself /…/.", which is in line with Sands et al. (1991), which among the reasons for small number of mentorships at the faculty-faculty level mentions a wide-spread opinion that a person holding a doctorate degree is capable of independent work and doesn't need additional learning.

Nevertheless, the senior researcher G doesn't look at the creation of new knowledge as a product of purely research activity, but as an effect of each contact with the assistant "/…/ it is created when we talk to the mentee /…/ when the mentee talks, explains, then the mentor has to have the ability to listen how he see it. And then the new knowledge is created, together."

The senior researcher I also argues for creation of new knowledge as a sum of the assistant's independent work and joint cooperation between the senior and junior researcher "/…/ the two of us, we define some guidelines, we direct the boat, and then he performs as much as possible of independent work. /…/ I don't have more than this amount of knowledge to give him, but we continue searching together, as this is the key to success – to keep searching together." The senior researcher H thinks similarly; in the process of creating new knowledge he emphasizes either the role of mentor or the role of assistant "/…/but you have to direct him, when it comes to new knowledge, into what he should be deeply engaged, why, how and then when he chooses some big section, then the mentor has to be the one to say 'Look, this would be important, this would be interesting, let's study this as well, let's work, this is waste, this doesn't make any sense.'" In line with this is also the experience of the junior researcher L, who says that he has gained new knowledge most often based on his own initiative and that regarding this he has been entirely free "/…/ if I got some ideas, I told this, we debated a little bit; perhaps the mentor, who has wider horizons, said 'Take a look as well at this book or that' /…/. Here, there haven't been any problems, I have never experienced that the mentor would say 'This - no'". The assistant M and her mentor first create new knowledge separately, and then they combine it into one entirety. "Actually, we create new knowledge separately, and then we build it in during conversation. /.../ we always tackle a thing separately, I prepare my part and she prepares her, then we get together and combine that into one entirety."

Among all interviewees, only the junior researcher K had very negative experience with the mentor and she says that between them there was neither creation of new knowledge nor knowledge transfer "/.../ there was absolutely no knowledge transfer. What I brought with myself, is what I used. I learned what I didn't know, and that's it. Catastrophe. We haven't created any new knowledge", which confirms that not everyone is suitable for a role of mentor (Walker and Scott, 1993) and that formally appointed mentors can be less interested in performing this function (Ragins and Cotton, 1991). The junior researcher N also has bad experiences with creation of new knowledge with a senior researcher, the reason for this being in the mentor's research inactivity "Regarding writing scientific articles, I don't have a close cooperation with my mentor, simply because she doesn't do any research or only a little bit." Also, the junior researcher O doesn't have experience with researching with her mentor "We haven't yet worked together on projects, so I can't talk about creation of new knowledge." The mentoring in the area of research encompasses helping junior researcher in achieving norms and standards as well as obtaining new knowledge, skills, behavior and values, which are required for his promotion (Keyser et al. 2008). In case of lack of the aforementioned, the junior researcher is deprived of knowledge which is necessary for promotion and successful career in the academia as well as for the development of new knowledge.

Mentoring as a Knowledge Transfer Process

It can be summarized from the given answers that knowledge transfer process between a senior and junior researcher runs mainly in a structured or unstructured way.

The researcher E believes that the role of senior researcher is also a transfer of knowledge onto the junior researcher; he says: "I cooperate as a partner .I help someone who studies to achieve better knowledge." The role of mentor as someone who transfers knowledge also to their colleagues see also the senior researcher E "It is in essence a transfer of knowledge and on the other side a direction how he/she should build himself/herself. Because a transfer of knowledge can be what you can give to him/her in some form, some knowledge that he/she doesn't have." The senior researcher J is of the similar opinion; he defines mentoring as leadership and transfer of knowledge on younger colleagues "I understand this as leading someone, helping him, giving him guidelines what should be done. If I have certain knowledge, to help him to achieve this. Not to tell him right away, but to give him the right starting points, so that he knows how to learn what he should learn." The senior researcher I also understands mentoring as a knowledge transfer process and at the same time emphasizes also the transfer of hidden knowledge and values "Apart from knowledge transfer, it is also a transfer of hidden knowledge, skills, behavior, values, way of thinking, and... especially values, which are lacking today." Most of the time mentors are mentioning what Nonaka (1994) has defined as a technical dimension of tacit knowledge, while the cognitive dimension that includes beliefs, ideas, paradigms, values, intuition, and mental models (Panahi, Watson, and Partridge, 2012) is considered only sporadically. The probable reason for this is that sharing the elements of cognitive dimensions requires more efforts and involves higher costs in terms of time than mentors are willing to spend. The latter is directly related to the reward that mentors can expect for their work (Cabrera and Cabrera, 2005).

The senior researcher G says that he can only transfer skills to his assistant "The only thing I can share are shortcuts, skills, purely technical skills which can make a way for mentee shorter. Only this can be shared, only this can be transferred." The senior researcher D emphasizes that knowledge transfer is always a two-way process "/.../ it isn't necessary to come always from my side to the assistant, but it is always two-way, as after all we are both doing similar stuff, we both have contacts with teachers, educators, students etc. During this work, certain conclusions arise, in particular during certain studies. And those conclusions, which one of us gets to know during our research, are debated." The assistant P also talk about two-way knowledge transfer between the senior and junior researcher; he describes cooperation with the mentor as "/.../ what we have done so far, we have been on somewhat same level, so we worked on the same problem and he shared with me what he found and vice versa. So, it was again somehow intertwined. There was a transfer from both sides /.../ somehow like this and we start in this way again." Wilkesman in Wilkesman (2011) talk about knowledge transfer as a reciprocal process, as they say someone can be an expert in one area and a novice in the other.

From the interviews, one can summarize mainly two ways of knowledge transfer between the senior and junior researcher, namely structured and unstructured knowledge transfer. The junior researcher N says that the knowledge transfer between her and her mentor is completely informal "Knowledge transfer is very informal; when we face a problem, or perhaps a new idea, for instance for some new thing /.../" The junior researcher M is of similar opinion and she says that she participates in knowledge sharing through a personal contact with her mentor "Knowledge transfer is through a conversation", while on the contrary the senior researcher J understands knowledge transfer process as a multi-phase process "For the field which is interesting to him, I give him starting points in the form of articles that he should study – this is the first phase – for him to study the literature, and the second stage is to describe what he has studied, so I can check whether he has understood /.../" in this case, one deals with transfer of explicit knowledge, which can be coded and easily transfer to another person (Dyck et al., 2005).

The senior researcher B also sees the knowledge transfer process in several phases and emphasizes the importance of an assistant's independent work "/.../ first of all, overview of situation, where we are – what is my knowledge, what is the knowledge of my assistant, and then the problem which we would like to study, what kind of problem is it, how much do we know about it, what is the level of my knowledge, what is the level of knowledge of my assistant, and then some kind of sending of new conclusions, new facts, and finally also guiding mentee into independent research /.../" The senior researcher F also understands knowledge transfer as a process of collaboration between the mentor and the mentee, whereby the assistant is mostly independent "/.../ I am interested in what he is doing and what he is reading, very, very often I say 'Send me the stuff you are writing.' 'Yes, it's not good yet, it's not...' It's not important, none of us is... the first text he wrote and this now is..., but this is a process, and in this process someone in helping you and is warning you 'Look, you could have done this differently; perhaps you could have, I don't know, take into consideration also this.'" The latter is in line with Taylor and Wright (2004), who say that a relationship which motivates new ideas and is based on learning from errors positively influences effective knowledge sharing.

Mentoring as Interpersonal Relationships

It can be seen from the answers of interviewees that the relationship between the senior and junior researchers is equivalent, with the exception of negative experience of the junior researcher K. Senior researchers emphasize that in the mentoring relationship in particular honesty and absence of fear are important.

The researchers understand the relationship mentor-mentee as an equal relationship. The junior researcher M says about the relationship with the mentor "I am very lucky. I wouldn't say that in our case we have a distinctive mentor-mentee relationship; this means that there is no feeling that someone is superior – subordinate /.../" and she defines the relationship with the mentor as "/.../ we are completely on friendship level /.../ I have a sense of equality." The junior researcher P also talks about equal relationship with the mentor "/.../we function as a team, we are equal." The senior researcher C is of similar opinion; he says about his assistant "We are partners, as we have same goals. /.../ we are equal /.../" The senior researcher D also talks about equal relationship between the senior and junior researcher; he says "/.../ we are equal; there is no things as he has lower title currently, this doesn't mean anything to me. She also has her opinions, her research, and her views/.../".

However, the junior researcher K had completely different relationship with the mentor; she says about her experience as assistant "/.../ exactly this 'little slave' prototype of assistant. /.../ It's when they throw you in the water and you should swim; if you save yourself, fine; if not, the next one comes in." The assistant N has similar experience of absence of mentor at the beginning of academic career; he says about it "This is one minus, that no one can introduce you, no one can tell you 'Come on, you too write something, so we can go to the conference together '. When you come to the university, you actually don't even know what a conference is."

Senior researchers in particular emphasize the role of honesty in the relationship between the mentor and mentee or between the senior and junior researcher; the senior researcher F thus says "Each man should be brought to the point where he gives maximum out of himself, and you have to be honest with him; so, you shouldn't be rough on him, discourage him, but simply tell him where his limitations are." Related to this, he adds "/.../ and I thinks that such open relationship with the mentee actually gives him as well some kind of courage /.../." Honesty towards the mentee is also emphasized by the senior researcher B, who says that the mentor must be ".../ correct to show and to let the mentee clearly know when he has achieved something, exceeded something or found something which is by its result a novelty and an achievement both for him as well as for his mentor /.../." The senior researcher G also talks about correctness and honesty; she says about a mentor "/.../ he has to be able to acknowledge and praise a mentee 'You did this well. Now, you say why you think it's good.' to acknowledge those steps where a movement occurs, when one feels that a professional-research movement occurs /.../." In addition to honesty, the senior researcher E emphasizes in particular absence of fear in the senior-junior researcher relationship "/.../ that someone dares to tell you something, that things are debated and that perhaps together we improve something." And he says about the relationship with his assistant "There is no that much fear, sometimes it can happen; rather, it is based primarily on exchange of different opinions /.../." The senior researcher I also agrees with this and says "The mentee has to dare, also to dare to oppose to the mentor and the mentor has to know how to cultivate this, by no means to oppress it. /.../" The importance of honest relationship and critical attitude is also emphasized by the senior researcher C who says the following about a junior researcher "He has to be creative he has to give added value, he can be critical and say 'I don't believe this'." The senior researchers' statements are in line with Detsky and Barelocher (2007), which claim that the mentees should not consent to the mentor, but instead express their opinion freely.

Mentorship and the Use of ICT

All sixteen interviewees agree that ICT plays very important role in their work; nevertheless, it cannot replace personal contact and in particular when it comes to transfer of tacit knowledge from one person to another (Harris and Lecturer, 2009; Hildrum, 2009; Falconer, 2006; Murray and Peyrefitte, 2007). All participants of the research use ICT primarily as a means of communication, namely the use of e-mails, while two junior researchers communicate with their mentors also via e-classroom, one senior and one junior researcher use web applications for working on common documents. Also, only one senior and one junior researcher use the video communication application. This limited consideration of ICT by participants can be partly attributed to the fact that the latter make sense of ICT within mentorship context primarily as a communication tool and less as a tool for the creation and transfer of knowledge. On the other hand, it can maybe relate to the issue of trust in the organization itself. Not only the use of various Web 2.0 tools can be unacceptable in some environments as they have a sign of avoidance of the "serious" work but there is also a question of privacy. These types of web tools cannot guarantee full anonymity of an individual which can have negative consequences for him in case when his expressed opinions are against the majority or against the opinions by the organization's leadership. The next question is whether an appropriate infrastructure is available for the wider ICT uptake. However, in the light of the answers given by the participants it seems that the mentors and mentee expect that their relationships take place as much as possible in a face-to-face manner.

The senior researcher J emphasizes today's key role of ICT and says "I think we're "dead" today without ICT, I see this even when we are left without electricity and the internet doesn't work. We can talk while we're having coffee, yes, about stuff, but on each meeting the computer is with us, something that belongs to ICT is present..." The senior researcher H is of similar opinion "Thanks God it exists. If we didn't have it, it would be much harder for us today." The senior researcher G also emphasizes the importance of ICT and ranks it as follows "Very high, 10, if it is 1 to 10." The junior researcher L acknowledges the importance of ICT for daily work and for collaboration with other colleagues; however, he describes it as "today, it is a necessary evil /.../"

All interviewees see ICT primarily as a tool for quick exchange of information and easier coordination of work as "The reason is simple – because of lack of time," says the senior researcher G. Both senior as well as junior researchers included into the research agree with this. This is confirmed also by the senior researcher E, who says "Since we don't have that much time, for all urgent things we constantly use computer," which confirms that in different higher education disciplines lack of time for a contact between senior and junior researchers occurs (Ehrich et al., 2004). Work-related obligations and corresponding lack of time for personal contacts between a mentor and mentee allocated a key role to ICT for communication and coordination of their collaboration and for exchange of information, which is also a conclusion of the senior researcher A when he says "This primarily makes coordination and communication easier /.../ significantly easier. This means you can be constantly in touch /.../ they simplified the speed of information flow, of coordination of meetings, continuous contact. /.../". The senior researcher B also emphasizes ICT as a tool for communication and at the same time he adds that it simplifies the work process and the process of obtaining information "/.../ first of all, this is an important way of communication, of simplification of work and it has been an important aspect recently. /.../ today, this enables significantly easier and faster work, obtaining of information /.../". The junior researcher O also says that he and his mentor use a computer "more for some quick information". The senior researcher J uses ICT for quick understanding with his assistant and says that "If we have to agree, for instance, about a meeting, we use mail, Skype to quickly agree about a time for meeting"; however, he points out a weakness of such communication "/.../ mail is asynchronous; you send and then you wait for an answer."

Only three researchers, of which two junior and one senior researcher, mention that they use other web applications as well with a mentor or mentee. Despite increasingly established use of e-learning, only two junior researchers mentioned it specifically as a communication channel between senior in junior researchers. The assistant N replaces communication with the mentor via e-learning "/.../ I publish what I have done during exercises and she looks it up there, so she doesn't call me on the phone, and she sees which copies I publish there"; the assistant P and his mentor occasionally have similar experience "/.../ he, for instance, publishes material intended for students, and then from there I take the information that I need in order to successfully conduct exercises". The senior researcher I says that with his assistant they use the application for sharing common files "Google drive, so we can have common documents and for teaching process and for exchange of research, we cannot be without it." The assistant M and her mentor use free-of-charge web applications, in particular when they write joint scientific articles "/.../ all Google, we even write articles via Google Sites, so that we have insight any time when one of us changes anything, everything is visible, so that we use this thing a lot /.../". At the same time, the assistant M adds that it hasn't been like that forever and that she introduced a mentor into such way of e-collaboration "Many times we use those things; however, this was one thing that I had to learn in order to even function" and as a consequence of lack of knowledge regarding ICT he emphasizes problems during transfer of knowledge. "/.../ if she mastered this thing, she could contribute something else as well, as she has a lot of other knowledge, and simply because she doesn't master those things, she gets stuck in one point." This confirms that the use of ICT in Slovenian formal and non-formal educational system is still low and shows the lack of informations about the opportunities that eLearning can bring (Ala-Mutka, Gaspar, Kismihok, Suurna, and Vehovar, 2010). The senior researcher H has also adapted to greater use of ICT; in it, he sees a possibility of communicating with younger generation "In case of someone who is 10 years younger than me, he is much more familiar with this and likes this way of work more, but I try to take this into consideration .../".

ICT successfully overcomes physical distance between individuals; however, primarily two persons included into the research use it for this purpose. The junior researcher M says that he and his mentor prefer to use it for communication than electronic mail "/.../ many times we say Thank you God for Skype. When she is somewhere and I am elsewhere, it doesn't work any other way! We prefer it over e-mails; we are both very happy that it exists. We use it tremendously." The senior researcher H also agrees that Skype is of essential importance when it comes to overcoming physical distance; he emphasizes this primarily with the following words "We also use Skype without any problem; I go abroad for a lecture and if there is anything urgent I say 'Wait for me to go online'. Here, there are no more limitations and it's good that it exists!" At the same time, the senior researcher H acknowledges that there are certain limitations for ICT communication, even though we are dealing with video communication "Personal approach has some other message, as when you talk to someone, you feel immediately if there is a problem or not, whether the communication is open or not, whether the source of a problem is totally elsewhere. Over the phone, you don't see this, via e-mail you don't see it, you see it on Skype if the camera is turned on, and even then it is questionable. ICT has great influence, but it doesn't give you that 100-percent realistic answer, while personal contact does this." The senior researcher J shares this opinion and he is critical regarding video conferences. "I don't have good experience with it. When we did department meetings, there was someone who couldn't join us and he was with us via Skype from Ljubljana. I don't know how much that man got out of it, how much did we, as for those of us who are in the same space the communication is different that with those who talk via Skype. He is automatically on another level /…/" and he also makes a conclusion that "/…/ in case of video conferences, some components of a debate are lost, you lose that feeling."

The interviewed academics agree that ICT has a big role in their work and that it enables easier cooperation among colleagues. However, they share the opinion that technology cannot replace personal contact. The senior researcher C thus says "I absolutely prefer personal contact. Personal contact is the most important thing in life. The quality of communication is in personal contact, in voice, in articulation, in enthusiasm." The senior researcher J is of similar opinion, when he says "/…/ of course, sometimes it doesn't work any other way. But if it's possible, I prefer personal contacts."

The senior researcher A points out the limitations of ICT, in particular when we are dealing with transfer of knowledge to a younger colleague "/…/ when it is necessary to explain certain research question, I can't do it via those contemporary information communication technologies /…/ when you send an electronic message, you cannot explain in it all aspects of certain problem, which you would like to tackle with him or solve it" and as a key disadvantage of electronic media he notices the possibility of creation of misunderstandings. "It is not understood that a subordinate or mentee understands this in the same sense as you wrote it, so in this area the tools are scarce". The senior researcher G also agrees with this and he says "/…/ when you see what happens, when the message is said, did you hurt a person, is he uncertain, does he understand what you are talking about, you can check understanding; on the other hand, through ICT you don't know anything, whether a person understands or not" and at the same time he sees point in a personal contact primarily regarding a mentee's professional development. "I would say that a person can develop professionally, as an expert only with face to face contact and that, if communication runs only via ICT, a person on the other side is still lonely. /…/ He develops only in this direct human relationship /…/." The senior researcher H acknowledges that when it comes to a process of transferring knowledge to an assistant, he prefers personal contact, primarily because of easier communication and because he understands that process as an important element of the mentor's attitude "/…/ if you show someone 'Look, I would do like this…' and you also argument why, as writing this in e-mail is a big waste of time and also rather unhuman. /…/ If, on the other hand, you tell someone 'Look, I would do this, this and that, because of this, this and that, someone will understand it as this, this and that…', this is actually also a stage of mentoring, because you direct him into the direction you want a joint project, joint matter to go. This is first. And second is that by doing so you are also transferring knowledge to him." The senior researcher B shares similar opinion, who also emphasizes personal contact in mentoring relationship. "If we are talking about mentoring, if we are talking about guiding a young person in the area of research, of a teaching colleague, we cannot work here without a personal contact. No way." The senior researcher I comes to a similar conclusion and he thinks that via face to face contact much more than knowledge is transferred "In case of 'face to face', transfer of other skills occur, which cannot be obtained in an isolated space /…/ for sure, some behavior patterns are transferred patterns of reacting, a candidate gets to know other tricks as well and perhaps he hears something from me that he otherwise wouldn't hear", and this more is a tacit knowledge which can only be transferred via personal contact and collaboration (Lam, 2000). This relationship between a person who gives knowledge and a person who receives knowledge cannot be replaced by any modern technology (Roberts, 2000).

Solutions and Recommendations

The study was conducted on a small research sample, but despite that fact it contributes to better understanding of the relationship between senior and junior researchers in the relation mentor- mentee. Regarding research questions which are related to the role of mentoring in higher education, based on the participants' answers we can conclude that we are dealing with equal relationship between a senior and junior researcher, whereby the key function of a senior researcher (mentor) is guidance and leading younger colleague on his/her professional path. In this context, a senior researcher motivates and supports assistant's critical thinking and individual expert development, he allows him independent work, if necessary, he corrects him and complements him and occasionally also »opens doors«. Some senior researchers emphasize a desire that their assistants would surpass them in a professional sense and they see added value of mentoring in this. However, the relationship between the mentor and the mentee seems marginal and superficial, as the mentee is occasionally left to himself. This situation can cause a high level of uncertainty and consequently stress which can result even in the abandonment of the academic career.

Answers related to the research question about transfer of knowledge and creation of new knowledge between senior and junior researchers indicate that mentors mostly generously transfer explicit knowledge to younger colleagues, whereby they advocate for mainly independent work of junior researchers even regarding this. Some point out that the transfer of knowledge is a two-way process of giving and receiving knowledge, in which both the senior and the junior researcher play the roles of provider and recipient of such knowledge.

For knowledge transfer, the majority of interviewees establishes that we are not dealing only with a transfer of explicit knowledge, but also for transfer of tacit knowledge (e.g. skills and values), which can only be transferred through collaboration and personal contact. Creating of new knowledge between senior and junior researchers usually occurs as a sum of individual work of both participants. The new knowledge created in this way becomes a starting point for further work together. Out of six interviewed junior researchers, half of them acknowledge that they haven't created new knowledge with the mentor, as they collaborate with him only in the teaching process. Senior researchers emphasize the autonomy of junior researchers as an opportunity for learning and professional development, while forgetting that, at the beginning of their career, novices usually feel lonely and lost in a work environment that they are not yet familiar with, as well as overwhelmed by their many new work tasks. Such working conditions can have a demotivational effect and result in dissatisfaction, more mistakes made and in some extreme cases, even in the end of one's academic career.

It can be concluded from the answers related to the research question on the role and importance of ICT for collaboration between senior and junior researcher that the most commonly used ICT tool is a computer or electronic mail, which serve above all for quick exchange of information and coordination of work. Minority of those included into the study use document sharing applications. Also, a minority uses tools for video-communication and e-learning. Senior and junior researchers share the same opinion that ICT is of key importance for conducting their work, but it cannot replace personal contact.

Based on the obtained results, we can conclude that the function of mentoring on the senior-junior researcher level is primarily guidance and leading younger colleagues and allowing them enough freedom for individual professional development. This can be observed also in the process of transferring and creating new knowledge, where senior and junior researchers work mostly independently. One can conclude from this that only a smaller part of the interviewees are aware of the importance of transferring tacit knowledge, which occurs primarily through close cooperation and personal contact. This lack of awareness means reducing possibility for transfer of wider specter of knowledge from the older to younger generation of researchers.

FUTURE RESEARCH DIRECTIONS

Despite the limited size of the research sample, the results obtained can serve as a starting point for further research in the field of cooperation between senior and junior researchers.

The first question refers to the effect of having a high degree of autonomy on a junior researcher's career. The survey findings indicate that junior researchers work largely independently. If a junior researcher has a high degree of autonomy, this means that the senior researcher in the role of mentor can primarily transfer explicit knowledge onto the junior researcher. On the other hand, tacit knowledge, which is a kind of sum of the senior researcher's previous academic experience, is not transmitted or is transmitted to a much lesser extent. Previous studies (Dreher and Ash, 1990; Scandura, 1992) have shown that mentoring has a positive impact on the mentee's career. Therefore, it would be interesting to study the influence of the mentor-mentee relationship on the career of a junior researcher, in cases where the junior researcher has limited contact with the senior researcher and thus a limited number of opportunities to participate in the process of transfer of tacit knowledge.

Only a small number of the senior researchers who participated in the study saw a relationship with their younger colleagues as an opportunity for a mutual transfer of knowledge and learning. Based on the results obtained, the question arises, whether senior researchers are open to receiving knowledge from junior researchers and in which areas junior researchers may assume the role of the person providing the knowledge, while senior researchers assume the role of recipient of the knowledge.

The last question arising on the basis of the results obtained refers to the collaboration between senior and junior researchers within scientific research projects. The results of the study showed that older and younger university professors cooperate with each other mainly in the implementation of pedagogical work and, to a lesser extent, in the field of scientific research, that is, work on research projects, where new knowledge is created. As stated in the available literature (Byrne and Keefe, 2002), research activity, in addition to teaching and community service, is one of the fundamental activities of higher education institutions. Therefore, future studies should focus on the scientific and research activities of universities and on cooperation between members of research groups and should determine the importance of mentoring in the realization of research mission of higher education institutions. Furthermore, the use of ICT should be transferred in the teaching process and teachers should be willing to adopt this tools even if students are often more proficient in their use than teacher (Bull et al., 2008). Some higher education institutions are already starting to use ICT to provide access to higher education to a broader public, to find innovative ways of teaching and learning, or to reduce costs (Reeves, 2003).

CONCLUSION

Mentoring as a way of collaboration and knowledge transfer between more and less experienced person is a common form of cooperation in higher education. Results of our study indicate that this form of collaboration is focused on the pedagogical process rather than the research process. In both cases, junior researchers mostly work independently. Hence, the transfer of knowledge between senior and junior researcher is mostly explicit while the transfer of tacit knowledge is limited.

Based on our findings we propose the following guidelines:

1. The collaboration between senior and junior researchers should be based on regular contact and common work, which would contribute to a greater transfer of explicit and also tacit knowledge between both researchers. Greater attention should be given to the transfer of tacit knowledge, i.e. the transfer of know-how, which needs a large degree of face-to-face contact.
2. Higher education institutions should support and encourage more collaboration and knowledge transfer between senior and junior researchers. Especially in the field of research as in addition to teaching and community service since it is one of their fundamental activities. Finally, the findings from the research process should be transmitted in the pedagogical process, where possible. It seems that higher education institutions should change their organizational structure and emphasize more the values such as cooperation, trust and respect in a way to stimulate transfer of knowledge among colleagues.

The survey results serve as a basis for further research in the field of mentoring between senior and junior researchers. Limitation of the research is the sample size, since it was carried out on a limited number of Slovenian senior and junior researchers. A similar study should be conducted in different academic environments, which would give a broader view on the characteristics of such kind of collaboration. The influence of junior researcher on the senior researcher also needs to be studied as well as the impact of the large degree of the junior researcher's autonomy on his/her career. At the same time it would be necessary to pay special attention to the participation of senior and junior researchers in the field of research.

REFERENCES

Abidi, S. S. R., Hussini, S., Sriraj, W., Thienthong, S., & Finley, G. A. (2009). Knowledge sharing for pediatric pain management via a Web 2.0 framework. In K.-P. Adlassnig et al. (Eds.), *Medical Informatics in a United and Healthy Europe* (pp. 287–291). IOS Press.

Ala-Mutka, K., Gáspár, P., Kismihók, G., Suurna, M., & Vehovar, V. (2010). Status and Developments of eLearning in the EU10 Member States: The cases of Estonia, Hungary and Slovenia. *European Journal of Education*, *45*(3), 494–513.

Alavi, M., & Leidner, D. E. (2001). Review: Knowledge management and knowledge management systems: Conceptual foundations and research issues. *Management Information Systems Quarterly*, *25*(1), 107–136. doi:10.2307/3250961

Allen, T. D. (2007). Mentoring relationships from the perspective of the mentor. In B. R. Ragins & K. Kram (Eds.), *The handbook of mentoring at work: Theory, research and practice* (pp. 123–147). Thousand Oaks, CA: Sage.

Andrews, K. M., & Delahaye, B. L. (2000). AT Influences on knowledge processes in organizational learning: The psychosocial filter. *Journal of Management Studies*, *37*(6), 797–810. doi:10.1111/1467-6486.00204

Baranik, L. E., Roling, E. A., & Eby, L. T. (2010). Why does mentoring work? The role of perceived organizational support. *Journal of Vocational Behavior*, *76*(3), 366–373. doi:10.1016/j.jvb.2009.07.004 PMID:20401322

Berk, R. A., Berg, J., Mortimer, R., Walton-Moss, B., & Yeo, T. P. (2005). Measuring the Effectiveness of Faculty Mentoring Relationships. *Academic Medicine*, *80*(1), 66–71. doi:10.1097/00001888-200501000-00017 PMID:15618097

Bhagia, J., & Tinsley, J. A. (2000). The mentoring partnership. *Mayo Clinic Proceedings*, *75*(5), 535–537. doi:10.1016/S0025-6196(11)64225-0 PMID:10807085

Biloslavo, R., & Prevodnik, M. (2010). Impact of Organizational Culture on Knowledge Management in Higher Education. In D. Harorimana (Ed.), *Cultural Implications of Knowledge Sharing, Management and Transfer: Identifying Competitive Advantage* (pp. 152–179). Hershey, PA: Information Science Reference. doi:10.4018/978-1-61520-965-1.ch418

Boice, R. (1992). *The new faculty member*. San Francisco: Jossey-Bass.

Borisoff, D. (1998). Strategies for effective mentoring and for being effectively mentored: A focus on research institutions. *Journal of the Association for Communication Administration*, *27*, 84–96.

Bull, G., Thompson, A., Searson, M., Garofalo, J., Park, J., Young, C., & Lee, J. (2008). Connecting Informal and Formal Learning Experience in tha Age of Participatory Media. *Contemporary Issues in Technology & Teacher Education*, *8*(2), 100–107.

Burke, R. J., & McKeen, C. A. (1997). Benefits of mentoring relationships among managerial and professional women: A cautionary tale. *Journal of Vocational Behavior*, *51*(1), 43–57. doi:10.1006/jvbe.1997.1595

Byrne, M. W., & Keefe, M. R. (2002). Building Research Competence in Nursing Through Mentoring. *Journal of Nursing Scholarship*, *34*(4), 391–396. doi:10.1111/j.1547-5069.2002.00391.x PMID:12501744

Cabrera, A., & Cabreta, E. F. (2002). Knowledge-sharing Dilemmas. *Organization Studies*, *23*(5), 687–710. doi:10.1177/0170840602235001

Cabrera, E. F., & Cabreta, A. (2005). Fostering knowledge sharing through people management practices. *International Journal of Human Resource Management*, *16*(5), 720–735. doi:10.1080/09585190500083020

Cawyer, C. S., Simons, C., & Davis, S. (2002). Mentoring to facilitate socialization: The case of the new faculty member. *International Journal of Qualitative Studies in Education*, *15*(2), 225–242. doi:10.1080/09518390110111938

Chao, G. T., Walz, P. M., & Gardner, P. D. (1992). Formal and informalmentorships: Acomparison on mentoring functions and contrast with non-mentored counterparts. *Personnel Psychology*, *45*(3), 619–636. doi:10.1111/j.1744-6570.1992.tb00863.x

Chong, S. A. (2009). Mentoring: Are We Doing It Right? *Annals of the Academy of Medicine, Singapore*, *38*(7), 643–646. PMID:19652858

Chua, A. (2001). Relationship between the types of knowledge shared and types of communications channels used. *Journal of Knowledge Management Practice*. Retrieved from http://www.tlainc.com/articl26.htm

Claro, D. P., de Oliveira Claro, P. B., & Hagelaar, G. (2006). Coordinating collaborative joint efforts with suppliers: The effects of trust, transaction specific investments and information network in the Dutch flower industry. *Supply Chain Management*, *11*(3), 216–224. doi:10.1108/13598540610662112

Cox, M. D. (1997). Long-term patterns in a mentoring program for junior faculty: Recommendations for practice. In D. DcZure (Ed.), To improve the academy (pp. 225–268). Stillwater, OK: New Forum Press and the Professional and Organizational Development Network in Higher Education.

Currall, S. C., & Judge, T. A. (1995). Measuring trust between organizational boundary role persons. *Organizational Behavior and Human Decision Processes*, *64*(2), 151–170. doi:10.1006/obhd.1995.1097

Darwin, A., & Palmer, E. (2009). Mentoring circles in higher education. *Higher Education Research & Development*, *28*(2), 125–136. doi:10.1080/07294360902725017

Davenport, T. H., & Prusak, L. (1998). *Working Knowledge*. Boston, MA: Harvard Business School Press.

Davenport, T. H., & Prusak, L. (2000). *Working Knowledge: How Organizations Manage What They Know*. Boston, MA: Harvard Business School Press.

De Vries, J. (Ed.). (2005). *More than the sum of its parts: 10 years of the Leadership Development for Women Programme at UWA*. Perth, Australia: The University of Western Australia.

DeTienne, K. B., Dyer, G., Hoopes, C., & Harris, S. (2004). Toward a model of effective knowledge management and directions for future research: Culture, leadership, and CKOs. *Journal of Leadership & Organizational Studies*, *10*(4), 26–43. doi:10.1177/107179190401000403

Detsky, A. S., & Barelocher, M. O. (2007). Academic Mentoring – How to Give It and How to Get It. *Journal of the American Medical Association*, *297*(19), 2134–2136. doi:10.1001/jama.297.19.2134 PMID:17507350

DiCicco-Bloom, B., & Crabtree, B. F. (2006). The qualitative research interview. *Medical Education*, *40*(4), 314–321. doi:10.1111/j.1365-2929.2006.02418.x PMID:16573666

Douglas, C. A. (1997). *Formal mentoring programs in organizations: An annotated bibliography*. Greensboro, NC: Center for Creative Leadership.

Dreher, G. F., & Ash, R. A. (1990). A comparative study of mentoring among men and women in managerial, professional, and technical positions. *The Journal of Applied Psychology*, *75*(5), 539–546. doi:10.1037/0021-9010.75.5.539

Duffy, T., & Cunningham, D. (1996). Constructivism: Implications for the design and delivery of instruction. In Handbook of research for educational telecommunications and technology (pp. 170-198). Academic Press.

Dyck, B., Starke, F. A., Mischke, G. A., & Mauws, M. (2005). Learning to Build a Car: An Empirical Investigation of Organizational Learning. *Journal of Management Studies*, *42*(2), 387–416. doi:10.1111/j.1467-6486.2005.00501.x

Dyer, J. H., & Nobeoka, K. (2000). Creating and managing a high-performance knowledge-sharing network: The Toyota case. *Strategic Management Journal*, *21*(3), 345–367. doi:10.1002/(SICI)1097-0266(200003)21:3<345::AID-SMJ96>3.0.CO;2-N

Ehrich, L. C., Hansford, B., & Tennent, L. (2004). Formal mentoring programs in education and other professions: A review of the literature. *Educational Administration Quarterly*, *40*(4), 518–540. doi:10.1177/0013161X04267118

Ensher, E. A., Thomas, C., & Murphy, S. E. (2001). Comparison of traditional, step-ahead, and peer mentoring on protégés' support, satisfaction, and perceptions of career success: A social exchange perspective. *Journal of Business and Psychology*, *15*(3), 419–438. doi:10.1023/A:1007870600459

Fagenson, E. A. (1989). The mentor advantage: Perceived career/job experiences of protégés versus non- protégés. *Journal of Organizational Behavior*, *10*(4), 309–320. doi:10.1002/job.4030100403

Falconer, L. (2006). Organizational learning, tacit information, and e-learning: A review. *The Learning Organization*, *13*(2), 140–151. doi:10.1108/09696470610645476

Farjoun, M. (2010). Beyond Dualism: Stability and Change As a Duality. *Academy of Management Review*, *35*(2), 202–225. doi:10.5465/AMR.2010.48463331

Flanagin, A. J. (2002). The elusive benefits of the technological support of knowledge management. *Management Communication Quarterly*, *16*(2), 242–248. doi:10.1177/089331802237237

Gardiner, M. (1999). *Making a difference: Flinders University mentoring scheme for early career women researchers*. Adelaide, Australia: Flinders University Staff Development and Training Unit.

Geng, Q., Townley, C., Huang, K., & Zhang, J. (2005). Comparative knowledge management: A pilot study of Chinese and American universities. *Journal of the American Society for Information Science and Technology*, *56*(10), 1031–1044. doi:10.1002/asi.20194

Gilbert, M., & Cordey-Hayes, M. (1996). Understanding the process of knowledge transfer to achieve successful technological innovation. *Technovation*, *16*(6), 301–312. doi:10.1016/0166-4972(96)00012-0

Gill, A. (2009). Knowledge management initiatives at a small university. *International Journal of Educational Management*, *23*(7), 604–616. doi:10.1108/09513540910990834

Gordeyeva, I. (2010). *Enterprise 2.0: theoretical foundations of social media tools influence on knowledge sharing practices in organizations* (Master's Thesis). Retrieved from http://essay.utwente.nl/59921/1/MA_thesis_I_Gordeyeva.pdf

Gupta, B., Iyer, L. S., & Aronson, J. E. (2000). Knowledge management: Practices and challenges. *Industrial Management & Data Systems*, *100*(1), 17–21. doi:10.1108/02635570010273018

Haldin-Herrgard, T. (2000). Difficulties in diffusion of tacit knowledge in organizations. *Journal of Intellectual Capital*, *1*(4), 357–365. doi:10.1108/14691930010359252

Haring, M. J. (1999). The case for a conceptual base for minority mentoring programs. *Peabody Journal of Education*, *74*(2), 5–14. doi:10.1207/s15327930pje7402_2

Harris, R. J. (2009). Improving tacit knowledge transfer within SMEs through e-collaboration. *Journal of European Industrial Training*, *33*(3), 215–231. doi:10.1108/03090590910950587

Healy, C. C., & Welchert, A. J. (1990). Mentoring Relations: A Definition to Advance Research and Practice. *Educational Researcher*, *19*(9), 17–21. doi:10.3102/0013189X019009017

Hendriks, P. (1999). Why share knowledge? The influence of ICT on the motivation for knowledge sharing. *Knowledge and Process Management*, *6*(2), 91–100. doi:10.1002/(SICI)1099-1441(199906)6:2<91::AID-KPM54>3.0.CO;2-M

Higgins, M. C. (2000). The more the merrier? Multiple developmental relationships and work satisfaction. *Journal of Management Development*, *19*(4), 277–296. doi:10.1108/02621710010322634

Higgins, M. C. (2001). Reconceptualizing mentoring at work: A developmental network perspective. *Academy of Management Rewiev*, *26*(2), 254–288.

Hildrum, J. M. (2009). Sharing Tacit Knowledge Online: A Case Study of e-Learning in Cisco's Network of System Integrator Partner Firms. *Industry and Innovation*, *16*(2), 197–218. doi:10.1080/13662710902764360

Hislop, D. (2002). Mission impossible? Communicating and sharing knowledge via information technology. *Journal of Information Technology*, *17*(3), 165–177. doi:10.1080/02683960210161230

Hislop, D. (2005). *Knowledge Management in Organizations*. Oxford: Oxford University Press.

Holmes, D. R., Hodgson, P. K., Simari, R. D., & Nishimura, R. A. (2010). Mentoring making the transition from mentee to mentor. *Circulation*, *121*(2), 336–340. doi:10.1161/CIRCULATIONAHA.108.798321 PMID:20083720

Hooff, B. V. D., & Weenen, F. D. L. V. (2004). Committed to share: Commitment and CMC use as antecedents of knowledge sharing. *Knowledge and Process Management*, *11*(1), 13–24. doi:10.1002/kpm.187

Hsia, T. L., Lin, L. M., Wu, J. H., & Tsai, H. T. (2006). A Framework for Designing Nursing Knowledge Management Systems. *Interdisciplinary Journal of Information. Knowledge and Management*, *1*, 14–21.

Inkpen, A., & Dikur, I. (1998). Knowledge management processes and international joint ventures. *Organization Science*, *9*(4), 454–468. doi:10.1287/orsc.9.4.454

Inkpen, A. C., & Currall, S. C. (2004). The coevolution of trust, control, and learning in joint ventures. *Organization Science*, *15*(5), 586–599. doi:10.1287/orsc.1040.0079

Ipe, M. (2003). Knowledge sharing in organizations: A conceptual framework. *Human Resource Development Review*, *3*(4), 337–359. doi:10.1177/1534484303257985

Jackson, V. A., Palepu, A., Szalacha, L., Caswell, C., Carr, P. L., & Inui, T. (2003). Having the right chemistry: A qualitative study of mentoring in academic medicine. *Academic Medicine*, *78*(3), 328–334. doi:10.1097/00001888-200303000-00020 PMID:12634219

Johannessen, J. A., Olaisen, J., & Olsen, B. (2001). Mismanagement of tacit knowledge: The importance of tacit knowledge, the danger of information technology, and what to do about it. *International Journal of Information Management*, *21*(1), 3–20. doi:10.1016/S0268-4012(00)00047-5

Johnsrud, L. K. (1990). Mentoring for administrative staff: Relationships that help and relationships that hinder. In K. M. Moore & S. B. Twombley (Eds.), *Administrative careers and the marketplace* (pp. 57–66). San Francisco: Jossey Bass.

Kaplan, B., & Maxwell, J. A. (2005). Qualitative research methods for evaluating computer information systems. In J. G. Anderson & C. E. Aydin (Eds.), *Evaluating the Organizational Impact of Healthcare Information Systems* (2nd ed., pp. 30–55). New York: Springer. doi:10.1007/0-387-30329-4_2

Keyser, D. J., Lakoski, J. M., Lara-Cinisomo, S., Schultz, D. J., Williams, V. L., Zellers, D. F., & Pincus, H. A. (2008). Advancing institutional efforts to support research mentorship: A conceptual framework and self-assessment tool. *Academic Medicine*, *83*(3), 217–225. doi:10.1097/ACM.0b13e318163700a PMID:18316865

Khan, O. J., & Jones, N. B. (2011). Harnessing tacit knowledge for competitive advantage: An internal social network approach. *Journal for International Business and Entrepreneurship Development*, *5*, 232–248. doi:10.1504/JIBED.2011.038031

Kitzinger, J. (1995). Introducing Focus Groups. BMJ. *British Medical Journal*, *311*(29), 299–302. doi:10.1136/bmj.311.7000.299 PMID:7633241

Koberg, C. S., Boss, R. W., Chappell, D., & Ringer, R. C. (1994). Correlates and consequences of protégé mentoring in a large hospital. *Group & Organization Management*, *19*(2), 219–239. doi:10.1177/1059601194192007

KPMG. (2000). *Knowledge management research report 2000*. KPMG Consulting. Retrieved April 5, 2011, from http://www.providersedge.com/docs/km_articles/KPMG_KM_Research_Report_2000.pdf

Kram, K. E. (1985). *Mentoring at work: Developmental relationships in organizational life*. Glenview, IL: Scott Foresman.

Kram, K. E., & Hall, D. T. (1996). Mentoring in a context of diversity and turbolence. In E. E. Kossek & S. Lobel (Eds.), *Managing diversity: Guman resource strategies for transforming the workplace* (pp. 108–136). Cambridge, MA: Blackwell.

Lai, I. L. A. (2005). Knowledge management for Chinese medicines: A conceptual model. *Information Management & Computer Security*, *133*(3), 244–255. doi:10.1108/09685220510602059

Lam, A. (2000). Tacit knowledge, organizational learning and societal institutions: An integrated framework. *Organization Studies*, *21*(3), 487–513. doi:10.1177/0170840600213001

Lawson, C., & Lorenz, E. (1999). Collective learning, tacit knowledge and regional innovative capacity. *Regional Studies*, *33*(4), 305–317. doi:10.1080/713693555

Lin, C., Kao, M., & Chang, K. (2010). Is more similar, better? Interacting effort of the cognitive-style congruency and tacitness of knowledge on knowledge transfer in the mentor-protègè dyad. *Asian Journal of Social Psychology*, *13*(4), 286–292. doi:10.1111/j.1467-839X.2010.01325.x

Lindmark, S. (2009). *Web 2.0: Where Does Europe Stand. JRC Scientific and Technical Reports, Institute for Prospectice Technological Studies (IPTS), Joint Research Center (JRC)*. Seville: European Commission.

Lumpkin, A. (2011, October). A model for mentoring university faculty. *The Educational Forum*, *75*(4), 357–368. doi:10.1080/00131725.2011.602466

Lundvall, B. A., & Johnson, B. (1994). The learning economy. *Journal of Industry Studies*, *1*(2), 22–42. doi:10.1080/13662719400000002

Martin, J. S., & Marion, R. (2005). Higher education leadership roles in knowledge processing. *The Learning Organization*, *12*(5), 140–151. doi:10.1108/09696470510583520

Marwick, A. D. (2001). Knowledge management technology. *IBM Systems Journal*, *40*(4), 814–830. doi:10.1147/sj.404.0814

Mayfield, M. (2010). Tacit knowledge sharing: Techniques for putting a powerful tool in practice. *Development and Learning in Organizations*, *24*(1), 24–26. doi:10.1108/14777281011010497

Mayor, E., & Cordey-Hayes, M. (2000). Knowledge translation: A new perspective on knowledge transfer and foresight. *Foresight*, *2*(4), 411–423. doi:10.1108/14636680010802762

Mesner-Andolšek, D. (2011). Knowledge hoarding or sharing. In V. Dermol (Ed.), *Knowledge as business opportunity: Proceedings of the Management, Knowledge and Learning International Conference 2011*, (pp. 167-176). Celje, Slovenia: International School for Social and Business Studies.

Meyers, H. W., & Smith, S. (1999). Coming home-mentoring new teachers: A school-university partnershipmto support the development of teachers from diverse ethnic backgrounds. *Peabody Journal of Education*, *74*(2), 75–89. doi:10.1207/s15327930pje7402_7

Mintzberg, H. (1983). *Structure in Fives: Designing Effective Organizations*. Englewood Cliffs, NJ: Prentice-Hall.

Moller, K., & Svahn, S. (2004). Crossing east-west boundaries: Knowledge sharing in intercultural business networks. *Industrial Marketing Management*, *33*(3), 219–228. doi:10.1016/j.indmarman.2003.10.011

Morgan, R. M., & Hunt, S. D. (1994). The commitmenttrust theory of relationship marketing. *Journal of Marketing*, *58*(3), 20–38. doi:10.2307/1252308

Morrison, E. W. (1993). Newcomer information seeking: Exploring types, modes, sources and outcomes. *Academy of Management Journal*, *36*(1), 557–589. doi:10.2307/256592

Murphy, G., & Salomone, S. (2013). Using social media to facilitate knowledge trabsfer in complex engineering environments: A primer for educators. *European Journal of Engineering Education*, *38*(1), 1–15. doi:10.1080/03043797.2012.742871

Murray, M. (1991). *Beyond the myths and magic of mentoring: How to facilitate an effective mentoring program*. San Francisco: Jossey-Bass.

Murray, S. R., & Peyrefitte, J. (2007). Knowledge Type and Communication Media Choice in the Knowledge Transfer Process. *Journal of Managerial Issues*, *19*(1), 111–133.

Newell, S., & Swan, J. (2000). Trust and interorganizational networking. *Human Relations*, *53*(10), 287–328.

Nonaka, I. (1994). A Dynamic theory of Organizational Knowledge Creation. *Organization Science*, *5*(1), 14–37. doi:10.1287/orsc.5.1.14

Nonaka, I., & von Krogh, G. (2009). Tacit Knowledge and Knowledge Conversion: Controversy and Advancement in Organizational Knowledge Creation Theory. *Organization Science*, *20*(3), 635–652. doi:10.1287/orsc.1080.0412

Oliver, R. (2001). Developing e-learningenvironments that support knowledge construction in higher education. In Working for Excellence in the E-conomy (pp. 407-16). Churchlands: We-B Centre.

Olson, R. K., & Connelly, L. M. (1995). Mentoring through predoctoral fellowships to enhance research productivity. *Journal of Professional Nursing*, *11*(5), 270–275. doi:10.1016/S8755-7223(05)80007-9 PMID:7593970

Osimo, D. (2008). *Web 2.0 in government: Why and how (Scientific and Technical Reports, Institute for Prospectice Technological Studies [IPTS], Joint Research Center [JRC])*. Seville: European Commission.

Ostroff, C., & Kozlowski, S. W. (1993). The role of mentoring in the information gathering process of newcomers during early organizational socialization. *Journal of Vocational Behavior*, *42*(2), 170–183. doi:10.1006/jvbe.1993.1012

Panahi, S., Watson, J., & Partridge, H. (2012). Social media and tacit knowledge sharing: developing a conceptual model. World Academy of Science, Engineering and Technology, (64), 1095-1102.

Pierce, G. (1998). Developing new university faculty through mentoring. *The Journal of Humanistic Education and Development*, *37*(1), 27–38. doi:10.1002/j.2164-4683.1998.tb00403.x

Polanyi, M. (1958). *Personal Knowledge Towards a Post-critical Philosophy*. London: Routledge and Kegan Paul Ltd.

Polanyi, M. (1966). *The Tacit Dimension*. London: Routledge and Kegan Paul Ltd.

Pololi, L., & Knight, S. (2005). Mentoring faculty in academic medicine: A new paradigm? *Journal of General Internal Medicine*, *20*(9), 866–870. doi:10.1111/j.1525-1497.2005.05007.x PMID:16117759

Ragins, B., & Cotton, J. L. (1999). Mentor functions and outcomes: A comparison of men and women in formal and informal mentoring relationships. *The Journal of Applied Psychology*, *84*(4), 529–550. doi:10.1037/0021-9010.84.4.529 PMID:10504893

Ragins, B. R., & Cotton, J. (1991). Easier said than done: Gender differences in perceived barriers to gaining a mentor. *Academy of Management Journal*, *34*(4), 939–951. doi:10.2307/256398

Raisanen, T., & Oinas-Kukkonen, H. (2008). A system architecture for the 7C knowledge environment. *Information Modelling and Knowledge Bases XIX*, *166*, 217 Retrieved from http://www.epd.aau.dk/uploads/media/ejc_7c_information_system_architecture_final.pdf

Ramanan, R. A., Phillips, R. S., Davis, R. B., Silen, W., & Reede, J. Y. (2002). Mentoring in medicine: Keys to satisfaction. *The American Journal of Medicine*, *112*(4), 336–341. doi:10.1016/S0002-9343(02)01032-X PMID:11893387

Ramani, S., Gruppen, L., & Kachur, E. K. (2006). Twelve tips for developing effective mentors. *Medical Teacher*, *28*(5), 404–408. doi:10.1080/01421590600825326 PMID:16973451

Reeves, T. C. (2003). Storm clouds on the digital education horizon. *Journal of Computing in Higher Education*, *15*(1), 3–26. doi:10.1007/BF02940850

Ribeiro, R., & Collins, H. (2007). The bread-making machine: Tacit knowledge and two types of action. *Organization Studies*, *28*(9), 1417–1433. doi:10.1177/0170840607082228

Roberts, J. (2000). From Know-how to Show-how? Questioning the Role of Information and Communication Technologies in Knowledge Transfer. *Technology Analysis and Strategic Management*, *12*(4), 429–443. doi:10.1080/713698499

Rollett, H. (2003). *Knowledge Management: Processes and Technologies*. Boston: Kluwer Academic Publishers. doi:10.1007/978-1-4615-0345-3

Sahay, B. S. (2003). Understanding trust in supply chain relationships. *Industrial Management & Data Systems*, *103*(8), 553–563. doi:10.1108/02635570310497602

Sands, R. G., Parson, L. A., & Duane, J. (1991). Faculty Mentoring Faculty in a Public University. *The Journal of Higher Education*, *62*(2), 174–193. doi:10.2307/1982144

Scandura, T. A. (1992). Menorship and career mobility: An empirical investigation. *Journal of Organizational Behavior*, *13*(2), 169–174. doi:10.1002/job.4030130206

Soekijad, M., & Andriessen, E. (2003). Conditions for knowledge sharing in competitive alliances. *European Management Journal*, *21*(5), 578–587. doi:10.1016/S0263-2373(03)00107-5

Song, D. (2009). The tacit knowledge-sharing strategy analysis in the project work. *International Business Research*, *2*(1), 83–85. doi:10.5539/ibr.v2n1p83

Srikanthan, G., & Dalrymple, J. F. (2002). Developing a Holistic Model for Quality in Higher Education. *Quality in Higher Education*, *8*(2), 215–224. doi:10.1080/1353832022000031656

Steininger, K., Ruckel, D., Dannerer, E., & Roithmayr, F. (2010). Healthcare knowledge transfer through a web 2.0 portal: An Austrian approach. *International Journal of Healthcare Technology and Management*, *11*(1), 13–30. doi:10.1504/IJHTM.2010.033272

Steinkellner, P. F., & Czerny, E. J. (2010, December). *Educating Managers for a Paradox World – Duality and Paradoxes in Management*. Paper presented at the International Conference on Management Learning 'Management Makes the World Go Around. Vienna, Austria.

Stenmark, D. (2001). Leveraging Tacit Organizational Knowledge. *Journal of Management Information Systems*, *17*(3), 9–24.

Stewart, B. M., & Krueger, L. E. (1996). An evolutionary concept analysis of mentoring in nursing. *Journal of Professional Nursing*, *12*(5), 311–321. doi:10.1016/S8755-7223(96)80011-1 PMID:8871697

Sveiby, K. E. (1997). *The new organizational wealth: Managing and measuring knowledge-based assets*. San Francisco, CA: Berret-Koehler.

Sveiby, K. E. (2001). A knowledge-based theory of the firm to guide in strategy formulation. *Journal of Intellectual Capital*, *2*(4), 344–358. doi:10.1108/14691930110409651

Swap, W., Leonard, D., Shields, M., & Abrams, L. (2001). Using Mentoring and Storytelling to Transfer Knowledge in the Workplace. *Journal of Management Information Systems*, *18*(1), 95–114.

Taylor, W. A., & Wright, G. H. (2004). Organizational readiness for successful knowledge sharing: Challenges for public secto managers. *Information Resources Management Journal*, *17*(2), 22–37. doi:10.4018/irmj.2004040102

Tippins, M. (2003). Implementing knowledge management in academia: Teaching the teachers. *International Journal of Educational Management*, *17*(7), 339–345. doi:10.1108/09513540310501021

Treleaven, L. (2004). A knowledge-sharing approach to organizational change: A critical discourse analysis. In H. Tsoukas & N. Mylonopoulos (Eds.), *Organizations as Knowledge Systems: Knowledge, Learning and Dynamic Capabilities* (pp. 154–180). New York, NY: Palgrave Macmillan.

Viator, R. E., & Scandura, T. A. (1991). A study of mentor-protégé relationships in large public accounting firms. *Accounting Horizons*, *5*, 20–30.

Walker, A., & Scott, K. (1993). Preparing for leadership in schools: The mentoring contribution. In B. J. Carwell & E. M. Carter (Eds.), *The Return of the Mentor: Strategies for Workplace Learning* (pp. 77–90). London: The Falmer Press.

Wasburn, M. H., & LaLopa, J. M. (2003). Mentoring faculty for success: Recommendations based on evaluations of a program. *Planning and Changing, 34*(3-4), 250-264.

Whitcomb, M. E. (2004). The future of academic medicine: Caereer development of junior faculty. *Academic Medicine*, *79*(3), 195–196. doi:10.1097/00001888-200403000-00001 PMID:14985190

Whitely, W., Dougherty, T. W., & Dreher, G. F. (1991). Relationship of career mentoring and socioeconomic origin to managers' and professionals' early career progress. *Academy of Management Journal*, *34*(2), 331–351. doi:10.2307/256445

Whitley, G. G., & Oddi, L. F. (1998). Graduate student-faculty collaboration in research and publication. *Western Journal of Nursing*, *20*(5), 572–583. doi:10.1177/019394599802000505 PMID:9775739

Wilkesmann, M., & Wilkesmann, U. (2011). Knowledge transfer as interaction between experts and novices supported by technology. *VINE: The Journal of Information and Knowledge Management Systems*, *41*(2), 96–112. doi:10.1108/03055721111134763

Wilkesmann, U., Fischer, H., & Wilkesmann, M. (2009). Cultural characteristics of knowledge transfer. *Journal of Knowledge Management*, *13*(6), 464–477. doi:10.1108/13673270910997123

Wind, J., & Main, J. (1999). *Diving change*. New York: The Free Press.

Wocial, L. D. (1995). The role of mentorsin promoting integrity and preventing scientific misconduct in nursing research. *Journal of Professional Nursing*, *11*(5), 276–280. doi:10.1016/S8755-7223(05)80008-0 PMID:7593971

Yi, J. (2006). Externalization of tacit knowledge in online environments. *International Journal on E-Learning*, *5*(4), 663–674.

Yoder, L. (1990). Mentoring: A concept analysis. *Nursing Administration Quarterly*, *15*(1), 9–19. doi:10.1097/00006216-199001510-00005 PMID:2234538

Zack, M. (1998). What knowledge problems can information technology help to solve. In *Proceedings of the Fourth Americas Conferences on Information Systems*, (pp. 644-646). Academic Press.

ADDITIONAL READING

Ambrose, L. (2003). Multiple mentoring: Discover alternatives to a one-on-one learning relationship. *Healthcare Executive*, *18*(4), 58–60. PMID:12841061

Antonelli, C. (2008). The new economics of the university: A knowledge governance approach. *The Journal of Technology Transfer*, *33*(1), 1–22. doi:10.1007/s10961-007-9064-9

Baranik, L. E., Roling, E. A., & Eby, L. T. (2010). Why does mentoring work? The role of perceived organizational support. *Journal of Vocational Behavior*, *76*(3), 366–373. doi:10.1016/j.jvb.2009.07.004 PMID:20401322

Bauer, J., & Kenton, J. (2005). Toward Technology Integration in the Schools: Why it isn't Happening. *Journal of Technology and Teacher Education*, *13*(4), 519–546.

Bearman, S., Blake-Beard, S., Hunt, L., & Crosby, F. J. (2007). New directions in mentoring. In L. T. Eby and T. D. Allen's (Eds.), The Blackwell handbook of mentoring: A multiple perspectives approach (pp. 375–395). Malden, MA: Blackwell Publishing

Bouquillon, E. A., Sosik, J. J., & Le, D. (2005). 'It's only a phase': Examining trust, identification and mentoring functions received across the mentoring phases. *Mentoring & Tutoring*, *13*(2), 239–258. doi:10.1080/13611260500105808

Chandler, D. E., Kram, K. E., & Yip, J. (2011). An ecological systems perspective on mentoring at work: A review and future prospects. *The Academy of Management Annals*, *5*(1), 519–570. doi:10.1080/19416520.2011.576087

Dutton, J. E., & Ragins, B. R. (Eds.). (2007). *Exploring positive relationships at work: Building a theoretical and research foundation*. Mahwah, NJ: Lawrence Erlbaum Associates.

Eby, L. T., Durley, J. R., Evans, S. C., & Ragins, B. R. (2006). The relationship between short-term mentoring benefits and long-term mentor outcomes. *Journal of Vocational Behavior*, *69*(3), 424–444. doi:10.1016/j.jvb.2006.05.003

Eby, L. T., & Lockwood, A. (2005). Protégés' and mentors' reactions to participating in formal mentoring programs: A qualitative investigation. *Journal of Vocational Behavior*, *67*(3), 441–458. doi:10.1016/j.jvb.2004.08.002

Ensher, E. A., & Murphy, S. E. (2011). The Mentoring Relationship Challenges Scale: The impact of mentoring stage, type, and gender. *Journal of Vocational Behavior*, *79*(1), 253–266. doi:10.1016/j.jvb.2010.11.008

Gibson, S. K. (2006). Mentoring of women faculty: The role of organizational politics and culture. *Innovative Higher Education*, *31*(1), 63–79. doi:10.1007/s10755-006-9007-7

Grosshans, O., Poczwardowski, A., Trunnell, E., & Ransdell, L. (2003). Senior faculty retrospectives on mentoring. *Journal of Health Education*, *34*(3), 146–153. doi:10.1080/19325037.2003.10603546

Jonson, A., & Kalling, T. (2007). Challenges to knowledge sharing across national and intra-organizational boundaries: Case studies of IKEA and SCA packaging. *Knowledge Management Research and Practice*, *5*(3), 161–172. doi:10.1057/palgrave.kmrp.8500139

Kasprisin, C. A., Single, P. B., Single, R. M., & Muller, C. B. (2003). Building a better bridge: Testing e-training to improve e-mentoring programmes in higher education. *Mentoring & Tutoring*, *11*(1), 67–78. doi:10.1080/1361126032000054817

Kirkup, G., & Kirkwood, A. (2005). Information and communications technologies (ICT) in higher education teaching—a tale of gradualism rather than revolution. *Learning, Media and Technology*, *30*(2), 185–199. doi:10.1080/17439880500093810

Marginson, S. (2006). Dynamics of national and global competition in higher education. *Higher Education*, *52*(1), 1–39. doi:10.1007/s10734-004-7649-x

Norris, D. M., Mason, J., Robson, R., Lefrere, P., & Collier, G. (2003). A Revolution in Knowledge Sharing. *EDUCAUSE Review*, *38*(5), 16–26.

Payne, S. C., & Huffman, A. H. (2005). A longitudinal examination of the influence of mentoring on organizational commitment and turnover. *Academy of Management Journal*, *48*(1), 158–168. doi:10.5465/AMJ.2005.15993166

Ragins, B. R., & Verbos, A. K. (2007). Positive relationships in action: Relational mentoring and mentoring schemas in the workplace. In J. E. Dutton & B. R. Ragins' (Eds.), *Exploring positive relationships at work: Building a theoretical and research foundation*. Mahway, NJ: Lawrence Erlbaum.

Reynolds, D., Treharne, D., & Tripp, H. (2003). ICT – the hopes and the reality. *British Journal of Educational Technology*, *34*(2), 151–167. doi:10.1111/1467-8535.00317

Ringberg, T., & Reihlen, M. (2008). Towards a Socio-Cognitive Approach to Knowledge Transfer. *Journal of Management Studies*, *45*(5), 912–935. doi:10.1111/j.1467-6486.2007.00757.x

Rosenfeld, B., & Martinez-Pons, M. (2005). Promoting Classroom Technology Use. *The Quarterly Review of Distance Education*, *6*(2), 145–153.

Sambunjak, D., Straus, S. E., & Marusic, A. (2006). Mentoring in academic medicine: A systematic review. *Journal of the American Medical Association*, *296*(9), 1103–1115. doi:10.1001/jama.296.9.1103 PMID:16954490

Scandura, T. A., & Williams, E. A. (2004). Mentoring and transformational leadership: The role of supervisory career mentoring. *Journal of Vocational Behavior*, *65*(3), 448–468. doi:10.1016/j.jvb.2003.10.003

Schulze, S. (2009). Mentoring in higher education to improve research output: An ethnographic case study. *Acta Academica*, *41*(4), 138–158.

Sosik, J. J., Godshalk, V. M., & Yammarino, F. J. (2004). Transformational leadership, learning goal orientation, and expectations for career success in mentor–protégé relationships: A multiple levels of analysis perspective. *The Leadership Quarterly*, *15*(2), 241–261. doi:10.1016/j.leaqua.2004.02.003

Stensaker, B., Maassen, P., Borgan, M., Oftebro, M., & Karseth, B. (2007). Use, updating and integration of ICT in higher education: Linking purpose, people and pedagogy. *Higher Education*, *54*(3), 417–433. doi:10.1007/s10734-006-9004-x

Suppiah, V., & Sandu, M. S. (2011). Organisational culture's influence on tacit knowledge-sharing behaviour. *Journal of Knowledge Management*, *15*(3), 462–477. doi:10.1108/13673271111137439

Teichler, U. (2004). The changing debate on internationalisation of higher education. *Higher Education*, *48*(1), 5–26. doi:10.1023/B:HIGH.0000033771.69078.41

Unwin, A. (2007). The professionalism of the higher education teacher: What's ICT got to do with it? *Teaching in Higher Education*, *12*(3), 295–308. doi:10.1080/13562510701278641

Wanberg, C. R., Welsh, E. T., & Hezlett, S. A. (2003). Mentoring research: A review and dynamic process model. *Research in Personnel and Human Resource Management*, *22*, 39–124. doi:10.1016/S0742-7301(03)22002-8

Wang, S., Tomlinson, E. C., & Noe, R. A. (2010). The role of mentor trust and protégé internal locus of control in formal mentoring relationships. *The Journal of Applied Psychology*, *95*(2), 358–367. doi:10.1037/a0017663 PMID:20230075

Watson, S., & Hewett, K. (2006). A Multi-Theoretical Model of Knowledge Transfer in Organizations: Determinants of Knowledge Contribution and Knowledge Reuse. *Journal of Management Studies*, *43*(2), 141–173. doi:10.1111/j.1467-6486.2006.00586.x

KEY TERMS AND DEFINITIONS

Explicit Knowledge: Knowledge that can be easily codified in words, pictures and numbers.

Junior Researcher: Usually a young and less experienced researcher involved in research projects.

Knowledge Transfer: A conscious or unconscious process where someone's knowledge is transferred to another person.

Knowledge: The capacity to act developed through formal learning, practical hands-on experience and socialization.

Mentee: A less experienced person who acquires knowledge and experience through a relationship with a more experienced person.

Mentor: Who on the basis of expertise and experience directs and guides a less experienced person.

Mentoring: A relationship between a more and a less experienced person aimed at supporting the personal and career development of both.

Senior Researcher: A high skilled person with years of experience in the research field.

Tacit Knowledge: Context-specific knowledge which is embedded in an individual's skills, habits, values and experience; knowledge difficult to communicate and transfer.

Section 2

Case Studies in Different Domains

This section provides case studies that show how different communities tend to separate creating different and incommensurable cultures. If this is quite easy to understand, considering different research domains; it is also important to recognize this trend within the same research domain when different communities are identified by different schools, paradigms, or traditions.

Chapter 7
Towards the Interoperable Data Environment for Facilities Science

Vasily Bunakov
Science and Technology Facilities Council, UK

Catherine Jones
Science and Technology Facilities Council, UK

Brian Matthews
Science and Technology Facilities Council, UK

ABSTRACT

The research enabled by national and international photon and neutron source facilities makes a key contribution to the modern scientific community supporting thousands of researchers around the world to explore and understand the structure of materials. In this chapter, the authors describe the characteristics of facilities science with a distinct common research lifecycle which requires the provision of facility-based centralized IT infrastructure and data management platforms. The chapter then considers how the nature of facilities science is changing and what opportunities this brings for a more cohesive approach to data management. It then goes on to consider investigation research objects and to formalize the aggregation of related and contextual information which is important for re-use of research outputs by those who were not directly involved in the original experiment. Finally, there is consideration of what infrastructure can support the use of investigation research objects.

INTRODUCTION

Today's scientific research is conducted not just by single experiments but rather by sequences of related experiments or projects linked by a common theme that lead to a greater understanding of the structure, properties and behaviour of the physical world. This is particularly true of research carried out on large-scale facilities such as neutron and photon sources, which are used for the detailed investigation of the structure of matter in areas such as pharmaceuticals, chemistry, mate-

DOI: 10.4018/978-1-4666-6567-5.ch007

Copyright © 2015, IGI Global. Copying or distributing in print or electronic forms without written permission of IGI Global is prohibited.

rial science or biology. Such large-scale facilities support many individual researchers and small research teams who share access to common, rare resources supported by specialist infrastructure. Facilities science combines the characteristics of "big science" (dedicated infrastructure) with those of "bench science" (many individual experiments).

Experiments at these facilities require complex computing support to convert the information collected at the facility into the knowledge which is interpretable by the scientific community. Additionally, experiments are increasingly not carried out in isolation, but require the combination of sequences of experiments for the full understanding of the sample structure. A further feature of this area is the extent to which the research community is shared across facilities. As a consequence the support required by scientists to manage their data and access the compute resources required for a complete result is getting more complex and large-scale.

Facilities experiments often require sophisticated software for data analysis and visualization, as well as for the experiments' design through simulation of experimental environment. The advances of computer science and information technology have made it possible to use supercomputers for the simulation of experiments themselves, and use simulation along with experimentation to get new insights into the nature of samples of materials under investigation. Computer simulations produce substantial amount of data that contributes to data management requirements.

Thus in the area of facilities science, there is a growing need for a comprehensive data infrastructure across facilities to enhance the productivity of research. In this chapter, we discuss the issues and approaches to providing a common open data infrastructure to support the users of facilities as they pass through and between different facilities. The issues which we consider include:

- The special nature of scientific collaboration and support within facilities science, and the requirements this nature poses for infrastructure support for the scientific community. This is built around a particular scientific workflow which provides the basis for a common data infrastructure.
- The changing landscape for facilities science as data rates from instruments increase, automation increases in the data infrastructure, and co-investigations undertaken across facilities mean that sharing and combination of data is required. These factors are affecting the established scientific practice. Further, common changes in the research policy that encourage open access to data for validation, sharing and reuse are also influencing facilities science, leading to a need for more effective data curation, publication and sharing.
- Curating and publishing information about experiments in context beyond the publication of data alone, using a notion of *Investigation Research Object* as a sharable persistent unit of discourse for facilities science. This also allows the publication and sharing of data within a Linked Open Data framework. This can point the way towards a framework for data curation and data sharing in facilities science which is required, in addition to technological considerations, to make the collaborative IT solutions realistically designed and sustainable through time.

We conclude with a future vision of an Open Data Infrastructure for facilities science across facilities interoperating with different user communities which use the large-scale facilities.

BACKGROUND

The class of large-scale facilities considered in this chapter supply intense beams of particles (such as photons or neutrons) generated from sources derived from fundamental physics experiments (such as synchrotrons, free-electron lasers, or nuclear reactors). These intense beams are used as "microscopes" to probe into the structure of matter at a micrometre (10^{-6} m) or nanometre (10^{-9} m) scale at a resolution and precision unavailable from more conventional instruments within university laboratories, making them a much sought-after resource for researchers studying the structure of matter. The scale and expense of constructing and operating these sources mean that they are beyond the scope of individual research institutions on their own, and they are typically provided as a pooled resource at a national level, often as part of national laboratories open for international research, for example the ISIS Neutron Spallation Source at the Rutherford Appleton Laboratory (https://www.isis.stfc.ac.uk/), the Soleil Synchrotron Source in France (http://www.synchrotron-soleil.fr/), the Advanced Photon Source at Argonne in the USA (http://www.aps.anl.gov/); or international establishments, for example the Institut Laue-Langevin (ILL) (https://www.ill.eu/) and the European Synchrotron Radiation Facility (ESRF) (http://www.esrf.eu/), both based in Grenoble, France. Costing in the order of 0.5-1.0 billion euros to construct they represent some of the largest investments in science infrastructure undertaken.

Historically, these facilities emerged as a by-product of big science experiments conducting fundamental research in physics. An example of this is synchrotron x-ray radiation which was seen as a loss of energy in particle accelerators until successful experiments proved its potential for the investigation of various samples in materials science and biology. Some of the facilities are still being used in dual mode for fundamental research and for shared use by visiting teams from a variety of research domains; the example of such a facility is the particle accelerator in Novosibirsk, Russia, that is used part time for the fundamental research in particle physics, and part time as a synchrotron x-ray source (Baier, 2006). Some of the facilities, for example the aforementioned ISIS, are the fruits of the creative reuse of instruments that are not competitive any more for fundamental research but are ideal as a core of shared facilities for applied science. Now it has become a common pattern in developing research infrastructures to build facilities that from their very inception are large shared instruments dedicated for experiments in materials science, biology, pharmaceutics and other fields of applied research; the Diamond Light Source (http://www.diamond.ac.uk/) or the future European Spallation Source (http://europeanspallationsource.se/) are the examples of such decidedly designed facilities.

As a consequence of being a shared resource, facilities have developed a particular way of operating that is quite different from that of traditional Big Science. Facilities provide a core of scientists and engineers to operate and maintain the machine, while the science is undertaken by many small teams of individual scientists based in external research institutions, who visit the facility for a period of allocated time to access the beam and the specialised detectors provided at the facility. Thus, facilities science has a mixture of the characteristics of "big science", including a large and long-term capital investment, permanent support staff, investment in computing infrastructure, rigorous procedures; and "bench science", with individual scientists and small research teams, limited research aims – the study of particular features of material samples, short-time scales – typically up the length of a PhD grant, and limited support infrastructure available at the home institution.

The neutron and photon source facilities serve an expanding user community of 25,000 to 30,000 scientists across Europe, and a much wider community across the world. Applications are numerous, for example: crystallography reveals the structures of viruses and proteins important for the development of new drugs; neutron scattering identifies stresses within engineering components such as turbine blades, and tomography can image microscopic details of the structure of the brain.

Industrial applications include pharmaceuticals, petrochemicals and microelectronics. These different applications typically use different techniques to measure the beam's interaction with an experimental sample, such as x-ray diffraction, computed tomographic imaging, small angle neutron scattering and many others, and thus require specialised detectors. Consequently, a single facility will support a number of instruments clustered around the source of the high-intensity beams; for example the ILL has some 40 instruments distributed around its nuclear-reactor which acts as a source of high-intensity neutrons.

The traditional approach of many of the facilities leaves handling of data generated by the detectors in experiments almost entirely to the individual instrument scientists and research teams. While this local responsibility is well handled in most cases, this approach in general has become unsustainable to guarantee the longevity and availability of precious and costly experimental data. Large-scale facilities are advanced scientific environments which have demanding computing requirements. Modern instruments can generate data in extremely large volumes, and as many instruments as possible are placed around target areas or beam-lines in order to maximize the output from the expensive neutron or synchrotron x-ray resource. Consequently, the data volumes are large and increasing, especially from synchrotron sources, where a single tomographic image can be of the order 100Gb, and the data throughput is very high resulting in Terabytes of data in a single day of operation for some detectors; thus the data management requires large-scale data transfer and storage.

The co-existence of diverse communities involved in building instruments and software and also the different academic communities and disciplines, has led to a proliferation in data formats and software interfaces. The data collected has a large number of parameters, measured both from the operating environment (e.g. temperature, pressure) and from the sample (e.g. angles from a scattering pattern) and this requires a multi-variate analysis, typically over several steps.

As a result of these trends and the increased capability of modern electronic detectors and high-throughput automated experiments, the facilities will soon produce a "data avalanche", beyond the scope of many users to be able to handle the data. As a consequence, infrastructures for efficient and sustainable data management and analysis are required so that the experimental results can be effectively used by the researchers who conducted experiments; users can easily share and combine their data, software and other resources, and can also meaningfully share them with the wider research community.

The instruments and experimental techniques may differ between facilities; the purpose of research may be pure scientific inquiry, or more practical in view of industrial applications. What is common across facilities science is a business model for servicing the facility users (researchers); the users' social habits, e.g. the accepted modes of managing research output, are less definitive but also important. These commonalities lay a foundation for a generalised data lifecycle in facilities science, as well as for common metadata models and information systems architecture to support common approach to data collection, handling, analysis, archiving, publication and access. Thus when considering how to provide infrastructure to support facilities-based science, it is helpful to consider the whole of the research lifecycle involved, from submitting applications for use of the facility, through sample preparation and instrument configuration and calibration, through data acquisition and storage, secondary data filtering, analysis and visualisation to reporting within the research community, informally and through formal publication. By taking an integrated approach, taking into account the provenance of the data (e.g. creation, ownership, history), the infrastructure can maximise the potential for science arising from the management of data and data context across the lifecycle. Such an approach can ensure that data do not remain just archives of bits but are included in the evolving process of collaborative knowledge acquisition.

THE FACILITY SCIENCE DATA LIFECYCLE

Independent of specific disciplines there is an acknowledged "life cycle" of research which covers distinct stages in the research process. The agents and relative importance of the steps will depend on the role of the agent being considered. Figure 1 illustrates a researcher's view of the process; a funder, a research organization, a publisher or a preservation institution will focus on different aspects of this life cycle and bring additional steps relating to their business process and requirements.

We discuss in detail the "daily life" of a facility, how scientists interact with the facility and its various actors, and how they develop their science to derive new knowledge. This analysis is a foundation for sophisticated data modeling and open infrastructure design that we consider later.

Overview of Facilities Lifecycle

The stages of a simplified facility's experimental lifecycle are illustrated in Figure 2 and can be summarised as follows.

Proposal: For the user who intends to conduct her research using a large facility, it quite often starts from looking into her research field, e.g. materials science, and identifying opportunities presented by different research facilities compared to the experimental equipment the researcher may immediately have at her disposal, e.g. in her own university department. The facilities are typically oversubscribed with more users applying for the time on the instruments (often known as "beamtime") than can be actually offered; the specific physical conditions that may be required for a

Figure 1. A researcher's view of a facilities lifecycle

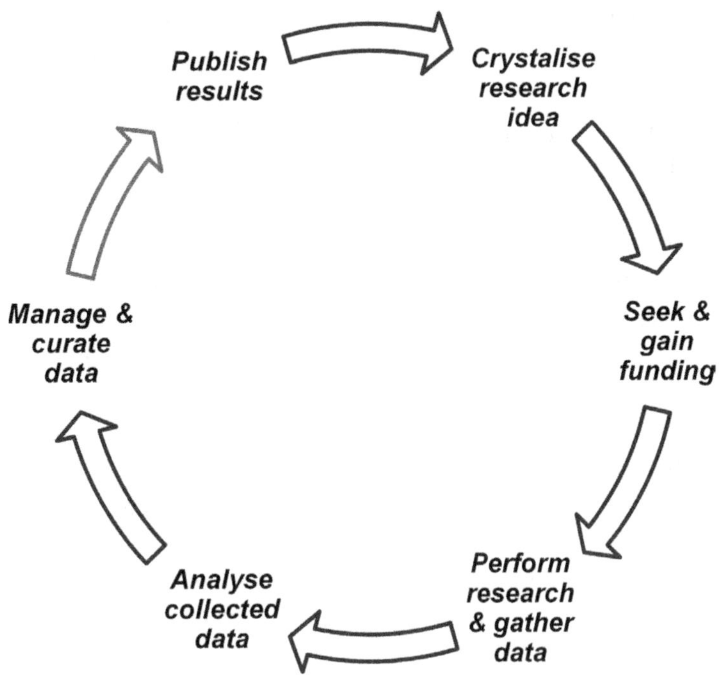

Figure 2. Generic research lifecycle in facilities science

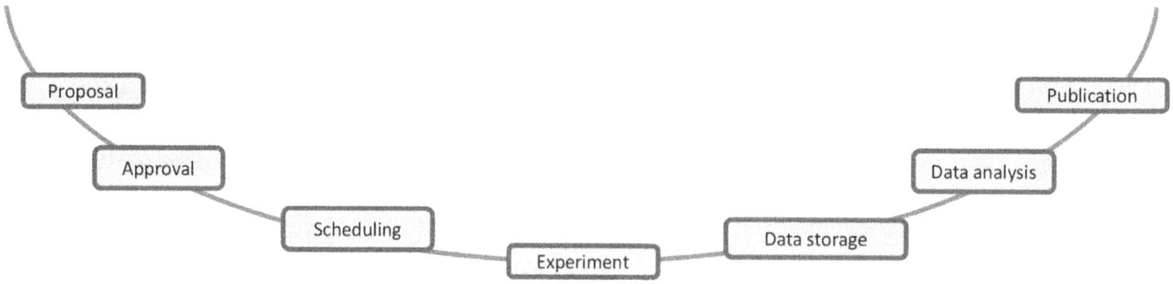

particular experiment, such as keeping samples in deep vacuum, or ability to apply substantial pressure to them, or parameters of the beam (energy, polarization) further contribute to the sometimes severe "niche" competition among facility users. The colleagues' opinion or their experience of dealing with a particular facility, especially discussions with individual instrument scientists, may play an important role which facility to actually choose for the experiment if there is a choice of comparable experimental techniques on more than one. Financial considerations may play role in making a decision, too, e.g. facilities that typically cover travel and living expenses, as well as the cost of beam time, may appear to be a more attractive choice even if the same sort of research can be conducted on the equipment in a local laboratory.

Then the user compiles a proposal for her research, according to the rules and formats imposed by a particular facility, applying to use a particular type of instrument on the facility for time to undertake experiments on particular material samples. This is lodged with the Facility.

Approval: Competition for the beam time may be harsh so facilities typically organize a weighted review of the research proposals they receive, considering their scientific merits and technical feasibility. Factors considered include:

- The research value of the proposal, that is the likely impact of the science derived from the use of the facility.
- The suitability of the facility experimental environment for the applicant's research needs, that is the likelihood of the chosen instrument being able to generate "good data" which can be used to derive a scientifically valid result.
- The availability of time slots required for the experiment.
- Risk assessment as the suggested experiment may require the use of hazardous samples or materials.

Scheduling: Successful proposals are allocated a time period for the use of an instrument within an operating cycle of the instrument and offer a user agreement to sign. Facilities typically require the registration of the researchers who will undertake the experiment, including any Doctoral students who are within the principal investigator's team. These experimenters will need to be trained in the use of the instrument and familiarized with the strict health and safety regimes which control their use. Further, the samples being analysed will also need to be described; these may include dangerous items (e.g. toxic, pathogenic or radioactive materials) or require special handling (e.g. cryogenic temperatures, or extremes of pressure), for which the facility will need to prepare. There may also be an agreement on the use of data, reflecting the policy that facility imposes on its users and may reflect the needs of other researchers and of the larger society; as an example, the policy of ISIS facility is that the results of non-commercial experiments conducted for public money should be released as open data after three years from the experiment date (ISIS, 2011).

Experiment: The user and her team will visit the facility to use the instruments for the allotted time, typically over a few days. The user will bring the samples and perhaps some additional equipment to the facility, and will work on the instrument, in conjunction with permanent facilities staff. Samples are placed in the beam, and during the exposure the beam particles are then blocked or deflected by the sample and then detected by banks of sensors arranged around the target area. These sensors then generate data on such parameters as particle counts, angle of deflection, time-of-flight of the particle, energy, or frequency and a series of measurements are taken. This may be an iterative process as the calibration and configuration of the instrument is adjusted; these are highly specialized and advanced machines which may require much adjustment in order to get good results, and not all samples may be equally suited to generating good data. Different instruments at the facilities have their own characteristics, but all have data acquisition software which will take data on the parameters of interest.

Data Storage: Data is aggregated into data sets associated with each experiment, stored in secure storage, within managed data stores in the facility, and systematically cataloged in local storage for further use. For the data collection that can be demanding in terms of data volume and velocity, facilities typically offer their computing and data archival services.

Data Analysis: The scientist takes the results of the experiments (the "raw data"), and carries out further analysis, as the results of the experiment are not usually ready for the immediate publishing or re-use. The data from the instruments often needs highly specialized interpretation to derive the required end result, typically a "picture" of a molecular structure, or a 3-D image of a nano-structure. There may also be data artefacts generated by the instrument itself which must be factored out by comparison with calibration date. Thus there may be a sophisticated data cleansing and data analysis as well as the involvement of other specialists such as theoreticians who "fit" a reasonable model, or a range of models, to the experimental results. It is common that experimentalists collaborate with theoreticians who they know well and who may reside in their home institutions across the globe. This analysis process may take a considerable amount of time (years is not unusual) and require intellectual input in refining algorithms.

Publication: Once a suitable scientific result has been derived from the data collected, then the scientist will report the results within journal articles that include results validation and often comparison to other results. The facility would like to be acknowledged, citing the instrument used, and informed of its publication, so that it can track the impact of the science derived from the use of its instruments.

Thus there is a Data Continuum from proposal to publication where data and metadata are managed together as a record of the experimental lifecycle of an experiment. This continuum is a basis for data modelling and various data management techniques applied by data management and information management specialists. Such themes as data visualization, data annotation, long-term digital preservation, data linkage and enrichment, are most prominent in this field.

The lifecycle of a facilities experiment illustrated by Figure 2 is thus a key concept for capturing the research output of a facility. We distinguish these facilities experiments from other types of experiment by calling them *Investigations*; an investigation is a set of experiments associated with the allocation of time on an instrument associated with a facilities proposal.

Actors Involved in the Lifecycle

Facilities science is in its nature a collaborative research effort with various stakeholders involved at the various stage of the lifecycle. The major actors involved in the lifecycle include:

The Experimental Team: A group of researchers who propose and undertake the experiment. These are typically from Universities or other publically funded research institutions, although a percentage will come from commercial use of the facility by industry. This team would typically be led by a *Principal Investigator* and would have expertise on the sample under examination within the experiment, its chemistry and properties. They may have some knowledge of the analytic technique being used to perform the experiment (e.g. crystallography, small-angle scattering, powder diffraction), but typically would not have detailed knowledge of the characteristics of the instrument, relying for this on assistance from the instrument scientist.

The User Office: A unit within the facility dedicated to managing external users of the facility. User Office staff and systems will typically register users, process their applications for beam-time, guide them through the process of visiting and using the facility, including managing any induction or health and safety processes, and collate information on the scientific outputs of the visit.

The Instrument Scientist: A member of the facility's staff who has specialist scientific knowledge of the capabilities of a particular instrument or beam-line and its use for sample analysis. They will typically advise and assist with the experiment on the instrument.

Other actors involved may include: *approval panels*, formed by scientific peers and charged with assessing proposals and allocating time on the instruments; *facility libraries*, which may collect information on resulting publications; *facility infrastructure providers* who maintain computing and data infrastructure within the facilities; and *facility operations staff* who manage the physical operation of the facilities, the moving of equipment, handling samples and chemicals, running the facility's source of beam. Further stakeholders in the use of the facility which should also be considered include: *collaborators*: researchers from different organizations, who join together for projects, with particular objectives and funding streams; *funders* who sponsor a particular research project or programme, and request a report; *suppliers*, that is experimental equipment, software and service providers; *research re-users,* including business and other consumers, perhaps of a non-commercial nature who may want to re-use the results of research; and *research output managers*, including journal publishers, data centre specialists, librarians, patent officers, which may be employed directly by the facility, or within other institutes which interact with the facility. Further, the facility itself has a *funding sponsor*, typically at a high level in government, which wishes to see that public investment is well spent in terms of research and ultimately economic impact in the use of the facility.

From the perspective of computing and data infrastructure, we can distinguish between *internal users* and *external users*, where internal users are facility staff including the user office managers and instrument scientists on the facilities staff who are within the facility's organisation and have daily interactions with the computing infrastructure, while external users are the end-users of the facilities. Facilities maintain very close working relationships with their user communities, through their normal operations, often working with the same experimental teams. Further, facilities have frequent consultative activities with external users, such as user group meetings, newsletters, mailing lists etc. Consequently, facilities have close knowledge of the needs and priorities of external users. Both groups are users of the computing and data infrastructures, the internal users using the infrastructures on a day-to-day basis, and the external users who interact with the infrastructure to expedite their work through the system and generating the results. Thus both of these groups have a stake in the data infrastructure.

The involvement of different stakeholders varies depending on the stage of research, e.g.

business and other consumers sit on the very end of the research lifecycle. There may be interesting interferences and even overlaps between different types of stakeholders; they should be thought of more as roles rather than separate human beings or groups. As an example, the team who install and tune experimental equipment on a synchrotron for the guest users may conduct their own experiments, too. Also an important notion is that facilities are themselves funders covering substantial parts of the total research costs. However, every stakeholder can be considered a part of the same research value chain based on a common infrastructure with typical patterns of the research conduct and of sharing its results.

SOCIAL AND COLLABORATIVE FACTORS

A number of social and collaborative factors characterise the changing nature of the scientific work undertaken within facilities. We discuss some of these factors here; then we consider how they affect the nature of the data modeling and data infrastructure required.

Changing Landscape of Facilities Science

The facilities business model, the technology and the patterns of the facilities users' social interactions are all interrelated and result in new challenges and new opportunities for the facility science stakeholders, specifically for data curators and information technology specialists.

A business model for user research on large facilities that emerged over the last 50 years has been influenced by recent developments. Instrumentation and data analysis have become more user friendly than in early days of facilities science. This has led, among other effects, to a lesser significance of the instrument scientists with a current trend of not including them as the authors of research papers; the estimate e.g. for biology papers is that about half of them do not now include any facility staff members as co-authors (Mesot, 2012).

The advances of instrumentation and Internet have also led to the emergence of specific services for research and industry such as the UK National Crystallography Service that allows users to send their samples for remote investigation according to one of the service plans (Coles & Gale, 2012). The sample exposure on a large facility like synchrotron radiation source may be just one of the experimental techniques included in the service plan so that users have a "seamless" interface for the multi-aspect investigation of a crystal substance submitted. The service provider then collects all the experimental data and supplies them to the user in pre-agreed formats. The facilities themselves have also started offering this sort of "express" service with the user personal presence not required for the conduct of experiment.

The users' attitude towards research may also have a significant influence on the research lifecycle and services in support of it. The experiments undertaken in these facilities are of growing complexity, and they are increasingly done by international research groups and many groups will combine data from experiments undertaken in more than one laboratory. The PaN-data collaboration (Photon and Neutron Data Infrastructure: http://www.pan-data.eu/) of 13 European facilities has sought to substantiate the anecdotal evidence that user frequently used more than one facility. In a exercise where the number users in common between facilities were estimated, it was shown that 22% of all users within Europe use more than one neutron or synchrotron radiation facility. Further, considerable numbers of users use both neutron and photon sources, gathering information which highlight different characteristics of the sample, and typically, 30-40% of the users of any of the photon or any of the neutron sources also use at least one other photon or neutron facility (PaNdata Collaboration, 2013). The reasons for this substantial level of facilities sharing are as often of a research nature as the characteristics of the experimental environment are different between facilities; the level of sharing is indicated in Table 1.

Another driver for change in data management and data curation is the emergence of new experimental techniques like neutron tomography, or using robots for manipulating multiple samples exposed to a synchrotron beam, or studies of dynamics of materials. The new techniques produce larger volumes of data making "Big Data" bigger than ever. This is causing significant problems as the traditional way of operating, whereby end users take a copy of the data with them to their home institution for further analysis is becoming impractical; the network bandwidth is not sufficient to practically download the data, and the home institute may not have the storage or compute capacity to effectively handle this data; the full implications of this are beyond the scope of this paper, but they need to be accommodated in the future data infrastructure provision. On the other hand, these techniques also raise potential opportunities for researchers to perform comparative and multi-aspect studies for the same samples using different experimental techniques, or using the same experimental technique for much wider variety of different samples. These trends appeal to providing a richer, well annotated and linked context for experimental data across different facilities, different experimental techniques and different sample types so that the mentioned research opportunities for comparative and multi-aspect studies could turn into reality.

Challenges and Opportunities for Knowledge Management in Collaborative Research Environment

The changing landscape of facilities science leads naturally to related challenges and opportunities in the knowledge management processes of the facilities research environment. We discuss some of these challenges to data and information infrastructures below.

The change of the instrument scientist's role who as mentioned does always receive a due credit for their job of preparing sophisticated experiments requires re-thinking of the attribution methods for research papers and other research outputs such as datasets. Facilities science may look at the developments such as role-based attribution in other fields of research (Marcos, Vara, & de Castro, 2012); this is just one example of how facilities' information departments could explore the new knowledge management culture elsewhere, and promote the best practices of it across facilities science. This example also shows that knowledge management is in fact a responsibility of everyone involved in research data lifecycle: the authors themselves, not any curation unit down the research

Table 1. Sharing the European neutron and photon facilities (as measured by PaNdata in 2011)

Total number of unique users	30873	
Using only neutrons	6719	21.7% of users
Using only photons	22021	71.3% of users
Using more than one facility	6863	22.2% of users
Using more than one photon source	4252	17.6% of all photon users
Using more than one neutron source	1734	19.6% of all neutron users

results distribution road, should be able to add the structured description of roles according to a reasonable metadata standard.

Information departments then can be seen as hubs or centres of expertise which monitor, refine, and communicate best practices of data curation and knowledge management for other stakeholders (research papers authors in the last example). The consistent and clearly formulated framework will make a collaborative data curation effort much better defined and communicated, and the best data curation and knowledge management practices more readily adopted by the research community. Supervision of various kinds of information through the research lifecycle will help then to create rich data aggregations and reproducible research workflows with contributions naturally made by different lifecycle stakeholders.

The next challenge and opportunity is presented by the emergence of research services such as the UK National Crystallography Service. This trend raises questions on the user management, research proposals management and data management in facilities science. Just one example is the future role and the content of data management policies that some facilities tend to impose on their users as a pre-condition for getting a facility resource for research. The policy may ask users to agree with the public release of their experimental data after a period of exclusive access (typically a few years), or contain the requirement to submit the list of resulting publications back to the facility user office. This works well in a traditional business model of facilities science but does not take into account the emergence of the service intermediaries who may need to be a subject of the data management policy, too, so that it becomes a multilateral agreement.

The data management policy format which is now just plain text is also questionable as it is not interpretable without a human; this will be likely not enough for the automated research proposals management and data release management across different facilities. The development of licences for data re-use, or the adoption of suitable ones could alleviate the problem but licences themselves might need a proper machine-oriented modelling for policy enforcement; the indication of what is possible in respect to structured modelling and automation of data licences can be seen in the recent formation of the Linked Content Coalition (www.linkedcontentcoalition.org) endorsed by the European Commission and some national governments. Again, information departments of large research facilities might consider borrowing the advanced practices and models of data licensing for their re-use in facilities science.

Another important consideration is the interoperability of metadata models and their actual implementations for different research facilities. A metadata model for facilities science that we call Core Scientific MetaData (CSMD) (see Matthews et al., 2009 and Matthews, Kourousias, Yang, & Griffin, 2012) is based on the research lifecycle as discussed and is implemented, with some modifications, in the ICAT software platform (Flannery et al., 2009).

The different stages of research lifecycle produce data artefacts (research proposals, user records, datasets, software, publications etc.) that are similar across research facilities so having a common metadata model like CSMD seems sensible. However, it may be applied differently by different facilities; there are several CSMD implementations in data catalogues across Europe by virtue of the ICAT platform but the model, and the actual use of its elements may vary among implementations. This may result in extra design and implementation overheads when we consider federated services for a few facilities (even when based on the same software platform), also there is no guarantee that once we have the federated solution agreed and implemented, it will be not affected sooner or later by the diverging business needs of different participants. The common data curation framework for facilities science might help to have these needs permanently monitored, properly communicated and effectively reconciled thus serving as a well-structured business analysis wrapper for technology solutions (Bunakov & Matthews, 2013).

An interesting development that may be considered a part of the emerging data curation framework but has exposed certain challenges, too, is the recent effort of minting Digital Object Identifiers for investigations performed on ISIS neutron facility (Wilson, 2012). Having permanent identifiers minted for particular investigations (experiments) should be enough for linking them to datasets and publications but in order to have a structured and linkable representation of a facility research environment other parts of it such as scientific instruments, experimental techniques, people, organizations, software, derived data sets etc. need minting or borrowing identifiers for them, too. There is currently no sustainability model for this activity, as well as for the steady production and support of landing Web pages where the permanent identifiers (all kinds of them) should ideally resolve into. The different aspects – modelling, technological, operational – of the permanent identifiers management should be an important part of the emerging data curation framework for facilities science.

We should also mention organizational barriers to sharing the content and the context of the research discourse: grant applications, facilities beam time applications (research proposals), the raw data collected, the research outputs, the models and the software used for data analysis or long-term digital preservation – all these components tend to be managed and published under separate ownership but can and should be interlinked and navigable in order to get the most of the impressive resource spent on the preparation and the actual conduct of facilities research. Linked Data might help here, and it proved to be a productive methodology for processing Big Data in some important research fields with industrial output such as drug discovery (Dumontier & Wild, 2012). There are even more advanced data modelling techniques for sharing the reproducible research workflows that are well accepted in some research domains, e.g. biology (Bechhofer et al., 2013). However, these techniques typically cover only certain parts of the larger research lifecycle that are immediately related to research work, with the Researcher as a major target of data linkage and data sharing. The needs of other stakeholders residing in education, industry, research management and funding, or policy making are underrepresented and do not have a consistent framework where all of them, along with the researchers and intermediary services, could fit in.

In the absence of a structured data curation framework, there is a risk that the information departments of large facilities are confined to supplying the technology solutions and IT services when their next role could be that of a conscious data curator and knowledge manager helping to increase data value across the entire research data lifecycle for the variety of stakeholders (Wilson, 2012); information technologies and services would be then a very important means to underpin the data curation role but not the end in themselves.

In order to adopt this new role, the information departments of large research facilities cannot entirely rely on the existing organizational structure as their role and actual influence in a larger research context is inevitably limited. What they can do is devise and elaborate a common framework for sharing the existing best practices across different organizational units and collaborative projects; the framework will also serve to bring the best practices from elsewhere for the adaptation to the needs of facilities science (Bunakov & Matthews, 2013). The projects, initiatives and working groups that the information departments are involved in will be a means to support certain "themes" in the common data curation framework. This should result in better opportunities for the organizational units and collaborative projects to interoperate, to reconcile their priorities, and to set common (and commonly understandable) goals.

Interaction with the User's Scientific Lifecycle

We cannot consider the facilities data lifecycle in isolation, but only in interaction with that of the user scientist, which is much more unpredictable. We give a simplified view of this lifecycle in Figure 3, derived from that of the Integrated Infrastructure for Structural Sciences (I2S2) project (I2S2 Partners, 2011). This is idealised; research rarely progresses in such straightforward well defined stages, with many interleavings and iterations of sample synthesis, experiments and data analysis for example, while literature review is an ongoing process. We also don't give a true impression of the time scales here; sample synthesis can take much skill and expertise in the laboratory, with many abortive and slow attempts to get a quality sample for use in the experiment, and much of the researcher's intellectual input may be in this stage. However the stages are all usually undertaken within a researcher's progress through a project and provide a conceptual starting point; the lifecycle for structural sciences such as crystallography is represented by Figure 3.

The facility's lifecycle above only interacts strongly with the user's lifecycle in a limited number of places, in the experiment execution and raw data collection (and not all the experiments may be at the facility, as there will be bench experiments, or experiments undertaken at smaller laboratory facilities). Thus only a relatively small part of lifecycle is under the direct control of the facility.

Once data is collected, there may also be a prolonged period of analysis, where data is repeatedly processed via various pieces of software, many of which will be home written, and via a process of iteration, data combination and

Figure 3. An idealised science lifecycle for the structural science

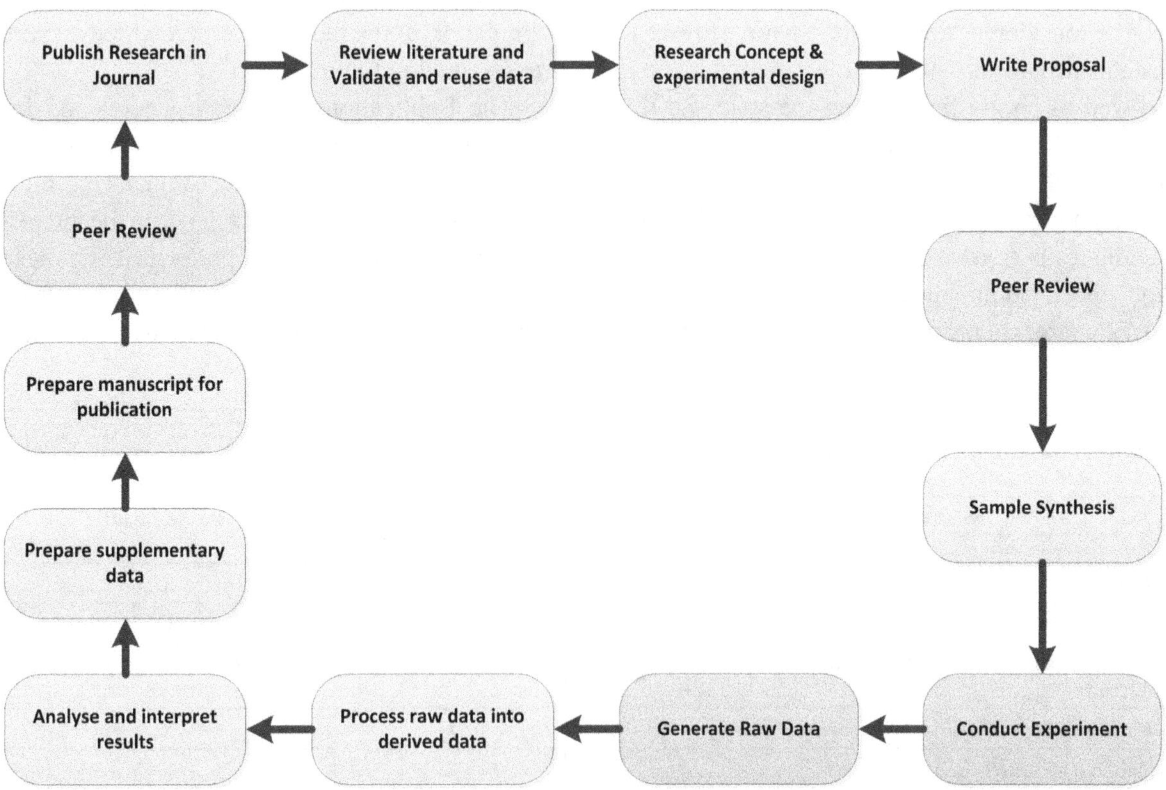

filtering, a final well-expressed result suitable for publication should, with good luck and the skill of the scientist, be derived. This is a highly complex process to track as this activity takes place out of the highly developed infrastructure of the facility; the complexity of this process is detailed in (Yang, Matthews, & Wilson, 2013). In that paper, the problem of supporting this process for a facility user is discussed, and the use of provenance to record the progress of analysis and make it available for reproduction is explored. We do not consider this further here, except to note that mechanisms for a future data infrastructure should support the process downstream of the data collection (see for example van Dam et al., 2011).

Entities in the Research Lifecycle

It is important to consider the types of entities that circulate in the research lifecycle. One of the popular entity types is represented by journal articles, e-prints, reports and other similar artefacts that irrespective of their physical manifestation can be considered "documents": they are the well-established means of research communication, as well as an aid for tracking the state and the trends of a research discourse. The "documents" have clear identity, allow review (of different kinds) and participate in citation networks; this facilitates performing the functions of the quality scholar communication and the measurable research tracking and evaluation; this also makes the "documents" valuable intellectual entities worth capturing in library catalogues, and sharing via advanced information services. Similarities between "documents" (publications) and investigations are outlined in Table 2.

The Investigations (that is the experiments associated with a facilities proposal) in facilities science have some essential features of their lifecycle similar to the document-like entities and are the natural candidates for being first-class members of a research discourse. The Investigation can in a way "cite" other Investigations: researchers applying for the beam time quite often refer to previously performed Investigations; the Investigation can "cite" research papers, and research papers can refer to Investigations they are based upon.

The items of software are another important type of information entities involved in the common knowledge management discourse on research facilities; the software can be "cited" in research papers, and is needed for the very interpretation and rendering of the experimental results. The major types of intellectual entities circulating in the facilities research lifecycle are represented in Figure 4.

The Publications (research papers) and Investigations can be considered parts of research discourse in a particular subject domain, e.g. biology, chemistry or materials science in the interest of which the experiment is performed. To make sense of these intellectual entities, one needs specialist knowledge in the subject domain. Datasets

Table 2. Similarities between publication and investigation

Feature/Aspect	Document (Publication)	Investigation
Is an intellectual entity	X	X
Is a subject of peer review	X	X (via proposal approval)
Can cite all significant related intellectual entities of a research discourse.	X	X (via research object – see below)
Citation chains (steps of discourse) observed	X	X
Universal identifiers "mints" available	X	X

Figure 4. Major intellectual entities of the facilities research lifecycle. The arrows represent a typical direction of the well-formed references ("citations").

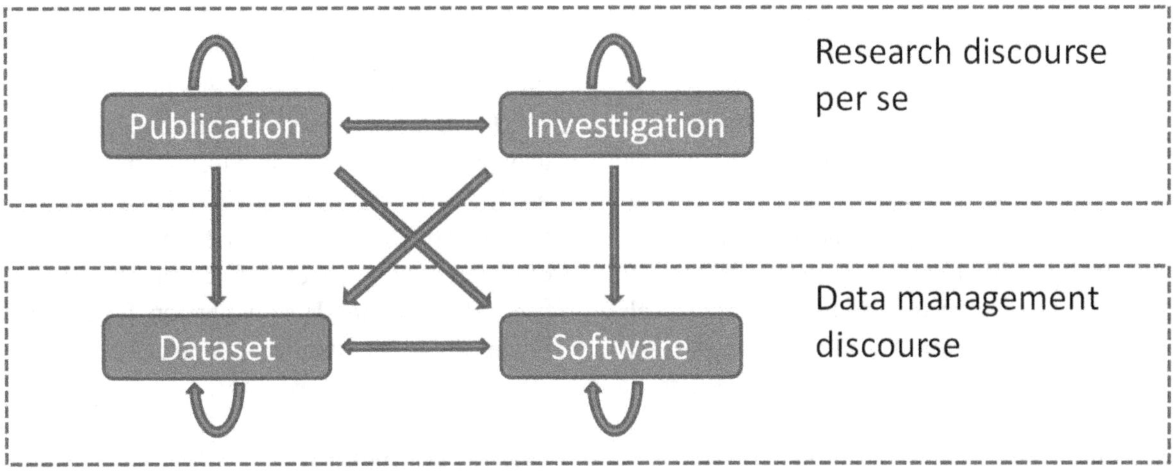

and Software both support the specialist research discourse and can be considered belonging to the data management discourse where the IT specialists and knowledge management professionals play the major role. Both layers are mutually interacting and interfering parts of a larger (common) research discourse in facilities science. Investigations that emerge close to the start of the facilities research lifecycle can be considered "seminal" intellectual entities for growing complex Research Objects around them; we elaborate on this in the following sections of this chapter.

DATA PRODUCTION AND KNOWLEDGE TRANSFORMATIONS

As discussed above, raw data is then streamed off via data acquisition and data management systems which collect, aggregate and move the data to short or long term storage to await further analysis. Traditionally, this process has been carried out with standard file systems; however, to cope with the increased complexity and volume of data, an ICAT data catalogue system has been developed which captures the information associated with a facilities investigation (Flannery et al., 2009). This cataloguing component, based on an information model capturing a view of a facilities investigation, provides a common point of gathering information about the experiment. This captures information on the experimental team and intent from the proposal system, and when the experimental visit takes place, will register the data sets, their locations in storage, and experimental parameters. This information is then exposed via an API, either for users to use for browsing and data download on or off site via a web interface, or else integrating with analysis tools and frameworks so that they can search for and access the data directly.

This existing infrastructure offers the core capability for managing data; we discuss how this can be extended to capture a more complete picture of the facilities investigation. We consider how facilities data management infrastructure can evolve to support a more complete view of the facilities experimental lifecycle described. Further, we also need to take into account the changing social landscape of facilities, which requires a more complete and transparent contextualization of the experiment, with appropriate accessibility and attribution.

Linking and Publishing

A research project produces many outputs during its lifetime; some of these are formally published and disseminated with a view to advancing the accepted knowledge about a topic, some of these will relate to the administration of the project and some to the steps in the process. They may include tools which capture the research process informally such as blogs and electronic laboratory notebooks; social media and dissemination such as Facebook, domain specific interaction sites and Twitter and maybe even alternative impact mechanisms described through altmetrics (www.altmetrics.org). With the changes in the environment due to Open Access publications and Open Data and the increasing expectation that results will be able to be validated by others, there is a growing need to enable the electronic navigation between these different outputs and to provide assurance to the persistence of these objects and their links.

It can be said that these objects and their relationships form a graph of connections and this will be the same regardless of who is intending to use them, but the relative importance of the objects depends on the viewpoint of the potential user, so the same underlying graph may be displayed in alternative views depending on the audience so for example the link to the funding grant from a piece of data may be important to the funder but not to the publisher of a journal article based on the data.

One method of ensuring the persistence of data is the concept of data publication and this is gaining traction in some disciplines to support the research process. This involves enabling public access to data with appropriate guarantees of integrity in the management and persistence of the data, and encouraging researchers to cite the use of the data within publications. The intentions behind data publication include: assigning credit and recognition to the collectors of data; encouraging the inspection of data by peers to assess the quality of the data, and to validate the assertions of scientific insights claimed in published articles arising from the analysis of the data; enabling the reuse of the data by other researchers to re-analyse to discover new insights and reportable results, thus furthering the value of the research which arises from the data collection. As a consequence, a number of different approaches and infrastructures have been advocated for data publication, for example Lawrence, Jones, Matthews, Pepler, & Callaghan, 2011 or Parsons & Fox, 2013.

Contextualisation

Some facilities, for example ISIS and ILL, are already using Datacite (www.datacite.org) Digital Object Identifiers (DOIs) in order to publish data arising from the experiments for citation and reuse. The choice of DataCite DOIs rather than other alternative systems for persistent identification was due in part to the trustworthiness and familiarity of CrossRef DOIs used in the publication world. The established use and trust of DOIs has transferred to the DataCite DOI system, which is backed by national libraries and other memory institutions around the world. While this it data citation is still at an early level of maturity, and norms for citation and bibliometrics have yet to be established, Datacite has gained acceptance by researchers so that they are prepared to use DOIs both for publishing their data and also in citing the use of other data.

The publication of the experimental data with the basic Dublin Core fields that DataCite expects provides a mechanism for the location and identification of specific data, however it does not guarantee that the data once found will be usable. For this to be achieved, it is important to collect the contextual information which enables the end-user to make sense and re-use the actual data file retrieved. This additional context depends on the academic discipline and the knowledge of the new user, but for Facility Science may include:

- Information about the Principle Investigator,
- Information on the proposal for the experiment including: title, scientific aims, prior art, funding sources,
- Information about the sample used,
- Information about the funding of the wider research project this experiment fits into,
- Which instrument has been used (and the date & technical details),
- What primary/raw data have been generated,
- What publications have been generated,
- What analysed data have been generated .

The two major approaches to data contextualization can be referred to as "predesigned" and "opportunistic". The "predesigned" approach relies more on a sophisticated, possibly multi-level metadata model that helps to discover data, see its relation to different organizational or intellectual entities, and finally interpret it in a meaningful way. Different actors involved in the research lifecycle just fill in the "placeholders" left in the large and universal model for them. The idealised Scientific Research Lifecycle Model (see Figure 3) is an example of such approach in structural sciences.

The "opportunistic" approach relies on simpler models populated with the actual information by collaborative effort of various participants of the research lifecycle so that individual descriptions of the specific parts of a common lifecycle can be used as "building blocks" for assembling the whole range of the sensible research activity "chains" or "networks" (Bunakov, 2013).

Contextual information is not a complete set at the point of creation of the data, it changes and adapts through the use and reuse of the data. Some of the contextual information is only available/known at certain points in time, for example when the experiment was performed, and some can be added or deduced after the event. What is appropriate contextual information is also dependent on the use that data is going to have; trying to replicate an experiment requires different information to say someone interested in the history of science looking at the experimental results.

Research Objects

The notion of Research Object (http://www.researchobject.org/) plays an important role in recent developments of data management for facilities science. The RO gathers information about research artefacts and their attributes and relationships. These are usually represented as Linked Data; thus RDF is used as the underlying model and representation, with a URI used to uniquely identify artefacts. As ROs are linked data objects, they can link into to the existing Linked Data cloud to provide additional context information and be managed by the standard tools of Linked Data and the Semantic Web. In addition they have the following characteristics:

- Standard vocabularies are used to represent research process relationships, such as workflow (workflow4ever http://www.wf4ever-project.org/), provenance (e.g. Prov-O http://www.w3.org/TR/prov-o/), and citation (e.g. cito http://purl.org/spar/cito/). The use of standard vocabularies encourages shared understanding, enables reuse and allows the use of tools which are tailored for their specialised semantics.
- A boundary is provided to the object as an aggregation, so we can determine membership of the research object; typically, OAI-ORE (http://www.openarchives.org/ore/) is used for this purpose.
- The whole research object can be identified via a URI, so its own history and attributes can be related as a first class research artefact in its own right.

The notion of the boundary of a RO is particularly important. A research artefact can be linked to a number of research artefacts. An investigator, workflow or instrument can participate in a number of investigations; a publication may use the output of several investigations to support its results. If this is represented as a simple web of linked data, then it would be difficult to distinguish which artefacts and relationships are members of which research object. We need a definition of a boundary to determine membership of the Research Object; OAI-ORE, with its notions of Aggregation and Resource Map provides such a boundary. Research Objects have been discussed in for example Bechhofer et al., 2013, Shaon et al., 2012, Belhajjame, 2012. Within the facilities context we have developed the concept of an Investigation Research Object, to link the information objects related to an experiment in one unit, representing the facility point of view, as illustrated in Figure 5. For more details see Matthews, Bunakov, & Jones, 2013.

One advantage of this approach is that there is no need to wait until a publication has been generated to start the aggregation of content, it can be started at the time the research artefact is created.

Activities at the start of the research workflow are within a single e-infrastructure provided by the facility and thus are amenable to being utilised as the initial implementation of this proposed framework. By following the lifecycle of a successful beamtime application, we can collect all the artefacts and objects related to it, with their appropriate relationships. As this is strongly related to allocation of the resources of the facility, this is a highly appropriate intellectual unit for the facility; facilities want to record and evaluate the scientific results arising from the allocation of their resources.

The Value of Investigation Research Objects

The use of IROs can enable a number of potential uses and benefits within an integrated open data research infrastructure.

Provide management for research artefacts beyond the experiment. Using IROs allows the

Figure 5. Investigation Research Object after a complete lifecycle

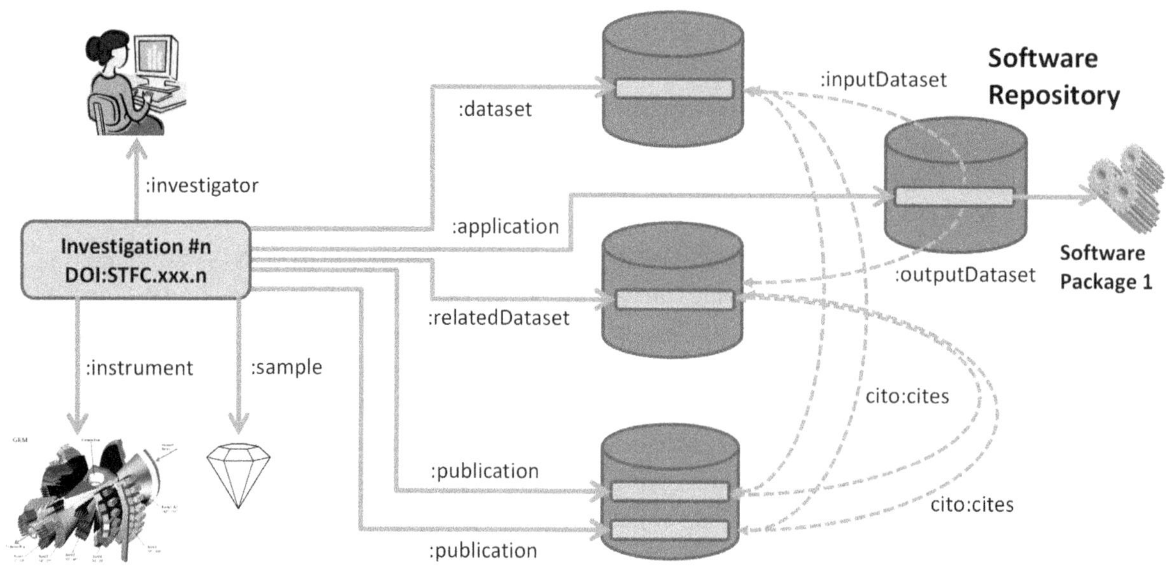

facility e-infrastructure to control the creation of the aggregation, and thus the resource map describing the Aggregation will be an authoritative collection of related research artefacts. We identify the aggregated resources using its own URI for the Aggregation (locking in the context). When other Resource references the individual Aggregated Resources or the Aggregation itself, it uses the IRO's URI. This helps to direct web traffic to the facility hosted services, as the Resource Map facilitates discovery and enable (point and click) access to the Aggregated Resources on or via facility server.

Further, exposing an Investigation DOI (Aggregation) as a compound research object allows us to reference all the related research artefacts using the same DOI. There is no need to register separate DOIs for its Dataset or Datafile as these objects can be referenced as an Aggregated Resource within the Investigation DOI (Aggregation).

Consistent presentation of the research artefacts can be supported, as the facility can use content negotiation to match what to return for a request, thus supporting multiple views and formats.

Facilitate e-Scholarship. The use of IROs provides a context to bind the distributed research artefacts together, with all Aggregated Resources are linked within the context of an Investigation, and unambiguously define the relationship between the resources within the Investigation. This can aggregate heterogeneous resources on the web independent of repository containment, including provenance info on the Aggregated Resources. This allows the traceability of research results to be followed, in both a human and machine interpretable manner, providing a basis for validation of results. Thus IROs provide the appropriate unit of research discourse for facilities science to support e-Scholarship, in analogy to published documents, as discussed above.

Further, publishing the IROs on the web improves discovery as they can be consumed by web crawlers, data mining applications, and be made available as part of the Linked Open Data network. The rich semantics and metadata should facilitate interpretation and re-use, while the use of ORE means that the Investigation and its related research artefacts is packaged under one URI, and thus the compound object can be referenced and annotated directly, to facilitate exchange. Usage and citation statistics can be gathered on an Investigation Object level, so we can gather more accurate statistics of use in context.

Support multiple viewpoints. Regardless of discipline there is an acknowledged "life cycle" of research, which is realised in many ways depending on the audience and purposes; a researcher, a funder, a research organization, a publisher or a preservation institution will focus on different aspects of this life cycle and bring additional contextual links relating to their business process and requirements. For these different stakeholders the central object to which context is added will be different as their world viewpoint is different, for example a publisher will want to establish links from the publication; a funder may wish to do the same for grants. We have described building the links to the investigation from our viewpoint as a facility which is responsible for the creation, discovery and curation of the investigation undertaken at the facility.

The use of research objects supports well this notion of different points of view. Publishing data within a linked open data context in particular makes notions of what constitutes a coherent viewpoint of relevant resources and relationships hard to capture. By providing boundaries and criteria for membership, research objects can support multiple points of view within one data infrastructure. Thus different stakeholders can construct, use and reuse the context relevant to them, and also be credited to the portion of the object which is appropriate to their contribution.

Support data publication. We would propose to use investigation research objects as the unit of publication for our facilities data. Thus we would identify investigation and their related resource maps by persistent identifiers, and use them to generate a landing page. This would be extensible to provide access to the research object in its entirety and include related entities to provide more information in context which could be accessed by other automated agents. Metadata associated with the DOI would need to be changed. Currently, the Datacite metadata field ResourceType supports Dataset and Collection (amongst others), neither of which is correct in this context (DataCite Metadata Working Group, 2011). We would propose that the list of allowed values for this field is extended to include the notion of experiment, study or investigation.

Using the notion of research object as a more open ended bounded object raises the notion of what exactly is being published persistently in this case. If we add additional information are we maintaining stability? Research Objects are well suited to notions of versioning, where we can relate objects together as they change, thus keeping the old boundary stable. Further, we would propose to have different levels of assurance in our case. The core information on the experiment (sample, instrument, parameters, raw dataset) would remain constant, with other information being secondary and subject to possible extension; this would made clear in the presentation.

Support data preservation. Shifting the focus from the data to the investigation makes the data preservation activity a more complex one, as it moves from activities relating to the preservation of a well-defined digital object to include not only the digital object but also activities to ensure that the complex linked data is packaged to maintain its integrity and meaning, and links still point to resolvable objects. For preservation purposes it is important that these links are permanent to ensure the integrity of the object.

Outstanding Challenges

There are outstanding challenges posed by the initial developments in preserving data through the concept of Investigation Research Objects.

Firstly there is the notion of immutability of the IRO: there are some which are immutable (experiment and raw data) –and there are some which are extensible (supplementary data & publications) – is there a point when these extensible items should be present? The Workflows4ever project Evolution model explored this and it will be interesting to see how this work can be transferred to a facilities context.

Secondly there is the issue of what changes mean that the IRO, with its unique identifier becomes so different that it needs to be considered to be a new entity with a new identifier. An example of this would be when the underlying experimental data is migrated from one format to another, is this same IRO or a different one? Should there be links to both versions even though the fact that a migration has occurred means that there was some preservation risk to the original data?

Using the notion of research object as a more open ended bounded object raises the question of what exactly is being published persistently in this case. If we add additional information are we maintaining stability? Research Objects are well suited to notions of versioning, where we can relate objects together as they change, thus keeping the old boundary stable.

TOWARDS AN OPEN DATA INFRASTRUCTURE FOR FACILITIES SCIENCE

As we have discussed, the user environment in facilities has evolved to support a much more collaborative community. Users move between facilities in search of the best result on the most suitable instrument, and will travel internationally to access these instruments. Indeed, with the provision of "express services", users may not even have to make visit in person, instead delivering their samples via specialist courier services to the facility, where they are subjected to standard procedures to generate a partially analysed data set which can be delivered to the user. Users also use different techniques on different instruments to study different characteristics of their sample, combining data from different sources to give a more complete picture of the sample's properties. Further, the analyses may take place in large, international collaborations, with a sequence of experiments at different instruments. Facilities science is international and collaborative.

Thus Facilities are increasingly sharing the same user communities, and sometimes the same science programmes, over the same samples, and this is a strong incentive for further development of their infrastructure. Facilities science can be made both more efficient and effective if the facilities themselves ease the collaborative process. Facilities can provide common infrastructure to support the user, so the user can experience a similar environment to support their experiments in each facilities, thus making the facility easier to access and speeding the user's journey through the facility, and also making the combination and analysis of results from different facilities more straightforward. Thus one of the major aims of the PaN-data consortium, mentioned above, is a common infrastructure for user management and data management.

The vision is for facilities to standardise and integrate research data infrastructures in order to establish a common and traceable pipeline for the scientific process from scientists, through facilities to publications. Each facility on its own provides a similar science lifecycle as described above, supporting the user through proposal to publication; while the support is different in the details of the systems provided, the overall process is similar. Thus each facility provides each own pipeline, as given in Figure 6, where each facility's investigation goes separately through the full science data lifecycle sequence, from generation of raw data to publication of results a series of common stages to generate a scientific result. By making

Figure 6. Current view of data pipeline at facilities

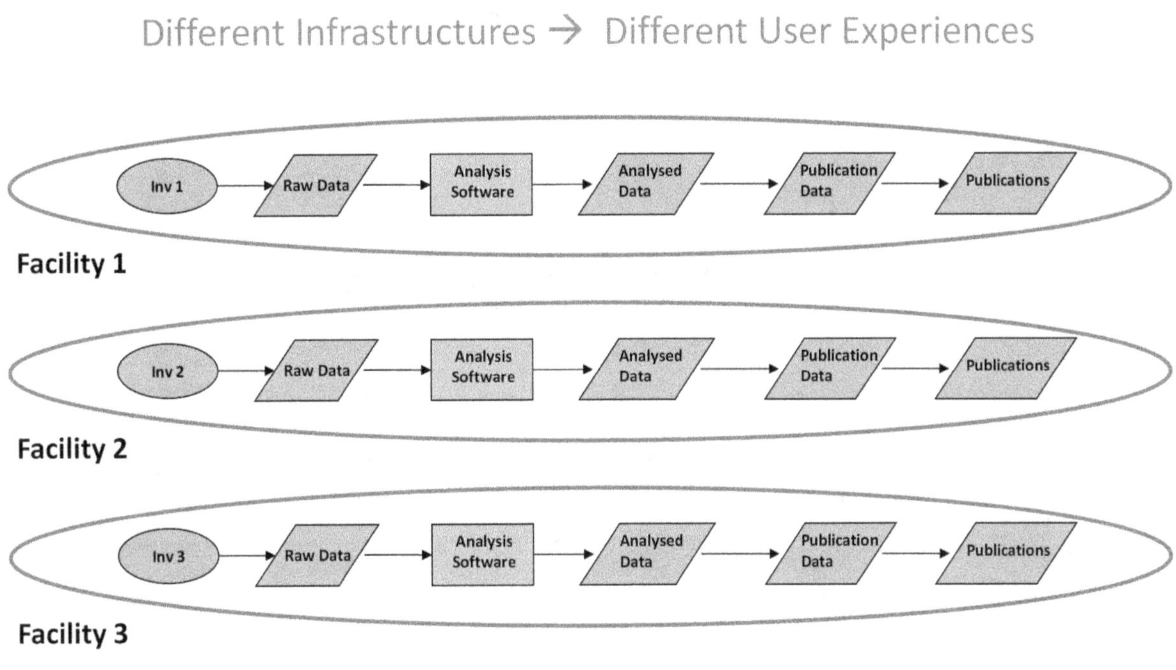

the systems supporting each stage interoperable, by using common data standards and APIs, by making software used at each facility compatible, and by federating the metadata used to record and make accessible the steps through the investigation lifecycle, the user environment can change to that given Figure 7, where the user experience is similar at each facility. Perhaps more importantly, the data can be accessed in common way between facilities, so that it can be published, aggregated and merged together to provide a unified view of the science.

At the heart of the vision is a series of federated systems, providing interoperable metadata which allow scientists to perform cross-facility, cross-discipline interaction with experimental and derived data, with near real-time access to the data. This will also deliver a common data management experience for scientists using the participating infrastructures particularly fostering the multi-disciplinary exploitation of the complementary experiments provided by neutron and photon sources. In the future view, a single user experience is enabled through the use of shared technologies at the different facilities. It is clear that the common data management scheme will offer many synergies and allow completely novel possibilities for aggregating, combining and analyzing data from a multitude of sources, using a variety of techniques to probe matter more deeply and completely.

Figure 7. Future view of the federated data pipeline at facilities

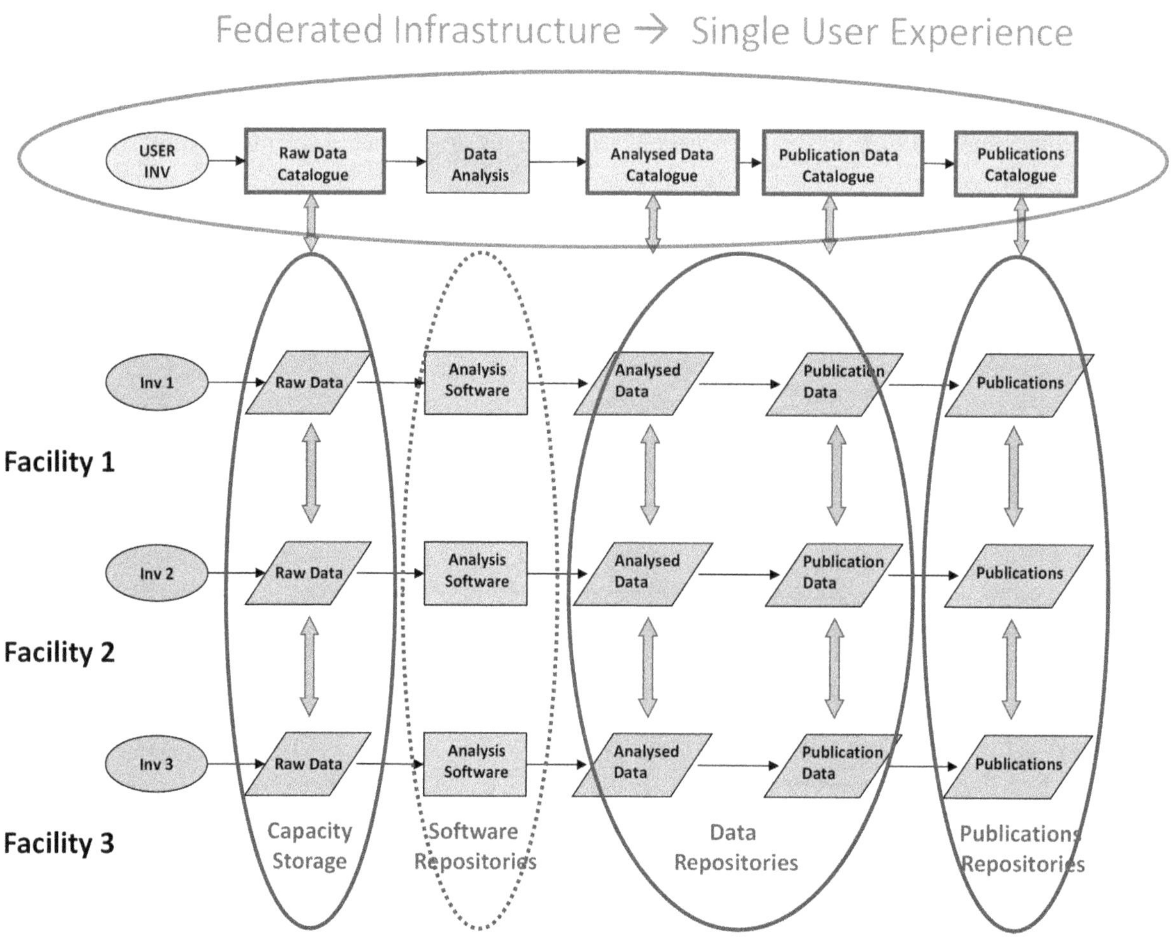

Such an infrastructure will be more reactive to the social and collaborative challenges outlined in this paper, reacting to the higher level of institutional support as data volumes increases with more systematic cataloguing, and supporting international collaboration. Within this infrastructure the notion of investigation research objects provides a unified model of the experimental process which supports multiple perspectives, for cross-facility's and cross-experiment tracking and provenance supporting validation and reproducibility, for data publication, and data preservation. The notion of IROs as presented above can naturally be used to links instruments, datasets, software, publications or other research artefacts, which are in other facilities outside the current investigation, using the existing linked data mechanisms. Common user management systems further this approach by making the association of items to users much more straightforward. This extends to linking to whole investigations, for a compound research object, formed of several individual facilities IROs. These compound research object represent the research undertaken from the point of view of the user scientist; this is represented in Figure 7 as the "User Investigation", although it is not a facilities investigation in sense we have defined above – it is unlikely to be associated with a particular facility, but rather represents a campaign of research, associated with a funding source, a publication, a research theme, or just on a common theme. The ability to accommodate multiple viewpoints over one common linked data graph structure is one of the strengths of the research object approach. However, as this user investigation object is not associated with any particular facility, it may not necessarily be constructed or held at any one facility. It would thus become a function of the federated infrastructure to represent, publish and preserve these objects; alternatively, this may become a function of external international bodies to provide a repository to hold and record these connections. For the development of a healthy ecosystem where Research Objects can circulate, the lessons learned from the years-long data modeling and data sharing effort in other branches of research should be taken into account, in particular MOLES metadata framework for environmental data (http://proj.badc.rl.ac.uk/moles) and research workflows for life sciences (Bechhofer et al., 2013). Yet the structure of the Research Objects for facilities science and the regime of their circulation are likely to be different owing to the different research lifecycle and the specific business model for operating facilities.

Finally, this infrastructure should as much as possible be open. National governments wish to maximise the science output from facilities to justify the high expense of these investments, and there are international initiatives like EUDAT (http://www.eudat.eu) and Research Data Alliance (http://rd-alliance.org/) that promote open data for research on the international level. Funders are increasingly recognising that data arising from experiments, such as the use of facilities, is a public good, with a greater potential economic impact derived from making it available for inspection and reuse than keeping it private. This is becoming recognised and data policies are being developed in facilities encouraging openness, such as the ISIS data policy discussed previously. Thus there is a need to maximise the use of data for the original data collectors, by capturing, organising and presenting it to them in a manner so that it can be analysed with the most up-to-date techniques, and not be a subject of unnecessarily repetition of the experiment through lost or poor quality data. Further, there is an increased recognition that output can be maximised by managing data for the long-term so that it can be reused by future scientists rather than re-doing the experiment. As we have discussed, the notion of investigation research object provides a suitable mechanism to publish and preserve the information collected and maintained in facilities computing infrastructure. However, this is the beginning of a process, and facilities need also to be responsive to the needs of their users. There is still much sensitivity within the user community around priority, intellectual property and control over release of information. Consequently, facilities need to act carefully to ensure that their community is reassured.

CONCLUSION

The business model of facility science discussed in this chapter has evolved over several decades, and has resulted in a thriving international community of researchers, engineers and information management professionals. This model combines the features of "big science", a dedicated large-scale operational infrastructure, with those of "bench science", with many individual researchers and experiments on a huge range of topics. The facility science research lifecycle presented in this paper has laid a foundation for the emergence and the evolution of data infrastructures and software platforms for managing and sharing data, supported by specific metadata models, programming interfaces and workflows. The landscape of facilities science is changing to a situation characterized by the increased sharing of facilities for the same research purpose or different aspects of it, by high level of experiment automation, and by the emergence of third-party services that are in some cases a "proxy" between a user and a facility. This motivates new developments in data modeling and data management that support the knowledge exchange among different stakeholders of the facilities research lifecycle. This is a challenge and an opportunity for information management professionals to redefine their role in the facilities research lifecycle, as well as in sharing the results of facilities research with other research domains and interested parties in a wider society. The ultimate goal is to make the knowledge more accessible for use and reuse, and thus increase the impact of science which arises from the use of expensive investments.

REFERENCES

I2S2 Partners. (2011). *I2S2 Idealised Scientific Research Activity Lifecycle Model.* Retrieved from http://www.ukoln.ac.uk/projects/I2S2/documents/I2S2-ResearchActivityLifecycleModel-110407.pdf

Baier, V. (2006). *Forty years of acting electron-positron colliders.* arXiv:hep-ph/0611201

Bechhofer, S., Buchan, I., De Roure, D., Missiera, P., Ainsworth, J., & Bhagata, J. et al. (2013). Why linked data is not enough for scientists. *Future Generation Computer Systems*, *29*(2), 599–611. doi:10.1016/j.future.2011.08.004

Belhajjame, K., Corcho, O., Garijo, D., Zhao, J., Missiera, P., Newman, D., et al. (2012). Workflow-Centric Research Objects: A First Class Citizen in the Scholarly Discourse. In *Proceedings of the ESWC2012 Workshop on the Future of Scholarly Communication in the Semantic Web (SePublica2012).* ESWC.

Bunakov, V. (2013). Core semantic model for generic research activity. In *Proceedings of 15th All-Russian Conference "Digital Libraries: Advanced Methods and Technologies, Digital Collections" (RCDL 2013).* Retrieved from http://purl.org/net/epubs/work/10938269

Bunakov, V., & Matthews, B. (2013). Data curation framework for facilities science. In *Proceedings of 2nd International Conference on Data Technologies and Applications (DATA 2013).* doi:10.5220/0004593302110216

Coles, S. J., & Gale, P. A. (2012). Changing and Challenging Times for Service Crystallography. *Chemical Science*, *3*(3), 683–689. doi:10.1039/c2sc00955b

DataCite Metadata Working Group. (2011). *DataCite Metadata Schema for the Publication and Citation of Research Data*. V2.2. Retrieved from http://schema.datacite.org/meta/kernel-2.2/doc/DataCite-MetadataKernel_v2.2.pdf

Dumontier, M., & Wild, D. (2012). Linked Data in Drug Discovery. *IEEE Internet Computing*, *16*(6), 68–71. doi:10.1109/MIC.2012.122

Flannery, D., Matthews, B., Griffin, T., Bicarregui, J., Gleaves, M. L., ... Kleese, K. (2009). ICAT: Integrating Data Infrastructure for Facilities Based Science. In *Proceedings of e-Science: Fifth IEEE International Conference on e-Science*. IEEE.

ISIS. (2011). *ISIS Data Management Policy*. Retrieved from http://www.isis.stfc.ac.uk/user-office/data-policy11204.html

Lawrence, B., Jones, C., Matthews, B., Pepler, S., & Callaghan, S. (2011). Citation and peer review of data: Moving towards formal data publication. *Int. Journal of Digital Curation*, *6*(2), 4–37. doi:10.2218/ijdc.v6i2.205

Marcos, E., Vara, J. M., & de Castro, M. (2012). Author order: What science can learn from the arts. *Communications of the ACM*, *55*(9), 39–41. doi:10.1145/2330667.2330680

Matthews, B., Bunakov, V., & Jones, C. (2013). *Investigations as research objects within facilities science*. Paper presented at the 1st Workshop on Linking and Contextualizing Publications and Datasets. Malta.

Matthews, B., Kourousias, G., Yang, E., & Griffin, T. (2012). Model of the data continuum in Photon and Neutron Facilities. *PaNdata ODI, Deliverable D6.1*. Retrieved from http://pan-data.eu/sites/pan-data.eu/files/PaNdataODI-D6.1.pdf

Matthews, B., Sufi, S., Flannery, D., Lerusse, L., Griffin, T., Gleaves, M., & Kleese, K. (2009). *Using a Core Scientific Metadata Model in Large-Scale Facilities*. Paper presented at the 5th International Digital Curation Conference. London, UK.

Mesot, J. (2012). A need to rethink the business model of user labs? *Neutron News*, *23*(4), 2–3. doi:10.1080/10448632.2012.725324

OAIS. (2012). Reference Model for an Open Archival Information System. *CCSDS 650.0-M-2 (Magenta Book) Issue 2, June 2012*. Retrieved from http://public.ccsds.org/publications/archive/650x0m2.pdf

PaNData collaboration. (2013). *Counting Users of the European Photon and Neutron Facilities 2012 Survey*. Retrieved from http://pan-data.eu/Users2012-Results

Parsons, M. A., & Fox, P. A. (2013). Is Data Publication the Right Metaphor? *Data Science Journal*, *12*(0), WDS32–WDS46. doi:10.2481/dsj.WDS-042

Shaon, A., Callaghan, S., Lawrence, B., Matthews, B., Osborn, T., Harpham, C., & Woolf, A. (2012). Opening Up Climate Research: A Linked Data Approach to Publishing Data Provenance. *International Journal of Digital Curation*, *7*(1), 163–173. doi:10.2218/ijdc.v7i1.223

van Dam, K., Li, D., Miller, S. D., Cobb, J. W., Green, M. L., & Ruby, C. L. (2011). Challenges in Data Intensive Analysis at Scientific Experimental User Facilities. In Handbook of Data Intensive Computing, (pp. 249-284). Springer. doi:10.1007/978-1-4614-1415-5_10

Wilson, M. (2012). Meeting a scientific facility provider's duty to maximise the value of data. In *DataCite Summer Meeting, Digital Research Data in Practice (DataCite2012)*. Retrieved from http://purl.org/net/epubs/work/62852

Yang, E., Matthews, B., & Wilson, M. (2013). Enhancing the Core Scientific Metadata Model to Incorporate Derived Data. *Future Generation Computer Systems*, *29*(2), 612–623. doi:10.1016/j.future.2011.08.003

ADDITIONAL READING

Ball, A. (2012) *Review of Data Management Lifecycle Models*. University of Bath. Retrieved from http://opus.bath.ac.uk/28587/1/redm1rep120110ab10.pdf

Bizer, C., Heath, T., & Berners-Lee, T. (2009). Linked Data - the story so far. *International Journal on Semantic Web and Information Systems*, *5*(3), 1–22. doi:10.4018/jswis.2009081901

Bourras, C., Giannaka, E., & Tsiatos, E. (2009). e-Collaboration Concepts, Systems and applications, in: Kock, N. (Ed.). Information Science Reference, E-Collaboration: Concepts, Methodologies, Tools, and Applications, Vol. 1.

Brase, J., & Farquhar, A. (2011). Access to Research Data. *D-Lib Magazine*, *17*(1/2). doi:10.1045/january2011-brase

Brooking, C., Shouldice, S. R., Robin, G., Kobe, B., Martin, J. L., & Hunter, J. (2009). Comparing METS and OAI-ORE for Encapsulating Scientific Data Products: A Protein Crystallography Case Study. *In 5th IEEE International Conference on e-Science (E-Science'09), Oxford, UK, 9-11 Dec 2009, 148-155*, doi:10.1109/e-Science.2009.29

Conway, E., Giaretta, D., Lambert, S., & Matthews, B. (2011). Curating Scientific Research Data for the Long Term: A Preservation Analysis Method in Context. *The International Journal of Digital Curation.*, *6*(2), 38–52. doi:10.2218/ijdc.v6i2.204

Coppola, M., Jégou, Y., Matthews, B., Christine Morin, C., Prieto, L. P., & Sánchez, O. D. et al. (2008). Virtual Organization Support within a Grid-Wide Operating System. *IEEE Internet Computing*, *12*(2), 20–28. doi:10.1109/MIC.2008.47

Corti, L., Van den Eynden, V., Bishop, L., & Woollard, M. (2014). *Managing and Sharing Research Data. A Guide to Good Practice*. SAGE Publications Ltd.

Crompton, S. (2012). Collaborative Information Management In Scientific Research Processes. In *8th IEEE International Conference on eScience (eScience 2012), Chicago, USA, 8-12 Oct 2012*. Retrieved from http://purl.org/net/epubs/work/64386

Giaretta, D. (2011). *Advanced Digital Preservation*. Springer. doi:10.1007/978-3-642-16809-3

Goble, C., Bhagat, J., Aleksejevs, S., Don Cruickshank, D., Michaelides, D., & Newman, D. et al. (2010). myExperiment: A repository and social network for the sharing of bioinformatics workflows. *Nucleic Acids Research*, *38*(2), W677–W682. doi:10.1093/nar/gkq429 PMID:20501605

Heath, T., & Bizer, C. (2011). Linked Data: Evolving the Web into a Global Data Space (1st edition). Synthesis Lectures on the Semantic Web: Theory and Technology, 1 (1), 1-136. Morgan & Claypool. doi:10.2200/S00334ED1V01Y201102WBE001

Higgins, S. (2008). The DCC Curation Lifecycle Model. *International Journal of Digital Curation*, *3*(1), 134–140. doi:10.2218/ijdc.v3i1.48

Lin, S., & Yen, E. (Eds.). (2011). *Data Driven E-Science: Use Cases and Successful Applications of Distributed Computing Infrastructures*. Springer. doi:10.1007/978-1-4419-8014-4

Matthews, B., Shaon, A., & Conway, E. (2012). How do I know that I have Preserved Software? In Delve, J., Anderson, D., Dobreva, M., Baker, D., Billenness, & C., Konstantelos, L. (Eds.). The Preservation of Complex Objects. Vol. 1. Visualisations and Simulations. The University of Portsmouth.

Pryor, G., Jones, S., & Whyte, A. (Eds.). (2013). *Delivering Research Data Management Services. Fundamentals of good practice*. Facet Publishing.

Rothenberg, J. (2000). Preserving Authentic Digital Information. In *Authenticity in a Digital Environment*. Washington, DC: Council on Library and Information Resources.

Shaon, A., Callaghan, S., Lawrence, B., Matthews, B., Woolf, A., Osborn, T., & Harpham, C. (2011). A Linked Data Approach to Publishing Complex Scientific Workflows. *In 7th IEEE International Conference on E-Science, Stockholm, Sweden, 5-8 Dec 2011*, 303-310. doi:10.1109/eScience.2011.49

Taylor, I., Deelman, E., Gannon, D., & Shields, M. (Eds.). (2007). *Workflows for e-Science: Scientific Workflows for Grids*. Springer. doi:10.1007/978-1-84628-757-2

KEY TERMS AND DEFINITIONS

Beamtime: Time slot allocated to investigation.

Facility User: Visiting researcher who conduct investigation using facility.

Facility: Large scientific instrument shared by individual researchers or small research teams.

Investigation: Series of "experiments" or "measurements" conducted by a facility user.

IRO (Investigation Research Object): Semantically rich aggregation of information resources that reflects different aspects and stages of investigation.

Principal Investigator: Visiting researcher who is responsible for the organization and conduct of investigation.

User Office: Facility business unit that registers users and allocates beamtime.

Chapter 8
Collaborative Seabed-Habitat Mapping:
Uncertainty in Sediment Data as an Obstacle in Harmonization

Sytze van Heteren
TNO - Geological Survey of The Netherlands, The Netherlands

Vera Van Lancker
Royal Belgian Institute of Natural Sciences, Belgium

ABSTRACT

This chapter discusses the uncertainties associated with sediment data, which are a commonly neglected obstacle in collaborative seabed-habitat mapping, complicating harmonization. Standardization of these data, although an important prerequisite in the harmonization of surrogate-based habitat maps, is not enough to achieve full map uniformity. Through time and between research institutes, data have been acquired, described, analyzed, and classified using a range of methods and protocols. The effects of differing error margins and indicative meanings remain hidden for most end users in environmental science and policy. The interpolation required to generate full-coverage data products from sedimentological point data creates additional uncertainty. Finally, small-scale habitat heterogeneity not captured by even the densest sampling places limits on the usability of sediment-based surrogates as well. Despite these limitations, there is much room for improvement when it comes to using surrogates in habitat mapping. Multiple visualizations generated from merged digital sediment datasets with flexible querying protocols are especially promising.

INTRODUCTION

There is international political momentum for the safeguarding and conservation of seabed biodiversity that is under increasing pressure from human activities such as fisheries, aggregate extraction, dredging and disposal of sediment, and wind-farm development. In this light, the extent and overall status of marine habitats must be assessed quantitatively. High-quality ecosystem and seabed-habitat maps are needed to spatially plan human activities and to designate marine protected areas at the most

DOI: 10.4018/978-1-4666-6567-5.ch008

Copyright © 2015, IGI Global. Copying or distributing in print or electronic forms without written permission of IGI Global is prohibited.

suitable locations. Effective integrated coastal and marine management compliant with the European Marine Strategy Framework Directive, Habitat Directive, and Water Framework Directive requires a transnational approach. Common, internationally agreed metadata and data standards are viewed as key elements in the harmonization of data products such as habitat maps, but adherence to these standards does not suffice in generating seamless habitat maps. Numerous acquisition and analytical protocols, each with their own uncertainties, preclude full harmonization. Their influence on the quality and indicative meaning of data, which differs from one dataset to the next, is difficult to quantify and usually not considered by data managers and map makers. Uncertainties related to sampling, subsampling, description, analysis and data interpolation form potential sources of error that will translate into mismatches in transnational sediment maps.

A specific challenge in harmonization concerns the common use of sediment-related surrogates in habitat mapping. Surrogates are biophysical variables that can be mapped with a quantifiable correspondence to the occurrence of benthic species and communities (Harris, 2012a). Since determining the distributions of key species of seabed habitats directly is costly, physical seabed parameters are frequently used as proxies of biological diversity and as indicators of habitat suitability. Abiotic variables, including morphological and sedimentological parameters, have been measured and monitored for many decades. Collection of morphological and sedimentological data is relatively easy and more cost-effective than the acquisition of data on seabed species and assemblages. In a stepwise, multi-faceted approach, various abiotic characteristics can be measured directly in the field or in the laboratory, documented by visual description, or deduced indirectly using remote-sensing techniques. The biological relevance of both the surrogates used and the divisions or classes defined within each surrogate is commonly unknown or merely assumed. Even where established quantitatively, the validity of observed biotic-abiotic relationships may not extend far beyond the area of consideration, as other variables may become more important. In short, potential habitat suitability as derived from a limited number of surrogates does not translate directly into actual habitat distribution.

Within the context of valorizing geological data for habitat mapping, Van Lancker & Van Heteren (2013) and Dolan et al. (2013) showed how sediment and terrain characterization respectively provide relevant knowledge about benthic habitats. They emphasized the need for standardization and harmonization, discussing principal methodological approaches and classification systems. Case studies presented as part of the two reviews illustrate how different data resolutions affect the granularity of sediment and terrain characterization. Here, we further elaborate on uncertainties related to sediment data, on seabed heterogeneity, and on up- and downscaling issues that need to be considered when merging adjacent regional data products. We illustrate how these issues may translate into mismatches linked in part to inconsistent definitions and vocabularies. Common standards, which would eliminate these inconsistencies, are a first prerequisite in unlocking the full potential of seabed-sediment databases. There is an equally important need, however, to consider differences in quality and indicative meaning related to a multitude of acquisition, description, analysis and interpolation methods and protocols.

STANDARDIZATION AND HARMONIZATION IN SEABED-HABITAT MAPPING

A seabed habitat is "a particular environment [on the seabed] that can be distinguished by its abiotic characteristics and associated biological assemblage, operating at particular, but dynamic spatial and temporal scales in a recognizable geographic area" (MESH Consortium, 2007). The mapping of seabed habitats should be considered as a process in which their distributions and extents are determined to create a full-coverage map with distinct boundaries that separate adjacent habitats or habitat complexes. A quality habitat map provides our best estimate of habitat distribution at a point in time, making use of the most up-to-date and complete knowledge available (MESH Consortium, 2007).

There is a subtle difference between standardization and harmonization, not only in habitat mapping but also in general. Standardization refers to conformity, providing a predefined, formal structure that may address data format as well as classification issues. Harmonization refers to consistency, ensuring that data from different owners can be merged and jointly processed. For harmonization in itself, adherence to official national or international standards is not a prerequisite. Common informal terminology and classification are sufficient. Both standardization and harmonization are needed, however, to make regional and particularly transnational seabed-habitat mapping feasible. Common standards are most easily adopted in new data-acquisition activities. Increased awareness about the added value of collaborative work, and of merged databases in particular, has resulted in numerous standardization initiatives that are increasingly coordinated in a European or even wider context. The transformation of legacy metadata, data and particularly data products (i.e. stored in old formats or computer systems that make access and processing difficult) toward harmonized and interoperable databases has been much more challenging. The rewards of complete interoperability are obvious. Common infrastructures and tools can be created for accessing, sharing, exchanging, analyzing and visualizing harmonized data and data products. Transnational and global infrastructures will allow the use of multiple or even numerous databases and information caches to accomplish common tasks.

SEDIMENT CHARACTERIZATION

Steps and Associated Uncertainties

Sediment is a mixture of particles with varying sizes and shapes. The spatial distribution of seabed-sediment types defined on the basis of these particle properties reflects an interplay of near-surface geology, morphology, hydrodynamics, sediment transport, environmental context, and even biota.

Sediment characterization is a stepwise, multifaceted activity. Various characteristics can be estimated visually, measured in the field or in the laboratory, or derived from (ground-truthed) proxies provided by acoustic data or video imagery. In developing surrogate maps for seabed-habitat studies, remote observation, sampling, sub-sampling, description, analysis, classification and even interpolation may all play a role (Figure 1). Each of these steps brings its own uncertainties and can take place in many different ways. The choice of methods is partly a function of equipment availability and cost, but also reflects suitability to particular field conditions or sediment types. Advantages and disadvantages of each combination of methods depend on the size and nature of the area studied.

A first step in sediment characterization is survey design, which can be optimized using sedimentological system and process knowledge. The aim is to capture all relevant spatial variability in sediment characteristics. In some cases, surveying or sampling should be done in a small sub-area (and upscaled later); in other instances, it is better to follow a gradient. The choices made will influence the end result, which will always include elements of interpolation and upscaling (Figures 2 and 3). The time and timing of data acquisition also matter. To obtain the maximum amount of information from samples, it is best to collect acoustic data first and then calibrate these using seabed-sediment samples. Ideally, the time between acoustic surveying and subsequent sam-

Figure 1. Elements of sediment characterization and mapping. The choice of elements is part of an overall research strategy determined on the basis of a few basic questions (left).

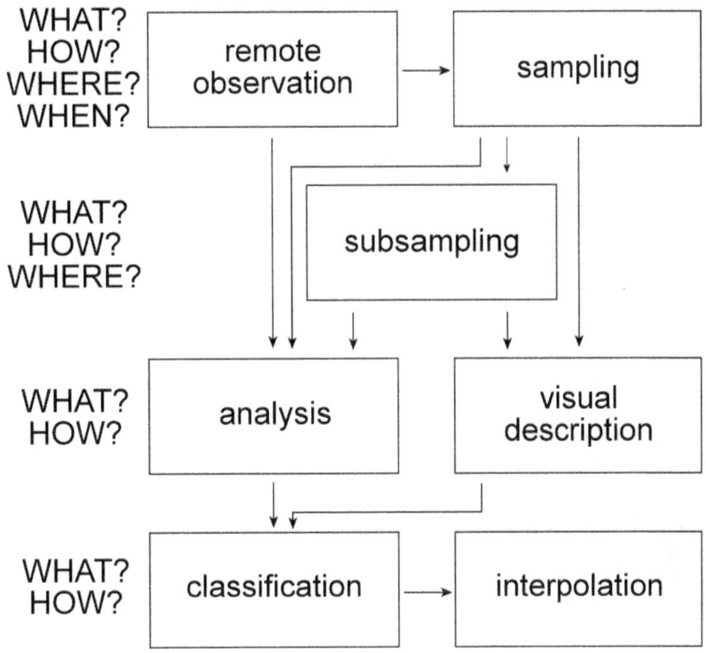

Figure 2. Characterizing seabed variability (left-hand panels) by interpolating partial-cover remote-sensing images (middle panels) or upscaling of smaller-scale full-coverage remote-sensing images (right-hand panels). For laterally continuous patchy areas, full-coverage data acquisition in small subsections may be more useful than surveying along grids (upper panels). For areas marked by broad-scale changes along a gradient, the reverse is true (lower panels).

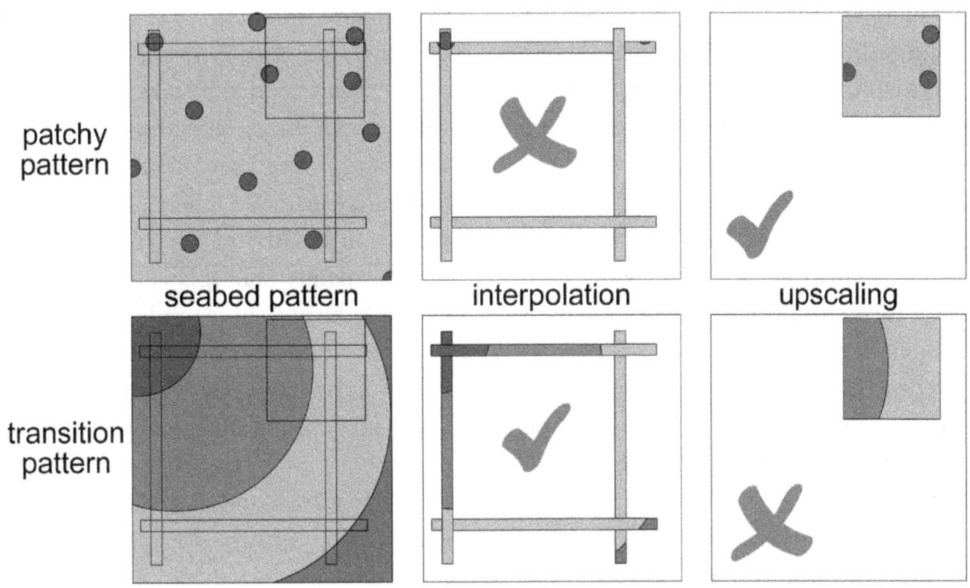

Figure 3. Characterizing seabed variability (left-hand panels) by interpolating widely distributed sediment samples (middle panels) or upscaling of smaller-scale sample clusters (right-hand panels). For laterally continuous patchy areas, closely spaced samples in small subsections may be more useful than samples along regular grids (upper panels). For areas marked by broad-scale changes along a gradient, the reverse is true (lower panels). The location and distribution of samples will determine how well the resulting data product captures reality.

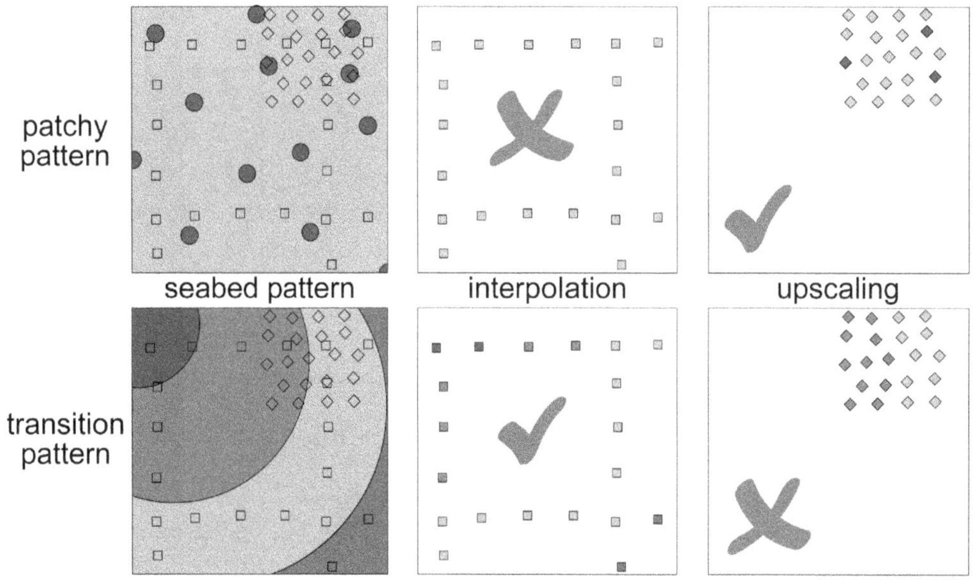

pling is kept to a minimum to reduce the influence of changing hydro-meteorological conditions on seabed properties as much as possible (cf. Renosh et al., 2014). The vintage (age) and the timing of data acquisition should be considered also when working on existing datasets, since seasonal and long-term dynamics change the character of the seabed sediment. Finally, sampling method is of influence. Overestimation of gravel content and distribution, for example, can be expected when samples have been collected with Van Veen grabs, used frequently during the early years of sedimentological seabed research. Large clasts may get trapped between the jaws upon closure. In that case, it is easy for finer-grained material to escape from the grab, resulting in preferential gravel enrichment.

Once samples have been taken, they can be described directly or be sub-sampled first. Sub-sampling is needed when samples are too voluminous for complete description or analysis. Generally, more material can be described visually than can be analyzed with field or laboratory instruments. For grain-size analysis and sediment composition, both relevant in habitat mapping, sub-samples need to be taken that are representative of both the entire sample and the top layer. Obtaining a representative sub-sample may be difficult, especially when a few coarse clasts are present in a matrix of fines or in case of stratified heterogeneous intervals sampled as single unit. The volume of a sub-sample needed to accurately describe or analyze a sample or sub-sample depends on its grain size. For muds and sands, less than 100 grams is sufficient to carry out accurate particle-size analyses. When gravel-sized particles (including shells and shell fragments) are present, tens of kilograms may be required (Head, 1980). It is commonly best to measure the coarser gravel fractions before sub-sampling.

Harmonizing Visual Descriptions and Laboratory Analyses

Descriptions and analyses of grain size provide qualitative or quantitative measures of the relative abundance of different fractions. Usually, end users are unaware of the subjectivity and other limitations inherent to the methods and protocols used. Overall, errors associated with laboratory analyses are systematic, and thus easier to quantify and correct than those made in visual descriptions. Nevertheless, it may be difficult to compare results obtained from one method with those from another. A good example relates to the measurement of grain size with various laboratory techniques, each producing weight or volume percentages for a number of size classes that may differ between methods. For individual samples, grain-size distributions analyzed by dry or wet sieving, photo analysis, optical counting, electro-resistance counting, settling tubes and laser diffraction have been shown to vary significantly (Figure 4). Commonly, the largest differences concern the silt and clay fractions, as observed in calibration studies (e.g., Konert & Vandenberghe, 1997; Rodríguez & Uriarte, 2009). These differences are due in part to behavioral aspects of particles that vary strongly in shape and chemical composition, but pre-treatment also plays a role. Fine particles may form aggregates upon drying in an oven, and naturally formed aggregates such as floccules may be broken up by various pre-treatment methods intended to separate the particles of unwanted aggregates formed during sample storage and preparation. For those methods that measure grain size directly, non-spherical particles may skew the results. Depending on the method used, irregularly shaped gravel, sand and silt grains are measured preferentially along either their longest or their narrowest axis. Platy particles cause even stronger differentiation of results. When indirect proxies of grain size are measured, such as light-reflection properties, translating the results into grain size is prone to error. Some conversion models routinely applied to translate laser measurements into volume fractions, for

Figure 4. Grain-size-distribution curves of identical samples are a function of sample preparation (with aggregation or disintegration of floccules and fecal pellets), instrumentation and protocols.

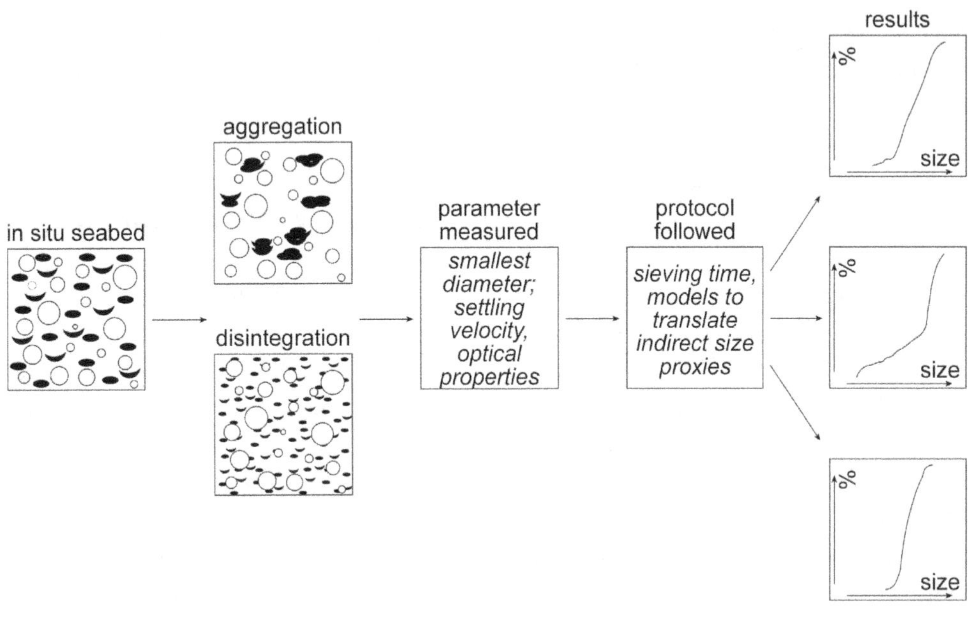

example, filter out the fine tails of the grain-size spectrum as noise, and thus underestimate mud content. With these examples in mind, it is easy to see why metadata and calibration studies are critical when merging and harmonizing multiple grain-size datasets.

A next source of uncertainty in harmonization relates to the simplification of grain-size data. When grain size is described visually, range, mean and median grain size, and sorting are the most common parameters listed. The most useful visual descriptions also list gravel, sand and mud percentages that should add up to 100%. Single descriptors of sediment size are easy to harmonize, but have limited value. Mean and median grain size, for example, are poor representations of mixed sediments, especially of those with bi- or multimodal particle-size distributions. In the past, many grain-size distributions measured in the laboratory were simplified to similar sets of key parameters, with associated data either discarded or stored separately as tables or graphs. State-of-the-art numeric sediment databases comprising quantitative analytical results reduce simplification, providing relatively complete descriptions of sediments (e.g., full grain-size distributions). They allow more flexibility in experimenting with different parameters or class boundaries. Still, harmonization may be complicated by a lack of information on the characterization methods and protocols used.

Classifying Sediment Data

Classification of sediment data can be seen as standardized simplification. It is a main cause of mismatches where national or regional databases for adjacent areas adhere to different standards. In the absence of clear natural breaks, there have been different definitions of terms used to describe grain-size classes and similarly differing definitions of terms used to describe the relative proportions of different grain-size fractions (Long, 2006). Attempts to agree on common international classifications have mostly failed. Although there have been efforts to propose a small number of sediment-classification systems with broad validity and relevance, many regional and national systems remain in use. In visual descriptions of sediment samples, for example, many scientists are bound by ever-changing national standards, usually in their native languages. Also, the focus of classification has never been on broad-scale harmonization, but rather on optimized schemes that work best for a particular place or purpose, creating differences even within individual countries. Even when the standards and protocols used to collect, describe and analyze data are clearly reported in the metadata, harmonization is rarely attempted. Calibration studies are missing, and database managers are daunted by a looming reclassification of enormous numbers of existing analyses and observations that are mostly still on paper rather than in a digital form, and available as text descriptions rather than standardized code. As long as databases were used separately, or merged only at a regional level, the multitude of classifications and languages used had only limited downsides. Discrepancies were either ignored or solved in practical approaches that were determined on a case-by-case basis. From a European or global perspective, however, such ad hoc adjustments are both unfeasible and undesirable. Coordinated transnational cooperation is needed to find common solutions and to distinguish perceived from real mismatches in datasets.

Commonly Used Terminology

The most widely and consistently used size categories of clastic sediment are gravel, sand, silt and clay, which may be subdivided into several subcategories by using modifiers such as coarse, medium and fine. Because it uses clear, universally recognized terminology and provides the maximum resolution achievable in visual descriptions, the Udden-Wentworth scale (Udden, 1914; Wentworth, 1922) has been adopted as the standard in academic research. It is not always easy to translate loosely defined text with familiar size categories and modifiers into the precise values and terms of the standardized Udden-Wentworth definitions. The Geological Survey of the Netherlands, for example, has switched between standards about ten times during the past 80 years (Figure 5). Metadata are incomplete at best, making it difficult to reconstruct which standard was used for which sample description.

For sediments consisting of different clastic components or size fractions, most sedimentologists in the international academic world use the classification systems developed by Shepard (1954) and Folk (1954). Shepard (1954) made a subdivision using a single ternary diagram with sand, silt and clay. Schlee (1973) later modified his scheme and added a second ternary diagram with gravel, sand, and mud (silt and clay). The most commonly used system today is the Folk classification, labeling sediment using the relative proportions of its gravel, sand and mud percentages. The Folk terms provide a lot of information without being a complete description, and are therefore very useful and understandable for a majority of end users. In Europe, the Folk classification has been slightly modified by the British Geological Survey and expanded by Flemming

Figure 5. Class boundaries for different types of sand, as used in the Netherlands during the past century (after Van der Meulen et al., 2003). Udden-Wentworth, German DIN 4022, and ISO 14688-1 classifications are shown for comparison.

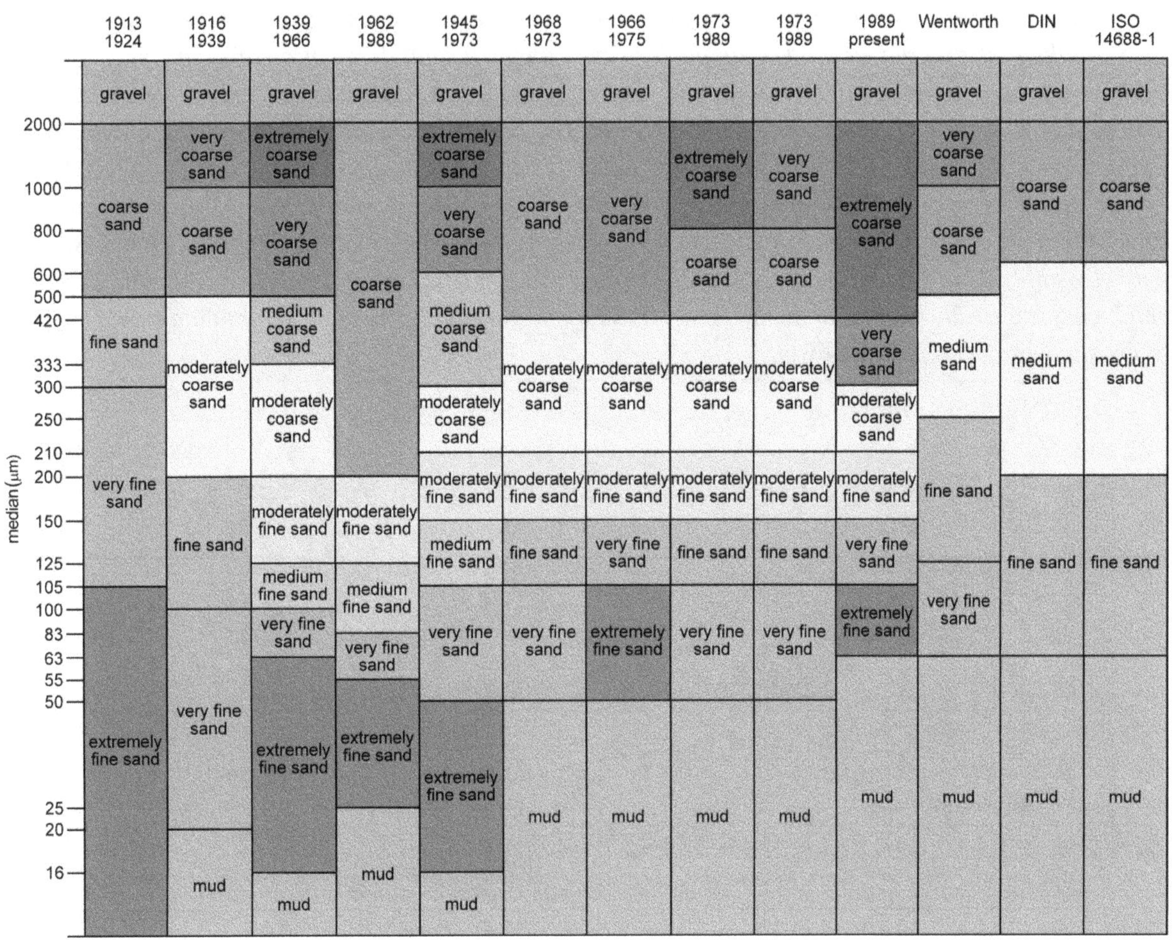

(2000) (Figure 6). In the gravel fraction, siliciclastic particles are sometimes distinguished from shells and shell fragments (bioclastic gravel) that add to the percentage of gravel but behave differently under the influence of waves and currents. Density and shape differences between siliciclastic and biogenic particles explain why two samples, similar in grain size, have different implications for bed shear stress, an important habitat parameter.

In the conversion from locally used terminology to the Folk classification, needed to harmonize visual descriptions and laboratory granulometry, two translation options are possible (Figure 6). When gravel, sand and mud percentages are listed, they can be plotted directly on the ternary diagram, giving a single class as output (in case of the example, gravelly mud). When information on gravel, sand and mud content has to be derived from code or from text fields (for example, moderately gravelly, clayey silt), the translation depends on the availability of the associated term definitions. When these definitions are available

Figure 6. Conversion from local terminology (left) to modified Folk classification (right; as used by the British Geological Survey in its 1:1,000,000 and 1:250,000 seabed-mapping program, with a further subdivision of gravel-free sand-mud mixtures as proposed by Flemming (2000) (see codes below the right-hand triangle)). Black dot in the Folk diagram on the right denotes a sample description for which gravel, sand and mud percentages are available. Shaded zone in the local classification diagram on the left (NEN5104 norm in the Netherlands), as transferred to the Folk diagram, denotes a sample description for which the gravel, sand and mud ranges can be derived from the terminology used. Dashed circle in the Folk diagram denotes a sample for which only a textual description of sediment type exists, without a proper definition of its meaning in terms of gravel, sand and mud percentage. In local terminology, (g) is slightly gravelly, g is moderately gravelly, gg is very gravelly, mG is muddy gravel, ssG is extremely sandy gravel, sG is very sandy gravel, (s)G is moderately sandy gravel and ((s))G is slightly sandy gravel. In modified Folk terminology, M is mud, (s)M is slightly sandy mud, sM is sandy mud, mS is muddy sand, (m)S is slightly muddy sand, S is sand, (g)M is slightly gravelly mud, (g)sM is slightly gravelly sandy mud, (g)mS is slightly gravelly muddy sand, (g)S is slightly gravelly sand, gM is gravelly mud, gmS is gravelly muddy sand, gS is gravelly sand, mG is muddy gravel, msG is muddy sandy gravel, sG is sandy gravel and G is gravel.

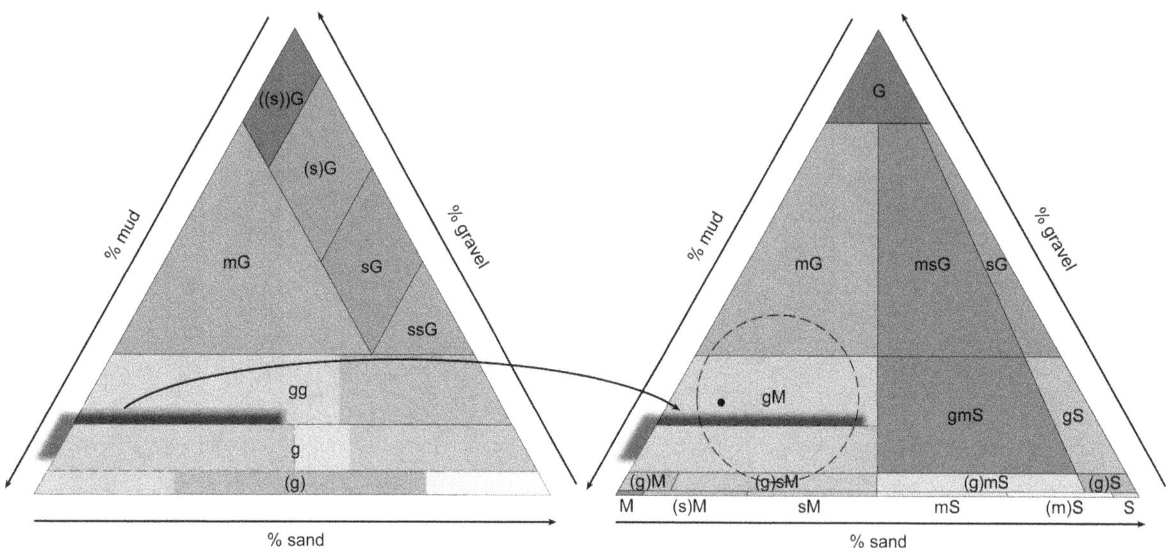

(for example, moderately gravelly means 5-15% gravel), mapping can be based on this constraining information. This translation is less precise than the classification based on numerically expressed relative proportions. When such definitions are not available, mapping relies on assumptions and is therefore even less accurate.

Interpolation

Maps and grids created to visualize seabed-sediment parameters rely on interpolation, except when they are derived from full-coverage acoustic data. Interpolation adds additional uncertainty to data products relying on described and measured seabed-sediment data that are already associated with various components of random and systematic error. Traditionally, maps were drawn manually, and interpolation-related uncertainty was reduced through incorporation of geological expertise (e.g., differentiation between sandbanks and swales). Nowadays, most maps are produced digitally, and interpolation is automated and optimized using information from parameters with a known relation to sediment type or grain size, such as bathymetry and morphology. Modern software allows the use of a variety of analytical tools to apply different interpolation techniques and to statistically compare the results of interpolated results with subpopulations of the dataset that were not used in the interpolation (Figure 7).

There are increasing efforts to link probabilities to the static values associated with digitally mapped sediment classes. Sequential Indicator Simulation (SIS; Goovaerts, 1997), for example, renders multiple, statistically equally probable simulations of the distribution of sediment type or grain-size class (cf. Stafleu et al., 2011). It uses sample descriptions and analyses as hard data and searches the neighborhood for all remaining

Figure 7. Interpolation of Belgian grain-size data through ordinary kriging (left) and through kriging with external drift (right) (Verfaillie et al., 2006). Kriging with external drift uses a densely sampled auxiliary variable (in this case water depth) to estimate a target variable. This method enabled differentiation between sediment distributions within swales and on sandbanks.

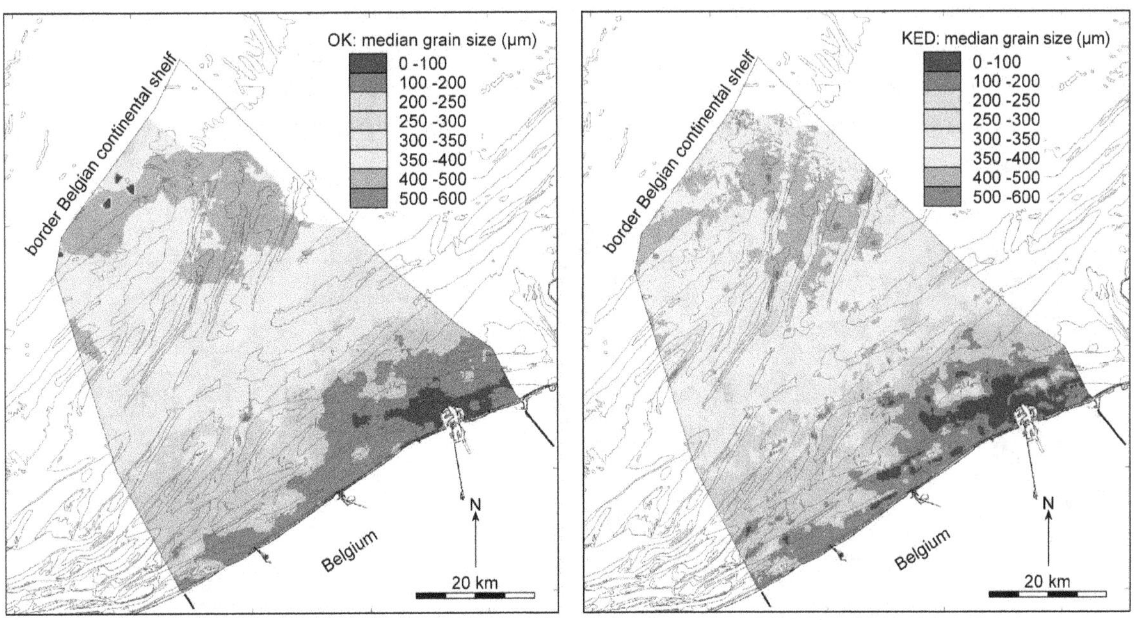

grid cells. From such simulations, generated with dedicated modeling software, probabilities of occurrence for each sediment type or grain-size class may be calculated. Such probability mapping provides an indication of model uncertainty. An alternative way to quantify and visualize interpolation-related uncertainty was presented by Lark et al. (2012). They used geostatistical interpolations of point observations on gravel, sand and mud percentages, obtained from about 30,000 seabed samples from the UK continental shelf, to calculate the probability of occurrence of sediment types. For the classification systems of Long (2006) and James et al. (2010), the local conditional probability of each sediment-texture class was mapped. For each grid cell, the probability of the class of maximum probability was also plotted, providing a powerful assessment of uncertainty.

Heterogeneity

An awareness of small-scale heterogeneity is highly important in seabed-sediment mapping and essential to understand the limitations of interpolation. The transition from sandy to muddy areas, for example, is not a simple spatial gradient that can be expressed easily on a map. Rather, it is marked by patchiness that may change quickly as conditions on the seabed and in the water column fluctuate. High-quality observations show that many gravel beds and mud layers are covered to varying degrees with sand veneers (Figure 8), adding to the complexity of detecting and monitoring their limits of occurrence. Patchiness and surficial veneers that are strongly different from underlying sediment are two forms of mixed sediment, a loosely defined term that also includes various mixtures of gravel, sand and mud that do not show specific lateral or vertical variability (Stevenson, 2011). In light of these multiple definitions and because of rapid temporal changes, mapping mixed sediment accurately and in a harmonized way is not possible. It is one of the strongest reminders of the fact that no map is 100% correct 100% of the time.

Mismatches and the Importance of Visualization

Usually, mismatches caused by differing acquisition, description, analytical and classification

Figure 8. Fine-scale patchiness in the swale of the Brown Bank tidal ridge in the southern North Sea (Van Dijk et al., 2012). Here, the seabed sediment generally consists of consolidated mud covered by a thin layer of sand. In some places, the mud lies directly at the surface. Left: box core; center: core; right: side-scan-sonar image.

methods are difficult to recognize from collated data, especially when they concern large databases. Many mismatches become apparent only when data are visualized in data products such as maps or models. Many of the largest discrepancies are associated with borders between neighboring countries or other government entities. They may be a result of different mapping practices, but also reflect differences in the times of data acquisition, as well as inaccuracies. When median grain size (d50) was mapped using sediment databases from British, Belgian and Dutch parts of the North Sea, a small but influential difference in definition of the plotted variable became obvious as soon as the individual maps were merged (Figure 9, inset). Belgian and Dutch researchers frequently list the median grain size of the sand fraction, in part to exclude the influence of primarily bioclastic gravel. British researchers include both (mostly siliciclastic) gravel and mud in their calculations of d50.

Figure 9. Cross-border incompatibility in an initial map of median grain size (d50) for British, Belgian and Dutch seabed sediment (Maljers et al., 2006; inset), as compared to final harmonized product showing d50 of the sand fraction (Verfaillie, 2008; large map). Different color scales are used for contrast. Median grain size in the inset increases from blue to red. The large red and orange areas that mark British waters do not continue in Belgian waters. In the inset, note that the Belgian data extend beyond the boundary of the Belgian part of the North Sea.

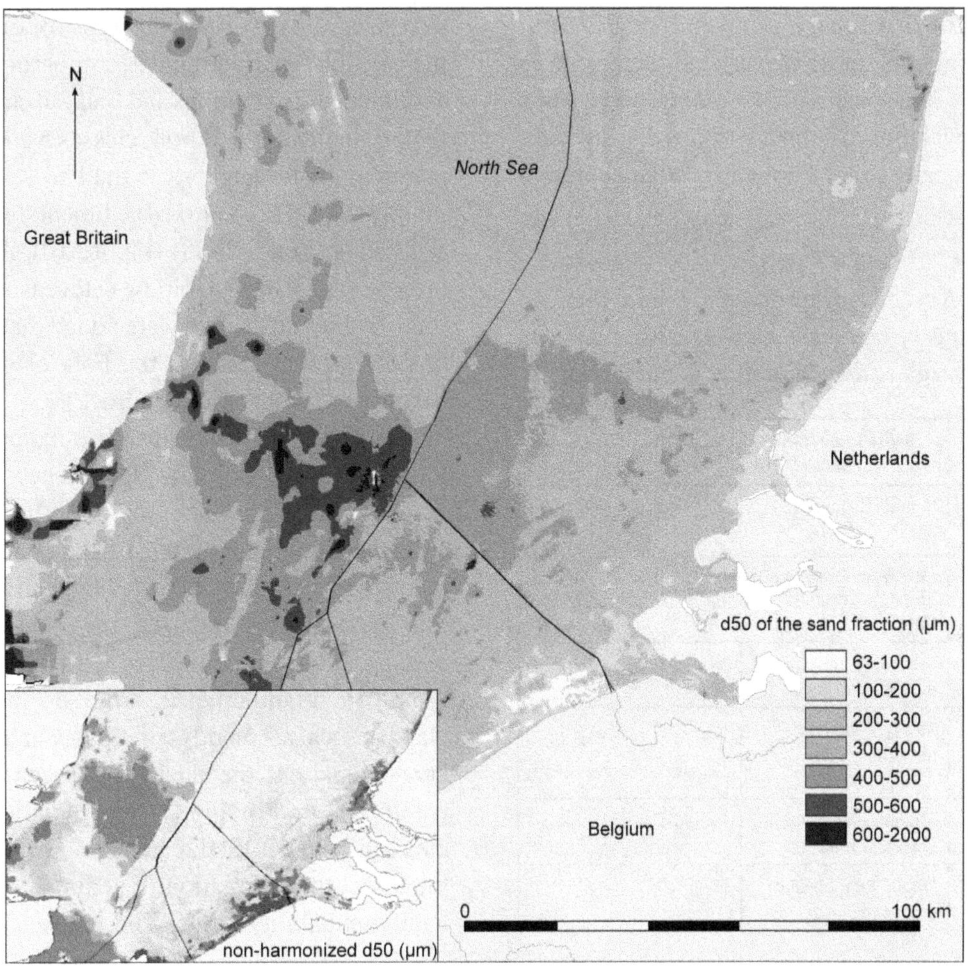

FROM SEDIMENT CHARACTERIZATION TO HABITAT MAPPING

Surrogacy in Habitat Mapping

In habitat mapping, surrogacy concerns those abiotic characteristics that best explain the species assemblage in a particular space and at a particular time. The indicative value of surrogates depends in part on the heterogeneity of the area under consideration. Grain size and sediment type can be used to predict (with some known probability and uncertainty) the occurrence of species assemblages in unexplored areas and between sparse biological observation points. Several surrogacy studies in which species assembly was analyzed against abiotic variables found sediment properties to be useful predictors of benthic communities (Harris, 2012b) (Table 1). However, seabed sediment is only one of many variables determining the ability of a species or assemblage to prosper. Although it helps to delineate zones that are sedimentologically suitable for a particular habitat, it cannot be used to predict the actual distribution of the associated species and assemblages within these potentially suitable areas. Most species are unable to fully exploit the so-called fundamental niches (*sensu* Hutchinson, 1957) associated with sediment.

Table 1. Key surrogacy studies quantifying the importance of sediment characteristics for the prediction of the distribution of benthic species

Study	Most Important Variables	Variability Explained (%)
Stevens & Connolly, 2004	% mud, distance to ocean	30
Beaman et al., 2005	Slope, % gravel, % $CaCO_3$	75
Post et al., 2006	% mud, % gravel, disturbance, depth	59
Beaman & Harris, 2007	Slope, % gravel, turbidity	62
Willems et al., 2008	Median grain size, % gravel, % mud	19/44/37
Gogina et al., 2010	Depth, total organic carbon	50/43

In broad-scale habitat mapping, harmonization of sediment-based surrogates is carried out at a rather general level. Sediment type as determined from gravel, sand and mud proportions has been used most frequently and successfully. Transnational examples are available from projects MESH (MESH Consortium, 2007) and EUSeaMap (Cameron & Askew, 2011). In these projects, a simplified Folk classification was defined as a surrogate for the habitat complexes at Level 3 of the European Nature Information System EUNIS (http://eunis.eea.europa.eu/), a hierarchical habitat classification covering the terrestrial and marine habitat types of Europe. On the basis of expert judgment, comparing species-assemblage distributions and Folk substrate maps, the originally defined Folk classes were grouped into four main categories: mud to sandy mud, muddy sand to sand, mixed sediments, and coarse sediments (Long, 2006) (Figure 10). As part of project EMODnet-Geology (Stevenson, 2011), three additional classes were added that were not or poorly represented in the Folk classification (diamicton, boulder and bedrock).

For habitat assessments on a national or regional level, broad-scale approaches with few mapped classes may not provide sufficient detail for end users. Where a finer subdivision is needed, it is better not to simplify the Folk classification. For maximum granularity, the modified Folk classification can be merged with size information from granulometric analyses (generating classes such as muddy fine sand and gravelly coarse sand). If the relation between sediment parameters and biological communities or species can be quantified, conversion models allow the direct transformation of sediment maps into habitat-suitability maps. The relevance of such

Figure 10. Simplified Folk classification as defined in MESH and EUSeaMap for broad-scale habitat mapping (right; Long, 2006), compared to the modified Folk classification as used by the British Geological Survey in its 1:1,000,000 and 1:250,000 seabed-mapping program (left; see caption of Figure 6 for terminology). The codes below the left-hand triangle represent a further subdivision of gravel-free sand-mud mixtures as proposed by Flemming (2000). Note the dominance of mixed sediments in the simplified classification.

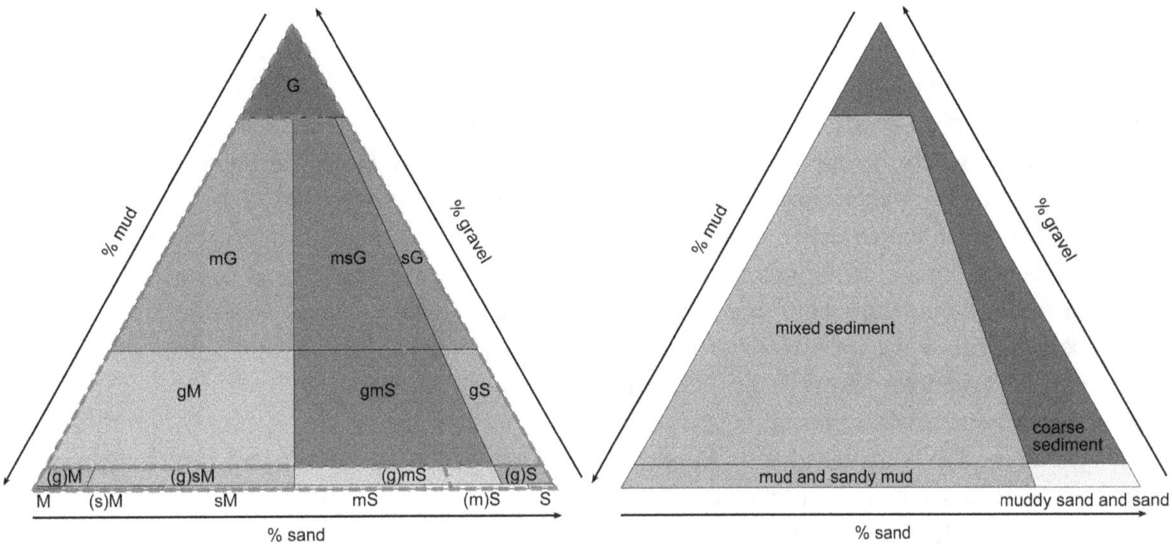

an approach has been demonstrated for Belgian waters, where median grain size and mud content have been found to best predict the occurrence of the main macrobenthic communities. This resulting habitat-suitability model (Degraer et al., 2008), which uses gridded datasets on d50 and mud content, was applied successfully farther north in the southern North Sea, up to the Dogger Bank (Verfaillie, 2008).

On an even smaller scale, discrepancies will still exist between point-based biological analyses and surrogacy-based interpretations. This is the case in areas with subtle sediment gradients (e.g., transitions from mud to sand), but also where lateral or vertical seabed heterogeneity is high. In these instances, both higher-resolution mapping and finer classification may be needed, requiring better coverage, more laboratory analyses of sediment, and sufficient ground-truthing with species identifications. In the habitat mapping of Belgian sand-bank areas, subtle spatial sediment differences were distinguished with the help of detailed terrain analyses (e.g., differentiating between sediments from swales, gentle/steep slopes and top zones). It proved possible to further link areas with subtly differing sediments and morphologies to the distribution of macrobenthic communities (Van Lancker et al., 2012). Some observed biotic-abiotic relationships, such as the preferential presence of dense aggregations of the tubeworm *Lanice conchilega* at the foot of tidal-ridge slopes, are being used to map and predict the distribution of ecosystem engineers, which contribute to seabed stability and coastal safety and are therefore considered an important asset in modern environmental management.

Propagation of Uncertainty Related to Sediment Characterization and Mapping

Links between sediment properties and benthos can be blurred by uncertainty related to sediment data and data products (Table 2). In this light, it is important to realize that poorly defined or apparently absent known relationships may be explained by imperfect sediment characterization and mapping. Conversely, apparently clear-cut relationships may be artefacts and their validity needs to be tested continuously. Even with the perspective of new, ground-breaking observation and analytical tools, finding and quantifying links will always be easier in broad-scale habitat transitions, well covered by samples and acoustics, than in fine-scale patchy areas with relatively few benthos analyses that need to be matched with sediment data from different locations and collected at different times. The roles of other abiotic parameters (e.g., sediment dynamics) should always be taken into account as well.

Confidence in sediment data and data products must be considered in a broad perspective, and include a valuation of sedimentological system and process knowledge. On the one hand, a confidence assessment indicates which limitations are associated with using available sediment characteristics as surrogates for benthic communities. On the other hand, it should provide a clear warning not to put too much weight to locally or regionally deduced links between sediment properties and biotic variables, especially when creating fine-scale maps. This is less of an issue in broad-scale habitat mapping, with small spatial details such as patchiness being filtered out of the final data products. Nevertheless, the danger of over-interpretation is present in all scales of habitat mapping.

Interpretation-Related Uncertainty

Quantifying relationships between sediment characteristics and biotic parameters is difficult, even when the same samples are subjected to

Table 2. Uncertainties associated with sediment data and data products

Step	Causes of Uncertainty	Actions Needed for Harmonization
Remote sensing	Survey design, method of acquisition (*incl.* calibration) and data processing, vintage, timing	Filtering data on the basis of methods and vintage
Sampling	Survey design, method of acquisition (*incl.* calibration) and data processing, vintage, timing	Filtering data on the basis of methods and vintage
Subsampling	Representativeness	Adherence to minimum sample sizes for sediments of different grain size
Description	Subjectivity, gross simplification	Description of standard parameters
Analysis	Pre-treatment, systematic error due to particle properties and imperfect conversion models, simplification	Intercalibration of granulometry methods and protocols, inclusion of methods and protocols in metadata
Classification	Simplification, inconsistency of standards, definition of class breaks, poor metadata	Development of mapping protocols for translating between standards, inclusion of standards in metadata
Interpolation	Spatial heterogeneity, temporal variability, poor system knowledge, validity of statistics, imperfect relation with external-drift variable	Use of standard methods, calculation of uncertainty
Interpretation	Subjectivity, poorly known surrogacy, poor system knowledge	Multidisciplinary and transnational collaboration and knowledge exchange

sedimentological as well as biological analyses. The complexity of marine ecosystems, governed by numerous abiotic and biotic processes and parameters, causes inconsistencies in correlations of sediment and biota between even the most closely spaced locations. In self-organization, for example, small-scale interactions between the components of an initially disordered system lead to larger-scale order. This process is prominently seen in the formation of clustered colonies out of individual tubeworms or bivalves (Dame, 2005). As a result of self-organization, adjacent sites with similar sediment characteristics may show enormous contrasts in biotic densities, with thousands of specimens in samples of colonized seabed and no specimens in samples collected outside the colonies. Clearly, sediment characteristics are surrogates of suitability, not of abundance. Differing successions of sediments in the benthos layer (upper half a meter of the sediment) also complicate the quantification of links between sediment and biota. The lateral distribution of certain benthic species, for example, may be related to shallow-subsurface layers that are different from the surface sediment. Usually, these layers are not shown on surface-sediment maps.

From a temporal perspective, seabed habitats are not a function of static and uniform conditions. They adjust, rarely instantaneously, to long-term, seasonal or even event-driven changes in the physical environment. Coarse sediments temporarily exposed at the surface by storm waves, for example, are covered by fines during fair-weather periods. Time of sample collection is critical. Some species need fluctuating dynamics for successful feeding and dwelling, and the favored habitat may change when larvae and juveniles develop into adults (e.g., Reiss & Kröncke, 2005).

In the ideal situation that all data acquisition and analysis have been carried out consistently, differences in data interpretation (translating data into information) can therefore still result in mismatches of data products. With so many uncertainties remaining, different researchers will continue to assign different indicative meanings to physical parameters or associated classes.

Semantics in Collaborative Work between Sedimentologists and Ecologists

State-of-the-art seabed-habitat mapping relies on the integration of data, information and knowledge from various disciplines, including geology, physical oceanography, hydrography, and marine ecology. It involves multiple levels of cooperation and harmonization between users from various backgrounds and organizations, each bringing his or her own unique expertise. Up until now, the focus of harmonization has mostly been on data-related aspects, addressing fixed parameters that are easily documented and standardized on a technical level. Associated semantics (vocabularies and ontologies), initially developed per discipline (for example in EU projects Geo-Seas and ODIP), are increasingly being exchanged between groups representing different disciplines. Semantic interoperability is still mostly discussed by technical experts, however, and has not been propagated to the research level. Common nomenclature, therefore, has not yet resulted in common understanding. The latter relies not just on the exchange of harmonized data but also on collaborative research, preferably aimed at some joint deliverable that requires semantic interoperability on an information and knowledge level. With the advent of Communities of Practice (Wenger et al., 2002; Fredericks, 2015, chapter 13 of this book), we can expect progressively more effective dialogs between disciplines, not just from a technical but also from a scientific and managerial point of view.

As demonstrated in this chapter, sediment parameterization by itself is prone to a lot of uncertainty, which typically remains underexposed in habitat mapping even though sediment sampling is standard practice in ecological studies. Much of this uncertainty can be highlighted if related metadata are recorded for all phases of data acquisition and sensor selection, and for post-measurement transformation of observations. SensorML and new vocabulary will enable this process, and will facilitate machine-to-machine integration of all aspects of datasets. Unfortunately, uncertainty is not (fully) included in the semantics of existing data. In this light, raising awareness on confidence issues associated with sediment data is a critical element in training a new generation of habitat mappers.

FUTURE RESEARCH DIRECTIONS

Although sediment characterization has proven valuable in broad- and fine-scale habitat mapping, the numerous small and large databases of sedimentological surrogates have plenty more to offer in habitat and habitat-suitability mapping, especially when combined with remote-sensing data and with newly collected information on benthic species. Standardization and harmonization are important, but they are only a first step in the optimization of surrogate-based mapping. One way forward in using sediment character to its full potential is a continued effort to find, quantify and explain relationships between abiotic and biotic parameters, including sediment dynamics, (bio)geochemical and geomechanical aspects, at various spatial and temporal scales. On a fine scale, this can be done by linking sediment and benthos analyses from single samples as part of local, project-related studies (Figure 11) carried out for a range of known field conditions. On a broad scale, samples collected for other purposes can be reused by creating multiple surrogate grids that may be compared to plotted biotic parameters as derived from samples or databases. Full-distribution grain-size data are particularly useful in creating such grids, and become very powerful in combination with associated uncertainties. To optimize the chance of finding least-biased relationships, benthos composition and distribution measured from seabed samples or observed by remote sensing can best be analyzed for locations where the occurrence of a certain sediment type has been mapped with the highest probability. Visualization of uncertainty, therefore, is critical.

As it becomes increasingly possible to digitally generate multiple visualizations for a single sediment dataset, researchers have an extra tool in hand to assess the relevance of surrogates and class breaks in habitat-mapping studies. Comparison of multiple renditions of habitat-suitability maps, created using slightly different sediment-based surrogacy parameters, with species or assemblage distributions mapped directly by marine ecologists can be used to choose one surrogate or classification over another. The most suitable surrogates are those that show maximum correspondence with biological data from point observations and remote sensing. This optimization process can be applied on various scales and resolutions, meeting the needs of different end users.

CONCLUSION

Common standards for metadata and data are seen as key elements in the harmonization of surrogate-based habitat maps, but adherence to these standards is not enough to realize the ambition of full map uniformity. The many methods and protocols for data acquisition, description, analysis and classification, usually with poorly defined error margins and indicative meanings, form an obstacle in our quest for seamless transnational habitat maps.

Many mismatches become apparent only once data are visualized in maps or models. The interpolation required to generate these full-coverage data products is an additional source of error,

Figure 11. Relationships between seabed morphology (data from multi-beam echo sounder; colors match water-depth profile showing a range between 18 (red) and 45 m (blue) below lowest astronomical tide), benthic assemblages, median grain size (d50, μm) and mud content (< 63 mm, %) across the Brown Bank tidal ridge in the southern North Sea (Van Dijk et al., 2012). For the benthos, yellow denotes high densities (1500-3179 ind/m²) of annelids and crustaceans; green denotes moderate densities (346-679 ind/m²) of annelids and crustaceans; blue denotes high densities (833-3808 ind/m²) of annelids, crustaceans, molluscs and some echinoderms; and red denotes very low densities (26-77 ind/m²) of various species.

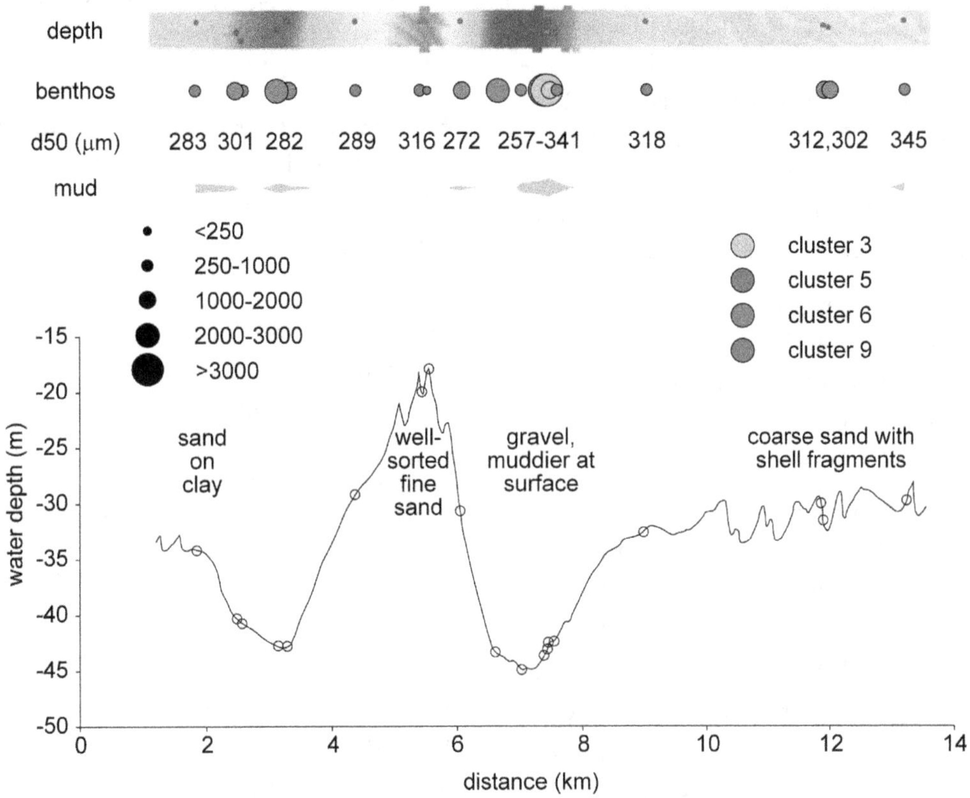

which can be quantified using various statistical approaches. An awareness of small-scale heterogeneity within even the smallest mapped units is crucial to understand both the possibilities and the limitations of using surrogates in habitat mapping.

Series of visualizations are instrumental in the selection of optimal class breaks, using maximum correspondence with plotted biological data such as species distribution. Even with this new approach, different researchers will continue to assign different indicative meanings to physical parameters or associated classes.

ACKNOWLEDGMENT

This chapter is a contribution to EU projects EMODnet Geology (MARE/2008/03; MARE/2012/10), Geo-Seas (FP7, Grant 238952) and ODIP (FP7, Grant 312492), and to Brain-be project TILES (Transnational and Integrated Long-term marine Exploitation Strategies), subsidized by Belgian Science Policy under contract BR/121/A2/TILES. We thank Ingemar Cato for his persistent reminder that mixed sediments cannot be captured in a single sediment class.

REFERENCES

Beaman, R. J., Daniell, J. J., & Harris, P. T. (2005). Geology–benthos relationships on a temperate rocky bank, eastern Bass Strait, Australia. *Marine & Freshwater Research*, *56*(7), 943–958. doi:10.1071/MF04306

Beaman, R. J., & Harris, P. T. (2007). Geophysical variables as predictors of megabenthos assemblages from the northern Great Barrier Reef, Australia. In B. J. Todd & H. G. Greene (Eds.), Geological Association of Canada Special Paper: Mapping the seafloor for habitat characterization (vol. 47, pp. 247–263). Academic Press.

Cameron, A., & Askew, N. (Eds.). (2011). *EUSeaMap - Preparatory Action for development and assessment of a European broad-scale seabed habitat map final report*. Retrieved January 27, 2014, from http://jncc.gov.uk/euseamap

Dame, R. F. (2005). Oyster reefs as complex ecological systems. In R. F. Dame & S. Olenin (Eds.), *The comparative roles of suspension-feeders in ecosystems* (pp. 331–343). Dordrecht, The Netherlands: Springer. doi:10.1007/1-4020-3030-4_19

Degraer, S., Verfaillie, E., Willems, W., Adriaens, E., Vincx, M., & Van Lancker, V. (2008). Habitat suitability modelling as a mapping tool for macrobenthic communities: An example from the Belgian part of the North Sea. *Continental Shelf Research*, *28*(3), 369–379. doi:10.1016/j.csr.2007.09.001

Dolan, M., Thorsnes, T., Leth, J., Alhamdani, Z., Guinan, J., & Van Lancker, V. (2013). Standardisation and harmonisation in seabed habitat mapping: role and added value of geological data and information. Part B: Terrain characterisation. *Geo-Seas report 10.5B*. Retrieved on April 7, 2013, from http://www.geo-seas.eu/content/content.asp?menu=0040028_000000

Flemming, B. W. (2000). A revised textural classification of gravel-free muddy sediments on the basis of ternary diagrams. *Continental Shelf Research*, *20*(10-11), 1125–1137. doi:10.1016/S0278-4343(00)00015-7

Folk, R. L. (1954). The distinction between grain size and mineral composition in sedimentary-rock nomenclature. *The Journal of Geology*, *62*(4), 344–359. doi:10.1086/626171

Fredericks, J. (2015). Persistence of knowledge across layered architectures. In P. Diviacco, P. Fox, A. Leadbetter, & C. Pshenichny (Eds.), *Collaborative Knowledge in Scientific Research Networks*. Hershey, PA: IGI Global Press.

Gogina, M., Glockzin, M., & Zettler, M. L. (2010). Distribution of benthic macrofaunal communities in the western Baltic Sea with regard to near-bottom environmental parameters. 2. Modelling and prediction. *Journal of Marine Systems*, *80*(1-2), 57–70. doi:10.1016/j.jmarsys.2009.10.001

Goovaerts, P. (1997). *Geostatistics for natural resources evaluation*. New York: Oxford University Press.

Harris, P. T. (2012a). Surrogacy. In P. T. Harris & E. Baker (Eds.), *Seafloor geomorphology as benthic habitat: GeoHAB atlas of seafloor geomorphic features and benthic habitats* (pp. 93–108). London: Elsevier Insights. doi:10.1016/B978-0-12-385140-6.00005-0

Harris, P. T. (2012b). Biogeography, benthic ecology, and habitat classification schemes. In P. T. Harris & E. Baker (Eds.), *Seafloor geomorphology as benthic habitat: GeoHAB atlas of seafloor geomorphic features and benthic habitats* (pp. 61–92). London: Elsevier Insights. doi:10.1016/B978-0-12-385140-6.00004-9

Head, K. H. (1980). *Manual of soil laboratory testing*. London: Pentech Press.

Hutchinson, G. E. (1957). Concluding remarks. *Cold Spring Harbor Symposia on Quantitative Biology*, *22*(0), 415–427. doi:10.1101/SQB.1957.022.01.039

James, J. W. C., Pearce, B., Coggan, R. A., Arnott, S. H. L., Clark, R., Plim, J. F., et al. (2010). The south coast regional environmental characterisation. *British Geological Survey Open Report OR/09/51*.

Konert, M., & Vandenberghe, J. (1997). Comparison of laser grain size analysis with pipette and sieve analysis: A solution for the underestimation of the clay fraction. *Sedimentology*, *44*(3), 523–535. doi:10.1046/j.1365-3091.1997.d01-38.x

Lark, R. M., Dove, D., Green, S. L., Richardson, A. E., Stewart, H., & Stevenson, A. (2012). Spatial prediction of seabed sediment texture classes by cokriging from a legacy database of point observations. *Sedimentary Geology*, *281*, 35–49. doi:10.1016/j.sedgeo.2012.07.009

Long, D. (2006). BGS detailed explanation of seabed sediment modified folk classification. In *MESH guide to marine habitat mapping*. Retrieved December 19, 2013, from http://www.searchmesh.net/default.aspx?page=1752

Maljers, D., Verfaillie, E., Gunnink, J., Van Heteren, S., & Van Lancker, V. (2006). Interpolating grain-size maps: transnational initiative between Belgium, Netherlands and UK. *MESH Technical Group Meeting*. Retrieved January 9, 2014, from http://www.searchmesh.net/default.aspx?page=1936

MESH Consortium. (2007). MESH guide to marine habitat mapping. *MESH, Mapping European Seabed Habitats*. Retrieved August 24, 2013, from http://www.searchmesh.net/default.aspx?page=1616

Post, A. L., Wassenberg, T. J., & Passlow, V. (2006). Physical surrogates for macrofaunal distributions and abundance in a tropical gulf. *Marine & Freshwater Research*, *57*(5), 469. doi:10.1071/MF05182

Reiss, H., & Kröncke, I. (2005). Seasonal variability of infaunal community structures in three areas of the North Sea under different environmental conditions. *Estuarine, Coastal and Shelf Science*, *65*(1-2), 253–274. doi:10.1016/j.ecss.2005.06.008

Renosh, P. R., Schmitt, F. G., Loisel, H., Sentchev, A., & Mériaux, X. (2014). High frequency variability of particle size distribution and its dependency on turbulence over the sea bottom during re-suspension processes. *Continental Shelf Research*, *77*, 51–60. doi:10.1016/j.csr.2014.01.024

Rodríguez, J. G., & Uriarte, A. (2009). Laser diffraction and dry-sieving grain size analyses undertaken on fine- and medium-grained sandy marine sediments: A note. *Journal of Coastal Research*, *25*, 257–264. doi:10.2112/08-1012.1

Schlee, J. (1973). U.S. Geological Survey Professional Paper: *Atlantic continental shelf and slope of the United States – sediment texture of the northeastern part (vol. 529L)*. Washington, DC: United States Government Printing Office.

Shepard, F. P. (1954). Nomenclature based on sand-silt-clay ratios. *Journal of Sedimentary Petrology*, *24*, 151–158.

Stafleu, J., Maljers, D., Gunnink, J. L., Menkovic, A., & Busschers, F. S. (2011). 3D modelling of the shallow subsurface of Zeeland, the Netherlands. *Netherlands Journal of Geosciences*, *90*, 293–310.

Stevens, T., & Connolly, R. M. (2004). Testing the utility of abiotic surrogates for marine habitat mapping at scales relevant to management. *Biological Conservation*, *119*(3), 351–362. doi:10.1016/j.biocon.2003.12.001

Stevenson, A. (Ed.). (2011). *EMODnet-Geology project final report*. Retrieved on December 30, 2013, from http://www.emodnet-geology.eu

Udden, J. A. (1914). Mechanical composition of clastic sediments. *Bulletin of the Geological Society of America*, *25*, 655–744.

Van der Meulen, M. J., De Lang, F. D., Maljers, D., Dubelaar, C. W., & Westerhoff, W. E. (2003). *Grondsoorten ent delfstoffen bij naam (Soil types and raw materials by name). Publicatiereeks Grondstoffen, 2003/16*. Delft, The Netherlands: Dienst Weg- en Waterbouwkunde. (in Dutch)

Van Dijk, T. A. G. P., Van Dalfsen, J. A., Van Overmeeren, R. A., Van Lancker, V., Van Heteren, S., & Doornenbal, P. J. (2012). Benthic habitat variations over tidal ridges, North Sea, the Netherlands. In P. T. Harris & E. Baker (Eds.), *Seafloor geomorphology as benthic habitat: GeoHAB atlas of seafloor geomorphic features and benthic habitats* (pp. 241–249). London: Elsevier Insights. doi:10.1016/B978-0-12-385140-6.00013-X

Van Lancker, V., Moerkerke, G., Du Four, I., Verfaillie, E., Rabaut, M., & Degraer, S. (2012). Fine-scale geomorphological mapping of sandbank environments for the prediction of macrobenthic occurrences, Belgian part of the North Sea. In P. T. Harris & E. Baker (Eds.), *Seafloor geomorphology as benthic habitat: GeoHAB atlas of seafloor geomorphic features and benthic habitats* (pp. 251–260). London: Elsevier Insights. doi:10.1016/B978-0-12-385140-6.00014-1

Van Lancker, V., & Van Heteren, S. (Eds.). (2013). Standardisation and harmonisation in seabed habitat mapping: role and added value of geological data and information: Part A: Sediment characterisation. *Geo-Seas report 10.5A*. Retrieved on April 7, 2013, from http://www.geo-seas.eu/content/content.asp?menu=0040028_000000

Verfaillie, E. (2008). *Development and validation of spatial distribution models of marine habitats, in support of the ecological valuation of the seabed*. (PhD thesis), Ghent University, Ghent, Belgium.

Verfaillie, E., Van Lancker, V., & Van Meirvenne, M. (2006). Multivariate geostatistics for the predictive modelling of the surficial sand distribution in shelf seas. *Continental Shelf Research*, *26*(19), 2454–2468. doi:10.1016/j.csr.2006.07.028

Wenger, E., McDermott, R., & Snyder, W. M. (2002). *Cultivating Communities of Practice: A guide to managing knowledge*. Boston: Harvard Business School Press.

Wentworth, C. K. (1922). A scale of grade and class terms for clastic sediments. *The Journal of Geology*, *30*(5), 377–392. doi:10.1086/622910

Willems, W., Goethals, P., Van den Eynde, D., Van Hoey, G., Van Lancker, V., & Verfaillie, E. et al. (2008). Where is the worm? Predictive modelling of the habitat preferences of the tube-building polychaete *Lanice conchilega*. *Ecological Modelling*, *212*(1-2), 74–79. doi:10.1016/j.ecolmodel.2007.10.017

ADDITIONAL READING

Blott, S. J., & Pye, K. (2006). Particle size distribution analysis of sand-sized particles by laser diffraction: An experimental investigation of instrument sensitivity and the effects of particle shape. *Sedimentology*, *53*(3), 671–685. doi:10.1111/j.1365-3091.2006.00786.x

Blott, S. J., & Pye, K. (2012). Particle size scales and classification of sediment types based on particle size distributions: Review and recommended procedures. *Sedimentology*, *59*(7), 2071–2096. doi:10.1111/j.1365-3091.2012.01335.x

Cooper, K. M. (2013). Setting limits for acceptable change in sediment particle size composition: Testing a new approach to managing marine aggregate dredging. *Marine Pollution Bulletin*, *73*(1), 86–97. doi:10.1016/j.marpolbul.2013.05.034 PMID:23806669

Cooper, K. M., Curtis, M., Wan Hussin, W. M. R., Barrio Froján, C. R. S., Defew, E. C., Nye, V., & Paterson, D. M. (2011). Implications of dredging induced changes in sediment particle size composition for the structure and function of marine benthic macrofaunal communities. *Marine Pollution Bulletin*, *62*(10), 2087–2094. doi:10.1016/j.marpolbul.2011.07.021 PMID:21872890

Eleftheriou, A. (Ed.). (2013). *Methods for the study of marine benthos*. Oxford: John Wiley & Sons. doi:10.1002/9781118542392

Folk, R. L. (1974). *The petrology of sedimentary rocks*. Austin, Texas: Hemphill Publishing Co.

Gao, S., Collins, M. B., Lanckneus, J., De Moor, G., & Van Lancker, V. (2003). Grain size trends associated with net sediment transport patterns: An example from the Belgian continental shelf. *Marine Geology*, *121*(3-4), 171–185. doi:10.1016/0025-3227(94)90029-9

Harris, P. T., & Baker, E. (Eds.). (2012). *Seafloor geomorphology as benthic habitat: GeoHAB atlas of seafloor geomorphic features and benthic habitats*. London: Elsevier Insights.

Huang, Z., McArthur, M., Radke, L., Anderson, T., Nichol, S., Siwabessy, J., & Brooke, B. (2012). Developing physical surrogates for benthic biodiversity using co-located samples and regression tree models: A conceptual synthesis for a sandy temperate embayment. *International Journal of Geographical Information Science*, *26*(11), 2141–2160. doi:10.1080/13658816.2012.658808

Jerosch, K. (2013). Geostatistical mapping and spatial variability of surficial sediment types on the Beaufort Shelf based on grain size data. *Journal of Marine Systems*, *127*, 5–13. doi:10.1016/j.jmarsys.2012.02.013

Nyberg, J., Lindberg, T., Holmberg, J., Zetterlund, S., & Stölen, L. K. (2007). Marine environmental data, availability and harmonization: BALANCE WP1 Final Report. *SGU-Report 2007:36*. Retrieved on April 7, 2013, from http://balance-eu.org/publications/index.html

KEY TERMS AND DEFINITIONS

Ecosystem Engineer: Any organism that physically, biologically or chemically creates or modifies its own habitat and thus influences the surrounding environment.

Habitat Suitability: Suitability of a habitat for selected species or assemblages, as determined on the basis of known biotic-abiotic relationships.

Indicative Meaning: Connotation of surrogate or proxy data for a parameter that has not been observed directly.

Legacy Data: Data stored in old formats or computer systems that make access and processing difficult.

Marine Habitat: Marine area marked by a dynamic set of physical conditions and biotic factors that jointly provide a suitable environment for one or more living plant or animal species.

Marine Strategy Framework Directive: European Directive introduced in 2008 to achieve or maintain Good Environmental Status in European seas.

Sediment: Mixture of naturally occurring particles with varying sizes and shapes, formed as a result of weathering and erosion, and subject to transportation by wind, water, ice, or gravitational processes.

Self-Organization: Process in which local interactions between various components of an originally disordered system result in some degree of overall order.

Surrogate: Proxy variable that can be mapped with a quantifiable correspondence to the occurrence of benthic species and communities.

Chapter 9
Knowledge Building and Computer Tools

Giuseppe M. R. Manzella
ETT SPA, Italy & Liguria Cluster of Marine Technology, Italy

ABSTRACT

The possibility of having free and open access to data, information, books, and conference papers allows each person to improve his/her comprehension and personal beliefs, so that he/she can embark on a social process of interaction with other people and with shared scientific and technological knowledge. This chapter describes some essential components of a knowledge-building environment based on open and free access to data. The internet-based knowledge-building environment also allows access to books, articles, and any publications that transform data into knowledge. The primary objective of the proposed knowledge-building environment is to document, store, and share all data and methods, making them available for review by other scientists or men of learning, who can then verify and reproduce the results.

INTRODUCTION

The diffusion of knowledge through books and use of the printing press, as well as the expansion of mass education through schools and universities, has accompanied the history of human evolution from the Renaissance to the present day. The current spread of new technologies and the emergence of the internet as a public network are carving out fresh opportunities to widen public knowledge and improve human life. In modern society, technologically-mediated information is constantly growing and permeates all aspects of civil society.

The link between *Scientia* (knowledge in Latin) and society has been cause for debate since the beginning of human civilisation. Socrates declared that he moved philosophy from heaven to earth. The 18th century saw the rise of reportage in *The Spectator,* a daily publication founded by Joseph Addison, as well as the so-called *Living Room Philosophy*, followed by Voltaire's *Philosophie pour les Dames* and Sade's *Philosophie dans le boudoir*. At the end of the 19th century, in Vienna, a journalistic presentation of information became popular that led Kraus (1909) to state: *The making of a journalist: no ideas and the ability to express them.* In 1976, the United States sociologist Daniel

DOI: 10.4018/978-1-4666-6567-5.ch009

Copyright © 2015, IGI Global. Copying or distributing in print or electronic forms without written permission of IGI Global is prohibited.

Bell (1976) introduced the notion of "information society" in his book *The Coming of Post-Industrial Society*, where he stated that the main axis of the new world society would be theoretical knowledge and he warned that knowledge-based services would be transformed into the central structure of a new economy and of an information-led society, where ideologies would become superfluous.

The present situation is very well represented by the opening stanza of T.S.Eliot's (1934) Choruses from "The Rock":

Where is the wisdom we have Lost in knowledge?
Where the knowledge we have Lost in information?

The question we should ask ourselves is: do we need to progress towards an information society or to a knowledge-based society? In the real world, information is disjointed and often deflected by commercial or political interests. They become social barriers that limit the possibility to move from *Scientia* towards an enhancement of cooperation, openness, and links to a knowledge-based society.

There is a virtuous circle in the scientific community that transforms data into information then into knowledge. This circle should be opened to move from fundamental to applied knowledge for a direct social benefit. The level of scientific literacy in the general public and the interest of young people in science should be raised, in order to usher in a new Renaissance that is not only restricted to a few privileged sectors of society, but expanded to society as a whole.

Knowledge is at the base of improvements in society and the economy. Knowledge-sharing (including information, skills, expertise) makes a group, an organisation and society as a whole more competitive. The spread of the internet and new communication technologies has provided opportunities to support 'communities of practice' (Lave and Wenger, 1991), i.e. groups of people with a common interest in specific fields. Knowledge management is driven by the need (inter alia) to make knowledge content available for new development cycles and for the supply of products and services to different communities. In any given community, easier access to knowledge can decrease the learning curve of new incomers and prevent the *'re-invention of the wheel'*, spawning new ideas for products and services.

Public engagement with science has become an almost obligatory passage point for science policy in Europe, even if its substantive forms and meaning still need development (Mejlgaard et al., 2012). A key issue for knowledge is innovation. The European Union has purposely tied its vision of scientific research to one of economic competitiveness through continuous technological innovation. In what is now called the "knowledge age," the health and wealth of societies depends increasingly on their ability to innovate.

UNESCO (2005), linked the concept of knowledge society to the Universal Declaration of Human Rights that includes: freedom of information, media pluralism, academic freedom, right to education, the right "to freely participate in the cultural life of the community, to enjoy the arts and to share in scientific advancement and its benefits" (Article 27, paragraph 1, of the Universal Declaration of Human Rights).

The general public, not just a specialised elite, needs to work creatively with knowledge. From a scientific and technological point of view, the knowledge society presents many challenges, as it requires the use of new technologies, including informatics, and the ability to use research results in a new and constructive manner.

This paper describes a simple knowledge-building environment that includes access to data and information residing in different information systems.

BACKGROUND: KNOWLEDGE BUILDING ENVIRONMENT

Knowledge Building may be defined as the production and continuous improvement of ideas of value to a community. Science is becoming, but is not yet an integral part of people's daily lives. Additional steps are necessary to build a society made up of scientifically literate individuals who have the ability to:

- Understand experiments and the reasoning behind basic scientific facts and their meaning;
- Ask, find, or determine answers to questions resulting from a curiosity about everyday experiences;
- Describe, explain, and predict natural phenomena;
- Read and understand articles about science in the popular press and engage in social conversation about the validity of the conclusions;
- Identify scientific issues underlying national and local decisions and express positions that are scientifically and technologically informed;
- Evaluate the quality of scientific information on the basis of its source and the methods used to generate it;
- Pose and evaluate arguments based on evidence and to apply conclusions from such arguments appropriately.

The primary objective of a knowledge-building environment is to document, store and share all data and methods, making them available for review and to provide the opportunity to verify and reproduce the results. Once created, data are valuable resources that can be used and re-used for future environmental, scientific, management and educational purposes. Data sharing facilitates new scientific inquiries, avoids duplicating data collection and provides rich real-life resources for well-defined objectives (e.g. environmental protection, education and training).

The purpose of Knowledge Building is to advance the frontiers of knowledge as perceived by the various components of society. "As perceived by ..." is an important qualification when Knowledge Building is undertaken in the social context. Identifying frontiers and judging what constitutes an advance are essential parts of Knowledge Building. Of course, frontiers perceived by children, for example, will differ from those perceived by professionals, but professionals may also disagree among themselves about where the frontier lies and what constitutes an advance. Dealing with such issues is part of the work of any knowledge-building group.

From a social point of view, the ability to connect discourses within and between communities opens up new possibilities for barrier-crossing and mutual support. The knowledge society is based on the so-called "Matthew effect" (the rich get richer): the more you know, the more you can learn. However, another important principle is that knowledge needs to be of value to people in their current lives, not merely banked against future needs.

KNOWLEDGE BUILDING, SCIENCE, AND SOCIETY

Science and technology continuously strive to improve existing knowledge with the participation of society as a whole. The progress of a society can be measured in hindsight by comparing knowledge available today with the past.

There are two ways in which substantial knowledge permeates society: as social practice (skilful actions) and as innovation base (creativity, change). A knowledge-building environment can stimulate questions and reveal emerging issues and can also enhance knowledge, improvement and development.

A knowledge-building environment needs multiple ways of representing and organising ideas and flexible ways to link them. It should provide tools for superordination, intended as the possibility to develop new ideas subsuming or integrating previously recorded ideas. Information and Communication Technology (ICT) is an important element of knowledge-building. It is an environment where ideas can be presented, discussed, revised, organised and combined.

Data, once collected, analysed, organised, and used by people with the necessary expertise, can be used again in social practice. Their 'good use' must be based on a technological and physical structure designed to facilitate collaborative knowledge building, decision-making and inference. This is called 'knowledge environment' and requires a 'knowledge-building environment' to address the needs of different communities in order to create knowledge and consider *People, Process and Technology* issues (Figure 1). Any implementation in one of the components has an impact to some extent, on all the others. Data and documentation are part of a powerful infrastructure, but they need to be established with *People* and *Process* implications and not just *Technology* (Lopez-Ferrer, 2015 see chapter 5 this book).

To build a knowledge-building environment, it is necessary to go beyond the *data* > *information* > *knowledge* flow and include elements that are mutually agreed upon by different communities. These elements must be '*interoperable*', based on '*standards*', '*common models*' and '*common vocabularies*'.

Interoperability and Standards

Interoperability is the capability of two or more systems of communicating and exchanging data. This is done, for example, with specific data formats and communication protocols. In most cases, interoperability is achieved 'post-facto', i.e. as a result of market dominance of particular products.

A standard should be an 'ab-initio' established norm or requirement with regard to a technical component of knowledge management. In reality, the process of developing technical standards includes spontaneous standardisation to produce 'de-facto standards'.

Data and Metadata Models

Data can be defined as a quantitative characteristic of an object. Data, information and knowledge differ mainly in abstraction, data being the least and knowledge the most abstract.

The main aim of data models is to support the development of information systems by providing data with a definition and a format. In the author's opinion, the most important effort now being made to develop a data model that can be applied to all kinds of data is the UNIDATA project (http://www.unidata.ucar.edu/software/thredds/current/netcdf-java/CDM/), which is providing the syntactic and semantic elements of a generic data file (Figure 2).

The UNIDATA Common Data Model (CDM) Feature Type layer turns CDM datasets into collections of 'feature type' objects (i.e. objects that can have a geographic location), and allows a user to extract subsets of the feature types "in coordinate space" i.e. using spatial and temporal bounding boxes.

Metadata are critical for recording data provenance, including processing steps and quality control certification (answers to the general questions: who, what, where, when, which, why, how). Metadata are traditionally found in the card catalogues of libraries. Today, they can be found in photographs, videos, web pages. International standards (e.g. ISO - International Organisation for Standardisation) apply to metadata. The definition of *metadata* given originally by Bagley (1968) was based on the ISO11179 standard: *data about the containers of data*.

A metadata model provides frames, rules, constraints and theories applicable and useful for the description of objects.

Common Vocabularies

Metadata and data formats are supported by common vocabularies. Their use in all metadatabases

Figure 1. The components of a knowledge management system

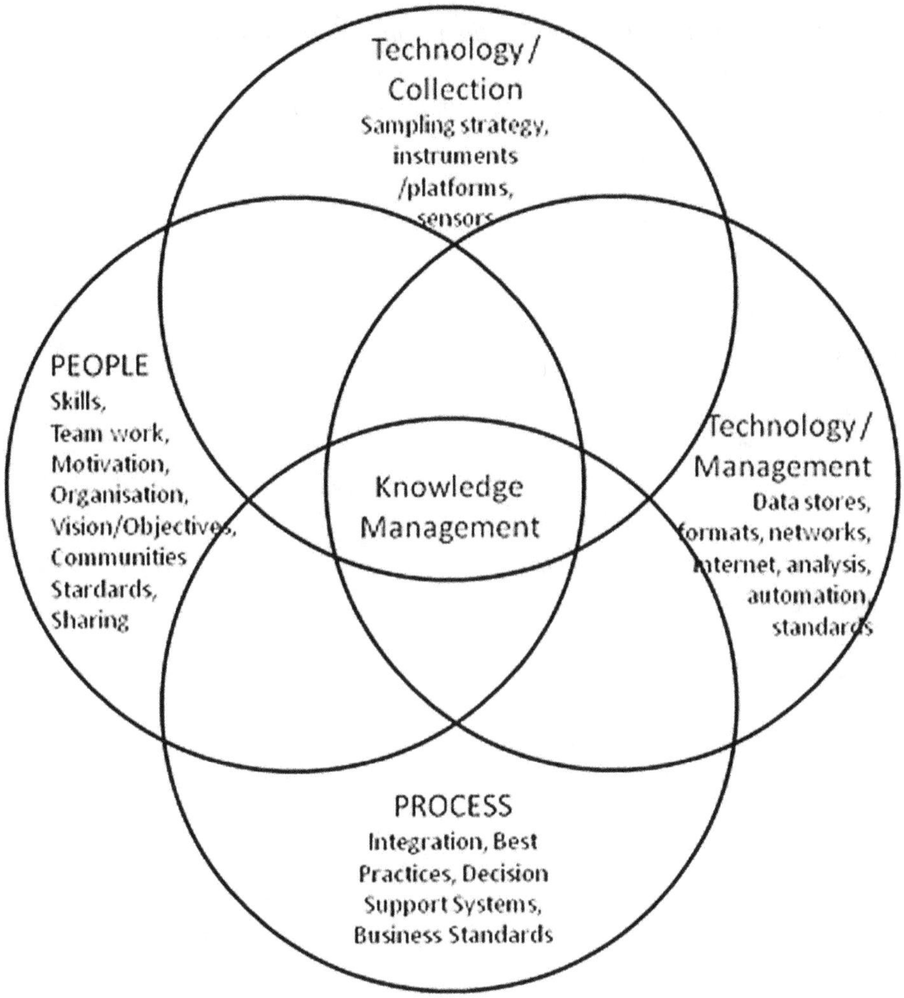

and data formats is an important prerequisite towards consistency and interoperability. Common vocabularies consist of lists of standardised terms that cover a broad spectrum of disciplines of relevance to the oceanographic and wider community. Using standardised sets of terms solves the problem of ambiguities associated with data mark-up and also enables records to be interpreted by computers. This opens up data sets to a whole world of possibilities for computer aided manipulation, distribution and long term reuse (http://www.seadatanet.org/Standards-Software/Common-Vocabularies).

SOLUTION PROPOSED: A SIMPLE KNOWLEDGE MANAGEMENT SYSTEM

A knowledge-building environment includes a physical workbench where something is being built, a component for sharing ideas and information. The physical workbench is composed of two main components:

- Data, collected using internationally agreed protocols for quality assurance, processed to transform raw data into physi-

Figure 2. The UNIDATA Common Data Model Feature Type. The digital object identifier (DOI) has been introduced as a component of the knowledge management system. For an optimal use of data this is not sufficient, additional information is needed on the means of data creation, purpose, time and date of creation, creator/author, standards used and location on a computer network. These and other elements (part of the knowledge management) are provided in the boxes on the right.

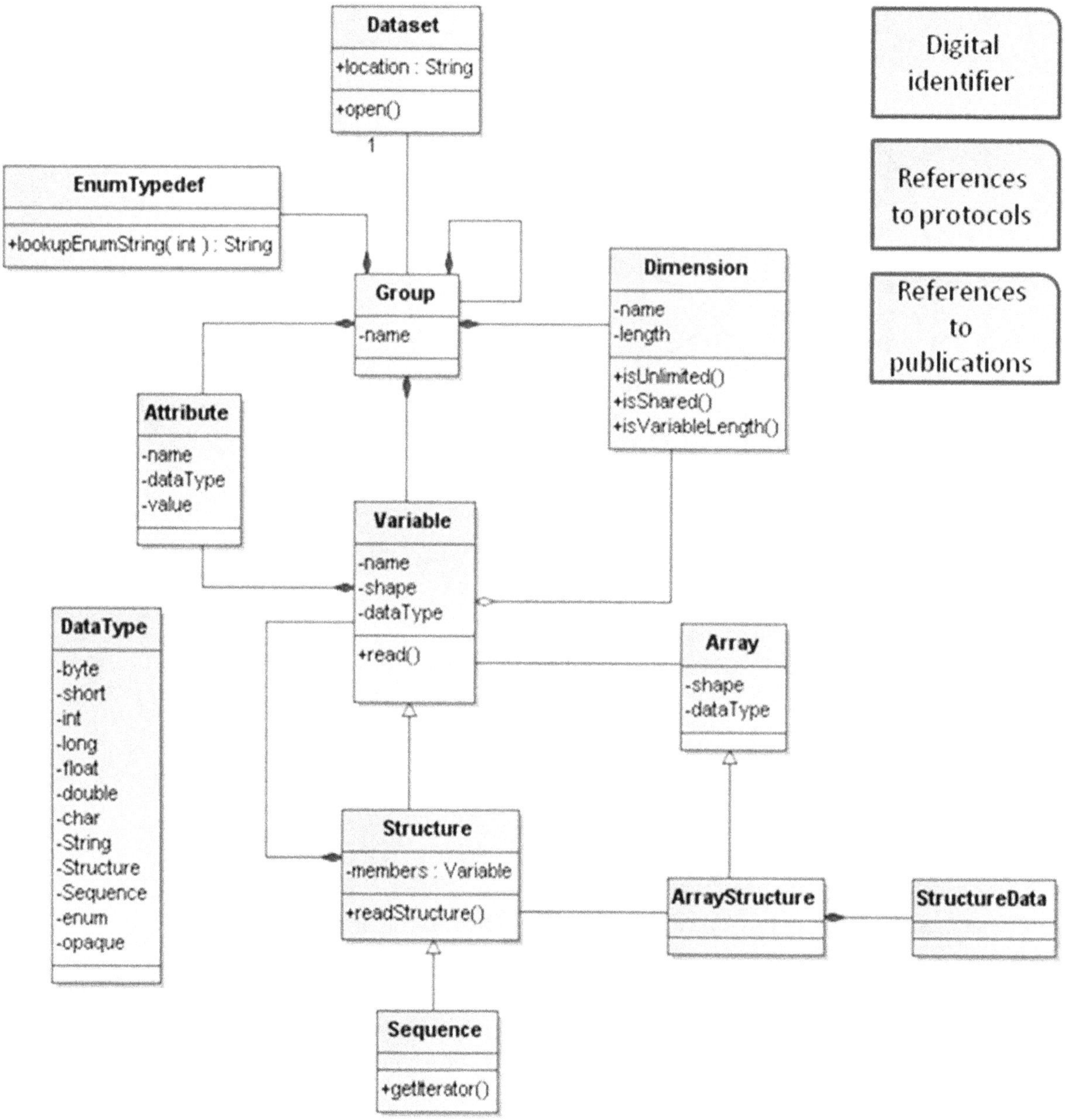

cal data, taking into account the relationship between parameters, validated with internationally agreed quality control protocols (information system/spatial data infrastructure - an open and free access to data/products and information related to data acquisition (including environmental conditions), instrumentation, protocols, for in situ and ex situ practices used for quality assessment and quality control and links to publications);

- Scientific interpretation of data, that starts with an analysis of reliability, data comparability/ compatibility and continues with the application of statistics; this is the content of publications (cross-references between data sets and bibliography/publications and/or protocols).

By introducing a third element, we have a simple but effective knowledge management system:

- Knowledge-building environment - a continuous learning process that allows the presentation of theories, comparison with observations, understanding and resolution of controversies.

Information Systems/Spatial Data Infrastructures

An information system is the combination of hardware and software used to store, process and distribute data. Information systems have evolved toward Spatial Data Infrastructures (SDI) enabling the discovery and delivery of spatial data from a data repository. SDI is the framework on which the INSPIRE directive is based. INSPIRE is the EU initiative to make spatial or geographical information more accessible and interoperable for a wide range of purposes supporting sustainable development. The basic components of an SDI are also part of the services required by INSPIRE:

- A software client to display, query, and analyse spatial data;
- A catalogue service for the discovery, browsing, and querying of metadata or spatial services, spatial datasets and other resources;
- A spatial data service allowing the delivery of data via the Internet;
- Processing services such as datum and projection transformations;
- A spatial data repository to store data.

To create and update spatial data, a Geographic Information System (GIS) software client or desktop is also required. On the base of the INSPIRE directive definitions, "spatial data" means any data with a direct or indirect reference to a specific location or geographical area. In this definition, there is a base for the passage from "simpler" Information Systems to SDI, since it is now recommended to access, manipulate or analyse data that can be mapped through a GIS. Technical standards have to be used in an SDI, as well as format and internet transfer standards defined by W3C (http://www.w3.org/). The use of ISO standards allows 'interoperability' among different SDI systems.

The ISO 19115 metadata standards are mostly used for catalogue purposes (the so- called 'discovery metadata'). The standard applies to:

- Cataloguing data (datasets), clearinghouse activities, full description of data,
- Geographic data, geographic data sets and geographical characteristics and properties of individual characteristics,

And define:

- Mandatory and conditional metadata sections, metadata entities, and metadata elements;
- The minimum set of elements (core metadata) required by the full range of metadata applications (data discovery, determining the relevance for use, data access, data transfer, and use of digital data);
- Optional metadata elements to allow for a more extensive standard description of geographic data, if required;
- A method for extending metadata to meet specialised requirements.

To facilitate the use and re-use of data, it is necessary to implement metadata services that can provide the following information:

- **What:** Describing the objectives, and/or the content of the project,
- **Why:** Explaining the reason and/or state-of-the-art before the project,
- **Where:** Describing the scientific rationale in which efforts are concentrated,
- **Who:** Describing the group of participants and/or the project participants and roles,
- **How:** Providing information and links to data collection methods and technologies.

In this case, in order to understand *why* a project and data collection was initiated and *how* it has been conducted, it is important to provide references to papers, protocols, and bibliography in general. At the same time, services that allow access to data must be provided:

- **What:** Listing the parameters that can be accessed,
- **Where:** Describing the geographical area where data were collected,
- **When:** Providing information on temporal coverage,
- **How:** Providing links to data.

One useful mandatory field in ISO19115 is the abstract describing the dataset, but it can also include additional elements providing important information. Services dedicated to parameters and units, geographical areas and instruments are provided by means of common vocabularies, which assure a certain level of interoperability. Other metadata services have been developed in Europe since the 1990s, giving overviews of European marine organisations, research projects and data sets, data acquisition and monitoring programmes (Schaap and Lowry, 2010).

- **EDMED:** An inventory of data sets collected during research cruises began in 1991. The *European Directory of Marine Environmental Data* describes data sets from a wide range of disciplines. It has established itself as a European standard for indexing and searching data sets relating to the marine environment.
- **EDMERP:** Research projects are described as metadata factsheets with their most relevant aspects. The primary objective is to support users in identifying interesting research activities and in connecting them to the research managers and organisations concerned throughout Europe. The *European Directory of Marine Environmental Programmes* covers marine research projects for a wide range of disciplines including marine meteorology; physical, chemical and biological oceanography; sedimentology; marine biology and fisheries; environmental quality; coastal and estuarine studies; marine geology and geophysics, etc.
- **EDMO:** The *European Directory of Marine Organisations* contains up-to-date addresses and activity profiles of research institutes, data holding centres, monitoring agencies, governmental and private organisations, that are in one way or another engaged in oceanographic and marine research activities, data & information management and/or data acquisition activities.

In practice, by properly organising the abovementioned services, the user is provided with some important information on data collection (Figure 3). However, these directories are not sufficient for an optimal use of data; there is still the need to include information on the protocols used for quality assessment, quality control, as well as on the accuracy and calibration of instruments. Ad-

Figure 3. The links between metadata services and data sets

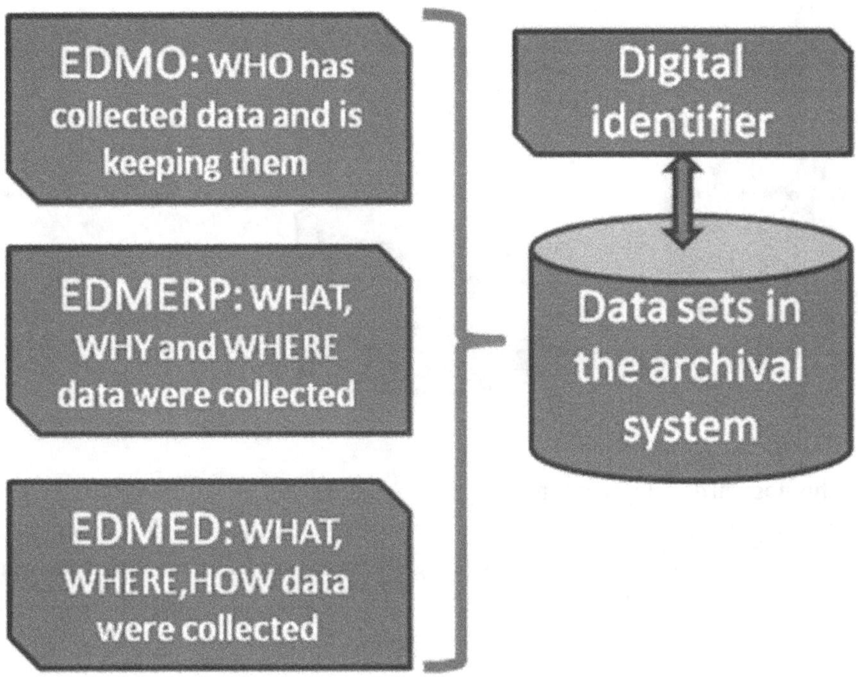

ditional information can avoid errors in the use of data. This can be achieved with an appropriate model of both data and metadata.

It must be underlined that moving from an information system to a spatial data infrastructure also requires the adoption of ISO19139 standards.

Cross-Reference Data: Publications

Data are described (metadata) and made accessible through specialised languages (i.e. XML http://www.w3.org/XML/). The inclusion of data set identifiers in metadata (Figures 2 and 3) could allow a cross-reference between data and publications and thus provide information on how data were used or, in turn, from publications move to the data (Figure 4). This element of the knowledge management system (cross-references between data sets and bibliography/publications and/or protocols) must be added to the Common Data Model presented in paragraph 3.2. Further information must be included in metadata to build a 'knowledge management' system. ISO 690 is the standard issued by the ISO, which specifies the elements that must be included in the bibliography of monographs and periodicals and related chapters, articles, etc. It provides a mandatory order for references and establishes conventions for the transcription and presentation of information derived from published sources.

FUTURE IMPLEMENTATION: PERSONAL COMPREHENSION VS. SOCIAL KNOWLEDGE

The future of a knowledge building environment is constituted by the third component of the knowledge management system. It is based on a learning model as a cycle of personal and social knowledge building proposed by Stahl (2000). Scientific and technological developments should become a part of normal human knowledge and activity, if necessary, in practical life choices. In other words,

Figure 4. The schematic links between data sets and protocols / publications. This is one element of the knowledge management system

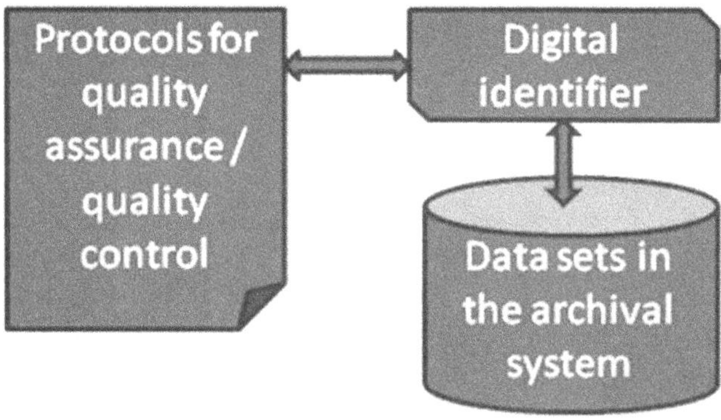

scientific and technological developments must be included in the continuous improvement of human knowledge, in order to increase the well-being of all societies. Scardamaglia and Bereiter (2003) suggested that in the absence of a pedagogical theory, learning-by-doing and apprenticeship are the methods to be applied in knowledge building. They distinguished between knowledge-building

Figure 5. The diagram of personal and social knowledge building. Elaboration of a schema proposed by Stahl (2000).

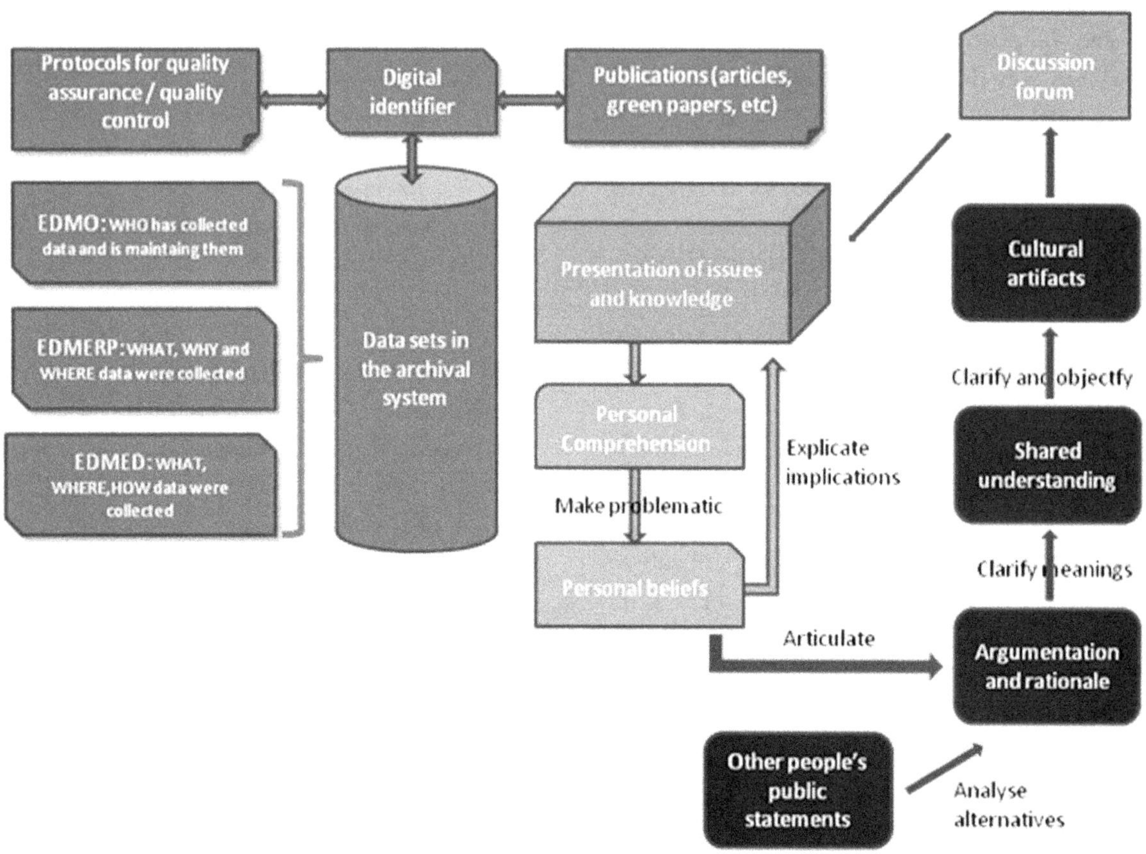

and learning, the latter being a personal process that results in changes in beliefs, attitudes and skills. In the context of this paper, learning is based on the previously presented knowledge management system; the phases of learning are presented in Figure 5.

The following elements are important for computer-based knowledge building:

- Presentation of a summary of research results.
- Links to high-quality textbooks.
- Supply of different levels of detailed information taken from generic content.
- Implementation of specific information sections.
- Integration of information for an interdisciplinary approach.

However, the tools presented above are not sufficient, as they must be based on formal education for a better understanding of the results. There is a need to share and transmit research results through meetings, conferences and not only through the web. The complete knowledge-building system is therefore based on both personal understanding and social discussion, as presented in Figure 5. An important component could include a mediated discussion forum where different ideas are presented. Every one participating to the forum must agree on some fundamental ethical issues and use the forum for learning and understanding both scientific concepts and each other, in order to find a common vision or pose new questions.

CONCLUSION

The motivation of this paper is the need of today's societies to share the results of scientific and technological research in order to improve the living conditions of human beings. The possibility of having free and open access to data, information, books and conference papers allows each person to improve his/her comprehension and personal beliefs, so that he/she can embark on a social process of interaction with other people and with shared scientific and technological knowledge. For a very long time, knowledge has been the exclusive domain of tightly-knit circles of intelligentsia. Today, the demand for democracy, the concept of openness and the gradual emergence of a public forum for knowledge, has fostered the spread of the ideas of universality, liberty and equality.

An important requirement in a knowledge-building environment is the establishment of shared understanding. This must be done by clarifying the meaning of used terms. A *glossary forum* can make explicit how different participants understand the terms they use. The glossary forum results in a group *glossary* of mutually agreed definitions. Such a glossary already represents a form of group knowledge. The glossary is, of course, subject to future debate and modification.

By providing access to data, it is possible to foster different *comparison perspectives*, in which we can view and contrast alternative perspectives and adopt or adapt ideas from other people's perspectives. The idea behind the comparison perspective is to allow a debate on data interpretations and foster convergence of thinking and sharing of insights or interpretations.

REFERENCES

Bagley, P. (1968). *Extension of programming language concepts*. Philadelphia: University City Science Center.

Bell, D. (1976). The Coming of the Post-Industrial Society. *The Educational Forum*, *40*(4), 574–579. doi:10.1080/00131727609336501

Eliot, T. S. (1935). The rock. In *The complete Poems and Plays 1909-1950*. San Diego, CA: Harcourt Brace & Company.

Kraus, K. (1909). *Sprüche und Widersprüche, Aphorismen*. Frankfurt, Germany: Bibliothek Suhrkamp.

Lave, J., & Wenger, E. (1991). *Situated Learning: Legitimate Peripheral Participation*. Cambridge, UK: Cambridge University Press. doi:10.1017/CBO9780511815355

Lopez-Ferrer, M. (2015). International funding and collaboration in Sea Level Rise research. In *Collaborative Knowledge in Scientific Research Networks*. IGI Global Press.

Mejlgaard, N., Bloch, C., Degn, L., Ravn, T., & Nielsen, M. W. (2012). *Monitoring Policy and Research Activities on Science in Society in Europe (MASIS) Final synthesis report*. Bruxelles, Belgium: European Commission.

Scardamaglia, M., & Bereiter, C. (2003). Knowledge Building. In *Encyclopedia of Education* (2nd ed., pp. 1370–1373). New York: Macmillan Publishing Company.

Schaap, D., & Lowry, R. (2010). SeaDataNet – Pan-European infrastructure for marine and ocean data management: Unified access to distributed data sets. *International Journal of Digital Earth*, *3*(1), 50–69. doi:10.1080/17538941003660974

Stahl, G. (2000). A Model of Collaborative Knowledge-Building. In FishmanB.O'Connor-DivelbissS. (Eds.), *Fourth International Conference of the Learning Sciences* (pp. 70-77). Mahwah, NJ: Erlbaum.

UNESCO. (2005). *Towards Knowledge Societies*. Paris, France: UNESCO Publishing.

ADDITIONAL READING

Babbie, E. (2010). *The Practice of Social Research*. Belmont, CA: Wadsworth.

Bereiter, C. (1994). Implications of postmodernism for science, or, science as progressive discourse. *Educational Psychologist*, *29*(1), 3–12. doi:10.1207/s15326985ep2901_1

Bereiter, C., Scardamalia, M., Cassells, C. & Hewitt, J. (1997) Postmodernism, Knowledge Building, and Elementary Science. The Elementary School Journal *97 (4), Special Issue,* 329-340.

Coro, G., Pagano, P., & Ellenbroek, A. (2013). Combining Simulated Expert Knowledge with Neural Networks to Produce Ecological Niche Models for Latimeria chalumnae. *Ecological Modelling*, *268*, 55–63. doi:10.1016/j.ecolmodel.2013.08.005

Diviacco, P. (2012). Addressing Conflicting Cognitive Models in Collaborative E-Research: A Case Study in Exploration Geophysics. In A. Juan, T. Daradoumis, M. Roca, S. Grasman, & J. Faulin (Eds.), *Collaborative and distributed e-research: innovations in technologies, strategies, and applications*. Hershey, PA: IGI Global. doi:10.4018/978-1-4666-0125-3.ch012

Gal, U., Yoo, Y., & Boland, R. J. (2004). "The Dynamics of Boundary Objects, Social Infrastructures and Social Identities," Case Western Reserve University, USA. *Sprouts: Working Papers on Information Systems 4 (11)*. Retrieved from http://sprouts.aisnet.org/4-11

Gottschalk-Mazouz, N. (2008). *Internet and the flow of knowledge: Which ethical and political challenges will we face?* In: *Wittgenstein and the Philosophy of Information. Proceedings of the 30th International Ludwig Wittgenstein-Symposium in Kirchberg, 2007* (pp. 215-232). Frankfurt: Ontos Verlag.

Holland, G., & Pugh, D. (Eds.). (2010). *Troubled waters: ocean science and governance*. Cambridge, UK: Cambridge University Press.

Laudon, K. C., & Laudon, J. P. (1988). *Management Information Systems* (2nd ed.). New York, NY: Macmillan Publishing Company.

Leigh, S. S., & Griesemer, J. R. (1989). Institutional Ecology, 'Translations' and Boundary Objects: Amateurs and Professionals in Berkeley's Museum of Vertebrate Zoology, 1907-39. *Social Studies of Science*, *19*(3), 387–420. doi:10.1177/030631289019003001

Manzella, G. M. R., Iona, S., & Nativi, S. (2010). Editorial "Earth and space science informatics: Informatics in oceanography". *Advances in Geosciences*, *28*, 1–2. doi:10.5194/adgeo-28-1-2010

Miller, M. (1987). Argumentation and cognition. In M. Hickman (Ed.), *Social and functional approaches to language and thought* (pp. 225–249). New York, NY: Academic Press.

Muukkonen, H., Hakkarainen, K., & Lakkala, M. (1999). Collaborative Technology for Facilitating Progressive Inquiry: the Future Learning Environment Tools. In C. Hoadley & J. Roschelle (Editors) *The proceedings of the CSCL '99 conference, December 12-15, 1999, Palo Alto,* pp. 406-415. Mahwah, NJ: Lawrence Erlbaum and Associates. doi:10.3115/1150240.1150291

Nativi, S., Manzella, G. M. R., Paolucci, F., Mazzetti, P., Pecci, L., Bigagli, L., & Reseghetti, F. (2006). Federated data bases for the development of an operational monitoring and forecasting system of the ocean: The THREDDS Dataset Merger. *Advances in Geosciences*, *8*, 39–47. doi:10.5194/adgeo-8-39-2006

Nativi, S., Mazzetti, P., Santoro, M., Boldrini, E., Manzella, G. M. R., & Schaap, D. M. A. (2010). CDI/THREDDS interoperability in the SeaDataNet framework. *Advances in Geosciences*, *28*, 17–27. doi:10.5194/adgeo-28-17-2010

Nonaka, I., & von Krogh, G. (2009). Tacit Knowledge and Knowledge Conversion: Controversy and Advancement in Organizational Knowledge Creation Theory. *Organization Science*, *20*(3), 635–652. doi:10.1287/orsc.1080.0412

Pijanowski, J. (2009). The Role of Learning Theory in Building Effective College Ethics Curricula. *Journal of College & Character*, *10*(3). doi:10.2202/1940-1639.1088

Roschelle, J., & Teasley, S. (1995). The construction of shared knowledge in collaborative problem solving. In C. O'Malley (Ed.), *Computer-supported collaborative learning*. Berlin, Germany: Springer Verlag. doi:10.1007/978-3-642-85098-1_5

Rossi, P. (2011). *La nascita della scienza moderna in Europa*. Bari, Italy: Laterza.

Scardamalia, M. (2002). Collective Cognitive Responsibility for the Advancement of Knowledge. In B. Smith (Ed.), *Liberal Education in a Knowledge Society*. Chicago, IL: Open Court.

Sensky, T. (2002). Knowledge management. *Advances in Psychiatric Treatment*, *8*(5), 387–396. doi:10.1192/apt.8.5.387

Serenko, A., & Bontis, N. (2004). Meta-Review of Knowledge Management and Intellectual Capital Literature: Citation Impact and Research Productivity Rankings. *Knowledge and Process Management*, *11*(3), 185–198. doi:10.1002/kpm.203

Serenko, A., Bontis, N., Booker, L., Sadeddin, K., & Hardie, T. (2010). A scientometric analysis of knowledge management and intellectual capital academic literature (1994- 2008). *Journal of Knowledge Management*, *14*(1), 3–23. doi:10.1108/13673271011015534

Stahl, G., Koschmann, T., & Suthers, D. (2006). Computer-supported collaborative learning: An historical perspective. In R. K. Sawyer (Ed.), *Cambridge Handbook of the Learning Sciences*. Cambridge, UK: Cambridge University Press.

Stanfield, J. H. (1985). The ethnocentric basis of social science knowledge production. *Review of Research in Education*, *12*, 387–415.

KEY TERMS AND DEFINITIONS

INSPIRE: European Union initiative to establish an infrastructure for spatial information in Europe that is geared to help to make spatial or geographical information more accessible and interoperable for a wide range of purposes supporting sustainable development.

Interoperability: The capability of systems and/or organizations to inter-operate. In informatics is the capability to exchange data (syntactic interoperability), with specified formats and communication protocols, or the ability to automatically interpret the information exchanged meaningfully and accurately in order to produce useful results as defined by the end users of different systems (semantic interoperability).

ISO: The International Organization for Standardization is an international standard-setting body composed of representatives from various national standards organizations.

Mark-Up: The terminology evolved from the "*marking up*" of paper manuscripts. In digital media this "mark" was replaced by tags, i.e. names surrounded by symbols.

Metadata: Information on data, describes the contents and context of data files. The main purpose of metadata is to facilitate in the discovery of relevant information. Metadata also helps organize electronic resources, provide digital identification, and helps support archiving and preservation of the resource.

Standard: It is usually a formal document that establishes uniform engineering or technical criteria, methods, processes and practices.

W3C: The World Wide Web Consortium (W3C) is the main international standards organization for the World Wide Web.

Chapter 10
Beyond Learning:
E-Learning Models for Scientific Research

Andrea Corleto
Italian National Agency for New Technologies, Energy and Sustainable Economic Development (ENEA), Italy

Veronica Tomassetti
Italian National Agency for New Technologies, Energy and Sustainable Economic Development (ENEA), Italy

ABSTRACT

The chapter analyzes the theme of collaboration in scientific research through Web 2.0 IT tools and a constructivist approach to knowledge, learning, and e-learning. It analyzes, in particular, the case of Sulcis CCS Summer School. Organized by ENEA, Sotacarbo, and the University of Cagliari, the school provided information and materials on technological aspects of Carbon Capture and Storage (CCS). This work might become the model for a new approach to education and research thanks to collaborative learning using the LMS Moodle. In this case study, the construction of knowledge moves from a teacher-centered model to one based on a community of practice. This model is naturally a "prototype," an experiment in progress, that needs to be refined through further experiences.

FRAMEWORK: SULCIS PROJECT

In late August 2012 the Carbosulcis miners organized a protest during which the Nurax'e Figus coal mine, in Gonnesa (province of Carbonia-Iglesias) was occupied. The workers demanded funding for the integrated project "carbon-mine" incorporating the capture and the geological storage of CO2. This brought to the attention of the public and all media, not only the pressing issue of unemployment in that area, but also the decisions in industrial development strategies that the Italian government is called to adopt.

Born under the pressure of Law number 99 of July 2009 with the intention to take advantage of funds provided by the European Commission NER 300 scheme, the "Sulcis Project" is an initiative of great challenge and technological innovation, planned in Sardinia by Sotacarbo company in close collaboration with ENEA, to develop, qualify

DOI: 10.4018/978-1-4666-6567-5.ch010

Copyright © 2015, IGI Global. Copying or distributing in print or electronic forms without written permission of IGI Global is prohibited.

and make economically competitive engineering solutions based on cutting-edge CCS technologies (Carbon Capture and Storage).

After going through various stages of adjustment, the project obtained the final go-ahead permission, at the beginning of August 2013, thanks to the signing of a Memorandum of Understanding, between the Ministry of Economic Development and the Sardinia Regional Government, for the development of a technological hub for clean coal in the Sulcis Iglesiente area. The Memorandum consists of two phases: the first is to plan the development of a research center that studies carbon dioxide capture and storage techniques, and the development of a 50 MWt experimental plant for CO2 capture. The second phase involves the construction of a power plant with CCS technology. In this context the idea of Sulcis CCS Summer School was envisioned. Launched by ENEA, Sotacarbo and the University of Cagliari in the summer of 2013, the school was the first step of an initiative that will proceed annually in a more comprehensive way and with a strong international connotation. The goal is to enable a permanent venue not only for research and training, but also to inform local and national administrators and stakeholders, to facilitate dialogue, increase confidence in these technologies and thus contribute to public acceptance.

THE RESEARCH FIELD: CARBON CAPTURE AND STORAGE

Global warming is already taking place and it is one of the biggest challenges mankind is facing. Failing to tackle it could have dramatic consequences. In order to have a reasonable chance of avoiding such consequences, the global average temperature must not increase by more than 2°C above the pre-industrial level. According to the Intergovernmental Panel on Climate Change (IPCC), this requires an overall 50 to 85% reduction of global greenhouse gas (GHG) emissions by 2050.

Achieving a 85% cut in global GHG emissions by 2050 seems to be possible, but it requires a tremendous effort to transform society into a low carbon emitting economy. This can only be achieved using a combination of solutions, as no single measure would be capable of reducing CO2 emissions on the massive scale required. This includes adopting measures for energy efficiency, a vast increase in use of renewable energies – and CO2 Capture and Storage (CCS).

Carbon Capture and Storage (CCS) is a technology that can prevent large amounts of carbon dioxide from being released into the atmosphere. The technology involves capturing CO2 produced by large industrial plants, compressing it for transportation and then injecting it deep into a rock formation at a carefully selected and safe site, such as depleted natural gas fields, deep saline aquifers and unmineable coal seams, where it is permanently stored.

Europe is recognised as one of the global leaders in the development of CCS technologies and the European Union adopted its CCS Directive in 2009 to help CCS become commercially viable, starting to recognise its potential as an important technology capable of reducing carbon emissions globally.

The CCS technology is currently being demonstrated worldwide at commercial scale, but without further study the role of CCS in meeting future climate goals will remain uncertain. CCS is expensive and energy intensive and it has been repeatedly stated that adding CCS to existing coal plants can lead to an increase in electricity prices, whilst reducing plant efficiency.

As a consequence, to cut down these costs and make the technology effective and performing it is essential to continue to study, test and research. Italian research centers and universities started specific studies and programmes regarding various aspects of CCS more than 10 years ago. Such initiatives have been financed through European Programmes and national funding. One of these initiatives is the realization in Sardinia Sulcis area of a medium size power plant, burning Sulcis coal and adopting CCS technologies.

There's no doubt that without policies to support these demonstrations, without environmental regulations and standards and without an engagement of local communities and citizens, CCS will not be able to play the large role here envisaged.

Engaging communities and citizens is what public awareness should do. Up to now research findings indicate that public awareness of CCS is low and that people know little about the option. To enable the public to form a fully considered opinion on a technology as complex as CCS, access to factual, comprehensive, and reliable information is required. For example, the public should be adequately informed as to how the implementation of a project may affect them. This is one of the goals of the Sulcis CCS Summer School: to create a link between the Community of Practice (CoP) represented by the School (and its participants at all levels), and the various stakeholders involved in the Sulcis area. This new "network of practice", thus constituted, would facilitate the spreading of correct information and its territorial growth, and the acceptance of the new technology by citizens, finally involved, avoiding the enforcement of political choices.

RESEARCH AND KNOWLEDGE TRANSFER ON INTERNET

Collaboration is recognized as essential to the advancement of scientific knowledge. The nature of co-production of knowledge in collaborative settings has been studied for some time, but little is known about how the process works in collaborations established and maintained through virtual organizations. With no doubt, Internet has changed the way people live, communicate and relate to each other and distances are no more considered a barrier for communication. The power of the World Wide Web and the increasing use of Internet has provided a network of infrastructures for collaborative technologies that supports communication between people and groups.

In the last years, all over the world thousands of people share their life on social networks, work together even if they are thousands of kilometers away from each other, and exchange tons of data and learn something new every day. Practically, Internet has reduced the distance among people and has encouraged collaboration since it tends, due to "spatial proximity", to generate more informal communications.

So, today, concepts like "online collaborative work" or "online collaborative learning" are well known and applied in daily practice in many working and learning environments.

Modern science and research are increasingly collaborative too, but it is not so easy to define tools or environments for online collaborative research. Two are the main problems to deal with.

- The first: Scientists and researchers carry out partnerships at a distance thanks to technologies such as e-mail, videoconferencing, shared whiteboards and centralized databases, but those tools do not form an organic set.
- The second: While information has become easy to transmit and share, knowledge is still difficult to transfer.

Collaboratories and the Science of Collaboration

Over the last 30 years researchers have made a number of large-scale attempts to build computer-supported scientific collaboration, often called *collaboratories*.

According to Wulf (1989) collaboratory is a "center without walls, in which the nation's researchers can perform their research without regard to physical location, interacting with colleagues, accessing instrumentation, sharing data and computational resources, accessing information in digital libraries". The concept of collaboratory, that considerably expanded from the earliest definitions, seems to be going in the right direction to solve the difficulties quoted in the previous paragraph.

In a workshop at the University of Michigan, in 2001, a group of scientists developed the following definition: "a collaboratory is an organizational entity that spans distance, supports rich and recurring human interaction oriented to a common research area, and provides access to data sources, artifacts and tools required to accomplish research tasks" (Olson et al., 2008). The definition, even if only slightly, shifts the focus on human interaction and a social dimension of research.

The social dimension becomes essential in the definition provided by Derrick Cogburn (2003):

A collaboratory is more than an elaborate collection of information and communication technologies; it is a new networked organizational form that also includes social processes; collaboration techniques; formal and informal communication; and agreements on norms, principles, values, and rules.

The Science of Collaboration (SOC) was a five year-project led by the National Science Foundation to study large-scale academic research collaborations across many disciplines. The goal of the SOC project was "to perform a comparative analysis of collaboratory projects, develop theory about this new organizational form, and offer practical advice to collaboratory participants and funding agencies about how to design and construct successful collaboratories" (Olson et al., 2008, p.55). The SOC project rated seven types of collaboratories:

- Shared Instrument,
- Community Data Systems,
- Open Community Contribution System,
- Virtual Community of Practice,
- Virtual Learning Community,
- Distributed Research Center,
- Community Infrastructure Project.

Leaving the others aside, we are interested in the Virtual Community of Practice collaboratory, that SOC defined as "a network of individuals who share a research area and communicate about it online. Virtual communities may share news of professional interest, advice, techniques, or pointers to other online resources". A limiting definition, compared to the potential of a real Community of Practice (CoP from here onwards), even if only virtual, that in our experience we have tried to capture and transfer, expanding its scope and objectives.

COLLABORATIVE LEARNING: WEB AND E-LEARNING 2.0

The concept of "Web 2.0" gained popularity in 2004 when O'Reilly Media and MediaLive hosted the first Web 2.0 conference: with this term they defined a new way of thinking about the creation and sharing of the resources of the Web. Compared to the static Web, the new Web 2.0 provided a gradual transition to the web imagined as a personal desktop. Web 2.0 was a synonym for broader self-created content sharing among users, such as photos, graphics, audio-files, etc. New search engines and SEO were developed and the first social networking systems took shape. The social software (as Myspace, Twitter, Linkedin, Facebook, etc.) is nowadays the core technology of web 2.0.

In the same year Stephen Downes used the term e-learning 2.0 to indicate a new model of learning based on the Web 2.0, in opposition to a conception of e-learning based on the transmission of knowledge, on the learning management system (LMS) and the learning object (LO).

The learning technology platform combined with social networking tools offers more natural and informal teaching situations, closer to everyday life-style.

This evolution reflects the change that has occurred in recent years in ICT and in teaching sciences. In this context, the objectivist concept has been replaced first by the constructivist concept and then by the current connectivist concept.

According to the objectivist concept, external reality is objective. Therefore, knowledge is considered to be a universal and objective reality, "mirrored" by a rational mind.

"The important metaphysical assumption of objectivism is that the world is real, it is structured, and that structure can be modeled for the learner" (Duffy & Jonassen, 1992, p.138). Learning consists of assimilating this reality.

This paradigm gave birth to behaviorist and neo-behaviorists approaches that have suggested educational technology based on a mechanistic view of the teaching-learning process (for example, the programmed instruction, the computer based training, the planning for goals, etc.).

According to the constructivist concept, knowledge is the result of a complex, multiple, subjective, socially and historically constructed product. Within a pedagogical sphere, constructivism redefines itself in socio-interactionist constructivism and emphasizes how the construction of meaning is negotiated and shared within a "community of interpreters". Merril has enumerated a few fundamental assumptions of a constructive vision of learning (Varisco, 1995, p.59):

- Knowledge is generated through experience;
- Learning is seen as a personal interpretation of the world;
- Learning is active, meaning it develops on the basis of experience;
- Learning is collaborative, because the final meaning is the result of multiple perspectives;
- Learning is situated (or anchored) because it happens in realistic settings;
- Assessment is integrated into the task and not an activity separate from it.

The social aspect – the collaborative learning - is essential: the subject carries out cognitive activities that he would not be able to perform alone. According to this theory, knowledge emerges as a shared activity.

In the context of educational technology, socio-interactionist constructivism has determined many learning environments; e.g. thanks to cognitive informatics Seymour Papert's proposals opened the way to the usage of active, constructive and social informatics tools. Significant learning, according to David Jonassen, would have the following six qualities (Varisco, 1995, p.62):

- Active, the student is responsible for his own results;
- Constructive, through the balance between the process of assimilation and accommodation;
- Collaborative, through the communities of learners, reciprocal teaching and the scaffolding and coaching offered by teachers;
- Intentional, as it actively and fully involves the student in the pursuit of cognitive goals;
- Conversational, because it deals with social processes, in particular those dialogical-argumentative;
- Contextualized, as the tasks of learning coincide with tasks in the real world;
- Reflexive, in that students organize what they have learned reflecting on the processes carried out and the decisions that were taken.

Connectivism redefines the concept of knowledge and learning through the principles of the Complex Networks Theory. In this perspective, the cognitive process is distributed across a network and is based on the recognition and interpretation of configurations. The cognitive process is deeply mediated by the diversity of the network and the strength of its connections; it relies on the adaptability of connection configurations, promotes the transfer of knowledge starting from the connections; adopts the speed of change of the domains of knowledge. (Del Moral Pérez, Cernea & Villalustre Martinez, 2011)

The various resources in Web 2.0 encourage the collaborative development of knowledge through four types of activities: collecting (search and store data), reflecting (select and create routes to information), connecting (enter virtual communities, create working groups) and publishing (sharing knowledge).

The rapid paradigm change so far outlined draws attention on the emergence of a pedagogical change that underlines the individual's ability to guide the complex process of building his own knowledge.

Facing phenomena like globalization, complexity, and acceleration that characterize the knowledge society, individuals must continually improve their knowledge, skills and competences. Moreover, individuals must manage their knowledge using basic skills such as creativity, analysis, ability to deal with complexity and ability in learning in a relationship.

Information technologies play an important role in this scenario because they can support and amplify the lifelong learning and act as tools for new cognitive skills, new forms of communication and collaboration among people. Olimpo underlines the fact that (Olimpo, 2010, p.13):

The first and perhaps the most important meaning of the use of educational technology in the knowledge society is precisely to create an environment in which individuals are naturally inclined to overcome their natural limitations and to expand the horizon of their concrete possibilities on a variety of fronts: access to information, knowledge building, communication and collaboration, expression, ability to deal with complexity, ability to give concrete and operational form to thought.

This transition is made particularly possible through the Web 2.0, going beyond the unidirectional nature of the first generation web. The Web 2.0 increases the social dimension of the network thanks to the many tools that can be defined as social software technologies and that facilitate the interaction among individuals, making them active creators of services and not just passive users.

The availability of tools such as blogs and wikis opens new pedagogical scenarios based on the integration of people through Internet and on the dialogue between opposite dimensions such as individual and social, formal and informal.

Social networking is the field of origin of the social climate and the basic culture of each virtual group, as it supports the motivation, the membership, the mutual understanding and the social grounding.

In formal contexts these elements are essential for the emergence of collaborative activities and should be promoted, through the technologies of Web 2.0 and appropriate methodologies, so that every individual can perceive the significance of his collaboration, can perceive himself as a useful resource and can therefore improve himself.

In an interesting model for collaboration among networks of people (Bonaiuti et al., 2006) social networking represents the dimension of informal e-learning that, transferred in more formal contexts, allows the overcoming of problems of online collaboration and first generation e-learning (such as the lack of a sense of belonging, the rapid decline in motivation for participation, etc.).

In this model, social networking combines the functions of organization and management collaboration, typically supported by the Learning

Management System, with those of reflection and metacognition, transverse technological tools supported by Web 2.0 technologies.

Web 2.0 creates an environment in which conditions and functions typical of the phenomenon of social networking - with the new tools and behaviors that it implies - can enhance the functions of traditional learning environments. (Bonaiuti et al., p.126).

Personal Learning Environment (PLE)

E-learning 2.0 envisages new learning scenarios. The paradigm shifts "from a perspective centered on the concept of standardized training proposal (or courses) targeted indiscriminately to a group of individuals, to the idea of subject-centered tools, where the integration of educational resources follows totally subjective dynamics" (Bonaiuti, 2006, p.63).

In this context, we observe a transition towards a personal learning environment (PLE). In a personal learning environment the use of technologies integrate lifelong learning, formal learning (intentional, structured and approved) and informal learning (unintentional, unstructured and not approved, connected to everyday activity of job, family and free time), self-organized learning and the development of virtual community. The approach of the PLE is learner-centered because the individual dynamically creates his own learning space, through the integration of all the opportunities offered by Web 2.0.

Bonaiuti explains:

It is in the PLE, which effectively acts as an alternative approach to e-learning platforms, that the subject provides an active construction of his training, largely made possible by the rich variety of resources and functions offered today by the Web, in which the subject puts together the many pieces of knowledge that represent the process of its development. It is on the Web, through the dense network of interactions (colleagues, community friendships, learning communities) that its multifaceted identity is substantiated and built. (Bonaiuti, 2006, p.72)

Within this system, the Learning Management System (LMS) is only one of the tools to support learning. The mashup allows to overcome an absence of flexibility.

A PLE puts together a wide range of the technological tools available. E.g.

- *Communication tools* (mailing list, chat, e-mail, web forum, blog, voice over IP, video conferencing, virtual classroom, wiki),
- *Data-sharing tools* (multimedia, maps, map-based services, social tagging, creative commons content, web applications, disk space on the network, e-book, file sharing, wikis),
- *Tools which support group processes* (shared calendar, planner, barometers, polls, sensors, social climate, conditional alert systems, virtual classroom, activities in platform, e-learning, e-learning lesson plan design, LAMS sequences, etc.),
- *Tools which support network-based individual work* (search engines, web application, web desktop, disk space on the network, podcasts, RSS, news feeds and aggregators, e-portfolio, e-books, e-journals, open access resources, open content, personal activities on platform, learning objects, test assessment, authoring tools),
- *Tools which foster networks of knowledge* (wikis, social networking, social tagging, blogs, research and building networks, FOAF - friends of friends, news, RSS, Atom, content syndication, collaborative writing, e-books, electronic magazines, open access resources, open content, group activities in the platform) .

- *Immersive simulation* (MOD, MOO, MUVE, role-playing games on the net, video conferencing, virtual reality, 3D simulation).

The concept of PLE requires a high level of personal maturity. Therefore it is more applicable in adult learning programs. Adults may find it useful to integrate formal and informal experiences; moreover they can compare their experiences with the ones of other individuals. This is the principle of mashup: mixing of contents and services, that mixes and re-use resources in different ways and with different functions. Despite the extensive debate that this has generated (for example, in learning quality assessment) there is no doubt that the "progressive hybridization of practices (moments of study, work, leisure), overlapping plans (formal/informal), methods (concepts of presence and distance are less and less distinguishable) and sizes involved (individual, group, social), already in action in our time, see the prospects for integration of e-learning 2.0 as a promising new path in the service of education". (Bonaiuti, 2006, p.103)

The experience of Sulcis CCS Summer School, described below, outlined a small model of PLE based on the integration of the web, social networks and LMS functionalities that could be a reference prototype for a future community of researchers and for the development of its scientific work.

SULCIS CCS SUMMER SCHOOL: BLENDED AND COLLABORATIVE LEARNING FOR RESEARCH AND EDUCATION ON CCS

The first edition of Sulcis CCS Summer School was held in July 23- 26 2013 at the Sotacarbo Research Center of Carbonia, Sardinia. Goals of Sulcis CCS Summer School are: the establishment of a permanent center for training and research on CCS technologies that gives continuity in time to teaching and experimentation; the establishment of a network of experts on the topic who work together exchanging know-how and experience, also on different projects; the establishment of a center for information on CCS; the dissemination of information to the public about this technology in order to promote public awareness.

Organized by ENEA, Sotacarbo and the University of Cagliari (Department of Mechanical Engineering, Chemical and Materials), with the essential support of other universities, research institutes and major companies in the sector, the 2013 School addressed about 30 undergraduate, graduate and PhD students. Frontal lectures and educational field activities offered the widest possible view on a very relevant issue: the reduction of carbon dioxide emissions in the atmosphere. The School provided information and materials on technological aspects of Carbon Capture and Storage (CCS), the latest developments of the research and its applications, the implementation of ongoing projects at European and international level.

The subject discussed required from the university students who attended the school, the acquisition and in-depth understanding of industry-specific skills (engineering, geology, chemistry, social sciences), as well as the acquisition of a multidisciplinary method in order to make a synthesis of the different disciplines, and to develop a comprehensive approach to the many aspects that the theme implies.

Training activities with lectures held by university professors, researchers and experts in the field were integrated and complemented with a website (www.sulcisccssummerschool.it), an e-learning course for specific training sessions and with the more widespread online social network tool, Facebook, taking into account that the age group addressed by the school, at least in the first instance, was that of young college students.

All the students of the summer school had access to the Moodle e-learning platform of ENEA Technical Unit for ICT (http://elearning.enea.it). There they had access to a specific course on CCS and they delivered the final report for the assessment of the acquired skills and the acknowledgment of two learning credits (CFU) and two research training credits (CFR) by the University of Cagliari.

The approach, therefore, was that of blended learning, a 'mixed' approach in which the traditional classroom activities are integrated and completed with e-learning courses.

With reference to the theoretical framework of social constructionism, the proposed teaching model emphasizes the active role of the student in the process of knowledge construction (renderings, models, answers, meanings, etc.). The online environment has been designed according to two lines of action: on the one hand, the building of virtual spaces that could facilitate the development of cognitive and metacognitive abilities; on the other, the attempt to facilitate the interaction and collaboration among participants in a Community perspective.

The first item was fulfilled through an intensive use of Moodle in its four basic functions: *management* (to track time and attendance, access management, etc.); *assessment* (quiz, vote, register, delivery tasks, etc.); *contents* (upload and organize learning objects, multimedia content, etc.) and *communication* (discussion forums, chat, email, etc.).

Documents, reports, articles, videos, publications and international regulations on CCS provided a working model and learning support. Moreover, the opening of educational materials to external sources had the goal of decontextualizing the cognitive process, making possible a metacognitive process of application of materials learned during the classroom lessons.

The second line of action emphasized the role of the community of students, trying to facilitate its set-up and the collaboration among its members. For this purpose, the interaction between participants was supported and encouraged through asynchronous and synchronous communication tools (web forums, chat rooms, instant messaging, wikis) and the link with the widespread social network Facebook (https://www.facebook.com/SulcisCcsSummerSchool). This space, in particular, was planned according to an informal e-learning perspective to promote informal contacts and to support students' sense of belonging. The goal was to stimulate the exchange of knowledge and information not only in a teacher-student perspective (top-down), but also among students themselves and in reverse from students to teachers (bottom-up). The attempt was to enhance the processes of construction and joint creation of "knowledge" – both in terms of combination and socialization of individual experiences –, and to empower the potential of collaborative learning, adding informal e-learning tools to formal-ones.

In this direction, the most significant and functional aspects generated by technology are (Varisco, 1995, p.62):

- *Knowledge construction* vs. its reproduction;
- *Conversation* vs. reception;
- *Articulation* vs. repetition;
- *Cooperation* vs. competition;
- *Critical reflection* vs. prescription.

Both sets of proposals had two objectives: in an initial phase the establishment of a learning community, made of different actors and tools, each one with its own cognitive perspective for the support of individual learning and of the community as a whole. The second step was the creation, at least at an embryonic level, of a true community of practice (as theorized by Wenger, better explained in the following) made of industry experts, researchers, students (the researchers of tomorrow) and stakeholders variously interested in the addressed theme.

The underlying theme was that of situated learning which has its essential feature in *legitimate peripheral participation*, namely the fact that the individual learns as a member of a community of practitioners and is fully involved in its socio-cultural practices. There can be no learning outside the community. Lave and Wenger (Lave & Wegner, 1991) explain:

The term "legitimate peripheral participation" provides a way to speak about the relations between newcomers and veterans as well as activities, identity, tools and community of knowledge and practice. It is concerned with the process by which newcomers become part of a community of practice. The will to learn is expressed and the meaning of learning is configured through the process by which a person becomes a full participant in a socio-cultural practice. This process includes, or rather subsumes, aware skills learning.

In this scenario, the use of web 2.0 tools was envisioned as a means for the development of certain skills and abilities that the connectionist perspective often puts in evidence for the current educational needs. This refers to the ability to select the information embedded in the community, to make connections among different nodes of information and to the choosing of relevant content by making relevant connections between ideas and concepts.

The proposed model showed its potential validity, because it led to the emergence of a small community of practice capable of directing its own learning process in a profitable way and to support mutual participation.

The scenario that could emerge, starting from this first group, is that of the establishment of a community of researchers that can address specific cognitive challenges through online interaction and collaboration. Reference tools are those of Web 2.0 and, in particular, their integration in a personal learning environment that, in our experience, might present itself already in terms of a single integrated area of Internet, Facebook and Moodle LMS resources. Each of them, in fact, was successfully integrated in the complex process of formal and informal learning, with an efficient use of all possibilities for communicating, sharing and supporting the processes of individual and group productivity.

FROM COLLABORATIVE LEARNING TO COLLABORATIVE RESEARCH: SULCIS CCS SUMMER SCHOOL'S COP

The combination of education and research - as in the case of Sulcis CCS Summer School here described - with new opportunities offered by e-learning 2.0, with an emphasis on building common knowledge, could have a disruptive impact on the patterns and processes of formal and informal learning, both at an individual and at an organizational level. This could provide also a new key to reinterpret the approach to scientific research in the desired direction of collaborative research. In this perspective, the paradigm of collaborative learning can be translated and adapted to a research environment, and extended in an even wider sense, pushing more on the concept of sharing. The challenge is to propose new methods and learning paths not only to deliver and share knowledge, but also to create and enhance collective learning processes. The exploration of the potential of those innovative forms for collaborative learning mediated by technology in complex environments (in functional and social-organizational terms) leads as a logical consequence to the concept of community of practice (CoP) in the Internet.

The theorist and researcher Etienne Wenger is the pioneer of research on CoPs and is considered a worldwide leading author on the subject. He is co-author, with Jean Lave, of the fundamental text "Situated Learning: Legitimate Peripheral Participation" (1991), in which the concept of communities of practice was outlined for the first time. In 1998 the author of Swiss origin published a text holding the same name, "Communities of Practice: Learning, Meaning, and Identity".

Etienne Wenger writes: "Communities of practice are groups of people who share a concern or a passion for something they do and learn how to do it better as they interact regularly" (Wenger, 1998). In line with the previous assumption, communities of practice are distributed and disseminated everywhere and are part of our daily lives. Each of us belongs to several CoPs, in some we are active members, in others we occupy a more peripheral role. Within each of the CoPs to which we belong, we assimilate and learn by interacting with others: our collaboration and knowledge sharing, within the community, modifies the production process and the attribution of meaning to contents and information. In CoPs, experience and collaboration are, in fact, revised and transformed (reified) in artifacts (symbols, procedures, rules, technology, products, jargon, etc.) through a process of renegotiation of meanings, deeply involved and situated in the context of the real problems of the social reality of reference.

Learning, therefore, becomes *situated* and develops as a result of:

- Involvement in activities,
- In specific contexts,
- In dealing with people.

In this way, the traditional concept of learning, considered a strictly individual path taking place separately from the practice, with a clear beginning and a clear end, is transformed. Learning is –according to Wenger – something different: it is a "social phenomenon" and effect of the situated experience. Learning and knowledge construction are seen from a "decentralized" perspective, as a social process based on experience, in which the acquisition of practices is simultaneous to that of social identity, to the belonging to a professional community and to the thorough knowledge of the organizational and relational rules of this community.

Scientific Community and the Construction of Knowledge in CoPs

"Communities are traditionally defined as collectivities of people who share values or beliefs, and whose social relations are affective relations, characterized by mutuality and emotional bonds, and who frequently interact" (Gläser, 2001). However, there are many exceptions to this definition, and the typical model of scientific community is a blatant example.

According to Merton, "scientific community originally referred to the worldwide collectivity of scientists, a collectivity that was assumed to be governed by a specific scientific ethos" (Merton, 1973). Scientists were assumed to produce new knowledge following the norms of communalism, universalism, disinterestedness and organized scepticism (Gläser, 2001). But the empirical observation of the behavior of the members of that community has repeatedly denied the beautiful theoretical aspiration quoted.

Khun's model seems to be the most fitting: "scientific communities are collectivities of scientists whose work is guided by a paradigm rather than by shared values and norms" (Khun, 1962). This model in particular contradicted the traditional concept of community, as it was no longer values that hold the community together, but knowledge, thus shifting the emphasis on the construction of knowledge, the knot that binds its participants mutually.

The sociology of science, since then and even today, questions and at the same time tries to give a modern definition of the "scientific community" which however, by its nature, suddenly changes with the changing of the resources and tools through which members that should be part of it interact. But, in line with Khun, the concept of community driven by the construction of knowledge is the one that better fits with that of Wenger's CoPs.

But what transforms a group of people who share an interest or a passion into a CoP? Wenger states that is the goal of mutual and imperishable bond between the two words of the "formula" to make the difference: learning. The aspiration to knowledge, to keep in line with the idea of scientific community formulated by Khun. Learning that configures itself as a result of active participation in the practices of one or more social communities to which we belong. It is the practice that defines the community and it is its "source of coherence" (Wenger, 1998), based on a link grounded on three assumptions:

- Mutual engagement,
- Joint enterprise,
- Shared repertoire.

Mutual engagement defines the practice, which exists because people are engaged in actions whose meanings they create and negotiate with each other. In other words, it is the social and relational fabric that allows to re-elaborate and attribute a meaning to the experience. And it is the agreement of mutual engagement that distinguishes the CoP from a network, a team or any other aggregate of people, because the development of the practice strengthens their functional relationships and induces the growth of individual skills.

Each member of a CoP finds a particular space and creates his own particular identity, which is further integrated and defined with the greater involvement in practice. The mutual engagement, therefore, does not result in homogeneity, but it creates relationships between people in different and complex ways and, if protracted, may become deeper than abstract affinities in terms of personal characteristics or social categories.

The negotiation of a joint enterprise is the second feature of the practice as a source of coherence for the community, since it binds and polarizes its activity. Joint enterprise:

- Is the result of a collective process of negotiation,
- Is defined by participants when they put it into effect,
- Creates among participants relations of mutual accountability.

However, because mutual engagement does not require uniformity, the joint enterprise does not require pure and simple approval. The enterprise, therefore, is not common because they all agree on everything, but because it is negotiated as a community.

Wenger emphasizes that the repertoire of a community, which is the source for the negotiation of meaning, is shared in a dynamic and interactive way. So the agreement, interpreted in the sense of literally shared meaning, does not influence the mutual engagement in the practice, nor the ultimate goal to reach. On the contrary, the diverging interpretations and misunderstandings are not just problems to be solved, but opportunities for the production of new meanings.

The occurrence of such conditions, of course, is created spontaneously but does not happen overnight. Some CoPs have existed for hundreds of years, as the community of artisans, others are less long-lived and can be incidentally formed. The development of a practice requires time, but what defines a CoP in its temporal dimension is not only a specific amount of time: if anything, is the mutual engagement, maintained over time in the joint pursuit of an enterprise, sufficient to share a meaningful learning. Ultimately, CoPs are shared histories of learning and this learning is accomplished by a shared doing: what community members are learning is their practice.

Learning is therefore not only a mental process or a simple focus on specific and explicit trainings or educational activities, but a full participation in a given social context. This underlines, once again, the role and the importance of experience in learning processes, to be understood as active involvement not only in a context, but especially in a social and relational fabric that allows to reprocess and assign a meaning to the experience.

CoPs in the Web 2.0 Era

Today, research on education is trying to assimilate some of the CoPs's crucial features in virtual learning environments. Internet, and the opportunities provided by new technologies, allow to broaden the concept of CoP and expand its value. Wenger's thesis about learning as a social and dialogic process of creation and development of joint meanings, in which the individual, as part of a group, gets support and motivation within the group itself, is enhanced through the application of Web 2.0 tools. However, in this sense, the additional value of 2.0 tools, in terms of community's effectiveness as a learning environment, must go beyond the mere provision of interconnectivity services, and must contribute to the definition of communication flows and modes of production. Technology must be truly collaborative, and capable of encouraging interaction among individuals. Furthermore, it must be capable of supporting real forms of collaboration, structured on a bond of interdependence between shared information, helping the process of joint learning. Just like in CoPs, where more social and relational aspects of the concept of community overlap those related to the "doing" that derives from practice, and the two aspects are deeply interrelated. A community of learning in virtual environments that does not stand on a specific practice, turns into a group or network of people interacting on topics and issues of different nature. Continuing along this line of thought, practice not supported by adequate social and relational fabric leads to isolation and exclusion from any possibility of collaboration and cooperation with others who are engaged in the same activity (Bettiol & De Pietro, 2003). A virtual learning environment, then, should be an enhanced model for an open and explorative production of meanings, focusing on situations where every individual has his space of action and is invited to take possession of the existing and go beyond it, in a constant process of individual and collective growth. This leads us to rule out from our analysis the wide variety of online communities that do not have a specific purpose of shared and participated building of knowledge. Attention must be focused on those models and tools capable of creating an environment for reflection and interpretation, and of giving a feedback that enriches the relationship between participants, so that the knowledge produced may be of greater value, allowing them at the same time to become creators and contributors within the new "learning networks".

How to Cultivate Communities of Practice

One of the characteristics of CoPs is their "informality", in that they arise around developed but not organized practices. CoPs exist in every organization and arise because their participants engage collectively in sets of recurring problems. Communities therefore do not appear in the exact moment that a project starts, or disappear with the end of a task. Communities have their own life cycle, which Wenger identifies in five stages, each one characterized by typical activities:

- **Potentiality:** People face similar situations without the benefit of a shared practice.
- **Coalition:** Members put themselves together and recognize their potential.
- **Activity:** Members engage themselves in the development of a practice.

- **Dispersion:** Members are no longer intensively engaged, but the community is still alive as a center of strength and knowledge.
- **Memory:** The community is no longer active, but people still remember it as a significant part of their identity.

Notwithstanding CoPs "natural" development, their evolution can be guided by a carefully structured project and it is useful and appropriate to identify experts to be involved, so that the community's work quality can be guaranteed. However, the learning plan cannot be based on the division of work between learners and teachers, or between those who organize learning and those who produce it. And it cannot be carried out by a totally separated community of educators or trainers. In this sense, Sulcis CCS Summer School configures itself as a structured project and relies on a "natural" community of experts and researchers in the field to share their knowledge and spread it as much as possible, even beyond the perimeter of the "restricted" CoP, as we will see further on. The experts are not only involved in the project, but they represent the frame that gave life to it and supports it, guaranteeing the project effectiveness. However, the school and the intended learning project, through the establishment of the CoP, goes beyond the distinction between students' and teachers' work, turning them into actors constantly interacting and exchanging roles in a process of "mutual" knowledge transfer.

According to Wenger, McDermott & Snyder (2002) to keep a CoP active it is necessary "to generate enough excitement, relevance, and value to attract and engage members". In other words, being able to create a sense of *aliveness*, a *vivaciousness* that stimulates the growth of the community and the interaction that makes it alive. In our case, except for a physical interaction limited to a few days, the tools to give vivaciousness are those of Web 2.0 and of informal e-learning: forums, chat rooms, instant messaging, wikis, social networks.

Each learning project requiring the presence of a CoP must, therefore, clearly take into account that a CoP is, by its very nature, spontaneous and self-directed and must be able to undertake the problem of how to support its realization through *vivaciousness*. To do this, Wenger, McDermott and Snyder (2002) set out seven principles:

- Design for evolution,
- Open a dialogue between inside and outside perspectives,
- Invite different levels of participation,
- Develop both public and private community spaces,
- Focus on value,
- Combine familiarity and excitement,
- Create a rhythm for the community.

Let's see them in detail, summarizing their description made by the three authors above mentioned.

Design for Evolution

Because CoPs are organic, designing them is more a matter of shepherding their evolution than creating them from scratch. The dynamic nature of communities is the key to their evolution. As the community grows, new members bring new interests and may pull the focus of the community in different directions. The purpose of a design is not to impose a structure but to help the community develop.

Open a Dialogue between Inside and Outside Perspectives

Good community design requires an insider's perspective to lead to the discovery of what the community is about, an understanding of the community's potential to develop and steward knowledge and, at the same time, it often takes an outside perspective to help members see the possibilities. Furthermore, it brings information from outside the community into the dialogue about what the community could achieve.

Invite Different Levels of Participation

Good community architecture invites many different levels of participation, but three are the main ones:

- A small *core* group of people who actively participate in discussions, take on communities project, identify topics to address and move the community along its learning agenda.
- An *active* group who attend meetings regularly and participate occasionally in the communities forums, without the regularity of the core group.
- A *peripheral* group who rarely participates.

Community members move through these levels: the key to good community participation and a healthy degree of movement between levels is to design community activities that allow participants at all levels to feel like full members.

Alive communities, whether planned or spontaneous, have a "coordinator" who organizes events and connects community members.

Develop Both Public and Private Community Spaces

Dynamic communities are rich with connections that happen both in the public places of the community – like meetings or web site – and the private space – the one-on-one networking of community members. The heart of a community is the web of "informal" relationships among community. The key to designing community spaces is to orchestrate activities in both public and private spaces that use the strength of individual relationships to enrich events.

Focus on Value

Communities thrive because they deliver value to the organization. Communities need to create events, activities and relationships that help their potential value emerge. Many of the most valuable community activities are the small, everyday interactions, informal discussions to solve problems, or one-to-one exchanges of information.

Combine Familiarity and Excitement

As communities mature, they often settle into a pattern of regular meetings, teleconferences, projects, web site use, and other ongoing activities. The familiarity of these events creates a comfort level that invites discussions.

At the same time, vibrant communities also supply divergent thinking and activity. For example, organizing conferences of experts with distinguished positions but different from those of the community. Conferences and workshops such as these bring the community together in a special way and thus facilitate a different kind of spontaneous contact between people, providing novelty and excitement that complements the familiarity of everyday activities.

Create a Rhythm for the Community

Vibrant CoPs have a rhythm. At the heart of a community is a web of enduring relationships among members: regular meetings, teleconferences, web site activity and informal lunches ebb and flow along with the heartbeat of the community. When that beat is strong and rhythmic, the community has a sense of movement and liveliness. The rhythm of the community is the strongest indicator of its aliveness.

Let us discuss the above mentioned seven points in the context of Sulcis CCS Summer School:

Design for Evolution: The school and its project try to channel a "stream" of knowledge and already existing techniques into the CoP's growth. Its evolution and growth should be measured through years, starting from the first edition of 2013, and web tools allow to go beyond the contingency of learning activities with face-to-face classroom interactions. The project's aim is to establish and support the development of the CoP, as a center for training and information on CCS.

Open a Dialogue between Inside and Outside Perspectives: The inside and implicit perspective is the design of the community, of its practice and of the new networks and synapses that its set-up could generate. The outside perspective is that of experts and researchers called to interact with the community in a stable or episodic way, but also of those who relate with the network of practice (see next pages) from a peripheral position.

Invite Different Levels of Participation: The different levels of participation are clear and obvious:

- **Small Core Group:** Experts, teachers and pupils of the school.
- **Active Group:** Those who interact with the core group for a limited period of time, occasionally or sporadically, without interrupting their interaction in a definitive manner, particularly because of the use of already mentioned computer tools.
- ***Peripheral* Group:** Those who come in contact with the community but are not part of it, standing on the sidelines of the network of practice.

Develop Both Public and Private Community Spaces: The proposed formal and informal learning and the joint creation of new knowledge, lessons, LMS forums and social network "public spaces"; the "private" space of informal relations within the virtual environment (instant messaging) or face-to-face.

Focus on Value: Is the expected goal, the creation of new events and activities with the effective participation of new community members. Future editions of the School will focus on a more "active" participation of students, researchers and experts, which should be not limited to lessons, interactions and joint practice.

Combine Familiarity and Excitement: It is the challenge and the goal to be achieved. Creating a sense of "community", trying to "institutionalize" it, and open the CoP to new ideas and new knowledge from outside. A process guaranteed by its very inclusive nature and by legitimate peripheral participation in the group's activities.

Create a Rhythm for the Community: This is still the weak side of the new CoP under construction. Assuring a strong and rhythmic beat is very difficult for several reasons including, first of all, the heterogeneous origins of its participants and the multidisciplinary nature of the topics. Because of this multidisciplinarity, discussions often take place among small groups. Of course, IT tools can address this problem and support community's rhythm.

Collaborative Learning, Communities and Networks of Practice in the Sulcis CCS Summer School

Once all the aspects that characterize a CoP and its 'life' have been defined, let us look at the process that may allow the transition from an experience of collaborative learning in virtual environments to the establishment, or attempted establishment, of a CoP.

If we assume that collaborating means working together and, therefore, distributing the tasks with the explicit intention of adding value to create something new, the elements that define the success of a collaboration, also in virtual environments, are:

- Competence among group's members and a shared and understood goal,
- Mutual respect and trust,
- The establishment and operation of shared spaces,

- Constant communication,
- Presence of formal and informal environments,
- Clear lines of responsibility, without restrictive boundaries,
- Acceptance that decisions should not be based on consensus and that physical presence is not essential,
- Selective use of people outside the group.

All these elements are clearly in line with what was previously said about CoPs and are, at least partially, also detectable in collaborative learning environments, under the form of computer tools that facilitate them. Particularly in our case, they are guaranteed by Moodle LMS's set of features and the extension of collaboration and learning beyond its borders, through the adoption of the widespread social network Facebook. Let us see in detail some of the most important and fundamental elements among those mentioned above, as they can be traced in the Summer School's teaching method:

- **The Goal, or Better, Shared and Understood Goals:** Those mentioned in the fifth paragraph of this chapter.
- **Establishment and Manipulation of Shared Spaces, Constant Communication and Presence of Formal and Informal Environments:** Shared and operable spaces are those of virtual environments (website, Facebook page and LMS), formal and informal, made available to all participants in the School (not only students, but also experts and teachers). Those virtual environments are constantly kept alive and active.

The simultaneous and multiple presence of formal and informal environments creates endless possibilities for interaction and the establishment of shared spaces. It increases the value not only of the subject (who, independently, develops his learning and accesses to the community through symmetrical relationships) but of the very community that, from then onwards, finds in technology and networking an opportunity and an additional tool to support processes of sensemaking and learning.

Similarly to what happens during a community practice, also learning in a collaborative virtual environment produces an enrichment of the shared repertoire and a modification of the joint enterprise, as well as a growth of the participants identity. And in creating a virtual learning community it is always necessary to give due consideration to these essential characteristics of a CoP. Encouraging an approach of this kind, through a mixed and complex system of work and learning tools, is the goal set by the teaching methodology identified for the Sulcis CCS Summer School.

In this context, the role played by the technology adopted - the LMS Moodle integrated with other tools and with a blended-learning approach - allows to organize a circuit enhancing distributed forms of knowledge, in a continuous reference and exchange among all the participants in the Summer School activities. The attempt is to convey a set of multidisciplinary skills connected to a practical activity. This is one step in a larger learning process of shared accumulation and acquisition of knowledge.

CoP, in this case, emerges as a tool to improve the efficiency of collaborative work and to enhance its "educational" impact. Wenger (1998) states, in fact, that learning involves an interaction between experience and competence and in CoPs the two things operate in close connection. In the Sulcis CCS Summer School the experience and competence of teachers and speakers acted in connection with the *newcomers* (those enrolled in Summer School), who accessed a community that they themselves contributed to make "lively" through the exchange and the interaction in the few days of face-to-face classrooms and, especially, in the practice carried out with social media tools and Moodle technology platform. In this way, an environment designed for collaborative learning expanded its boundaries to become, at least in an embryonic form, a CoP.

The methodological approach chosen, therefore, matches with a social interpretation of the mechanisms activating learning processes. It is exactly in the role played by the *newcomer* (in our case the student enrolled in Summer School), seeking recognition as a member of the community, that learning originates. This does not coincide banally and solely with the ability to learn concepts in order to become capable of performing a given task in an appropriate manner, but also in having internalized language, values and "norms" that characterize the community itself. Being familiar with specific scientific terminology, with an e-learning environment, sharing opinions and knowledge with experts, mastering new concepts with the critical capacity to put them in discussion. Sharing, without distinction, formal and informal learning activities.

In this context, learning is characterized not only as a transfer and assimilation of concepts, theories and information, but as a process of participation, belonging, engagement, inclusion and development of a new identity, which refers to the concept of *"legitimate peripheral participation"* developed by Lave and Wenger (Lave & Wenger, 1991), that is the process by which *newcomers* become part of a CoP. It is a structured process through which the "novices" absorb and are absorbed in a practical activity, characterized by a specific regime of participation, which allows them to contribute actively and legitimately to the performance of tasks, but being only marginally responsible of the final product.

Through their particular participation, newcomers approaching a social context defined by structured practices and identities, begin to deal with the knowledge and tacit culture of the group in order to gain understanding of them and to be recognized as legitimate members. The gradual transition from the status of newcomers to that of members, from a peripheral position to a central one, is related not only to the acquisition of technical skills typical of the group's practice, but also with the assimilation of its customs, its language and its culture. Through the legitimate peripheral participation in the group's activities, the newcomer gradually takes possession of the available social knowledge (techniques, tradition, tacit culture characterizing the group's experience) and, as soon as his ability is recognized as appropriate and relevant to the whole culture, in its turn, he becomes an expert like the others and an effective member of the group. A valid paradigm, needless to say, both in training and in research, especially when it aims to be collaborative.

The experience of Sulcis CCS Summer School intersects the CoP tipology with that of the so-called networks of practice and, in this sense, also provides a model for collaborative scientific online research. We have extensively said that about CoPs, but here it might be useful to highlight the feature that distinguishes them from networks of practice: their cohesion, their nature of consistent groups of individuals who know each other and work together. On the contrary, networks of practice are weakly connected communities of practice, in which the majority of members does not know the others, even though having common practices and knowledge. Connections among members are mostly indirect and participants are coordinated through a third-party: coordination and communication are quite explicit. The networks of practice are of considerable extent, a feature strengthened and increased by information technologies, but the degree of reciprocity is relatively low, and their members do not interact directly in a substantial manner.

Sulcis CCS Summer School, de facto, sums up in a single experience the two situations. It establishes a CoP, starting from a core group of experts and researchers on CCS, an already existing research field, and calls students to join it and begin to collaborate on a shared practice, structuring the construction of a new identity through learning.

At the same time, with Web 2.0 technological tools, creates around the CoP a managed and coordinated network of practice, including students and new researchers, and that can be potentially extended to the whole group of stakeholders related to the project (institutions, organizations, companies, schools, universities) and serves as a "buffer" and filter with the CoP's core. Thus, the question of public awareness and social acceptability of the new technology and its potential impact on the Sulcis area, finds a new transfer channel which, by its nature, tends to include and bind also the participants at the periphery of the network of practice.

Features of this new on-line CoP are: sharing of mutually defined practices, aggregation around an idea and shared tasks, organization according to the activities and needs that emerge, structure made of geographically distributed members and, aspect that interests us the most, use of web technologies and, specifically, e-learning platforms and tools to collaborate, communicate, train and develop knowledge, in which the level of examination depends on the degree of penetration and participation in the CoP.

CONCLUSION: CoPs FOR COLLABORATIVE E-RESEARCH

Analysis of Data

The data reported in the following, although without in any way claiming to be complete and exhaustive, show the use of the tools available for online activities.

In the period from 1 July to 30 September 2013, the Summer School course's page on the e-learning platform (http://elearning.enea.it/course/view.php?id=150), was viewed 856 times, with 162 unique pageviews (see Figures 1 and 2); it should be noted that although the students enrolled in the Summer School were only 30, the course was also open to teachers and all ENEA employees.

The course was divided into 8 sections and its documents received the following number of openings/views:

- **Documents:** 104 views.
- **Reports, Articles and Publications:** 95 views.
- **Regulations:** 18 views.
- **Sulcis Technological Hub:** 34 views.
- **Video:** 32 views.
- **Teaching Material - First Day:** 183 views.
- **Teaching Material - Second Day:** 91 views.
- **Teaching Material - Third Day:** 60 views.

Figure 1. Visits report. Source: http://www.google.com/intl/it/analytics/.

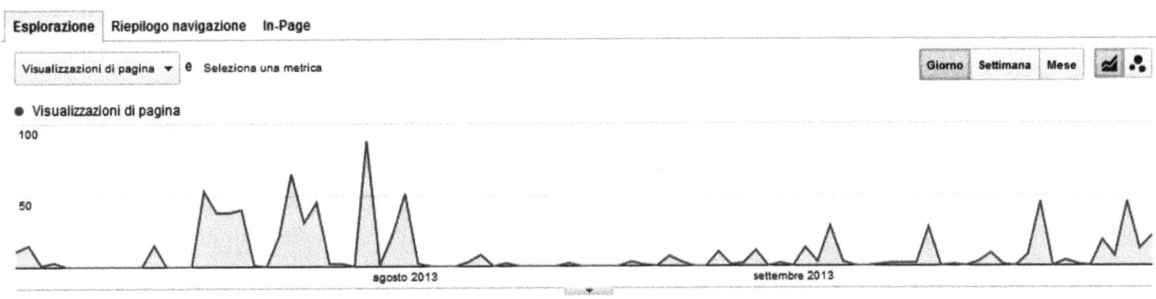

Figure 2. Visits and unique pageviews. Source: http://www.google.com/intl/it/analytics/.

Dimensione principale: Pagina Altro

Traccia righe | Dimensione secondaria | Tipo di ordinamento: Predefinito

	Pagina	Visualizzazioni di pagina	Visualizzazioni di pagina uniche
		856 % del totale: 9,52% (8.988)	162 % del totale: 3,47% (4.664)
	1. /courseMiew.php?id=150	856(100,00%)	162(100,00%)

The e-learning platform Forum was divided into five thematic areas relating to:

- About the Summer School,
- Platform usability,
- Netiquette,
- Educational Topics,
- Delivery of final reports.

Forum threads had 238 views, but the number of comments, answers and questions posted by students was low compared to the amount of visits made. This finding is in line with results coming to light from the analysis of Facebook Insights data reported below, compared to the level of online participation in educational activities.

The analysis of the statistics of the Facebook Sulcis CCS Summer School page provides some useful information to evaluate the role and importance of the use of communication tools and non-formal learning.

We took into account Facebook Insights statistics from 1 to 31 July 2013, which represent the range of maximum visibility and participation in the initiative and include the days when the Summer School actually took place.

The first fact to be pointed out is the high number of page views (3.806) and the total number of people who saw at least one post on the page (101). Viral reach, i.e. the number of people who saw the page or one of its posts from a news shared by a friend, is 215.

The posts on the page were clicked by an average of 37 people and specifically:

- **Photo View:** 36 people.
- **Video Play:** 20 people.
- **Other Clicks (News, Links, Documents):** 55 people.

Finally, 17 people shared the page news, posting comments, sharings, answers and "likes".

The analysis of the above data shows clearly that using a social network to increase the visibility of the School and to disseminate information about it was a successful choice, both for the high number of views obtained and the ease with which news published on the page spread, in a "viral" way, through Facebook contacts. It is quite noticeable that a lot more people than those who truly attended the school were informed of the event.

The average of contents actually clicked by Facebook users is in line with the number of members and participants, at various levels, in the School's activities and highlights the potential role of social networks to convey educational contents in an open and non-formal approach.

It must be pointed out the lack of participation and effective interaction on the entered contents, resulting from the low number of people who shared or commented the page's news. This data, however, must be interpreted taking into account that, in the month of July, the blended-learning course provided a direct interaction during the classrooms affecting the indirect and online virtual interaction.

At the end of the Sulcis CCS Summer School the students were asked to fill in a final evaluation questionnaire. The survey focused on four points:

- Overall perception of the experience,
- Teaching,
- Organization and services,
- E-Learning environment.

For each question a rating scale from 1 (little) to 5 (very much) was given. In addition, some open questions were added (opinions, suggestions, criticisms). The perception of the learning experience was generally found to be more than satisfactory. The highest scores were: fulfilment of expectations (59%), improvement of knowledge about the topic (71%), impact on future career (59%).

Percentages of best rating higher than 50% were reached in the evaluation of the e-learning environment. Two aspects deserved special attention:

- The added value that such a proposal may provide to a traditional educational process (71% of the participants gave the highest rating),
- Ease of use of the Moodle platform (highest rating for 59% of answers).

At this point in our discussion, it is quite natural to establish a link between the "Sulcis CCS Summer School model" and a community model which works for collaborative research, and especially for the one that makes use of IT tools.

This model is naturally a "prototype", an experiment in progress. Some aspects need to be revised, but this is part of a natural process, when someone has to "expose" oneself and take a challenge with others and shows a natural initial resistance. Nevertheless the establishment of a CoP is a good foundation on which we can build in the future, assuming that the "Center for coal technologies" and Sulcis CCS Summer School will find the essential support in institutions.

If, according to Khun, the scientific community is driven by the building of knowledge, and knowledge is built in a process of practice and social interaction (just like in Wenger and also in the paradigm of collaborators explained in section 3) the pattern of CoP seems to comply perfectly with scientific research. Summer School students took part in a "prototype-experience" that might be used as a model for their scientific activity in the future. If learning, which is knowledge, is built more effectively through essential collaboration in the practice of the Community, so the scientific research can take advantage of the mutuality. In this sense, because of its multidisciplinarity, CCS can be considered an ideal "training ground" for collaboration. The geologist who deals with CO2 confinement must necessarily interact with the engineer who is responsible for CO2 capture; the Sulcis CCS Summer School's young geologist had a number of IT tools to get in touch with his young colleague preparing for an engineering degree. He exchanged information and files via Facebook, watched videos on the website, used the e-learning platform's wiki to acquire a new vocabulary and deepen his knowledge in the subject. He participated and gave a sense of "aliveness" to the Community well before and well beyond the classes. In addition, his on the job interaction with professors and experts provided the dimension and the physical perception of belonging - even if only in a limited way - to the community to which he was called to be part of.

REFERENCES

Bettiol, M., & De Pietro, L. (2003). Comunità di pratica e organizzazione: Un rapporto da scoprire. *T D Tecnologie Didattiche*, *30*, 53–60.

Bonaiuti, G. (2006). E-learning 2.0, Il futuro dell'apprendimento in rete, tra formale e informale. *I quaderni di Form@re, n.6.*

Calvani, A., Bonaiuti, G., & Fini, A. (2008). Lifelong Learning: quale ruolo può svolgere l'e-learning 2.0? *Journal of e-Learning and Knowledge Society, 4,* 47-56.

Cantoni, L., Botturi, L., & Succi, C. (2007). eLearning, Capire, progettare, comunicare. Milano, Italy: Franco Angeli.

Cinque, M. (2010). *Agire creativo – Teoria, formazione e prassi dell'innovazione personale*. Milano, Italy: Franco Angeli.

Cogburn, D. L. (2003). HCI in the so-called developing world: What's in it for everyone. *Interaction*, *10*(2), 80–87. doi:10.1145/637848.637866

Del Moral Pérez, M. E., Cernea, D. A., & Villalustre Martinez, L. (2011). Scenari connettivisti e progetto di Learning Object adattabili alla diversità cognitiva. *TD Tecnologie Didattiche*, *19*(2), 102–111.

Duffy, T. M., & Jonassen, D. H. (1992). *Constructivist and the technology of instruction: A conversation*. Hillsdale, NJ: Lawrence Erlbaum Associates.

Esposito, G., & Mantese, G. (2003). *E-learning: una guida operativa*. Milano, Italy: Franco Angeli.

Fini, A. (2008). E-Learning 2.0: natura e sviluppo di una comunità. Esame di un caso. *Journal of e-Learning and Knowledge Society, 4*, 61 – 69.

Gabrielli, G. (2006). *Conoscenza, apprendimento, cambiamento. La gestione dei programmi di knowledge e change management*. Milano, Italy: Franco Angeli.

Gläser, J. (2001). *Producing Communities as a Theoretical Challenge*. Paper presented at the TASA 2001 Conference. Sydney, Australia.

Kuhn, T. (1962). *The Structure of Scientific Revolutions*. Chicago: University of Chicago Press.

Lave, J., & Wenger, E. (1991). *Situated learning. Legitimate peripheral participation*. Cambridge, UK: University of Cambridge Press. doi:10.1017/CBO9780511815355

Merton, R. K. (1973). *The Normative Structure of Science. In The Sociology of Science*. Chicago: The University of Chicago Press.

Olimpo, G. (2010). Società della conoscenza, educazione, tecnologia. *TD Tecnologie Didattiche*, *50*, 4–16.

Olson, G. M., Zimmerman, A., & Bos, N. (2008). *Scientific Collaboration on the Internet*. Cambridge, MA: The MIT Press. doi:10.7551/mitpress/9780262151207.001.0001

Rossi, P. (2009). *Tecnologia e costruzione di mondi*. Roma, Italy: Armando Editore.

Trentin, G. (2000). *Dalla formazione a distanza alla comunità di pratica attraverso l'apprendimento in rete.* Retrieved from http://www.tdmagazine.itd.cnr.it/files/pdfarticles/PDF20/Trentin.pdf

Trentin, G. (2001). *Dalla formazione a distanza all'apprendimento in rete*. Milano, Italy: Franco Angeli.

Varisco, B. M. (1995). Paradigmi psicologici e pratiche didattiche con il computer. *TD Tecnologie Didattiche*, *7*, 57–68.

Wenger, E. (1998). *Communities of Practice. Learning, Meaning and Identity*. Cambridge, UK: Oxford University Press. doi:10.1017/CBO9780511803932

Wenger, E., McDermott, R., & Snyder, W. (2002). *Cultivating Communities of Practice: A Guide to Managing Knowledge*. Cambridge, MA: Harvard Business School Press.

Wulf, W. A. (1989). The national collaboratory: A white paper. In T*owards a national collaboratory: Report of an invitational workshop at the Rockefeller University*. Washington, DC: National Science Foundation.

KEY TERMS AND DEFINITIONS

Carbon Capture and Storage (CCS): Is a technology that can prevent large amounts of carbon dioxide from being released into the atmosphere.

Collaborative Learning: Learning that takes place in a peer-oriented environment. The development of collaborative tools such as web conferencing, instant messaging, email, etc.

Community of Practice: An informal group that shares knowledge in an informal and mutually respectful manner.

Constructivism: A learning theory that states that learning is an active, contextualized process of constructing knowledge rather than acquiring it. The construction of knowledge is based on personal experiences.

E-Learning 2.0: A new model of learning based on the Web 2.0, in opposition to a conception of e-learning based on the transmission of knowledge, learning management systems (LMS) and learning objects (LO).

E-Learning: Short for electronic learning is a overall term that refers to all types of training that occurs on a digital medium.

E-Research: The application of advanced information and communication technologies to the practice of research.

Learning Management System: (LMS): Software application or Web-based technology used to plan, implement, and assess a specific learning process. LMS allows create assignments, distribute materials, hold online classes, create tests and grade students' work.

Moodle: Learning Management Systems (LMS) open source.

Personal Learning Environment (PLE): Tools, communities, and services that constitute the individual educational platforms that learners use to direct their own learning and pursue educational goals.

Web 2.0: Web 2.0 is a synonym for broader self-created content sharing among users, such as photos, graphics, audio-files, etc. New search engines and SEO were developed and the first social networking systems took shape. The social software (as Myspace, Twitter, Linkedin, Facebook, etc.) is nowadays the core technology of web 2.0.

Chapter 11
Collaborative Knowledge in Catchment Research Networks:
Integrative Research Requirements for Catchment Systems Science

Christopher (Kit) J. A. Macleod
James Hutton Institute, UK

ABSTRACT

Catchments are increasingly used to structure collaborative knowledge to advance understanding and management of terrestrial and freshwater natural resources. To produce collaborative knowledge of catchment systems requires greater levels of integrative research based on networks of researchers, policymakers, and practitioners. To enable increased levels of integrative research depends on development and application of digital technologies to improve collection, use, and sharing of data, and devising new knowledge infrastructures. Catchment observatories that integrate existing and novel physical, social, and digital networks of knowledge infrastructures are urgently needed. This chapter presents three leading international examples of collaborative networks of catchment researchers. In particular, the digital infrastructures they have developed to support collaborative knowledge in catchment research networks. This chapter provides future research directions required for the production, sharing, and use of collaborative knowledge in catchment research networks based on catchment systems science.

INTRODUCTION

There is a need to improve the production, sharing and use of collaborative knowledge of catchment systems through networks of researchers, policy makers and practitioners. This requires greater levels of systems based integrative research. Over a decade ago, Falkenmark (2001) stated that the greatest water related problem was our inability to link water, food and environmental security. Current mismatches in the demand for and supply of local to regional resources is leading to water, food and energy insecurity. Solutions are required that integrate interactions between water security (Bogardi et al., 2012), food security (Godfray et al., 2010) and energy security (Bazilian et al., 2011). These solutions need to be built on systems nexus based thinking (Hoff, Iceland, Kuylenstierna, &

DOI: 10.4018/978-1-4666-6567-5.ch011

Copyright © 2015, IGI Global. Copying or distributing in print or electronic forms without written permission of IGI Global is prohibited.

te Velde, 2012). We require systems based approaches to understand and manage these complex unstructured 'messy' socio-ecological situations (Ackoff, 1974). Systems based approaches are well established in several disciplines that range from systems biology, operational research to earth systems science. Lessons learned from these approaches include: the importance of problem structuring; the need to integrate qualitative and quantitative approaches; the need for network based approaches; and the requirement to increase the sharing and use of data (Macleod, 2010). These lessons highlight the importance of enabling integrative research in scientific research networks to produce, use and share collaborative knowledge. Catchments are increasingly used to structure our collaborative knowledge to improve our understanding and management of terrestrial and freshwater resources under environmental change.

In parallel to the growing realization that greater levels of collaborative knowledge in scientific research networks are required, a digital revolution has been taking place. This has been driven primarily by the emergence of distributed networks of computers and standards-based interoperability. "The Internet has changed not only the way society operates, but also how scientists work. There has also been a step change in the quantity of data that scientists routinely amass and need to analyse" (Thorpe, 2009). More recently, Nielsen (2012) in his book 'Reinventing discovery' sets out how the Internet is fundamentally changing how we are carrying out collaborative scientific research. How knowledge is generated and used are being changed by the presence and use of the Internet (Weinberger, 2011). Developments in web 2.0 internet technology have enabled a wider range of applications that involve contributions of data from networks of researchers and non-researchers e.g. (Frame et al., 2009; Rinner, Kessler, & Andrulis, 2008). Current developments with the semantic web (web 3.0) are enabling greater discovery and sharing of linked data (Hall, De Roure, & Shadbolt, 2009). There are calls for a greater understanding of collective knowledge systems, where the semantic web meets the social web (Gruber, 2008). A vision for the Internet is where we can create social machines (web 4.0) that enable people e.g. researchers, to be creative, whilst the machines carry out more of the administrative tasks (Hendler & Berners-Lee, 2010). Catchment systems based integrative research must harness developments in digital technologies to enable networks of researchers and non-research stakeholders produce, share and use collaborative knowledge to better understand and manage catchment systems.

The objective of this chapter is to present the status and research needs for greater levels of systems based integrative research for the production, sharing and use of collaborative knowledge in catchment research networks. To enable increased levels of integrative research depends on development and application of digital technologies to improve collection, use and sharing of data and devise new knowledge infrastructures. This chapter focuses on the requirements for catchment observatories that integrate existing and novel physical, social and digital networks of knowledge infrastructures. To support this focus, I present three leading international examples of collaborative networks of catchment researchers and their development of catchment observatories. In particular, the digital infrastructures they have developed to support collaborative knowledge in catchment research networks. I conclude by discussing the future research directions required for greater levels of production, sharing and use of collaborative knowledge in catchment research networks based on catchment systems science.

STATUS OF CATCHMENT RESEARCH NETWORKS

Catchment Research Networks Span Multiple Research Disciplines, Policy, and Practice

To increase the production, sharing and use of collaborative knowledge of catchments requires catchment systems science that spans research, policy and practice to understand and manage catchments from the plant to global scales. Catchment systems science involves the study of surface water hydrological catchments framed as complex socio-ecological systems with cross-scale dominating processes. Collaborative knowledge production requires integrative research that can be interdisciplinary or transdisciplinary research depending on the research questions. Developing collaborative knowledge of catchments requires systems based approaches based on key concepts of scale, connectivity and indicators of thresholds in state or process (Macleod & Ferrier, 2011). There are broad communities of multi and interdisciplinary researchers studying catchments in addition to more narrowly focused disciplines, for example: integrated water resources management (Cortner & Moote, 1994; Jewitt, 2002); adaptive water management (Dewulf, Francois, Pahl-Wostl, & Taillieu, 2007; Pahl-Wostl, 2007; Pahl-Wostl, Lebel, Knieper, & Nikitina, 2012); and catchment science (Collins & Ison, 2010; Jordan, Haygarth, Shortle, & Harris, 2012; MacKinnon & Tetzlaff, 2009; Tetzlaff et al., 2007).

Catchments are an example of complex adaptive systems, in which higher levels of patterns emerge from smaller scale processes (Levin, Grenfell, Hastings, & Perelson, 1997). Catchment system science integrates existing perspectives of landscapes (Wu & Loucks, 1995) and riverscapes (Thorp, Thoms, & Delong, 2006) as hierarchical patch dynamics. Catchments are composed of mosaics of rural and urban land use and networks of water bodies, the former determined by physical factors that include climate, topography and soil and a range of historical and current social factors that include ownership and commodity prices. Temporal and spatial networks of water bodies in catchments are influenced by climate, hydrogeology, geomorphology and landuse. There are a wide range of established scientific disciplines that focus on particular structures, processes or scales in catchments and the wider landscape e.g. hydrologists study the quantity and/ or quality of water at one or more spatial scales in catchments. Each discipline tends to understand how catchments function from a particular perspective, for example the experience and knowledge of each researcher will influence how they perceive in what ways water moves through particular catchments (Bracken et al., 2013).

Surface water catchments are increasingly used as biophysical units for natural resource management. In the EU, the Water Framework Directive (2000/60/EC) and Floods Directive (2007/60/EC) necessitate catchment based approaches. Catchment management requires increased levels of integration between groups of: natural and social scientists; land and water users; land and water managers; and planners and policy makers across spatial scales (Macleod, Scholefield, & Haygarth, 2007). Greater levels of integrative research are required at the science-policy interface (Macleod, Blackstock, & Haygarth, 2008; McGonigle et al., 2012). There needs to be improved interaction between the policy demands for research and the scientific push from researchers (Holmes & Clark, 2008). With an appreciation if the interaction involves opening up or closing down science-policy-management dialogues (Macleod et al., 2008). Integrating water and land management in catchments under environmental change requires systems based approaches that: can be predicted, monitored and evaluated; are based on common flexible frameworks e.g. ecosystem services; enable scientific credibility to be combined with practical usefulness; and improve the linkages between data-models-evidence-policy through place based studies (Macleod & Haygarth, 2010).

Examples of Catchment Research Networks and the Need for Integrative Research

Catchment based networks of researchers, policy makers and practitioners are increasingly set up to understand and manage catchments. For example, researchers monitor and manipulate individual catchments and small networks of catchments as part of nationally or internationally funded research projects. Catchments are widely used for long term studies of the impacts of environmental change on the quantity and quality of water e.g. (Langan et al., 1997; Likens & Bormann, 1974; Lovett, Likens, Buso, Driscoll, & Bailey, 2005; Wegehenkel, 2003). They are important sentinels of environmental change, especially lakes and reservoirs (Williamson, Saros, & Schindler, 2009) as they integrate temporal and spatial inputs from watersheds and airsheds (Schindler, 2009). Examples of national and international catchment research networks include: the Global Lake Ecological Observatory Network (GLEON) (Hanson, 2007); the recently completed North-Watch project (Tetzlaff, Carey, & Soulsby, 2013); the ongoing Demonstration Test Catchment project in England (Outram et al., 2013); the Shale Hills Critical Zone Observatory in the US (Takagi & Lin, 2011); and the TERENO observatories in Germany (Zacharias et al., 2011). In the North-Watch network, space for time substitutions enabled hypotheses to be tested on how catchments may respond to changes in climate forcings (Tetzlaff et al., 2013). Since catchments in the north (above 50°) are particularly vulnerable to climate change, due to the influence of snow on the hydrological cycle and the importance of 0°C as a critical environmental threshold (Tetzlaff et al., 2013). Advances in high temporal resolution data collection and analysis are providing new insights into catchment functioning (Jordan, Arnscheidt, McGrogan, & McCormick, 2007; Milne, Macleod, Haygarth, Hawkins, & Lark, 2009; Outram et al., 2013). An international workshop on aquatic ecosystem sustainability concluded that integrative research was required that would enable research communities to access locally collected data, integrate this data with existing data catalogs, and reproducibly and efficiently process the data and models (Harmon et al., 2011). To increase production, sharing and use of collaborative knowledge from catchment systems science networks requires greater levels of systems based integrative research.

INTEGRATIVE RESEARCH REQUIREMENTS FOR COLLABORATIVE KNOWLEDGE IN CATCHMENT RESEARCH NETWORKS

Greater levels of systems based integrative research are required to produce, share and use collaborative knowledge from catchment research networks. Integrative research, as defined by Winder (2003) covers both interdisciplinary (between research disciplines) and transdisciplinary (between researchers and non-researchers) research projects (Tress, Tress, & Fry, 2005). Integrative research is based on disciplinary research. Integrative research provides additional challenges and benefits to carrying out narrowly focused disciplinary research. These relate to the extra time and communication that is required, but also the attitudes of researchers and their knowledge and skills for carrying out integrative research. In particular, it requires time to develop shared understanding between a team of researchers and in the case of transdisciplinary research, non-research stakeholders. Communication is vital, not only to develop this shared understanding at the start of a project, but also during the research and at the end. Collaborative eScience research teams are often impeded by a lack of understanding of each other's research interests to generate interdisciplinary research questions (Pennington, 2011). These collaborative teams require excellent management and leadership to deliver integrative research. Though integrative research takes place in teams, teams are comprised of individuals. These individuals need to have the knowledge, skills and attitude to balance their disciplinary expertise and ability to carry out integrative research. "Professionals who understand domain science, software best practices, and the issues of social interfacing would be enormously beneficial in almost any area of knowledge" (Edwards et al., 2013). Collaborative knowledge in catchment based networks requires greater levels of systems based integrative research involving networks of researchers, policy makers and practitioners. In the following sections I present challenges to carrying out greater levels of integrative research and the potential for digital solutions.

Challenges to Collaborative Knowledge Production in Catchment Research Networks

Carrying out integrative research in collaborative knowledge networks is challenging. There are knowledge, team and technical challenges to carrying out greater levels of integrative research (Kragt, Robson, & Macleod, 2013). Knowledge challenges include different epistemologies of researchers and the need to develop and share ontologies. Team challenges include differences in disciplinary methods, values and language and the extra time required for communication, management and leadership of team based integrative research. Technical challenges can include the lack of interoperability and reuse of data and models. Related to these are three main interoperability challenges for greater levels of integrative research: technological, semantic and organizational (Nativi, Mazzetti, Craglia, & Pirrone, 2013). To enable increased levels of integrative research depends on development and application of digital technologies to improve collection, use and sharing of data and devise new knowledge infrastructures.

Digital Technologies to Improve Collection, Sharing, and Use of Data in Catchment Research Networks

Integrative research to produce collaborative knowledge requires development and application of digital technologies to generate and share data. Data and its metadata are central to scientific research. Science is becoming data intensive. Researchers need to adjust how they work based on: the data available; accessing and storing of data in spatial data infrastructures (SDIs); requirements for greater levels of data sharing and reuse; importance of metadata; and how data is used to advance knowledge and decision making about natural resource management under global environmental change (Table 1). Increases in the volume, variety, velocity and veracity of data from sensors and model simulations are now available via the Internet to help scientists understand the relationships between society and their environment (Sellars et al., 2013). There is great interest in utilizing advances in big data and the associated cyberinfrastructure (Dumbill, 2013; Hampton et al., 2013; Vitolo, Elkhatib, Reusser, Macleod, & Buytaert, 2015). The large number of smaller datasets and their management must also be considered, as small quantities of data does not equal simpler management (Akers, 2013). Observations and how they are organized and manipulated are vital to many fields of science including hydrology and water resources research (Horsburgh, Tarboton, Maidment, & Zaslavsky, 2008). A "data first" approach has been called for, to improve integrative earth systems science that focusses on improving conventions, formats and consistency in data rather than develop interfaces and analysis tools to improve modelling (Parsons, 2011). A lot of data is not being captured (it is runoff) and re-used (Borgman, 2012). A recent survey of data custodians highlighted that a wide range of data is at risk of being lost, because of a lack of available time and skills for data preservation (Thompson, Robertson, & Greenberg, 2013). Current levels of scientific data sharing and reuse are low outside the fields of astronomy and genomics (Borgman, 2012). Preservation of data is also a major challenge, especially since digital data is a vital resource for education, research, policy and the wider knowledge economy (Blue Ribbon Task Force on Sustainable Digital Preservation and Access, 2010). To address the need for greater capture and preservation of data there are profound changes in requirements for scientists to deposit and share data. For example, research funders increasingly require data management plans.

Sharing and reuse of data needs to become standard practice to address the scientific challenges associated with global environmental change (Table 1). Sharing of data has been shown to increase the citation rate of publications (Piwowar, Day, & Fridsma, 2007). But sharing and citing of data needs to become more common in the natural (Costello, 2009) and social sciences (Mooney, 2011). There are technological and sociological challenges to increasing the level of data sharing amongst scientists that include standardization of methods and the lack of metadata, before more open data will reside in federated repositories (Reichman, Jones, & Schildhauer, 2011). There are a range of data identification technologies including DOIs that are favored by many commercial publishers and URIs/URLs, but none deal with data reformatting (Duerr et al., 2011). "When scientists from two or more disciplines work together on related problems, they often face what we call 'science friction'. As science becomes more data-driven, collaborative and interdisciplinary, demand increases for interoperability among data, tools, and services. Metadata – usually viewed simply as 'data about data', describing objects such as books, journal articles, or datasets – serve key roles in interoperability. Yet we find that metadata may be a source of friction between scientific collaborators, impeding data sharing" (Edwards, Mayernik, Batcheller, Bowker, & Borgman, 2011). Metadata is vital to enable data discovery and there are greater numbers of metadata resources being produced and used by researchers e.g. the EU-funded WISER project (Water bodies in Europe: Integrative Systems to assess Ecological status and Recovery)

Table 1. Themes and challenges/requirements for knowledge infrastructures summarized from Edwards et al. (2013)

Theme	Challenges/Requirements
Theme 1 how are knowledge infrastructures changing?	Knowledge in perpetual motion; Shifting borders of tacit knowledge and common ground; Complexities of sharing data across disciplines and domains; New norms for what counts as knowledge; Massive shifts in publishing practices, linked to new modes of knowledge assessment; Challenges to traditional educational institutions; Navigating across scales of space and time, and rates of change; Standards and ontologies.
Theme 2 how do knowledge infrastructures reinforce or redistribute authority, influence and power?	Approach this problem as a design opportunity; Create a professional cadre at the interface between scientists and developers; Use the participatory design movement as a model for work practices.
Theme 3 how can we best study, know and imagine today's (and tomorrow's) knowledge infrastructures?	Create and nourish mechanisms for large-scale, long-term research; Build interdisciplinary collaborations across natural and social sciences; Develop comparative analysis techniques for studying large-scale, long-term data; Create sustainable, shareable data archives; Build better software for qualitative work; Integrate qualitative work with statistical techniques and social network analysis; Imagine new forms of cyber scholarship.

(Schmidt-Kloiber, Moe, Dudley, Strackbein, & Vogl, 2013). Standards-based services from SDI are gradually becoming available and used by researchers, but often their architecture needs to be improved to enable greater bottom up use of data services (Diaz, Granell, Gould, & Huerta, 2011). Publishing data requires information on the provenance of the methodology and who collected it, where and when, to enable interpretation and trust that enables reproducibility and reuse (Bechhofer et al., 2010). Science will benefit from the data fusion capacity of linked data, but requires a publication model based on the notion of research objects, that are 'units of knowledge' (Bechhofer et al., 2010). More recently, Castelli *et al.* (2013) called for 'scientific communication infrastructure' to bridge the gap between research digital libraries and data centres, to improve discovery, linking and publication across the research process. Automated ways are required to quality control the data and track its provenance to increase data integrity (Campbell et al., 2013). Porter et al. (2012) reviewed challenges of using sensors to study lakes that include information management and analytical challenges from the massive deluge of data.

Digital Technologies to Advance Knowledge Infrastructures in Catchment Research Networks

Collaborative knowledge is dependent on integrative research, but also on the physical, social and digital knowledge infrastructures to enable networks of data, information and knowledge to be collected, processed, stored, shared and reused (Figure 1). Science is a dynamic process of understanding how the world works through

Figure 1. Integration of domain data, information and knowledge, social science and advances in digital technologies for systems based integrative research

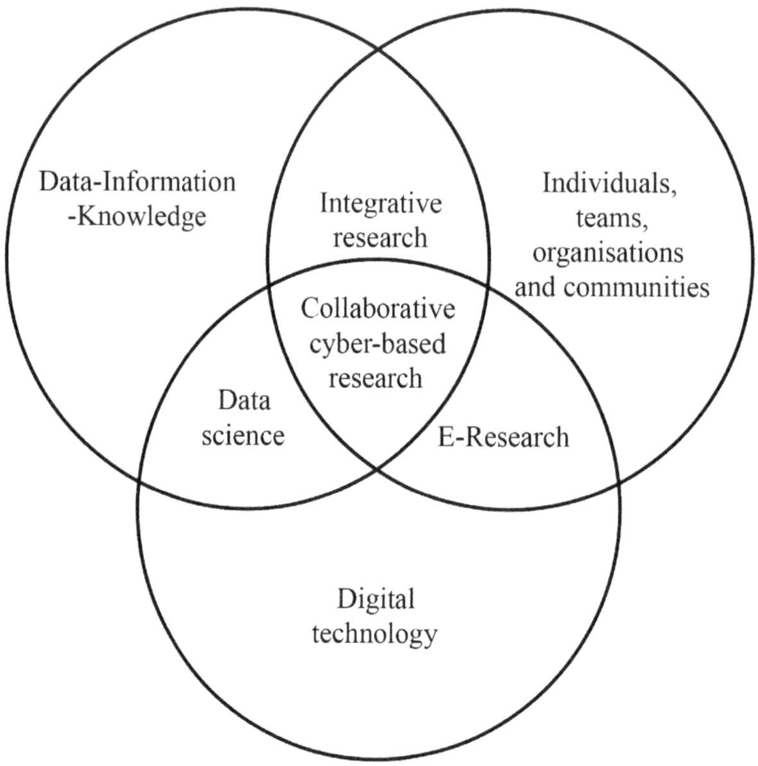

theoretical and empirical studies and building that information and knowledge into powerful and coherent frameworks. Empirical data becomes information, when through context we add meaning. The generation, use and sharing of knowledge is often viewed as a series of processes between the tacit knowledge people hold and how this is externalised into explicit knowledge e.g. papers and models, and then internalised by others into tacit knowledge.

There are a range of models of the theory and practice of knowledge management, these are covered in depth by Dalkir (2011). The traditional way of viewing relationships between data, information, knowledge and decision making is as a pyramid, made famous by Russell Ackoff in his 1988 presidential address to the International Society for General Systems Research (1989), where data is filtered through information to knowledge. In the past, we reduced the complexity of the world so we could understand it in a paper-based system, but that limited our capacity for knowledge generation, sharing and application. With the development of the Internet, Weinberger (2011) argues that the methods (i.e. network) we have available for understanding the world and our interactions with it have changed radically. A network based perspective of the relationships between data, information and knowledge is now required.

Digital technologies and the growth in team based integative research are drastically changing how collaborative knowledge is developed and used. Edwards (2010) defined knowledge infrastructures as "robust networks of people, artifacts, and institutions that generate, share, and maintain specific knowledge about the human and natural worlds." Knowledge infrastructures and our understanding of them are changing rapidly with increased development and use of Internet and Web digital technologies: how people create, share and dispute knowledge (Edwards et al., 2013). An National Science Foundation (NSF) funded workshop looked at: 1) how are knowledge infrastructures changing?; 2) how do knowledge infrastructures reinforce or redistribute authority, influence, power and control?; and 3) how can we best study, know and imagine today's knowledge infrastructures? A summary of the main points they raised are in Table 1. In the context of this chapter, one of the largest changes in knowledge infrastructure is due to the complexities of sharing data across disciplines and with non-researchers. This change relates to the need for knowledge infrastructure that will enable longer term and larger scale collaborations between natural, social and computer scientists with shareable data archives. To address these requirements for knowledge infrastructure to support catchment research networks, there is a need for catchment observatories that integrate existing and novel physical, social and digital networks of knowledge infrastructures (Figure 1).

Developing Catchment Observatories for Collaborative Knowledge in Catchment Research Networks

Catchment observatories (sometimes called platforms) are needed that can integrate our physically based understanding of their functioning and related manipulation and monitoring, along with our understanding of the social processes controlling their use and management through to the digital technologies required to integrate existing physical, social and digital knowledge infrastructures and support systems based integrative research. There is growing scientific and wider societal interest in the development and use of virtual observatories and community science platforms to provide and integrate physical, digital and social infrastructures to support scientific research (Gloria, McGuinness, Luciano, & Zhang, 2013; Gurney et al., 2011; McGuinness, 2012). These virtual observatories and community science platforms are the scientific social machines that will provide the knowledge infrastructures for global environmental change research in the 21st century. Virtual observatories originated in the astronomy community to enable data archiving, sharing and analysis between professional and amateur astronomers, along with development of standards-based software to enable novel science e.g. (Chilingarian et al., 2009). In ecology, established long term networks of field sites are increasingly incorporating social science to go beyond traditional ecological research, to understand landscapes as multi-scale socio-ecological systems (Haberl et al., 2006; Redman, Grove, & Kuby, 2004). For example, the newly established US NSF funded NEON project is providing a new network of place based sites and cyberinfrastructure for advancing regional and continental scale ecological science (Schimel, Hargrove, Hoffman, & MacMahon, 2007). The process of setting up river basin (catchment) observatory has no set recipe, but requires long-term collaborative development of problem focused integrative teams of researchers and stakeholders (Quinn, Hewett, Muste, & Popescu, 2010).

The social networking infrastructure of observatories needs to ensure that research is carried out efficiently and effectively amongst the diverse range of individuals, groups and institutions involved (Figure 1). This includes: researchers carrying out the research; non-research stakeholders involved in the research; and users of the data, information and knowledge. Pennington (2008) highlighted that there is often still a lack of effective collaborative research as we need to better understand how we can produce the environments and interactions for effective interdisciplinary research. As greater levels of research becomes team based, there is a need to learn from management sciences and the newly developing field of 'science of team science' (Syme, 2008) and its multi-level perspective (Börner et al., 2010). Team working requires good leadership with leaders having three key skill sets: cognitive (including clear vision of the project's aims and objectives); structural (e.g. gaining support and facilitating linkages); and processual (ensuring productive interactions) (Lyall, Bruce, Tait, & Meagher, 2011). A vision for long term networks of science platforms is required as they provide researchers with valuable opportunities to collaborate (Mauz, Peltola, Granjou, Van Bommel, & Buijs, 2012). However, there are challenges to carrying out research based on networks of observatories and platforms at the individual, site and network levels. For example, through interviews, workshops and surveys with graduate students it was apparent that there were cultural, institutional and logistical barriers to them carrying out their research and differences in their knowledge and engagement with the LTER network (Romolini, Record, Garvoille, Marusenko, & Stuart Geige, 2013).

To enable greater levels of collaborative knowledge production, sharing and use there are an increasing number of digital technologies including web based resources supporting the development and application of digital catchment observatories (Horsburgh et al., 2009; Kunkel et al., 2013; Vitolo et al., 2015). Development, use and governance of spatial data infrastructure (SDI) to support the environmental sciences are well established (De Man, 2006). Increasing efforts to develop and subsequently use community-based geospatial cyberinfrastructure is improving people's access to data and information (De Longueville, 2010). Geospatial cyberinfrastructure "refers to the infrastructure that supports the collection, management and utilization of geospatial data, information and knowledge for multiple science domains" (Yang, Raskin, Goodchild, & Gahegan, 2010). A review of the future requirements for more effective geospatial cyberinfrastructure stressed the need for: 1) understanding social heterogeneity in relation to problem/situation awareness; 2) improving how we can transform data collected to the information required; 3) greater use of semantic approaches for more effective knowledge exchange; 4) iterative development of geospatial cyberinfrastructure through engagement with stakeholders; 5) involvement of social and domain scientists to advance citizen science approaches; 6) use of cloud computing to geospatial cyberinfrastructure to provide open and transparent platforms; and 7) a geospatial cyberinfrastructure research agenda that includes a wider set of stakeholders, including those from academia, governments and their agencies, NGOs and the wider public (Yang et al., 2010). Advances in cyberinfrastructure and its accessibility to researchers are enabling development of digital catchment environments that integrate *in situ* and remotely sensed measurements with databases and computational models (Muste et al., 2010). It is now possible to develop cyberinfrastructure platforms to support understanding and management of key catchment processes. For example, Muste et al. (2013) assessed the amount of soil erosion from agricultural land in Clear Creek Catchment in eastern Iowa, USA based on the CUAHSI Hydrologic Information System (HIS). Web based catchment observatories for local and national scale catchment modelling have been piloted (Buytaert, Baez, Bustamante, & Dewulf, 2012; Gurney et al., 2011). These catchment systems science observatories are still in their infancy.

INTERNATIONAL EXAMPLES OF DIGITAL INFRASTRUCTURE PROJECTS SUPPORTING COLLABORATIVE KNOWLEDGE IN CATCHMENT RESEARCH NETWORKS

In this section I present three leading international examples of collaborative networks of catchment researchers and their development and use of digital infrastructure to advance collaborative knowledge production, sharing and use. These include examples from North America (NSF funded CUAHSI HIS) and from Europe (UK NERC funded EVOp and the German Helmholtz Association Centers funded TERENO/TEODOOR). These exemplars all supported advancing collaborative knowledge in catchment research networks through the development of catchment observatories. However, each project had: different objectives; applied and developed a range of digital technologies; had differing levels and timescales of funding; and varying numbers of disciplinary experts involved in the catchment research networks.

NSF Funded CUAHSI Hydrologic Information System (HIS)

Over the past decade a consortium of hydrologists and software engineers funded through the NSF in the US have developed an impressive stack of informatics infrastructure to support hydrological science. The aim of the CUAHSI HIS project was "to enhance hydrologic science by facilitating user access to more and better data for testing hypotheses and analyzing hydrologic processes" http://his.cuahsi.org/project.html. This involved developing the HIS open source cyberinfrastructure system for sharing hydrologic data as well as supporting the wider hydrologic information community, through holding regular national and regional meetings. Horsburgh et al. (2011) set out the "architecture and functional requirements for an environmental observatory information system that supports collection, organization, storage, analysis, and publication of hydrologic observations." To improve the accessibility of CUAHSI HIS to smaller research networks around the world, Conner et al. (2013) recently developed a light weight Linux version using open source and free software enabling research groups with little technical expertise or funding to access the benefits of CUAHSI HIS.

Digital Infrastructure

Architecture and Web Services: The architecture of CUAHSI HIS was based on a service-orientated architecture (SOA) linking data discovery (HIS Central), data publication (HydroServer) and data access (HydroDesktop and other clients). With metadata, search and data services joining these components, called WaterOneFlow based on the original WaterML.

Data Standards, Semantics and Provenance: Horsburgh et al. (2008) developed and applied the Observations Data Model for the storage and access of point observational data for multiple researchers. The design of the Observations Data Model enabled a systematic solution to data management that removed many of the semantic and syntactic differences between datasets. To meet the challenges of creating large amounts of scientific data, Horsburgh et al. (2009) developed a method based on controlled vocabularies to enable federated network of data across a national network of environmental observatories.

Applications: HydroDesktop is a geographical user interface tool that is part of the HIS, with the purpose to improve data discovery and access though novel use of web services and extendable interface to help reduce time to navigate different data repositories with a range of data formats and interfaces (Ames et al., 2012). This can be extended by plugins e.g. HydroR is a R statistical computing environment plugin that enables scientists and engineers to discover data and model it in the same location (Horsburgh & Reeder, 2014). Recently CUAHSI HIS has been linked with the Open Modelling Interface (OpenMI) to benefit from its capacity to support integrated modelling and allow the HIS to focus on hydrologic data management and visualization (Castronova, Goodall, & Ercan, 2013).

Use in Catchment Observatories to Support Collaborative Knowledge in Research Networks

The CUAHSI HIS has been widely used in North America and increasingly around the world e.g. (Kadlec & Ames, 2011). The HIS has been applied in individual research catchments and in national networks, for example, Muste et al. (2013) presented an assessment of soil erosion from agricultural land in Clear Creek Catchment in eastern Iowa, USA and Horsburgh et al. (2011) demonstrated its use in a national network of observatory sites.

NERC Funded Environmental Virtual Observatory Pilot (EVO)

The Environmental Virtual Observatory pilot (EVO) project http://www.evo-uk.org/ ran between 2010 and 2013 and included 33 researchers from 12 environmental research centers and institutes across the UK (Gurney et al., 2011). The aim was to develop a cloud based test bed for exploring scenarios concerning catchment understanding and management. This involved assessing the potential of cloud technologies to make environmental monitoring and decision making more efficient, effective and transparent to the whole community at the local catchment level and at the national level. A project advisory board representing potential end users and stakeholders from academia, industry and government agencies guided the development of the pilot. In addition to the local and national level catchment exemplars, the team explored broad issues relating to security, intellectual property rights, data licensing, and the development and testing of the web technology stack. A review of web technologies for environment modelling can be found in Vitolo et al. (2015), including an evaluation of the digital technologies tried and used in the development of the EVO web based observatory.

Digital Infrastructure

Architecture and Web Services: The EVO used a set of web applications to access, use and visualize a range of catchment data and models. A RESTful architecture was chosen rather than SOA, due to its ease of use and existing RESTful infrastructure.

Data Standards, Semantics and Provenance: This pilot project was based on a range of fairly small datasets available in plain text or CSV format. Model results were generated in an XML format to enable interoperability amongst components and addition of metadata. Data were stored in a PostgreSQL database.

Applications: The models (Topmodel, and related FUSE ensemble of hydrological models and a biogeochemical export coefficient model) were written in R and calls to the models were made through PyWPS package (an implementation of the Open Geospatial Consortium (OGC) Web processing standard) and the RPy2 connector. Web based visualization was based on Google maps and the Flot library. An experimental online workflow editor using Taverna was developed.

Use in Catchment Observatories to Support Collaborative Knowledge in Research Networks

The EVO was only a pilot project. At the local level, its potential was applied and evaluated in three main catchments with a range of local stakeholders. At the national level, hydrological and biogeochemical model based applications were develop as proof of concept (Freer et al., 2012; Greene et al., submitted; Greene et al., 2012).

Helmholtz Association Centers Funded TERENO/TEODOOR

In 2008 a network of Terrestrial Environmental Observatories (TERENO) http://teodoor.icg.kfa-juelich.de/overview-de was established in Germany as part of an interdisciplinary research network with the aim to study the impacts of land use change and climate change (Zacharias et al., 2011). The focus was on environmental monitoring of the states and fluxes from the subsurface to the lower atmosphere. To manage the collection, sharing and use of large amounts of data, a distributed SDI was created, TEreno Online Data RepOsitORy (TEODOOR) (Kunkel et al., 2013). The TEODOOR data portal enabled the local data infrastructure to communicate using OGC compliant web services.

Digital Infrastructure

Architecture and Web Services: A range of standard OGC web services including Web Coverage Services, Web Feature Services, Web Map Services, Web Processing Services and Web Catalogue Services were used to enable the processing and exchange of data and metadata.

Data Standards, Semantics and Provenance: Each of the catchment observatories collected data from field based sensors which was then processed by a time series management system (TSM 2.0) and stored in a PostGIS database. They extended the Observation Data Model schema from CUAHSI HIS. Metadata profiles based on EU-INSPIRE Directive were produced. The metadata were published by a Catalogue Service, so users could discover the data.

Applications: A data visualization application based on the OpenGeoSys framework was tested and is under further development. The TERENO data portal brought together data published through the OGC web services and used a PLONE content management system to allow users to upload and share a range of resources.

Use in Catchment Observatories to Support Collaborative Knowledge in Research Networks

All four of the TERENO terrestrial environmental observatories use the TEODOOR distributed geodata infrastructure. These range from lowland catchment observatories in northern and central Germany to an upland catchment observatory in the Bavarian Alps. Six Helmholtz Association Centers are actively involved and an impressive range of researchers in these catchment research networks.

FUTURE RESEARCH DIRECTIONS

Integrative Research Requirements for Catchment Observatories Based on Catchment Systems Science

In this chapter, I have set out that greater levels of systems based integrative research are required for the production, sharing and use of collaborative knowledge in catchment research networks. There are challenges to carrying out integrative research. Catchment research networks, which are often dominated by natural scientists, need to learn from other disciplines, including the social and management sciences, how to carry out greater levels of systems based integrative research. There needs to be greater levels of international scientific community based development of the required digital informatics infrastructure to support the use of observatories in catchment research networks. This will enable linking of networks and observatories not only in each individual project or in each country, but internationally, which is required to further our understanding of how environmental change is affecting catchment management and the functions catchments provide to society. Advances in the development and application of digital technologies in catchment observatories will help integrate existing physical, social and digital knowledge infrastructures. We need to learn from existing catchment based research initiatives like CUAHSI HIS and GLEON. Collaborative knowledge of catchment systems is required to develop and link solutions to water, food, environment and energy security. This collaborative knowledge needs to span research, policy and practice networks. Catchments are complex adaptive systems and there is still a large amount of research required to understand cross-scale catchment processes under environmental change. This integrative research requires networks of researchers building and using catchment observatories to serve as the knowledge infrastructure of the 21st century. International initiatives are needed to enable systems based integrative research and development of catchment research network infrastructure. Examples of recent international initiatives include Future Earth and the IAHS Panta Rhei. Future Earth (http://www.futureearth.info/) aims to increase the linkages between natural and social sciences to study global environmental change. Whereas the IAHS Panta Rhei is aimed more specifically at understanding the dynamic interactions between human systems and the water cycle, which is essential for understanding and managing water security (Montanari et al., 2013). These international initiatives along with national and regional funding calls e.g. European H2020 Infrastructure, and the growing number of bottom up observatory initiatives will help further develop the theory and practice of catchment observatories and integrate existing and novel physical, social and digital knowledge infrastructures.

CONCLUSION

Collaborative knowledge of catchments requires greater levels of systems based integrative research involving networks of researchers, policy makers and practitioners. Integrative research is team based and requires excellent management and leadership, along with extra time to develop shared understandings and enable effective communication across scientific disciplines, policy makers and practitioners. To enable increased levels of integrative research depends on the development and application of digital technologies to improve collection, use and sharing of data and devise new knowledge infrastructures. Catchment observatories that integrate existing and novel physical, social and digital networks of knowledge infrastructures are urgently needed. Catchment systems science is required to increase the levels of integrative research and to develop and apply these open source digital technologies to enable catchment observatories to serve effectively as 21st century knowledge infrastructures, supporting international catchment research networks.

REFERENCES

Ackoff, R. L. (1974). *Redesigning the Future: Systems approach to societal problems*. New York: Wiley.

Ackoff, R. L. (1989). From data to wisdom. *Journal of Applied Systems Analysis*, *16*, 3–9.

Akers, K. G. (2013). Looking out for the little guy: Small data curation. *Bulletin of the American Society for Information Science and Technology*, *39*(3), 58–59. doi:10.1002/bult.2013.1720390317

Ames, D. P., Horsburgh, J. S., Cao, Y., Kadlec, J., Whiteaker, T., & Valentine, D. (2012). HydroDesktop: Web services-based software for hydrologic data discovery, download, visualization, and analysis. *Environmental Modelling & Software*, *37*(0), 146–156. doi:10.1016/j.envsoft.2012.03.013

Bazilian, M., Rogner, H., Howells, M., Hermann, S., Arent, D., & Gielen, D. etal. (2011). Considering the energy, water and food nexus: Towards an integrated modelling approach. *Energy Policy*, *39*(12), 7896–7906. doi:10.1016/j.enpol.2011.09.039

Bechhofer, S., Ainsworth, J., Bhagat, J., Buchan, I., Couch, P., Cruickshank, D., et al. (2010). Why linked data is not enough for scientists. In *Proceedings of IEEE Sixth International Conference on e-Science*, (pp. 300-307). IEEE. doi:10.1109/eScience.2010.21

Blue Ribbon Task Force on Sustainable Digital Preservation and Access. (2010). *Sustainable Economics for a Digital Planet: Ensuring Long-Term Access to Digital Information*. Final Report of the Blue Ribbon Task Force on Sustainable Digital Preservation and Access.

Bogardi, J. J., Dudgeon, D., Lawford, R., Flinkerbusch, E., Meyn, A., & Pahl-Wostl, C. etal. (2012). Water security for a planet under pressure: Interconnected challenges of a changing world call for sustainable solutions. *Current Opinion in Environmental Sustainability*, *4*(1), 35–43. doi:10.1016/j.cosust.2011.12.002

Borgman, C. L. (2012). The conundrum of sharing research data. *Journal of the American Society for Information Science and Technology*, *63*(6), 1059–1078. doi:10.1002/asi.22634

Börner, K., Contractor, N., Falk-Krzesinski, H. J., Fiore, S. M., Hall, K. L., & Keyton, J. etal. (2010). A Multi-Level Systems Perspective for the Science of Team Science. *Science Translational Medicine*, *2*(49), 49cm24. doi:10.1126/scitranslmed.3001399 PMID:20844283

Bracken, L. J., Wainwright, J., Ali, G. A., Tetzlaff, D., Smith, M. W., Reaney, S. M., & Roy, A. G. (2013). Concepts of hydrological connectivity: Research approaches, pathways and future agendas. *Earth-Science Reviews*, *119*(0), 17–34. doi:10.1016/j.earscirev.2013.02.001

Buytaert, W., Baez, S., Bustamante, M., & Dewulf, A. (2012). Web-Based Environmental Simulation: Bridging the Gap between Scientific Modeling and Decision-Making. *Environmental Science & Technology*, *46*(4), 1971–1976. doi:10.1021/es2031278 PMID:22260091

Campbell, J. L., Rustad, L. E., Porter, J. H., Taylor, J. R., Dereszynski, E. W., & Shanley, J. B. etal. (2013). Quantity is Nothing without Quality: Automated QA/QC for Streaming Environmental Sensor Data. *Bioscience*, *63*(7), 574–585. doi:10.1525/bio.2013.63.7.10

Castelli, D., Manghi, P., & Thanos, C. (2013). A vision towards Scientific Communication Infrastructures. *International Journal on Digital Libraries*, *13*(3-4), 155–169. doi:10.1007/s00799-013-0106-7

Castronova, A. M., Goodall, J. L., & Ercan, M. B. (2013). Integrated modeling within a Hydrologic Information System: An OpenMI based approach. *Environmental Modelling & Software*, *39*(0), 263–273. doi:10.1016/j.envsoft.2012.02.011

Chilingarian, I., Cayatte, V., Revaz, Y., Dodonov, S., Durand, D., & Durret, F. etal. (2009). A Population of Compact Elliptical Galaxies Detected with the Virtual Observatory. *Science*, *326*(5958), 1379–1382. doi:10.1126/science.1175930 PMID:19797624

Collins, K. B., & Ison, R. L. (2010). Trusting emergence: Some experiences of learning about integrated catchment science with the Environment Agency of England and Wales. *Water Resources Management*, *24*(4), 669–688. doi:10.1007/s11269-009-9464-8

Conner, L. G., Ames, D. P., & Gill, R. A. (2013). HydroServer Lite as an open source solution for archiving and sharing environmental data for independent university labs. *Ecological Informatics*, *18*, 171–177. doi:10.1016/j.ecoinf.2013.08.006

Cortner, H., & Moote, M. (1994). Trends and issues in land and water resources management: Setting the agenda for change. *Environmental Management*, *18*(2), 167–173. doi:10.1007/BF02393759

Costello, M. J. (2009). Motivating online publication of data. *Bioscience*, *59*(5), 418–427. doi:10.1525/bio.2009.59.5.9

Dalkir, K. (2011). *Knowledge management in theory and practice* (2nd ed.). MIT.

De Longueville, B. (2010). Community-based geoportals: The next generation? Concepts and methods for the geospatial Web 2.0. *Computers, Environment and Urban Systems*, *34*(4), 299–308. doi:10.1016/j.compenvurbsys.2010.04.004

De Man, W. H. E. (2006). Understanding SDI; complexity and institutionalization. *International Journal of Geographical Information Science*, *20*(3), 329–343. doi:10.1080/13658810500399688

Dewulf, A., Francois, G., Pahl-Wostl, C., & Taillieu, T. (2007). A framing approach to cross-disciplinary research collaboration: Experiences from a large-scale research project on adaptive water management. *Ecology and Society*, *12*(2).

Diaz, L., Granell, C., Gould, M., & Huerta, J. (2011). Managing user-generated information in geospatial cyberinfrastructures. *Future Generation Computer Systems-the International Journal of Grid Computing-Theory Methods and Applications*, *27*(3), 304–314. doi:10.1016/j.future.2010.09.002

Duerr, R. E., Downs, R. R., Tilmes, C., Barkstrom, B., Lenhardt, W. C., & Glassy, J. etal. (2011). On the utility of identification schemes for digital earth science data: An assessment and recommendations. *Earth Science Informatics*, *4*(3), 139–160. doi:10.1007/s12145-011-0083-6

Dumbill, E. (2013). Making sense of big data. *Big Data*, *1*(1), 1–2. doi:10.1089/big.2012.1503

Edwards, P. N. (2010). *A Vast Machine: Computer Models, Climate Data, and Politics of Global Warming*. MIT Press.

Edwards, P. N., Jackson, S. J., Chalmers, M. K., Bowker, G. C., Borgman, C. L., & Ribes, D. etal. (2013). *Knowledge Infrastructures: Intellectual Frameworks and Research Challenges*. Ann Arbour, MI: Deep Blue.

Edwards, P. N., Mayernik, M. S., Batcheller, A. L., Bowker, G. C., & Borgman, C. L. (2011). Science friction: Data, metadata, and collaboration. *Social Studies of Science*, *41*(5), 667–690. doi:10.1177/0306312711413314 PMID:22164720

Falkenmark, M. (2001). The greatest water problem: The inability to link environmental security, water security and food security. *International Journal of Water Resources Development*, *17*(4), 539–554. doi:10.1080/07900620120094073

Frame, I., Austen, K. F., Calleja, M., Dove, M. T., White, T. O. H., & Wilson, D. J. (2009). New tools to support collaboration and virtual organizations. *Philosophical Transactions of the Royal Society A: Mathematical, Physical and Engineering Sciences, 367*(1890), 1051-1056.

Freer, J. E., Clark, M. P., Odoni, N. A., Coxon, G., McMillan, H. K., Souvigent, M., et al. (2012). *Understanding the main uncertainties in hydrological impact ensembles of regional climate models predictions for large catchments in the UK.* Paper presented at the American Geophysical Union 45th Annual Fall Meeting. San Francisco, CA.

Gloria, M. J. K., McGuinness, D. L., Luciano, J. S., & Zhang, Q. (2013). Exploration in web science: instruments for web observatories. In *Proceedings of the 22nd international conference on World Wide Web companion*. Academic Press.

Godfray, H. C. J., Beddington, J. R., Crute, I. R., Haddad, L., Lawrence, D., & Muir, J. F. etal. (2010). Food security: The challenge of feeding 9 billion people. *Science*, *327*(5967), 812–818. doi:10.1126/science.1185383 PMID:20110467

Greene, S., Johnes, P. J., Bloomfield, J. P., Reaney, S. M., Lawley, R., & Freer, J. etal. (Manuscript submitted for publication). Improving national capability in nutrient flux modelling using a high resolution geospatial framework. *Environmental Modelling & Software*.

Greene, S., Johnes, P. J., Freer, J., Bloomfield, J. P., Reaney, S. M., Macleod, C. J. A., & Odoni, N. A. (2012). *The Environmental Virtual Observatory (EVO): can cloud-based modelling provide new understanding of nutrient cycling processes from catchment to national scale?* Paper presented at the 17th International Nitrogen Workshop: Innovations for Sustainable use of Nitrogen Resources. Wexford, Ireland.

Gruber, T. (2008). Collective knowledge systems: Where the Social Web meets the Semantic Web. *Web Semantics: Science, Services, and Agents on the World Wide Web*, *6*(1), 4–13. doi:10.1016/j.websem.2007.11.011

Gurney, R., Emmett, B., McDonald, A., Blair, G., Buytaert, W., Freer, J., et al. (2011). *The environmental virtual observatory: a new vision for catchment science*. Paper presented at the American Geophysical Union, Fall Meeting. New York, NY.

Haberl, H., Winiwarter, V., Andersson, K., Ayres, R. U., Boone, C., & Castillo, A. etal. (2006). From LTER to LTSER: Conceptualizing the socioeconomic dimension of long-term socioecological research. *Ecology and Society*, *11*(2), 13.

Hall, W., De Roure, D., & Shadbolt, N. (2009). The evolution of the Web and implications for eResearch. *Philosophical Transactions of the Royal Society a-Mathematical Physical and Engineering Sciences, 367*(1890), 991-1001.

Hampton, S. E., Strasser, C. A., Tewksbury, J. J., Gram, W. K., Budden, A. E., & Batcheller, A. L. etal. (2013). Big data and the future of ecology. *Frontiers in Ecology and the Environment*, *11*(3), 156–162. doi:10.1890/120103

Hanson, P. C. (2007). A grassroots approach to sensor and science networks. *Frontiers in Ecology and the Environment*, *5*(7), 343–343. doi:10.1890/1540-9295(2007)5[343:AGATSA]2.0.CO;2

Harmon, S. G., Guo, Q., Hanson, P., Hofmann, A., Jones, B., Knoblock, C., et al. (2011). *ISI Technical Report Number ISI-TR-674 October, 2011*. Information Sciences Institute, University of Southern California.

Hendler, J., & Berners-Lee, T. (2010). From the Semantic Web to social machines: A research challenge for AI on the World Wide Web. *Artificial Intelligence*, *174*(2), 156–161. doi:10.1016/j.artint.2009.11.010

Hoff, H., Iceland, C., Kuylenstierna, J., & te Velde, D. W. (2012). Managing the Water-Land-Energy Nexus for Sustainable Development. *UN Chronicle*, *49*(1&2).

Holmes, J., & Clark, R. (2008). Enhancing the use of science in environmental policy-making and regulation. *Environmental Science & Policy*, *11*(8), 702–711. doi:10.1016/j.envsci.2008.08.004

Horsburgh, J. S., & Reeder, S. L. (2014). Data visualization and analysis within a Hydrologic Information System: Integrating with the R statistical computing environment. *Environmental Modelling & Software*, *52*(0), 51–61. doi:10.1016/j.envsoft.2013.10.016

Horsburgh, J. S., Tarboton, D. G., Maidment, D. R., & Zaslavsky, I. (2008). A relational model for environmental and water resources data. *Water Resources Research*, *44*(5), n/a. doi:10.1029/2007WR006392

Horsburgh, J. S., Tarboton, D. G., Maidment, D. R., & Zaslavsky, I. (2011). Components of an environmental observatory information system. *Computers & Geosciences*, *37*(2), 207–218. doi:10.1016/j.cageo.2010.07.003

Horsburgh, J. S., Tarboton, D. G., Piasecki, M., Maidment, D. R., Zaslavsky, I., Valentine, D., & Whitenack, T. (2009). An integrated system for publishing environmental observations data. *Environmental Modelling & Software*, *24*(8), 879–888. doi:10.1016/j.envsoft.2009.01.002

Jewitt, G. (2002). Can integrated water resources management sustain the provision of ecosystem goods and services? *Physics and Chemistry of the Earth*, *27*(11-22), 887-895.

Jordan, P., Arnscheidt, A., McGrogan, H., & McCormick, S. (2007). Characterising phosphorus transfers in rural catchments using a continuous bank-side analyser. *Hydrology and Earth System Sciences*, *11*(1), 372–381. doi:10.5194/hess-11-372-2007

Jordan, P., Haygarth, P., Shortle, G., & Harris, R. (2012). Catchment science and policy for agriculture and water quality. *Environmental Science & Policy*, *24*, 1–3. doi:10.1016/j.envsci.2012.09.009

Kadlec, J., & Ames, D. (2011). Design and Development of Web Services for Accessing Free Hydrological Data from the Czech Republic. In J. Hřebíček, G. Schimak, & R. Denzer (Eds.), *Environmental Software Systems. Frameworks of eEnvironment* (Vol. 359, pp. 581–588). Springer Berlin Heidelberg. doi:10.1007/978-3-642-22285-6_63

Kragt, M. E., Robson, B. J., & Macleod, C. J. A. (2013). Modellers' roles in structuring integrative research projects. *Environmental Modelling & Software*, *39*(0), 322–330. doi:10.1016/j.envsoft.2012.06.015

Kunkel, R., Sorg, J., Eckardt, R., Kolditz, O., Rink, K., & Vereecken, H. (2013). TEODOOR: A distributed geodata infrastructure for terrestrial observation data. *Environmental Earth Sciences*, *69*(2), 507–521. doi:10.1007/s12665-013-2370-7

Langan, S. J., Wade, A. J., Smart, R., Edwards, A. C., Soulsby, C., & Billett, M. F. etal. (1997). The prediction and management of water quality in a relatively unpolluted major Scottish catchment: Current issues and experimental approaches. *The Science of the Total Environment*, *194*, 419–435. doi:10.1016/S0048-9697(96)05380-6

Levin, S. A., Grenfell, B., Hastings, A., & Perelson, A. S. (1997). Mathematical and computational challenges in population biology and ecosystems science. *Science*, *275*(5298), 334–343. doi:10.1126/science.275.5298.334 PMID:8994023

Likens, G. E., & Bormann, F. H. (1974). Linkages between terrestrial and aquatic ecosystems. *Bioscience*, *24*(8), 447–456. doi:10.2307/1296852

Lovett, G. M., Likens, G. E., Buso, D. C., Driscoll, C. T., & Bailey, S. W. (2005). The biogeochemistry of chlorine at Hubbard Brook, New Hampshire, USA. *Biogeochemistry*, *72*(2), 191–232. doi:10.1007/s10533-004-0357-x

Lyall, C., Bruce, A., Tait, J., & Meagher, L. (2011). *Interdisciplinary Research Journeys: Practical Strategies for Capturing Creativity*. London: Bloomsbury. doi:10.5040/9781849661782

MacKinnon, D., & Tetzlaff, D. (2009). Conceptualising scale in regional studies and catchment science–towards an integrated characterisation of spatial units. *Geography Compass*, *3*(3), 976–996. doi:10.1111/j.1749-8198.2009.00231.x

Macleod, C. J. A. (2010). *What can we learn from systems based approaches: from systems biology to earth systems science?* Paper presented at the 2010 International Congress on Environmental Modelling and Software, Modelling for Environment's Sake, Fifth Biennial Meeting. Ottawa, Canada.

Macleod, C. J. A., Blackstock, K. L., & Haygarth, P. M. (2008). Mechanisms to improve integrative research at the science-policy interface for sustainable catchment management. *Ecology and Society*, *13*(2).

Macleod, C. J. A., & Ferrier, R. C. (2011). Temperate grasslands in catchment systems: the role of scale, connectivity and thresholds in the provision and regulation of water quality and quantity. In G. Lemaire, J. Hodgson, & A. Chabbi (Eds.), *Grassland productivity and ecosystem services* (pp. 229–238). CABI. doi:10.1079/9781845938093.0229

Macleod, C. J. A., & Haygarth, P. M. (2010). Integrating water and agricultural management under climate change. *The Science of the Total Environment*, *408*(23), 5619–5622. doi:10.1016/j.scitotenv.2010.02.041 PMID:20304465

Macleod, C. J. A., Scholefield, D., & Haygarth, P. M. (2007). Integration for sustainable catchment management. *The Science of the Total Environment*, *373*(2-3), 591–602. doi:10.1016/j.scitotenv.2006.12.029 PMID:17239423

Mauz, I., Peltola, T., Granjou, C., Van Bommel, S., & Buijs, A. (2012). How scientific visions matter: Insights from three long-term socio-ecological research (LTSER) platforms under construction in Europe. *Environmental Science & Policy*, *19*, 90–99. doi:10.1016/j.envsci.2012.02.005

McGonigle, D. F., Harris, R. C., McCamphill, C., Kirk, S., Dils, R., Macdonald, J., & Bailey, S. (2012). Towards a more strategic approach to research to support catchment-based policy approaches to mitigate agricultural water pollution: A UK case-study. *Environmental Science & Policy*, *24*, 4–14. doi:10.1016/j.envsci.2012.07.016

McGuinness, D. L. (2012). *Community science-the next frontier*. Paper presented at the AGU. San Francisco, CA.

Milne, A. E., Macleod, C. J. A., Haygarth, P. M., Hawkins, J. M. B., & Lark, R. M. (2009). The wavelet packet transform: A technique for investigating temporal variation of river water solutes. *Journal of Hydrology (Amsterdam)*, *379*(1-2), 1–19. doi:10.1016/j.jhydrol.2009.09.038

Montanari, A., Young, G., Savenije, H. H. G., Hughes, D., Wagener, T., & Ren, L. L. etal. (2013). "Panta Rhei—Everything Flows": Change in hydrology and society—The IAHS Scientific Decade 2013–2022. *Hydrological Sciences Journal*, 1–20.

Mooney, H. (2011). Citing data sources in the social sciences: Do authors do it? *Learned Publishing*, *24*(2), 99–108. doi:10.1087/20110204

Muste, M., Kim, D., Arnold, N., Whiteaker, T., Just, C., & Kruger, A. (2010). Digital catchment inception using community project components. *Proceedings of the ICE-Water Management*, *163*(1), 13–25. doi:10.1680/wama.2010.163.1.13

Muste, M. V., Bennett, D. A., Secchi, S., Schnoor, J. L., Kusiak, A., & Arnold, N. J. etal. (2013). End-to-End Cyberinfrastructure for Decision-Making Support in Watershed Management. *Journal of Water Resources Planning and Management*, *139*(5), 565–573. doi:10.1061/(ASCE)WR.1943-5452.0000289

Nativi, S., Mazzetti, P., Craglia, M., & Pirrone, N. (2013). The GEOSS solution for enabling data interoperability and integrative research. *Environmental Science and Pollution Research International*, 1–16. PMID:24243262

Nielsen, M. A. (2012). *Reinventing Discovery: The New Era of Networked Science*. Princeton, NJ: Princeton University Press.

Outram, F., Lloyd, C., Jonczyk, J., Benskin, C. M., Grant, F., & Dorling, S. etal. (2013). High-resolution monitoring of catchment nutrient response to the end of the 2011-2012 drought in England, captured by the demonstration test catchments. *Hydrology and Earth System Sciences Discussions*, *10*(12), 15119–15165. doi:10.5194/hessd-10-15119-2013

Pahl-Wostl, C. (2007). Transitions towards adaptive management of water facing climate and global change. *Water Resources Management*, *21*(1), 49–62. doi:10.1007/s11269-006-9040-4

Pahl-Wostl, C., Lebel, L., Knieper, C., & Nikitina, E. (2012). From applying panaceas to mastering complexity: Toward adaptive water governance in river basins. *Environmental Science & Policy*, *23*(0), 24–34. doi:10.1016/j.envsci.2012.07.014

Parsons, M. A. (2011). Making data useful for modelers to understand complex Earth systems. *Earth Science Informatics*, *4*(4), 197–223. doi:10.1007/s12145-011-0089-0

Pennington, D. D. (2008). Cross-Disciplinary Collaboration and Learning. *Ecology and Society*, *13*(2).

Pennington, D. D. (2011). Bridging the Disciplinary Divide: Co-Creating Research Ideas in eScience Teams. *Computer Supported Cooperative Work-the Journal of Collaborative Computing*, *20*(3), 165–196. doi:10.1007/s10606-011-9134-2

Piwowar, H. A., Day, R. S., & Fridsma, D. B. (2007). Sharing detailed research data is associated with increased citation rate. *PLoS ONE*, *2*(3), e308. doi:10.1371/journal.pone.0000308 PMID:17375194

Porter, J. H., Hanson, P. C., & Lin, C.-C. (2012). Staying afloat in the sensor data deluge. *Trends in Ecology & Evolution*, *27*(2), 121–129. doi:10.1016/j.tree.2011.11.009 PMID:22206661

Quinn, P., Hewett, C., Muste, M., & Popescu, I. (2010). Towards new types of water-centric collaboration. *Proceedings of the ICE-Water Management*, *163*(1), 39–51. doi:10.1680/wama.2010.163.1.39

Redman, C. L., Grove, J. M., & Kuby, L. H. (2004). Integrating social science into the long-term ecological research (LTER) network: Social dimensions of ecological change and ecological dimensions of social change. *Ecosystems (New York, N.Y.)*, *7*(2), 161–171. doi:10.1007/s10021-003-0215-z

Reichman, O. J., Jones, M. B., & Schildhauer, M. P. (2011). Challenges and Opportunities of Open Data in Ecology. *Science*, *331*(6018), 703–705. doi:10.1126/science.1197962 PMID:21311007

Rinner, C., Kessler, C., & Andrulis, S. (2008). The use of Web 2.0 concepts to support deliberation in spatial decision-making. *Computers, Environment and Urban Systems*, *32*(5), 386–395. doi:10.1016/j.compenvurbsys.2008.08.004

Romolini, M., Record, S., Garvoille, R., Marusenko, Y., & Stuart Geige, R. (2013). The Next Generation of Scientists: Examining the Experiences of Graduate Students in Network-Level Social-Ecological Science. *Ecology and Society*, *18*(3), art42. doi:10.5751/ES-05606-180342

Schimel, D., Hargrove, W., Hoffman, F., & MacMahon, J. (2007). NEON: A hierarchically designed national ecological network. *Frontiers in Ecology and the Environment*, *5*(2), 59–59. doi:10.1890/1540-9295(2007)5[59:NAHDNE]2.0.CO;2

Schindler, D. (2009). Lakes as sentinels and integrators for the effects of climate change on watersheds, airsheds, and landscapes. *Limnology and Oceanography*, *54*(6), 2349–2358. doi:10.4319/lo.2009.54.6_part_2.2349

Schmidt-Kloiber, A., Moe, S. J., Dudley, B., Strackbein, J., & Vogl, R. (2013). The WISER metadatabase: The key to more than 100 ecological datasets from European rivers, lakes and coastal waters. *Hydrobiologia*, *704*(1), 29–38. doi:10.1007/s10750-012-1295-6

Sellars, S., Nguyen, P., Chu, W., Gao, X., Hsu, K.-, & Sorooshian, S. (2013). Computational Earth Science: Big Data Transformed Into Insight. *Eos, Transactions, American Geophysical Union*, *94*(32), 277–278. doi:10.1002/2013EO320001

Syme, S. L. (2008). The science of team science - Assessing the value of transdisciplinary research. *American Journal of Preventive Medicine*, *35*(2), S94–S95. doi:10.1016/j.amepre.2008.05.017 PMID:18619409

Takagi, K., & Lin, H. (2011). Temporal dynamics of soil moisture spatial variability in the Shale Hills Critical Zone Observatory. *Vadose Zone Journal*, *10*(3), 832–842. doi:10.2136/vzj2010.0134

Tetzlaff, D., Carey, S., & Soulsby, C. (2013). Catchments in the future North: Interdisciplinary science for sustainable management in the 21st Century. *Hydrological Processes*, *27*(5), 635–639. doi:10.1002/hyp.9707

Tetzlaff, D., Soulsby, C., Bacon, P., Youngson, A., Gibbins, C., & Malcolm, I. (2007). Connectivity between landscapes and riverscapes—a unifying theme in integrating hydrology and ecology in catchment science? *Hydrological Processes*, *21*(10), 1385–1389. doi:10.1002/hyp.6701

Thompson, C. A., Robertson, W. D., & Greenberg, J. (2013). Where Have All the Scientific Data Gone? LIS Perspective on the Data-At-Risk Predicament. *College & Research Libraries*, crl13–crl500.

Thorp, J. H., Thoms, M. C., & Delong, M. D. (2006). The riverine ecosystem synthesis: Biocomplexity in river networks across space and time. *River Research and Applications*, *22*(2), 123–147. doi:10.1002/rra.901

Thorpe, A. (2009). Environmental eScience. *Philosophical Transactions of the Royal Society A: Mathematical, Physical and Engineering Sciences, 367*(1890), 801-802.

Tress, G., Tress, B., & Fry, G. (2005). Clarifying integrative research concepts in landscape ecology. *Landscape Ecology*, *20*(4), 479–493. doi:10.1007/s10980-004-3290-4

Vitolo, C., Elkhatib, Y., Reusser, D., Macleod, C. J. A.,, & Buytaert, W. (2015). Web technologies for environmental Big Data. *Environmental Modelling & Software*, 63, 185-198.

Wegehenkel, M. (2003). Longterm evaluation of land use changes on catchment water balance--a case study from North-East Germany. *Physics and Chemistry of the Earth Parts A/B/C*, *28*(33-36), 1281–1290. doi:10.1016/j.pce.2003.09.004

Weinberger, D. (2011). *Too big to know: rethinking knowledge now that facts aren't facts, experts are everywhere, and the smartest person in the room is the room*. New York: Basic Books.

Williamson, C. E., Saros, J. E., & Schindler, D. W. (2009). Sentinels of Change. *Science*, *323*(5916), 887–888. doi:10.1126/science.1169443 PMID:19213905

Winder, N. (2003). Successes and problems when conducting interdisciplinary or transdisciplinary (= integrative) research. In B. Tress, G. Tress, A. van der Valk & G. Fry (Eds.), Potential and Limitations of Interdisciplinary and Transdisiplinary Landscape Studies (pp. 74-90). Wageningen: Delta Series 2. Alterra Green World Research and Wageningen University.

Wu, J. G., & Loucks, O. L. (1995). From balance of nature to hierarchical patch dynamics: A paradigm shift in ecology. *The Quarterly Review of Biology*, *70*(4), 439–466. doi:10.1086/419172

Yang, C. W., Raskin, R., Goodchild, M., & Gahegan, M. (2010). Geospatial Cyberinfrastructure: Past, present and future. *Computers, Environment and Urban Systems*, *34*(4), 264–277. doi:10.1016/j.compenvurbsys.2010.04.001

Zacharias, S., Bogena, H., Samaniego, L., Mauder, M., Fuß, R., & Pütz, T. etal. (2011). A Network of Terrestrial Environmental Observatories in Germany. *Vadose Zone Journal*, *10*(3), 955–973. doi:10.2136/vzj2010.0139

ADDITIONAL READING

Blue Ribbon Task Force on Sustainable Digital Preservation and Access. (2010). *Sustainable Economics for a Digital Planet: Ensuring Long-Term Access to Digital Information*. Final Report of the Blue Ribbon Task Force on Sustainable Digital Preservation and Access.

Borgman, C. L. (2012). The conundrum of sharing research data. *Journal of the American Society for Information Science and Technology*, *63*(6), 1059–1078. doi:10.1002/asi.22634

Börner, K., Contractor, N., Falk-Krzesinski, H. J., Fiore, S. M., Hall, K. L., & Keyton, J. etal. (2010). A Multi-Level Systems Perspective for the Science of Team Science. *Science Translational Medicine*, *2*(49), 49cm24. doi:10.1126/scitranslmed.3001399 PMID:20844283

Campbell, J. L., Rustad, L. E., Porter, J. H., Taylor, J. R., Dereszynski, E. W., & Shanley, J. B. etal. (2013). Quantity is Nothing without Quality: Automated QA/QC for Streaming Environmental Sensor Data. *Bioscience*, *63*(7), 574–585. doi:10.1525/bio.2013.63.7.10

Edwards, P. N., Jackson, S. J., Chalmers, M. K., Bowker, G. C., Borgman, C. L., & Ribes, D. etal. (2013). *Knowledge Infrastructures: Intellectual Frameworks and Research Challenges*. Ann Arbour: Deep Blue.

Gruber, T. (2008). Collective knowledge systems: Where the Social Web meets the Semantic Web. *Web Semantics: Science, Services, and Agents on the World Wide Web*, *6*(1), 4–13. doi:10.1016/j.websem.2007.11.011

Hendler, J., & Berners-Lee, T. (2010). From the Semantic Web to social machines: A research challenge for AI on the World Wide Web. *Artificial Intelligence*, *174*(2), 156–161. doi:10.1016/j.artint.2009.11.010

Holmes, J., & Clark, R. (2008). Enhancing the use of science in environmental policy-making and regulation. *Environmental Science & Policy*, *11*(8), 702–711. doi:10.1016/j.envsci.2008.08.004

Horsburgh, J. S., Tarboton, D. G., Maidment, D. R., & Zaslavsky, I. (2011). Components of an environmental observatory information system. *Computers & Geosciences*, *37*(2), 207–218. doi:10.1016/j.cageo.2010.07.003

Kragt, M. E., Robson, B. J., & Macleod, C. J. A. (2013). Modellers' roles in structuring integrative research projects. *Environmental Modelling & Software*, *39*(0), 322–330. doi:10.1016/j.envsoft.2012.06.015

Kunkel, R., Sorg, J., Eckardt, R., Kolditz, O., Rink, K., & Vereecken, H. (2013). TEODOOR: A distributed geodata infrastructure for terrestrial observation data. *Environmental Earth Sciences*, *69*(2), 507–521. doi:10.1007/s12665-013-2370-7

Montanari, A., Young, G., Savenije, H. H. G., Hughes, D., Wagener, T., & Ren, L. L. etal. (2013). "Panta Rhei—Everything Flows": Change in hydrology and society—The IAHS Scientific Decade 2013–2022. *Hydrological Sciences Journal*, 1–20.

Nielsen, M. A. (2012). *Reinventing Discovery: The New Era of Networked Science*. Princeton, NJ: Princeton University Press.

Pennington, D. D. (2011). Bridging the Disciplinary Divide: Co-Creating Research Ideas in eScience Teams. *Computer Supported Cooperative Work-the Journal of Collaborative Computing*, *20*(3), 165–196. doi:10.1007/s10606-011-9134-2

Porter, J. H., Hanson, P. C., & Lin, C.-C. (2012). Staying afloat in the sensor data deluge. *Trends in Ecology & Evolution*, *27*(2), 121–129. doi:10.1016/j.tree.2011.11.009 PMID:22206661

Reichman, O. J., Jones, M. B., & Schildhauer, M. P. (2011). Challenges and Opportunities of Open Data in Ecology. *Science*, *331*(6018), 703–705. doi:10.1126/science.1197962 PMID:21311007

KEY TERMS AND DEFINITIONS

Catchment Observatory: Linked physical, social and digital knowledge infrastructure that are required to support catchment systems science to be carried out by catchment research networks.

Catchment/Watershed: Areas of land which all surface water run-off flows into single points (usually the sea).

Geospatial Cyberinfrastructure: The digital infrastructure that supports the collection, management and utilization of geospatial data, information and knowledge for multiple science domains.

Integrative Research: Includes both interdisciplinary (between research disciplines) and transdisciplinary (between researchers and non-researchers) research projects.

Knowledge Infrastructure: Robust networks of people, artifacts, and institutions that generate, share, and maintain specific knowledge about the human and natural worlds.

Section 3

Formalization of Syntax and Semantics of Static Knowledge

After providing theoretical contributions and case studies confronting communities and domains, what follows is related to practical proposals. We recognized that there can be essentially two approaches in this, namely: those that rely on the possibility that knowledge can be made explicit and therefore formalized and those that instead, considering the limitations imposed by tacit knowledge and by the social aspects of scientific research, try to propose solutions where knowledge is left implicit but where its management can anyway be supported by IT systems. If knowledge is eligible for formalization, two paths can be identified, namely: one based on static knowledge formalization, such as semantics, controlled vocabularies, and Web ontologies (this section), and one based on dynamic knowledge formalization (the following section).

Chapter 12
Loose Integration of Local Information to Generate Collaborative Marine Science Knowledge

Adam Leadbetter
British Oceanographic Data Centre, NERC, UK

Robert Arko
Columbia University, USA

Cynthia Chandler
Woods Hole Oceanographic Institution, USA

Adam Shepherd
Woods Hole Oceanographic Institution, USA

Roy Lowry
British Oceanographic Data Centre, NERC, UK

ABSTRACT

This chapter focuses on improved access to marine science data, enabling researchers to generate new information and knowledge products. The history of controlled vocabulary developments in marine sciences, from paper publications to the Semantic Web, is explored in detail. This history is being furthered through the publication of Linked Open Data, meaning: the publication of clearly identifiable entities; a simple, universal mechanism for retrieving resources; a generic graph-based data model; and publishing explicit relationships to other resources. Progress towards Linked Open Data for marine science is reported in this chapter. As shown by the Data-Information-Knowledge ecosystem, the approach of "small pieces of data loosely joined" provides presentation and organisation to data, which creates information. The use of query endpoints to integrate this information from multiple locations into a knowledge base, which required active collaboration between cooperative partners to truly generate new knowledge and to address emerging science questions, is described.

DOI: 10.4018/978-1-4666-6567-5.ch012

Copyright © 2015, IGI Global. Copying or distributing in print or electronic forms without written permission of IGI Global is prohibited.

INTRODUCTION

Data collected during the course of oceanographic research expeditions have traditionally been archived close to the data originator: either in institutional or project based Data Assembly Centres (DACs) or on a national scale in the network of National Oceanographic Data Centres (NODCs) supported by the Intergovernmental Oceanographic Commission of UNESCO's International Oceanographic Data and Information Exchange (IODE). This tradition led to the creation of several attempts to integrate the data from the DACs and NODCs either in portals or multinational data systems. Until recently, data were collected for one purpose, but research paradigms are shifting and expectations along with them for increased access to data and desire to integrate multidisciplinary data from disparate sources. While it is important to curate data close to the point of origin, it is important to 'set data free' and with sufficient documentation that the data can be interpreted accurately by others. A possible solution to providing sufficient documentation is through the use of Sensor Observation Services (SOS, http://www.opengeospatial.org/standards/sos). Fredericks (2015, see chapter 13 this book) presents a vision of semantic SOS, and this is to be extended in the ongoing European Commission SenseOCEAN project. We present here the results of a recent collaborative network in loosely coupling data systems using Semantic Web and Linked Open Data methodologies in order to generate new information and knowledge within the ocean science domain and in support of this new research paradigm, concentrating on the integration of multidisciplinary data from a range of sources. These methodologies are explained in detail by Leadbetter (2015, see chapter 2 this book).

Controlled vocabularies are collections of terms that may be used to populate a given element in a structured metadata document or a data file. The controlled vocabulary concept has been an essential part of the British Oceanographic Data Centre's (BODC) data infrastructure from its initial design in the late 1970s to the present day. However, the way in which controlled vocabularies have been managed has changed considerably, with development from loosely managed basic code tables into an integrated semantic resource with formal content governance and sophisticated serving technology based on internationally-agreed knowledge management standards. In this section of the chapter we will present the lessons that have been learned in building and maintaining these vocabularies through time and governing the content within communities.

In 2012, version 2.0 of the Natural Environment Research Council Vocabulary (NERC) Server (NVS) which is used to deliver these vocabularies was launched (Leadbetter, Lowry, & Clements, 2013b). This release underpins the rest of the work presented in this chapter.

Current practice in BODC controlled vocabulary management now underpins the semantic infrastructure of many oceanographic data management systems across the world, such as SeaDataNet in Europe (Schaap & Lowry, 2010), the Rolling Deck to Repository (R2R) programme (Arko, Chandler, Clark, Shepherd, & Moore, 2012) and the Biological and Chemical Oceanography Data Management Office (BCO-DMO) project (Chandler, Groman, Allison, Wiebe, & Glover, 2013) in the US. The NVS, BCO-DMO and R2R linkages are all implemented using Linked Open Data techniques (Leadbetter, Arko, Chandler, Shepherd, & Lowry, 2013a) which means that the collaboration has been driven through a "bottom-up" desire to integrate data sources to produce new knowledge, and not from a "top-down" imposition of standards. Recently, the Integrated Marine Observing System in Australia has begun a process of mapping its vocabularies to the NVS. A major benefit of this "bottom-up" approach is that we have maintained and respected the local naming conventions in use, and have created the integrated knowledge space through a mapping of naming terms, not a replacement. In this section we examine in detail the process of connecting the datasets and information sources: including the technical difficulties encountered and their solutions and the practices of both physical and virtual communication involved in creating an intercontinental network of this type. Other initiatives which have taken a similar approach include the Marine Metadata Interoperability project (http://marinemetadata.org/) which has its own Ontology Registry and Repository, with mappings from that resource to the NVS.

In practise, there exist two paradigms for linking metadata and data to a controlled vocabulary. The first is to respect the local vocabulary used, maintaining the local terms while creating a mapping between those terms and the standard vocabulary used to bridge the disparate communities. The second is for a data provider to become authoritative only for the data they provide, and to not disregard their own local vocabularies in favour of the content from a community standard vocabulary. Both of these approaches require users to navigate the vocabularies in order to create the relationships between either local vocabularies or (meta)data files and the community standard resource. Until now, much of this work has been done by hand, but there is ongoing research into the development of software tools to assist the mapping and linking process.

Once the desired controlled vocabulary entry has been located, a unique and persistent reference to it must be records, either embedded in the data file or in the metadata record or alongside its corresponding local vocabulary entry. In the general SeaDataNet system, this takes the form of a Uniform Resource Name, which is in interpreted to a Uniform Resource Locator (URL) allowing web-enabled applications to read the controlled vocabulary entry from the RDF/XML document returned. Referencing the URL directly is the more common, and more extensible technique as this allows use of controlled vocabulary terms published anywhere on the Semantic Web, not just within the SeaDataNet system.

Once the (meta)data are published online, this linkage results in the metadata and data conforming to the Linked Data specification: the resources are available on the World Wide Web; addressable by a URL; the URL resolves to something meaningful; and the document returned by the URL contains linkages to external documents. In an ocean sciences context, Linked Data may be used to federate data management questions between different data archiving centres; validate Web Processing Service chains; facilitate data discovery and data aggregation; allow integration of ocean sciences data with other environmental sciences data in a holistic, Earth Systems Science approach; semantically annotate metadata records and data files from the point of data collection; and associate datasets with journal publications, methodologies and broadcast media output.

BACKGROUND

The International Oceanographic Data and Information Exchange (IODE) has been in place since the 1960s. The IODE system forms a worldwide service oriented network consisting of Designated National Agencies, National Oceanographic Data Centres, Responsible National Oceanographic Data Centres and World Data Centres. During the past 50 years, Intergovernmental Oceanographic Commission Member States have established over 80 oceanographic data centres in as many countries. This network has been able to collect, control the quality of, and archive millions of ocean observations, and makes these available to Member States.

With the advance of oceanography from a science dealing mostly with local processes to one that is also studying ocean basin and global processes, researchers depend critically on the availability of an international exchange system to provide data and information from all available sources. Additionally, scientists studying local processes benefit substantially from access to data collected by other Member States in their area of interest. The economic benefit of obtaining data through exchange as opposed to collecting all of the data by oneself is significant.

In Europe, the National Oceanographic Data Centres have been connected by the SeaDataNet project, in which a regional portal gives access to data collected by countries throughout the region in a single portal, giving the data increased context and impact. In the United States, two examples of Data Archiving Centres, or Designated National Agencies, of importance to the work described in this chapter are given below.

The Biological and Chemical Oceanography Data Management Office (BCO-DMO) was created in late 2006 to address the need in the ocean ecosystem research community for improved access to marine research data. The Biological and Chemical Oceanography Sections and the Division of Polar Programs Antarctic Organisms & Ecosystems Program at the U.S. National Science Foundation fund the BCO-DMO to support the data management needs of the investigators funded by those NSF programs and to ensure that research data resulting from those awards make it into the public domain and are preserved in a permanent archive. In the course of serving in this role, the BCO-DMO recognized, through conversations with researchers, that simply putting the data online and ensuring persistent archive was inadequate to support the needs of marine ecosystem researchers. Additional functionality was needed to improve discoverability of resources (e.g. data sets and supporting documentation), and to make the data truly accessible, additional documentation was needed to enable accurate re-use and integration of the data.

The nascent NSF EarthCube initiative (http://earthcube.org/) in the United States has funded a series of domain workshops to engage a broad range of geoscience researchers. One message that has emerged clearly from those workshops is that while the research community favours a model of distributed, thematic data repositories that are responsive to their respective communities, they now want to see greater connectivity between those repositories (http://earthcube.ning.com/page/past-end-user-workshops/). A small collaborative effort was funded by NSF in 2013 to explore the promise of Linked Open Data (Bizer, Heath, & Berners-Lee, 2009) for the ocean science domain, with the expectation that the lessons learned could be applied across the geosciences. Two of the initial data repositories in the collaboration are those managed by the BCO-DMO and the R2R project, the latter also created and supported by NSF funding. The initial strategy is based on several essential notions:

- Two or more repositories contain complementary content,
- The repositories publish out the content for which they are authoritative,
- The content is exposed as Linked Open Data (Leadbetter et al., 2013a; Bizer et al., 2009),
- To the extent possible, values in the published content are uniquely identified by Uniform Resource Identifiers (URIs) rather than free text,
- The repositories agree to match their local terms to term URIs from an authoritative source.

Effective implementations of Linked Data are bolstered by sound, stable URIs for local content. After careful considerations, BCO-DMO decided on a Uniform Resource Identifier (URI) pattern for local resources of

http://lod.bco-dmo.org/id/concept/instance

e.g. http://lod.bco-dmo.org/id/dataset/3391

The 'lod' subdomain was chosen for representation of BCO-DMO content as 'Linked Open Data' or 'L', 'O', 'D'. Using a new subdomain for linked data URIs ensures that collisions with existing subdomains such as 'www' do not occur. This is a resource-oriented architecture, where use of the HTTP Accept header on http://lod.bco-dmo.org/id/ URI requests, allows BCO-DMO to respond appropriately with RDF/XML, Turtle, HTML, or JSON-LD representations of the same local resource. The final condition is based on lessons learned by BCO-DMO staff members through conversations with marine research scientists.

In 2006, the BCO-DMO was created by merging the formerly independent, project-specific data management offices that served the United States Joint Global Ocean Flux Study (US JGOFS, see Buesseler, 2001) and United States Global Ocean Ecosystem Dynamics (US GLOBEC, see Fogarty & Powell, 2002) research programs. Lessons learned during those large, coordinated research programs included the observation that researchers, especially those in collaborative research programs, establish and adopt shared terminology to describe the instrumentation, methodology and measurements that emerge as being common to the research program (Glover et al., 2006). In addition to reporting this observation when comparing the US JGOFS and the Long Term Ecological Research Network (LTER, see Hobbie, Carpenter, Grimm, Gosz, & Seastedt, 2003) communities, Baker and Chandler (2008) also acknowledged the importance of local vocabularies to a collaborative research community. However, it has also been recognized that as research scope has progressed from local to global, it is important to consider strategies to improve the efficacy of data discovery and subsequent understanding by a broader community beyond the scope of the initial project (Baker & Chandler, 2008; Michener, Porter, Servilla, & Vanderbilt, 2011).

Over the years, BCO-DMO staff members developed data access and delivery systems to support data discovery, access, display, assessment, integration, and export of data resources. The modular system design enables strategic integration of emerging technologies as they are deemed sufficiently mature for implementation in the BCO-DMO architecture. The goal is to enhance resource discovery for a broad community of researchers and range of clients while maintaining the integrity of content available from the BCO-DMO catalog and respecting the local vocabularies that have become familiar to communities of practice. BCO-DMO and R2R are working together with computer scientists at Rensselaer Polytechnic Institute, Wright State University, University of Maryland, Baltimore County and Marymount University to explore ways in which Semantic Web technologies could improve their existing data curation and access systems. Thus far several promising areas have been identified:

1. Cross-repository comparison as a way to validate complementary content and ultimately improve accuracy;
2. Improved data entry efficiency at 'downstream' repositories;
3. Enhanced user experience because local content is extended by resources from the complementary repository.

Early on in the research process, it became evident that for a Linked Open Data approach to be successful for our purposes, globally unique identifiers from authoritative sources would be helpful if not required for the main concepts that describe the classes of repository resources as well as the instances. For content curated by R2R and BCO-DMO, several of the most important concepts are:

- *Deployment* (e.g. a cruise on an oceanographic research vessel, or other platform type from which instrumentation and sampling devices are deployed);
- *Instrument*, including a vast array of sampling device types ranging from a relatively simple temperature sensor to complex, multi-sensor, integrated sampling systems designed to record simultaneous measurements from the marine environment; and

- *Parameter*, the collection of terms used to describe the observations and measurements reported in a data set.

As stated earlier, researchers often have local terminology they use to describe these same concepts in their own research, and controlled vocabulary lists were developed to achieve greater consistency within the BCO-DMO data system, and facilitate search interfaces. However, to make the BCO-DMO managed data accessible to a wider audience we investigated the availability of community-wide controlled vocabularies for our concepts and lists of local terms.

Following the recommendations published at the Marine Metadata Interoperability (MMI) Platform site for choosing a controlled vocabulary (Graybeal, 2009), BCO-DMO and R2R identified the NERC Vocabulary Server (NVS; Leadbetter et al., 2013b) as meeting all of the recommended criteria. Because the NVS terms also include Simple Knowledge Organization System (SKOS) to establish the relationships between terms within and between vocabulary lists, interfaces built by consumers such as BCO-DMO can take advantage of those semantics to build hierarchical arrangements of term lists organized as logical categories. Displaying the vocabulary terms as logical categories inherited from the NVS hierarchies assists with the data entry phase as well as data discovery.

The value of continuous, reliable record-keeping in oceanographic research is well-established (NRC Ocean Studies Board, 2000). R2R was created in 2009 with funding from the U.S. National Science Foundation to address the need for consistent, fleet-wide documentation, preservation, and dissemination of environmental sensor data routinely acquired ("underway" data) by U.S. academic, oceanographic research vessels. R2R is a broadly collaborative program that works closely with the University-National Oceanographic Laboratory System (UNOLS) Office, vessel technicians and schedulers at the fleet operating institutions, the U.S. National Oceanic and Atmospheric Administration (NOAA) National Data Centers, and many partner facilities that provide products and services in support of oceanographic research.

R2R collects essential information about each expedition that includes vessel identifier and operator; cruise identifier, project title, and research program (e.g. Climate Variability and Predictability, Geodynamic Processes at Rifting and Subducting Margins, and Hawaii Ocean Time-series); port stops and dates; science party names, institutions, and roles; funding agency and award identifier; sensor identifier, classification, manufacturer, and model; cruise reports and event logs; and data file manifest. The R2R data model is shown in Figure 1. Selected data sets are assessed for quality using automated statistical methods, and a suite of quality-controlled standard products are created that include ship track navigation and (for selected vessels, in partnership with the Shipboard Automated Meteorological and Oceanographic System initiative) real-time near-surface data.

Information about expeditions is published online by R2R in multiple formats including ISO 19115 XML metadata records suitable for long-term archiving, Open Geospatial Consortium Web Map and Feature Services suitable for Geographical Information Systems clients, and as Linked Open Data suitable for Semantic Web applications. The decision to deploy Linked Open Data was made early in the program, in response to feedback from data managers at partner facilities who sought easily-accessible, reliable, globally-unique identifiers for each vessel, cruise, sensor, and dataset in the R2R inventory. The D2RQ open-source software package (http://d2rq.org/) was deployed to map R2R's existing relational database into a collection of Resource Description

Figure 1. Linked Open Data model for the Rolling Deck to Repository (R2R) Program, showing primary concepts and relationships

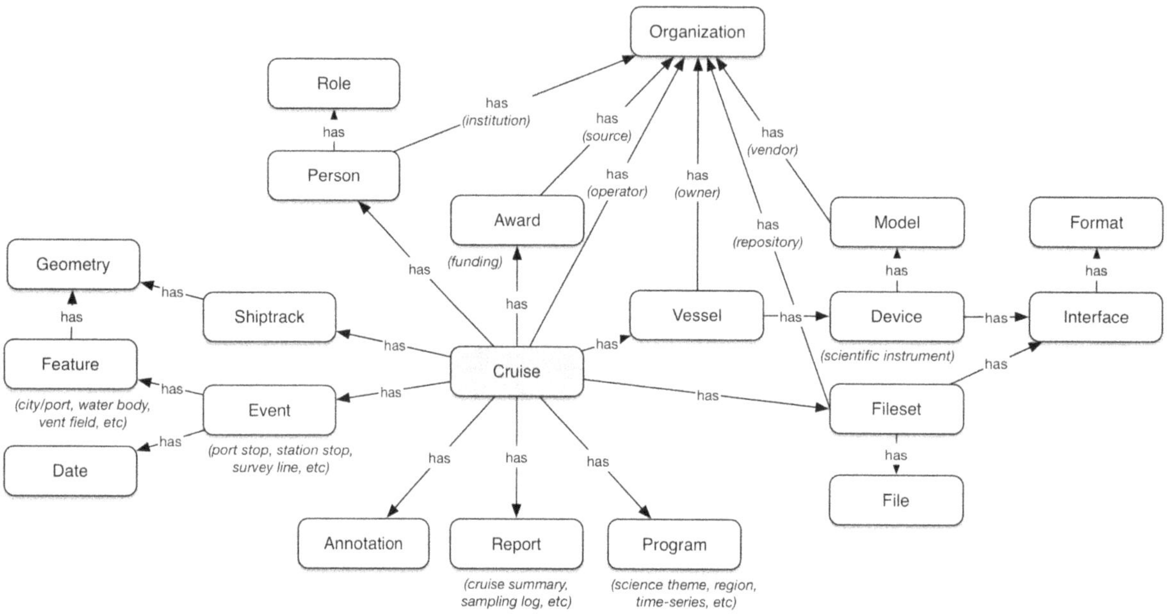

Format (RDF) resources on-the-fly, providing a "virtual triplestore" that delivers satisfactory performance without requiring separate storage or synchronization. Each resource is assigned a Uniform Resource Identifier (URI) using the pattern

http://data.rvdata.us/id/concept/instance

e.g. http://data.rvdata.us/id/cruise/MGL1212

The D2RQ package also provides a web service endpoint to query the resources using the SPARQL query language, and a Pubby Web browser to visually explore the resources.

R2R's deployment strategy for Linked Open Data includes mapping every key concept to a community-standard controlled vocabulary where such exists, and as of March 2014 this is roughly 7,000 links from R2R's database to the NVS. Work to date has focused on the NVS, which uses SKOS to organize terms in hierarchical collections with multilingual descriptions (Leadbetter et al., 2013b). For every term, NVS provides a URI that can be embedded in R2R's Linked Open Data resources, and dereferenced by Semantic Web applications to provide definitions and relationships with other terms. The ability to easily add and update terms, within a trusted governance structure, allows R2R to undertake vocabulary mapping in a flexible and efficient manner.

To accommodate a variety of clients, R2R publishes Linked Open Data for each vessel's trajectory (cruise track) in multiple formats. The geospatial extent of the track is published using the W3C Basic Geo Vocabulary for WGS84 Latitude/Longitude, allowing simple queries with point coordinates. A detailed approximation of the cruise track is published using the OGC GeoSPARQL Vocabulary (Perry & Herring, 2012), which accommodates a full range of geospatial object types including Points, Linestrings, and Polygons from the ISO 19125 Simple Features specification. Additional formats including JSON and thumbnail graphics are planned.

THE INFORMATION-KNOWLEDGE GLUE: COLLABORATIVE KNOWLEDGE ENCODING IN CONTROLLED VOCABULARIES

Controlled vocabularies have been developed to allow for consistent metadata and data annotation within the ocean science community through IODE since the 1980s (UNESCO, 1987). This development has allowed the exchange of data and information between the types of data stores described above in a science that has traditionally been expensive to undertake (due to the cost of mounting expeditionary research cruises) in sparsely sampled environmental domains. In recent times, the controlled vocabularies have been published to Semantic Web and Linked Data standards, allowing their content to be used in highly technological knowledge building applications.

The whole point of controlled vocabularies is that their content is controlled by some form of governance process. This governance process can be very informal, for example much of the current BODC controlled infrastructure was developed by individuals operating in isolation. It can also be extremely formalised, for example with the International Standards Organisation (ISO) where changes are proposed to and voted on by formally constituted committees. However, it can all too easily become totally anarchic, such as the situation within BODC during the 1990s where students were given tools with an 'add term to vocabulary' button.

Principles of Good Vocabulary Content Governance

In a worst case scenario for vocabulary governance, content governance sinks to the level wherein anyone could ask for anything to be done and their wish would be granted. Technical governance falls to the level in which multiple copies of some vocabularies are held by different organisations. Even worse, due to the lack of content governance in such cases, vocabularies would be independently maintained and likely to evolve in a manner similar to "Galapagos finches", meaning evolution into vocabularies that were similar but significantly different, and so of little use for enabling semantic interoperability.

Throughout the European Commission SEA-SEARCH project, NERC DataGrid, and European Commission SeaDataNet project a lot of energy was expended by BODC and the International Council for the Exploration of the Sea (ICES) to rectify a situation similar to that which was described in the above paragraph. Some recommendations from this experience follow in the sections below.

Many controlled vocabularies are made up of concepts that are classifications for other concepts or groups of real-world objects. There are an infinite number of ways to classify a given set of objects with differing numbers of objects in each category. Coarse-grained classifications have a small number of categories each containing a large number of objects. Fine-grained classifications have a much larger number of categories, each containing a smaller number of objects. Vocabulary content governance needs to determine the appropriate granularity for a given classification vocabulary based on an assessment of its usage.

The classic view of collection granularity is based on a hierarchical model. The hierarchy is built from several collections, each holding concepts of a similar granularity ranging from coarse to fine, and linked together by mappings. An example is the hierarchy formed by the SeaDataNet Disciplines (http://vocab.nerc.ac.uk/collection/P08/current/), SeaDataNet Agreed Parameter Groups (http://vocab.nerc.ac.uk/collection/P03/current/), BODC Parameter Discovery (http://vocab.nerc.ac.uk/collection/P02/current/), and BODC Parameter Usage vocabularies (http://vocab.nerc.ac.uk/collection/P01/current/).

A more modern viewpoint is to allow variable granularity within a single collection held together by internal mappings. The model chosen depends upon two factors. The first is the level of control required over metadata producers, who if permitted will inevitably use the coarsest-grained concepts available because it's easier. The second is the nature of the human interface in which the vocabularies will be used in discovery tools. If these are based on search criteria selection using drop-down lists then a classical hierarchy works better. For interfaces based on thesaurus browsing then mixed granularity works better.

There is another reason for including variable granularity within a single collection that is driven by the domain interest of the anticipated user community. For example, if the target user community solely comprised focused nutrient chemists then a classification like nitrate, nitrate, phosphate, silicate, urea, nutrients, non-nutrient chemistry, physical data, geological data, and biological data could be perfectly acceptable. The main problem with the BODC Parameter Discovery and SeaDataNet Agreed Parameter Groups vocabularies is that their target community has changed dramatically over time as usage extended from BODC to SeaDataNet to the European Commission Geo-Seas project and then to United Kingdom's Marine Environment Data and Information Network (MEDIN).

The fundamental definition of a controlled vocabulary is 'A set of terms that may be used to populate a particular element in a data or metadata model.' A strong data model should have a definition for every element, which may be converted into a set of semantic tests that should be passed by the concepts in its controlled vocabulary. For example, the SeaDataNet data model defines 'platform type' as 'A grouping term for vehicles, structures or organisms capable of bearing instruments or tools for the collection of physical, chemical, geological or biological samples or data'.

This may be translated into two semantic tests:

- Is it a group of vehicles, structures or organisms?
- Can it sensibly carry oceanographic sensors or samplers?

These tests may then be applied to candidate concepts for the controlled vocabulary. For example, the term 'submarine' passes on both counts (although there may be other issues such as granularity), but 'RRS Discovery', 'bus', 'seaweed' and 'salinity' all fail as they are because they are either individual vehicles (not oceanographic platforms), or they match neither of the selected criteria.

Note that it is possible for data model governance to change the semantic definition of a model element. For example, the platform type definition was extended to include organisms when the practice of attaching sensors to marine animals was introduced. However, this process needs to be carefully controlled. In particular, the process needs to be done explicitly and not implicitly through vocabulary extension (otherwise known as shoe-horning). If it isn't then the vocabulary will very rapidly become semantically impure.

There are many examples of semantically impure vocabularies that have resulted from a lack of clear thinking exacerbated by weak content governance resulting in decisions being made hastily under pressure. For example the BODC instrument category vocabulary (http://vocab.nerc.ac.uk/collection/C11/current/) includes concepts like 'Seismic survey' and 'Bottom temperature measurements' that are clearly not instrument groupings no matter what definition is used for 'instrument'.

Another path to vocabulary semantic impurity is through the use of vocabulary population to fix defects in data models. For example, it is all too easy to sort out a one-to-one relationship that should have been one-to-many by adding concepts to the controlling vocabulary that are lists. There are some examples of this in the full ICES ship code list (http://vocab.nerc.ac.uk/collection/C17/current/deprecated/) that have been excluded from the ship code list used by SeaDataNet (http://vocab.nerc.ac.uk/collection/C17/current/accepted/) such as "Route: Folkestone – Boulogne" and "Multiple Ships".

During the development of controlled vocabularies mistakes will inevitably be made. One of the most common is the accidental duplication of concepts, particularly with large controlled vocabularies. Another common problem with classification vocabularies is that they become finer grained as they evolve because the governance fails to realise that an extension request is a small part of a bigger picture. Sheer size can also become an issue, particularly for vocabularies that are used to populate drop-down lists in applications.

One way of fixing duplicates is to simply delete one of them. Good content governance practice should only allow this if all usage of the vocabulary is known and under the governance's control. One example is the BODC internal vocabularies that are included in NVS. If this good practice is broken then deletions will break systems in user organisations. Great care needs to be taken by governances to ensure rather than assume that all usage is under their control.

An alternative to concept deletion is concept deprecation, in which the concept is flagged as no longer recommended and tagged with information on the concept that should be used instead. This is now fully supported in NVS V2. However, the consequences of deprecation need to be thought through carefully and good governance practice for deprecation is "If in doubt - don't", especially for vocabularies that could potentially be used for mark up in large numbers of physical files (e.g. the BODC Parameter Usage Vocabulary). The recommended fix for duplicates where deprecation isn't allowed, such as the BODC Parameter Usage Vocabulary, is to map the duplicates as synonyms.

The NVS data model, in common with many other controlled vocabularies, includes an identifier often referred to as a code that represents the concept. Experience has shown that these should be semantically opaque, which ideally means random collections of characters. However, use of mnemonics to preserve the comprehension in small vocabularies is acceptable.

What is not recommended are codes that may be parsed to give meaning. The classic example of bad practice is the structured ICES ship code (now abandoned in policy if not practice) that comprises a 2-byte country code plus two bytes to identify the ship within country. This allowed a maximum of 1296 ships per country. When this was exceeded by the USA, the governance at the time solved the problem by setting up multiple country codes for the USA. They are still there in the Intergovernmental Oceanographic Commission country codes (http://vocab.nerc.ac.uk/collection/C18/current/deprecated/), which is one of the reasons for recommending a move to ISO country codes.

The NVS data model allows for a plaintext definition up to 4000 bytes long to be assigned to every concept. Good content governance should ensure that definitions are well thought out, clear, easy to understand and especially that they deliver additional information that is not available from the concept labels. Even ISO standards have fallen short of the mark here with many of the ISO19115 code lists having definitions that just repeat the preferred label.

The fundamental entity in the World Wide Web Consortium's Simple Knowledge Organisation System (SKOS, see Miles and Bechhofer, 2009) is the 'concept', which is an atomic piece of knowledge. This naturally maps onto the list entry in the NVS V1 data model, which had the attributes in the following bulleted list. Consequently, the concept is the fundamental entity in the NVS V2 data model, which has the following attributes:

- **An Identifier:** An alphanumeric string, sometimes referred to as a code, that is used to represent the concept to software agents. Concept identifiers are guaranteed

unique within the controlled vocabulary and ideally should be random strings devoid of semantic content.
- **A Preferred Label:** A string used to represent the concept to humans.
- **An Alternative Label:** A different string used to represent the concept to humans. In NVS the alternative label is generally used for an abbreviation to be used in applications where space is a problem.
- **A Definition:** A clear and unambiguous (and therefore often a long string of text) statement of the semantics associated with the concept.
- **A Status:** An indication of where the concept lies within the ISO19135 concept evolution model. Only two stages from this ISO19135 model - accepted and deprecated - are currently included in NVS V2.
- **A Version:** A simple monotonic integer starting at 1 that indicates the number of times the concept attributes have been changed since the concept was first created. The number of changes is the version number minus 1.
- **A Publication Date:** The timestamp of the last change to the concept attributes.

The data model includes two types of concept aggregation that are mapped to aggregation concepts in SKOS. The first aggregation type is the well-established idea of a 'controlled vocabulary' or 'code table'. This has been formally defined in NVS V2 as an aggregation of the concepts that may be used to populate a specified field in a data or metadata model and has been mapped to the 'concept collection' in SKOS.

Note that for historical reasons - NVS grew out of a system for code tables - the relationship between concepts and concept collections is strictly one-to-one. In other words, every concept needs to belong to one and only one collection. Indeed, the collection identifier forms a part of the address of a concept, which is somewhat at odds with current thinking that expects concepts to be independently addressed and shared between collections.

The second type of concept aggregation is based on the idea of a 'thesaurus'. This may be thought of as a 'concept soup' interlinked by a simple set of semantic relationships (broader, narrower, same and related). A small number of these concepts, known as 'top concepts' are declared as entry points. This is formally defined in NVS V2 as any group of concepts for which at least one is declared as a top concept. These thesauri have been mapped to 'concept schemes' in SKOS.

Whilst any mixture of concepts is possible, it obviously makes sense to include concepts that are semantically related - say those that specify a given discovery facet like 'platform type' - and to have top concepts that are semantically coarse-grained. However, it is also possible to build thesauri that are semantically more complex with multiple facets, for example 'parameters', 'platform types' and 'instrument types'. Providing appropriate mappings between the facets are in place, such a thesaurus is able to support semantically-enabled, multi-faceted discovery. Thesauri of this type provide us with an extremely powerful tool for quickly building semantic hierarchies (for example the SeaDataNet device catalogue, http://vocab.nerc.ac.uk/scheme/SDNDEV/), subsetting large vocabularies, or both.

The NVS V2 data model is summarised in Figure 2 and an example SKOS node from NVS 2 is given in Figure 3.

Mappings from the accepted version of SKOS were used, with narrowMatch converted to narrower and broadMatch converted to broader. The conversions for exactMatch and minorMatch were less straightforward because either no direct equivalents existed in SKOS or there were none in scope for the SKOS concept collection. However, the way in which minorMatch had been used in practice allowed it to be converted to SKOS related. The solution to the problem of no SKOS equivalent in scope for exactMatch was

Figure 2. A schematic representation of the data model used within version 2.0 of the NERC Vocabulary Server. A SKOS concept registered on the NERC Vocabulary Server is associated with one, and only one, SKOS concept collection from which its Uniform Resource Locator (URL) is derived. However, the concept may form part of multiple SKOS concept schemes.

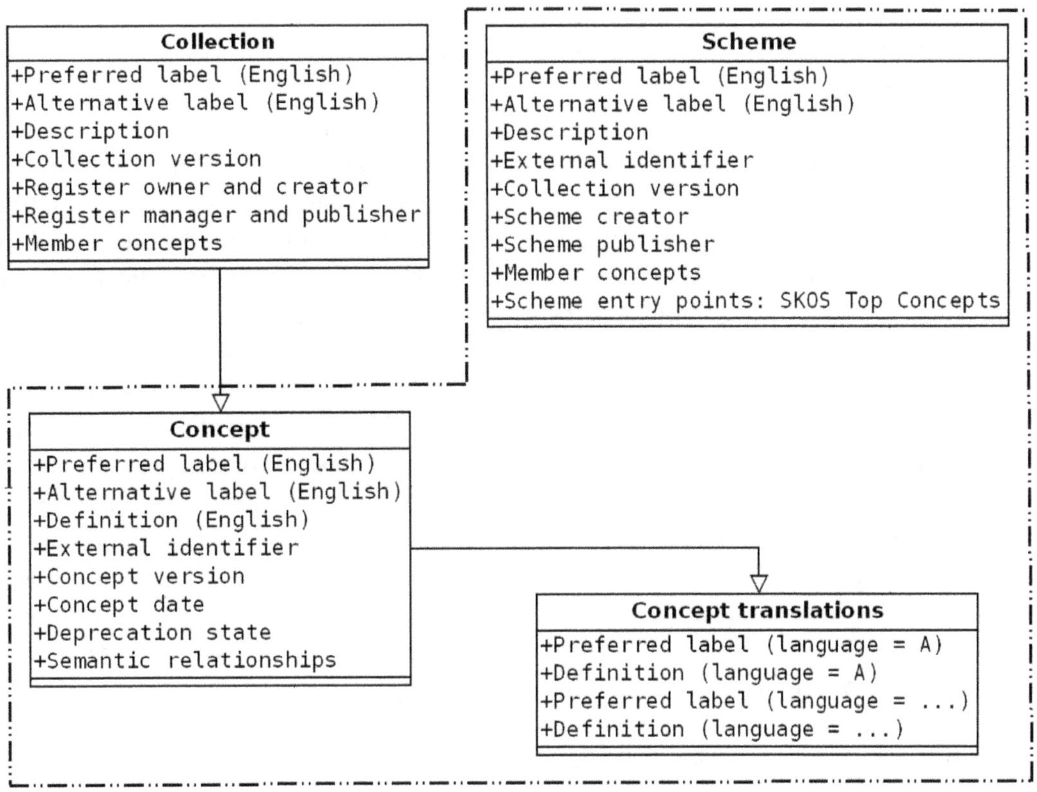

Figure 3. An example SKOS node from the NERC Vocabulary Server (http://vocab.nerc.ac.uk/collection/P07/current/CFSN0023/)

```
▼<rdf:RDF xmlns:rdf="http://www.w3.org/1999/02/22-rdf-syntax-ns#"
  xmlns:skos="http://www.w3.org/2004/02/skos/core#" xmlns:dc="http://purl.org/dc/terms/"
  xmlns:rdfs="http://www.w3.org/2000/01/rdf-schema#" xmlns:owl="http://www.w3.org/2002/07/owl#"
  xmlns:void="http://rdfs.org/ns/void#">
  ▼<skos:Concept rdf:about="http://vocab.nerc.ac.uk/collection/P07/current/CFSN0023/">
     <skos:prefLabel xml:lang="en">air_temperature</skos:prefLabel>
     <skos:altLabel/>
   ▼<skos:definition xml:lang="en">
       Air temperature is the bulk temperature of the air, not the surface (skin) temperature.
     </skos:definition>
     <dc:identifier>SDN:P07::CFSN0023</dc:identifier>
     <skos:notation>SDN:P07::CFSN0023</skos:notation>
     <owl:versionInfo>1</owl:versionInfo>
     <dc:date>2006-09-26 18:12:43.0</dc:date>
     <skos:note xml:lang="en">accepted</skos:note>
     <owl:deprecated>false</owl:deprecated>
     <skos:broader rdf:resource="http://vocab.nerc.ac.uk/collection/P02/current/CDTA/"/>
     <owl:sameAs rdf:resource="http://vocab.nerc.ac.uk/collection/P04/current/G220/"/>
     <skos:related rdf:resource="http://vocab.nerc.ac.uk/collection/P06/current/UPKA/"/>
     <skos:broader rdf:resource="http://vocab.nerc.ac.uk/collection/P25/current/ATEMP/"/>
     <owl:sameAs rdf:resource="http://vocab.nerc.ac.uk/collection/I01/current/0.0.0/"/>
     <owl:sameAs rdf:resource="http://vocab.nerc.ac.uk/collection/P64/current/G220/"/>
     <void:inDataset rdf:resource="http://vocab.nerc.ac.uk/.well-known/void"/>
   </skos:Concept>
  </rdf:RDF>
```

to use sameAs from the OWL namespace, which is not without its own problems (Halpin, Hayes, McCusker, McGuinness, & Thompson, 2010).

Whilst the NVS2 code has been written more flexibly, allowing use of the full set of predicates from the NERC DataGrid vocabulary of term relations (http://vocab.nerc.ac.uk/collection/C60/current/), this cannot be fully implemented until NVS V1 is withdrawn. Usage statistics indicate that this could be some time.

Another limitation of the V1 mapping implementation was that both subject and object needed to be terms held in the NVS concept repository. In other words, mappings could only be made between NVS concepts and not to concepts in other resources offering concepts addressed by Uniform Resource Locators (URLs). This was overcome by the addition of an "external" mapping table to the schema. An example of how this has been used in the context of the Climate and Forecast (CF) conventions is shown in Table 1.

The issues with NVS V1 versioning described above had to be addressed in V2. Although the ultimate V2 objective is to replace the V1 list and superlist containers with just collections, it was considered more prudent to decouple this process from the resolution of version synchronicity issues. Consequently, the first stage of NVS V2 version implementation was to synchronise the version number of all NVS V1 lists to the version number of their parent superlist and put measures in place to ensure that once synchronisation was achieved it was retained.

In NVS V2, each concept carries its own version number and publication date, which are trigger-managed fields in the database. Each time any concept attribute is modified, the trigger archives the record as it was, increments the version number and sets the last modification/publication timestamp to the current time. A consequence of this and the daily publication of concept collections and schemes is that several changes to – and hence republications of – concepts may occur during the lifetime of a single concept collection version. The next version of the collection only includes the latest version of each concept and so some concept versions will never be included in a published version of a collection.

This was a deliberate choice because it means that any prior version of any concept is represented by a single record in the database, which makes direct access to specified concept versions much easier to implement in future versions of NVS. The price paid is that assembling prior versions of

Table 1. Example semantic relationships stored within the NERC Vocabulary Server between Climate and Forecast (CF) Standard Names, the General Multilingual Environmental Thesaurus (GEMET), BODC's Units vocabulary and DBpedia. The CF Standard Names are related to GEMET and BODC Units through the skos:related property, BODC Units is related to DBpedia through both the skos:related (the top four examples) and owl:sameAs (final example) properties.

CF Standard Name	GEMET	BODC Units	DBpedia
tendency_of_atmosphere_mass_content_of_ammonia_due_to_emission_from_savanna_and_grassland_fires	Grass fire	kg/m^2/s	Kilogram; Square_metre; second
tendency_of_atmosphere_mass_content_of_ammonia_due_to_emission_from_forest_fires	Forest fire	kg/m^2/s	Kilogram; Square_metre; second
tendency_of_atmosphere_mass_content_of_carbon_dioxide_due_to_emission	Emission	kg/m^2/s	Kilogram; Square_metre; second
tendency_of_atmosphere_mass_content_of_carbon_monoxide_due_to_dry_deposition	Dry deposition	kg/m^2/s	Kilogram; Square_metre; second
air_temperature		Degrees Kelvin	Kelvin

concept collections is more complicated, but this problem has already been solved in NVS V1 and the solution has been carried over to V2.

In NVS V2, every concept collection is identified by a 3-byte identifier, such as P01 for the BODC Parameter Usage Vocabulary, P02 for the BODC Parameter Discovery Vocabulary and P25 for the NETMAR Aggregation Parameters Vocabulary. The identifiers for concepts are constructed as unregistered Uniform Resource Names (URNs), with the following syntax:

Namespace_identifier:*collection_identifier*::*concept_identifier*

- The namespace identifier 'SDN' used in NVS V1 has been retained.
- The collection identifier is the 3-byte identifier as per the superlist identifier in NVS V1.
- The two adjacent colons signify a null field. This is retained for backward compatibility with NVS V1.
- The concept identifier is a unique label within the collection.

A URN example is SDN:P01::TEMPPR01. Note this always refers to the current version of the concept. URNs for collections may specify the version of the collection. For example, SDN:P01:15: or (for the current version) SDN:P01::.

The URNS may be converted by simple string slicing into URLs of the form:

http://vocab.nerc.ac.uk/collection/collection_identifier/current/concept_identifier

For example:

http://vocab.nerc.ac.uk/collection/P01/current/TEMPPR01/

NVS resolves these URLs into World Wide Web Consortium Resource Description Framework (RDF, http://www.w3.org/RDF/) XML documents conforming to the SKOS standard.

In addition to concepts and concept collections, NVS V2 delivers thesauri as SKOS concept schemes that are identified using mnemonic names up to 80-bytes long, but generally much shorter. These are incorporated into URLs as follows:

http://vocab.nerc.ac.uk/scheme/scheme_name

For example,

http://vocab.nerc.ac.uk/scheme/SDNDEV

Within NVS, concept schemes are considered to be less formal aggregations of concepts than collections. Consequently, they are considered transitory and are not versioned.

Concept deprecation has been a problem with NVS. The model developed for Version 1 involved setting up a list for the deprecated concepts in the same superlist as its accepted siblings. For example the NVS V1 lists for the CF Standard Names were P071 for the accepted Standard Names and P072 for the deprecated Standard Names. In the back office, deprecation was implemented by changing a database field from 1 to 2. This didn't work well because concept deprecation ended up changing the Uniform Resource Identifiers (URIs) - both URN and URL - for the concept, which caused chaos both internally in BODC for vocabulary management and wherever the deprecated concept had been used externally. Guidance on the concept that should replace the deprecated concept was included as a mapping relationship in the RDF XML.

For NVS V2, the database expanded to include a status field that may be set to either accepted or deprecated as appropriate, leaving the identifier untouched. This is supported by the addition of a table that clearly identifies the concept (only one) that replaces the deprecated concept. Note that it is perfectly possible for the replacement concept to be subsequently deprecated and therefore a deprecation chain is set up. The concept status, the replacement concept and any concepts replaced by a concept are reported in the NVS V2 SKOS documents.

NVS V2 supports storage and presentation in the SKOS documents of concept preferred labels and definitions in multiple languages. The concept identifier, language identifier and text in that language are stored in a separate, multilingual database table.

Software Tooling for Collaborative Information Integration

To apply this encoded knowledge, which has been collaboratively built by the content governance authorities, it is necessary to map the terms used locally to describe data, to the terms in the community standard resource. In scientific data management, there are two essential software components that must be considered when attempting to map local entities to vocabulary terms with human intervention - the entity mapping platform and a user interface (UI). At the most basic level, the entity mapping platform assists in match compilation between the two datasets and is driven by mechanisms that attempt to model human decision-making processes of matching similar entities. Likewise, the basic function of the UI is to enable quick, intuitive, human interaction with these compiled matches for investigation and annotation.

In the BCO-DMO case study, mapping entities between local instrument and parameter vocabularies and SeaDataNet vocabularies served by the NERC Vocabulary Server was a time-consuming process for a human - a BCO-DMO data manager. In the early days of mapping BCO-DMO entities, the entity mapping platform was not automated, required human direction and was very crude - two web browser windows, one for each vocabulary at BCO-DMO and SeaDataNet. A BCO-DMO data manager would start with a local term in one browser, directed at the BCO-DMO web page for that term, and attempt to discover a match in the other browser, directed at SeaDataNet SKOS RDF/XML representations at the NERC Vocabulary Server. Discovery occurred by using the browser's document search feature to find relevant words in the SKOS documents. Since the SeaDataNet SKOS vocabularies are constructed hierarchically, the best match for a BCO-DMO term could be at any level of that hierarchy. For the data manager this means that at any level of the hierarchy, they must be searching for three different representations of a potential match - an exact definition, a broader definition and a narrower definition. This exercise proved to be a complex, time-consuming task. When the best match was discovered, the data manager could finally use a UI, a content management system web form, to record a match by storing the SeaDataNet term's URI in a BCO-DMO database.

The next iteration of software components to aid BCO-DMO mapping to SeaDataNet involved a set of Drupal web content management framework modules (https://github.com/ashepherd/vocab-mappings) to serve as the entity mapping platform. The modules sought to provide two functions - an interactive, visual representation of a vocabulary and a simple text string search over vocabularies. The platform required its own local copy of a vocabulary, and did so, by ingesting RDF/XML for a vocabulary. The local copy of a vocabulary could be updated with a newer version through a batch process that could be started manually. These vocabularies could be visually represented in HyperText Mark-up Language (HTML), and in the case of SKOS, could leverage the hierarchy of the skos:narrower and skos:broader properties to create collapsible tree structures in a browser. These collapsible trees made the human process of discovery simpler because the BCO-DMO data manager could now logically navigate from high-level, coarse descriptive terms in SeaDataNet down through to more refined, specific instance terms for instruments and parameters. Apart from the visual representation, a simple text string search could discover related terms in one or a number of pre-selected vocabularies. The search component's aim was to get as close to a direct match in the selected vocabularies as possible. From here, discovery started from the most refined terms and, if possible, allowed a data manager to traverse up a hierarchical vocabulary, if necessary. For example, in the case of instrument vocabularies, the name of an instrument might not indicate at all what type of measurements it makes or what function it serves. The instrument named *SBE 911plus* (a common instrument type in oceanography, used to measure conductivity, water temperature and depth, see http://www.seabird.com/products/spec_sheets/911data.htm) provides little indication of how best to classify it. In this case, attempting a search of *SBE 911plus* may result in a direct match or give some indication of the instrument's nature. If no meaningful results are returned, a data manager could attempt to mine a description of the instrument either stored in the BCO-DMO vocabulary or even on the Internet. Keywords extracted from this search might aid in using the Drupal module's search tool from the other perspective - the top-down approach. When a matching term was found, the data manager could use the same UI from the first iteration to record matches. While this Drupal module made the initial process for BCO-DMO data managers faster, it still wasn't making recommendations or providing a mechanism for a data manager to help the system learn how to make better recommendations. Another drawback was the software's inability to make recommendations about an existing coarse match that could be refined given new data when the local copy of a vocabulary was updated with fresher content.

At this point, BCO-DMO understood two things: 1) entity matching is a complex, time-consuming task for a human, and 2) the task does not scale for human operators as the number of local terms increases over time. Leveraging a machine's capability to assist the human process seemed the only sustainable way to commit to entity matching as a data management practice for BCO-DMO. After discovering the Silk entity matching tool (http://wifo5-03.informatik.uni-mannheim.de/bizer/silk/), the BCO-DMO attempted to leverage Silk's ability to learn entity linking rules based on positive and negative links to automate entity mapping recommendations - a practice called genetic programming. Positive links are links between two datasets that assert a match whereas negative links are assertions that two entities do not match. Since BCO-DMO already had a set of positive links, the Silk tool seemed promising. Based on Isele and Bizer (2013), the Silk tool had been used with successful results in pairing entities by calculating a confidence score based on a formula defined by its user. However, in the case of BCO-DMO and SeaDataNet vocabularies, Silk didn't have the natural language processing capabilities required to make informed decisions based on the entity descriptions available at BCO-DMO and SeaDataNet. Although Silk is powerful when the entity properties are close in distances as calculated by Levenshtein, Jaccard, and geospatial algorithms, a different mechanism was needed for BCO-DMO - an algorithm that could model human decision-making processes.

Partnering with Rensselaer Polytechnic Institute, BCO-DMO sought to develop a sustainable mapping platform with an automated recommendation mechanism that could learn based on human-asserted positive and negative matches, but more importantly model human-based decisions. This partnership began to explore the use of Bayesian networks and pointwise mutual information (PMI) as a framework for decision-making. The resulting entity mapping platform ingests vocabulary RDF data from a file or a SPARQL Protocol and RDF Query Language (SPARQL, http://www.w3.org/TR/sparql11-overview/) endpoint. When a user wants to map between two vocabularies, a mapping routine can be created on the platform and identified by a unique Internationalized Resource Identifier (IRI). A routine identifies both a source vocabulary and a destination vocabulary. For each entity in the source vocabulary, a confidence score is calculated against each entity in the destination vocabulary. Given a certain threshold for valid confidence scores, which can be defined per routine or at runtime, each match and its corresponding confidence score are stored in a triplestore in a graph identified by the routine's IRI. The triplestore is made available to the user through a SPARQL endpoint that provides the ability to query a relevant routine's result set by the routine's IRI. This capability allows UIs to be developed for annotating the result set and preserving data relevant to the mapping exercise. Locally, at BCO-DMO, a UI accesses the derived graph at the entity mapping platform and presents a web page listing of all the potential mappings for a given routine. For each potential mapping a BCO-DMO data manager, can assert one of three things - a positive match, a negative match, or undecided. These annotations are then stored locally with provenance. This provenance allows for versioning the entity matches enabling an audit trail as better mappings at the same or a new vocabulary are derived in the future.

BCO-DMO also decided to try this type of mapping approach to match its cruise resources (records of research vessel deployments) to records for the same resources in the R2R cruise catalogue, and to highlight a different aspect of entity matching. In the example of pairing BCO-DMO instruments and parameters to the SeaDataNet vocabularies, the emphasis was on discovering similarity. In mapping cruises between BCO-DMO and R2R, the emphasis lies on identifying exactness. The cruise metadata records in both repositories have these properties in common: cruise identifier (as assigned by the vessel operator), vessel name, vessel code (as assigned by the International Council for the Exploration of the Sea, ICES), start and end date, chief scientist name, and since both R2R and BCO-DMO map their vessel identifiers to the same community vocabulary, the community vessel URI. Because these properties are understood, in structure and data type, a tool such as Silk becomes useful. With Silk, an algorithm can be defined that measures the proximity of these individual properties. These measurements can be aggregated and weighted accordingly to compute a score that identifies for the user a quality of exactness. BCO-DMO is using Drupal to manage its metadata, and Drupal runs on the PHP: Hypertext Processor (PHP) language. Since PHP already defines a Levenshtein algorithm, BCO-DMO used the Drupal framework to build the matching tool between its cruises and the R2R cruise catalogue. Since the driving factor is exactness the algorithm can be unforgiving in cases of mismatches between properties. However, if the algorithm is too unforgiving there is a lost opportunity to use exactness matching techniques to validate content correctness. Therefore, the cruise matching algorithm approach followed the following rules:

1. Compute Levenshtein distance between cruise identifiers.
2. Ignore all potential matches where cruise identifier closeness was less than 85%.
3. Calculate the Levenshtein distance between cruise start dates, end dates, vessel names, vessel codes, the reported cruise chief scientist, and a shared community URI.
4. Aggregate the results based on user-defined input.

Before the algorithm is run, the user is allowed to specify how important certain cruise properties are in determining exactness as shown in Figure 4. The most important property is the cruise identifier, and because of that, its calculated distance should inform whether to proceed. Since cruise identifiers tend to be less than ten characters long, 85% seemed a reasonable threshold for deciding whether to pursue a potential match. If

Figure 4. A web page presented to a BCO-DMO data manager for tuning the cruise matching algorithm

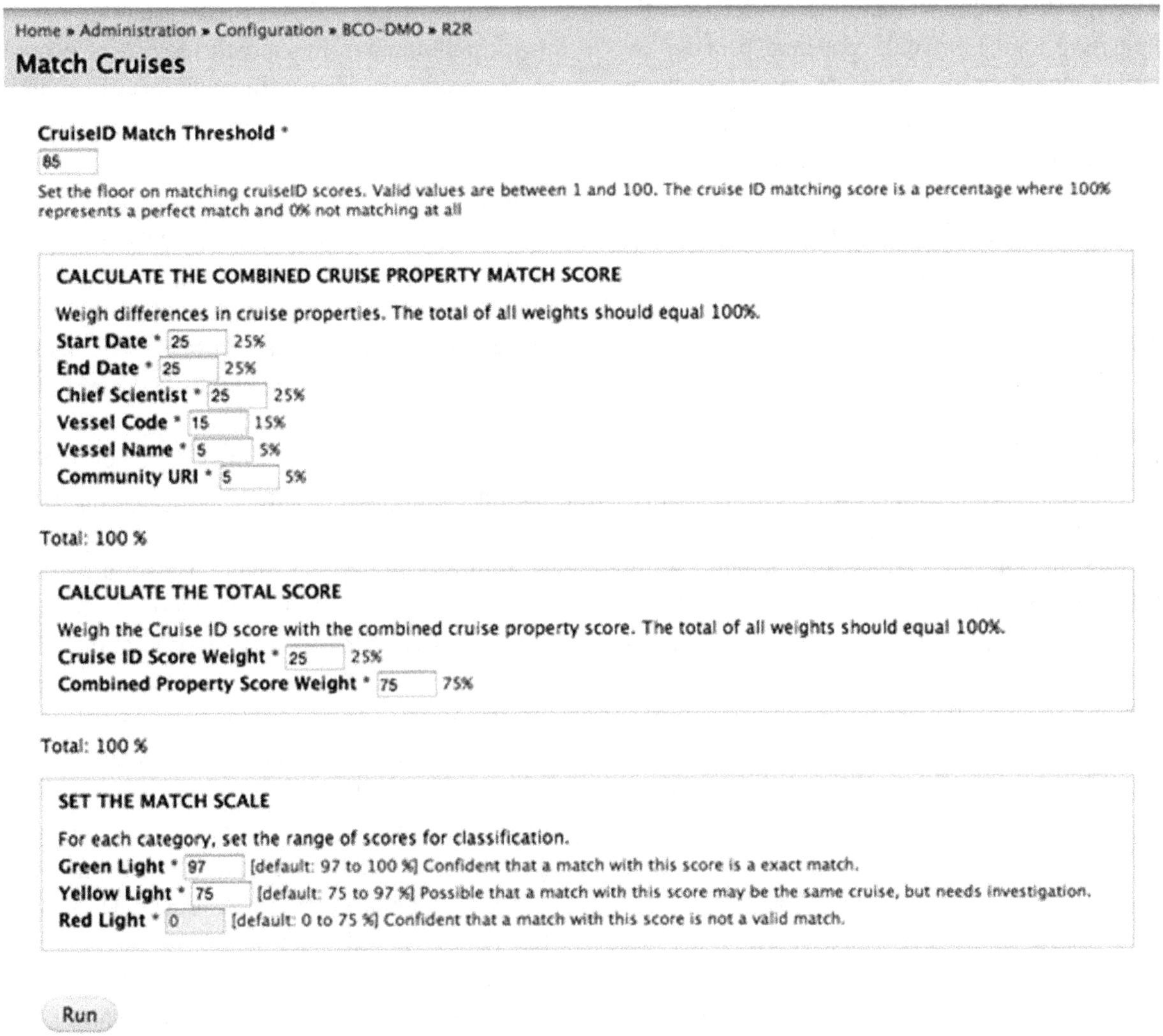

the threshold is set to 100%, the algorithm only returns potential matches where the cruise identifiers are exactly the same. The problem with this approach is that it doesn't account for errors in the entered cruise identifiers, resulting in such errors not being discovered. If a cruise identifier score is greater than 85%, the Levenshtein distance is then calculated for the rest of the properties and score values weighted according to the user inputs. The defaults at BCO-DMO weigh start, end date, and chief scientist the strongest because they are the most volatile properties. Since there is no community accepted way to universally identify researchers, the chief scientist comparison is performed on the string value of person's name. Comparatively, since the cruise identifier has matched adequately, there is a strong likelihood that the vessel name, code and community URI also adequately match. To better determine if there are mismatches in cruise metadata, weighing the volatile properties higher brings any total combined score for a cruise match down. Finally, the cruise identifier score is aggregated with the weighted sum of the other cruise property scores to determine an overall combined cruise score. In this aggregation, the default is to weigh the cruise identifier only 25% and the aggregated parameter score as 75%. This weighting considers that the matching algorithm only got here because the cruise identifier was adequate enough to begin with. Therefore, it is desirable for the algorithm to be unforgiving with the other parameters by weighting it much higher. This brings overall scores down which helps BCO-DMO data managers identify cruises that should score high match values and determine which cruise properties may be in error, if any.

FUTURE RESEARCH DIRECTIONS

As shown by the Data-Information-Knowledge ecosystem (Diviacco, 2015, see chapter 18 this book; Fredericks, 2015, see chapter 12 this book; and Leadbetter, 2015, see chapter 2 this book), the Linked Open Data approach of "small pieces loosely joined" (Weinberger, 2002) provides presentation and organisation to data, which creates information. Through the use of federated SPARQL endpoints this information can be integrated from multiple locations to create a knowledge base. This knowledge base has required continued discourse between the partners in order to truly generate new knowledge, which at the time of writing is much to do with answering data management questions (e.g. which datasets have been collected by cruises in the R2R project which are of interest to BCO-DMO). However, there are emerging science questions that can be answered through this technique as well. These emerging science questions include research in broad topics such as ocean acidification, oxygen minimum zones, biodiversity studies, and coastal zone management, and more focused investigations such as a recent effort by the environmental data centres of the UK's Natural Environment Research Council to use Linked Open Data techniques to link geological, hydrological and oceanographic data to track pollutants from source to sea (McCormick & Heaven, 2013). Emerging work in the semantic geospatial domain (http://www.w3.org/2001/sw/wiki/GeoSPARQL) will allow the exploration of data from overlapping cruise tracks from different sources. Related to this, the Ocean Data Interoperability Platform (http://www.odip.eu) is promoting a standard method for reporting summaries of research cruises across Europe, the USA and Australia encoded using controlled vocabularies published on the NVS, and explorations are underway as to how these may be represented as an ontology in the Semantic Web sense. Finally, as presented at the first meeting of the United Kingdom Ontology Network (http://dream.inf.ed.ac.uk/events/ukont-13/ukon_workshop_2012.html), the British Broadcasting Corporation (BBC) drives their website from a number of ontologies. Whilst initially developed for the sports division in order to track the matches, teams and players of the 2010 FIFA World Cup, this work was extended to create a programme ontology (http://purl.org/ontology/po/) that has an instance for every episode of every series broadcast on BBC television and radio stations.

Conforming to the Linked Open Data approach these instances all have unique, persistent URIs and are also mapped to other resources on the Semantic Web, in particular to topic categories from the DBpedia resource (http://dbpedia.org/) in the subject property. As scientific funding organisations are continually looking for new and innovative ways to create impact for their data, or to communicate their research outputs to the general public, leveraging the Linked Open Data approach to achieve this through links to the national broadcaster seems a rich area to explore further.

CONCLUSION

There are several existing, high quality, Linked Open Data resources published in the oceanographic domain. In the case of BCO-DMO and R2R, the two repositories manage related oceanographic data resources, such as research cruises and their datasets, but from different perspectives and with different goals. However, by linking their metadata to a common vocabulary such as the NERC Vocabulary Server, these repositories can now discover each other's related data through links to this common resource. Beyond the obvious advantage of increased resource discovery, this capability has made data management practices such as data validation, much easier to accomplish. For instance, using the SPARQL language, the two repositories can efficiently match their related cruise metadata to the other repository's cruise metadata through their common link to NERC Vocabulary Server. The SPARQL query will derive matches, and help to assess the accuracy of shared common metadata values. From these derived matches and assessments, data managers can quality control their own cruise metadata based on the results from the other repository. As the fundamental resources (platform, cruise, instrument, dataset) published by BCO-DMO and R2R are gradually enhanced with related links for relevant research programs, investigators, journal publications, funding awards, and gazetteer features, users will be guided efficiently to additional content that is likely to be of interest to their research focus.

As described earlier, the Linked Open Data Cloud has been a popular visualisation of the links between high quality Linked Open Data resources. In a similar manner, the Linked Ocean Data Cloud is an application of the ideas introduced recently by Blaumer (2013), in the Clean Energy Linked Open Data Cloud and links to the German Library Linked Open Data Cloud. Executing a World Wide Web search for similar micro-Linked Open Data clouds quickly reveals the existence of a Linguistic Linked Open Data cloud. We have assembled the relevant Linked Open Data resources of which we are aware, and have developed the diagram that can be seen in Figure 5. Where possible, URI links have been built between resources in order to expand the scope of the diagram as much as possible at this early stage. The diagram indicates those sources currently publishing high level Linked Open Data in RDF and those currently publishing to a lower Linked Open Data level.

All this is not to say that deploying a Linked Open Data resource is not without its difficulties. Deploying an effective Linked Open Data resource means that the URIs that are used to identify objects available from a repository should be well designed, in order to allow developers and clients to intuitively navigate the content. The United Kingdom government provides good advice on design patterns for URIs for Linked Open Data (Tennison, 2010), which has been widely adopted in many countries.

Another important consideration is the persistence of a URI, or the persistence of the data represented by the URI. If over time a URI becomes invalid, it is a fairly trivial matter for a software agent that is navigating the Linked Open Data cloud to ignore the response. However, if a human user has spent a long time building the links between one resource and another only to find that the URIs are no longer valid at one source, the result is a degradation of trust through a drop in the perception of competence or integrity of the service provider (Leadbetter, 2015, see chapter 2 this book). Or, if a person browsing through Linked Open Data resources discovers that several URIs in one namespace no longer represent anything, they are unlikely to trust that source in the long

Figure 5. A diagram illustrating the Linked Ocean Data Cloud comprising distributed sources of data, information and services available as a range of Linked Open Data levels

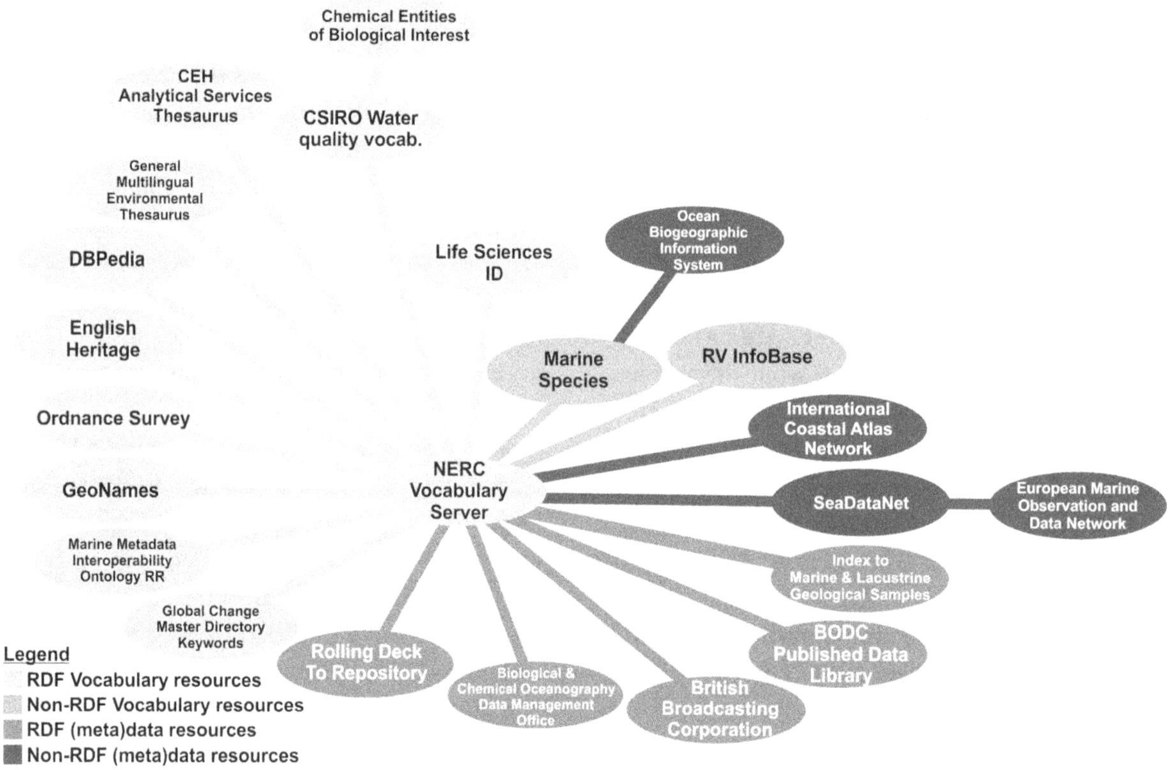

term. In building these networks it is important to remember that while the approach may be rooted in information technology there are many issues derived from human values that must still be respected. As outlined in the section on future research directions, the methodologies given above are beginning to be applied in answering science questions, moving them beyond the data management realm.

REFERENCES

Arko, R., Chandler, C., Clark, P., Shepherd, A., & Moore, C. (2012). *Rolling Deck to Repository (R2R): Linking and integrating data for oceanographic research.* Paper presented at 2012 Fall Meeting, American Geophysical Union. San Francisco, CA.

Baker, K., & Chandler, C. (2008). Enabling long-term oceanographic research: Changing data practices, information management strategies and informatics. *Deep-sea Research. Part II, Topical Studies in Oceanography*, *55*(18-19), 2132–2142. doi:10.1016/j.dsr2.2008.05.009

Bizer, C., Heath, T., & Berners-Lee, T. (2009). Linked Data - The story so far. *International Journal on Semantic Web and Information Systems*, *5*(3), 1–22. doi:10.4018/jswis.2009081901

Blaumer, A. (2013). *The LOD cloud is dead, long live the trusted LOD cloud.* Retrieved April 15, 2014 from http://blog.semantic-web.at/2013/06/07/the-lod-cloud-is-dead-long-live-the-trusted-lod-cloud/

Buesseler, K. (2001). Ocean biogeochemistry and the global carbon cycle; an introduction to the US Joint Global Ocean Flux Study. *Oceanography (Washington, D.C.), 14*(4), 5. doi:10.5670/oceanog.2001.01

Chandler, C., Groman, R., Allison, D., Wiebe, P., & Glover, D. (2013). One approach to ensuring long-term access to ocean research data. *Bollettino di Geofisica teorica ed applicata, 54* (suppl. 1), 213-5.

Diviacco, P. (2015). Maps, graphs and annotations as Boundary Objects in Knowledge Networks, Distributed Cognition and Collaborative e-Research. In P. Diviacco, P. Fox, C. Pshenichny, & A. Leadbetter (Eds.), *Collaborative Knowledge in Scientific Research Networks*. Hershey, PA: IGI Global.

Fogarty, M., & Powell, T. (2002). An overview of the U.S. GLOBEC Program. *Oceanography (Washington, D.C.), 15*(2), 4–12. doi:10.5670/oceanog.2002.17

Fredericks, J. (2015). Persistence of Knowledge Across Layered Architectures. In P. Diviacco, P. Fox, C. Pshenichny, & A. Leadbetter (Eds.), *Collaborative Knowledge in Scientific Research Networks*. Hershey, PA: IGI Global.

Glover, D., Chandler, C., Doney, S., Buesseler, K., Heimerdinger, G., Bishop, J., & Flierl, G. (2006). The US JGOFS data management experience. *Deep-sea Research. Part II, Topical Studies in Oceanography, 53*(5–7), 793–802. doi:10.1016/j.dsr2.2006.01.018

Graybeal, J. (2009). *Choosing and Implementing Established Controlled Vocabularies*. Marine Metadata Interoperability Project. Retrieved April 16, 2014 from http://marinemetadata.org/guides/vocabs/cvchooseimplement

Halpin, H., Hayes, P., McCusker, J., McGuinness, D., & Thompson, H. (2010). When owl:sameAs isn't the same: an analysis of identity in Linked Data. *Lecture Notes in Computer Science, 6496*, 305–320. doi:10.1007/978-3-642-17746-0_20

Hobbie, J., Carpenter, S., Grimm, N., Gosz, J., & Seastedt, T. (2003). The US Long Term Ecological Research program. *Bioscience, 53*(1), 21–32. doi:10.1641/0006-3568(2003)053[0021:TULTER]2.0.CO;2

Iseler, R., & Bizer, C. (2013). Active learning of expressive linkage rules using genetic programming. *Journal of Web Semantics, 23*, 2–15. doi:10.1016/j.websem.2013.06.001

Leadbetter, A. (2015). Examining trust in collaborative research networks. In P. Diviacco, P. Fox, C. Pshenichny, & A. Leadbetter (Eds.), *Collaborative Knowledge in Scientific Research Networks*. Hershey, PA: IGI Global.

Leadbetter, A., Arko, R., Chandler, C., Shepherd, A., & Lowry, R. (2013a). Linked Data: An oceanographic perspective. *Journal of Ocean Technology, 8*(3), 7–12.

Leadbetter, A., Lowry, R., & Clements, O. (2013b). Putting meaning into NETMAR – the open service network for marine environmental data. *International Journal of Digital Earth*. doi:10.1080/17538947.2013.781243

McCormick, T., & Heaven, R. (2013). *Geoscientific vocabularies and Linked Data at the British Geological Survey – progress and pragmatism*. Paper presented at 2013 Fall Meeting, American Geophysical Union. San Francisco, CA.

Michener, W., Porter, J., Servilla, M., & Vanderbilt, K. (2011). Long term ecological research and information management. *Ecological Informatics, 6*(1), 13–24. doi:10.1016/j.ecoinf.2010.11.005

Miles, A., & Bechhofer, S. (2009). *SKOS Simple Knowledge Organization System Reference*. World Wide Web Consortium. Retrieved April 16, 2014 from http://www.w3.org/TR/2009/REC-skos-reference-20090818/

NRC Ocean Studies Board. (2000). *50 years of ocean discovery: National Science Foundation 1950-2000*. Washington, DC: The National Academies Press.

Perry, M., & Herring, J. (Eds.). (2012). *GeoSPARQL - A geographic query language for RDF data*. Open Geospatial Consortium. Retrieved April 16, 2014 from http://www.opengeospatial.org/standards/geosparql

Schaap, D., & Lowry, R. (2010). SeaDataNet – Pan-European infrastructure for marine and ocean data management: Unified access to distributed data sets. *International Journal of Digital Earth*, *3*(Supplement 1), 50–69. doi:10.1080/17538941003660974

Tennison, J. (2010). *Creating URIs*. Retrieved April 16, 2014 from http://data.gov.uk/resources/uris

UNESCO. (1987). GF3: A General Formatting System for geo-referenced data: Volume 2: technical description of the GF3 format and code tables. Paris: Intergovernmental Oceanographic Commission.

Weinberger, D. (2002). *Small pieces loosely joined: a unified theory of the web*. New York, NY: Perseus Books.

KEY TERMS AND DEFINITIONS

Controlled Vocabulary: A controlled vocabulary provides a way to organize knowledge for subsequent retrieval. Vocabularies are used in subject indexing schemes, subject headings, thesauri, taxonomies and other forms of knowledge organization systems. Controlled vocabulary schemes mandate the use of predefined, authorised terms that have been preselected by the designer of the vocabulary, in contrast to natural language vocabularies, where there is no restriction on the vocabulary.

Data Archiving Centre: Data Archiving Centres (DACs) provide secure, long-term storage for research data. DACs should make available clear, searchable information on their data holdings, by the generation and publication of metadata. DACs should also form the first point of contact for expertise in the management of research data for domain scientists.

Linked Data, Linked Open Data: Linked Data is a technique that uses the World Wide Web to connect related data that was not previously linked, or uses the Web to lower the barriers to linking data currently linked using other methods. In practice, Linked Data often uses the Resource Description Framework model and Web addresses (as Uniform Resource Locators, or URLs) to achieve these linkages. Linked Open Data is an extension of these concepts to data which are freely available to all parties.

Ontology: In computer science and information science, an ontology formally represents knowledge as a hierarchy of concepts within a domain, using a shared vocabulary to denote the types, properties and interrelationships of those concepts.

Resource Description Framework: RDF is a standard model for data interchange on the World Wide Web. The RDF data model is based upon the idea of making statements about resources in the form of subject-predicate-object expressions, known as triples. A classical example of the subject-predicate-object triple is the sky-has colour-blue.

Semantic Web: The Semantic Web is the World Wide Web of data built on the Resource Description Framework model as a foundation for publishing and linking data.

Simple Knowledge Organization System (SKOS): SKOS is an area of work within the World Wide Web Consortium developing specifications and standards to support the use of knowledge organization systems (KOS) such as thesauri and controlled vocabularies within the framework of the Semantic Web.

Thesaurus: In the context of information retrieval, a thesaurus (plural: "thesauri") is a form of controlled vocabulary that seeks to dictate semantic manifestations of metadata in the indexing of content objects. A thesaurus serves to minimise semantic ambiguity by ensuring uniformity and consistency in the storage and retrieval of the manifestations of content objects.

Chapter 13
Persistence of Knowledge across Layered Architectures

Janet Fredericks
Woods Hole Oceanographic Institution, USA

ABSTRACT

In this chapter, a model demonstrating methods for integration of semantic technologies within observational data services is described. Implementation of the model captures knowledge about data provenance where it is best understood and also enables its persistence across architectural layers through the use of standards-based technologies. Domain experts can build upon the semantic layer to create meaningful ontologies. Brokering services can utilize the ontologies for automated mediation of terms and translation between standards-based technologies. Research communities will be enabled to operate within their own framework, utilizing their familiar, specialized terminology and tools. The role of communities of practice is explored relating to knowledge management across layered architectures. Implementation of semantic technologies within Web-based data and brokering services will minimize the operational barriers to data discovery and access and provide mechanisms that enable the formation of collaborative environments that will facilitate repeatable, well-documented research.

INTRODUCTION

Since the advent of the web and the subsequent ability to openly share and access data, there have been significant advances in creating, adopting and promoting standards to describe data provenance, requiring information such as: key words, ownership details, parameter units and geospatial-temporal coverage. The use of standards-based mechanisms is enabling access to machine-harvestable data along with the basic information describing it. But, are we missing an opportunity to do more than describe data and methods of access? Can we push the technologies and their implementation further to enable the capture of and machine-to-machine harvesting of information sufficient to also describe how data are created? The ability to understand observational provenance will provide a foundation of trust in our global data assets.

In the workshop report on "The Future of Scientific Knowledge Discovery in Open Networked Environments" (National Research Council, 2012), Paul Edwards describes the need to extend

DOI: 10.4018/978-1-4666-6567-5.ch013

Copyright © 2015, IGI Global. Copying or distributing in print or electronic forms without written permission of IGI Global is prohibited.

the types of available knowledge required to ensure trust and add value to data as they become available through open networked environments. He summarizes institutional impediments including the fact that participants are remote versus local, as well as noting the pervasive problem that agents along the digital life cycle do not consider it their responsibility to document data provenance and assume that they cannot know what will be needed by future unknown users. In Diviacco (2015, see chapter 1 this book), the complexity of understanding and knowledge relating to science within this environment is discussed, along with relevant levels of communication of knowledge, such as syntactic, semantic and pragmatic meaning.

With the wide adoption of web technologies, researchers are now able to conduct research remotely providing opportunities for collaborations internationally and across disciplines (Juan, Daradoumis, Roca, Grasman & Faulin, 2012). A collaborative environment enables participants to build a workspace to address common goals or problems by either sharing access to data or processes or both. It may evolve as a community with needs to access data from outside its own community or it may have data and want to share tools to utilize it. Researchers may want data to be dynamic or may share static data. Collaboration can focus on building data resources for sharing or building tools and workflows to be applied to disparate data sets.

Often researchers have to perform tedious tasks when accessing other's data, such as reformatting, subsampling, and interpretation of terms and conventions. Data providers typically serve data collections that require researchers to download volumes of data and to create specialized tools to access and subset the data, putting an onerous burden on their time and physical resources. Then there is the task of dealing with conversions of data formats and determining what is meant by each parameter in the set, assuring the understanding of units and conventions. The discovery and integration of data from outside a community's area of expertise can also be prone to error and omission. Each community may have a set of tools for processing, visualization or workflows that are to be applied to numerous data sets. Collaboration environments let scientists and decision makers focus on their respective research and analysis, rather than on the task of data discovery and transfer mechanisms.

The main concept of a collaborative environment is to create a methodology or workspace that facilitates ease of operation for a group of researchers or decision makers. Once the framework is built, the functionality can provide repeatable and documented results for use within the community. Either the data can change for utilizing the same workflows on different data (new location, different time, etc.) or the collected data are to be used in addressing a particular area of interest with different approaches. But, in all cases, the environment must provide access to knowledge about the data and processes that are being made available.

Considerable effort is being made throughout the world to bring data out of our offices and into the global arena. Also, researchers may benefit from sharing data with numerous data centers. For example, providing data to the U. S. National Oceanic and Atmospheric Administration (NOAA) National Ocean Data Center (NODC) not only provides a mechanism to expose one's data to the world, but NODC also archives data. The same data may be shared with the NOAA National Data Buoy Center (NDBC), where the focus is on quality control and real-time access and delivery, but data are not archived. Data provided through standards-based frameworks should be easily integrated into our national and international data centers as well as the evolving systems of systems, such as Global Earth Observation System of Systems (GEOSS – Butterfield, Pearlman & Vickroy, 2008), the National Science Foundation Ocean Observatory Initiative (OOI – Chave et al., 2011) and NOAA Integrated Ocean Observing System (IOOS - Willis, 2012). National and international efforts primarily focus on interoperating across data centers, where much of the knowledge about observational provenance is not readily available. This also leaves the onus on researchers to offer data by requiring reformatting of data and installation of specialized software, an activity for which they are seldom funded.

Integrative data collection is typically handled by mid- to large-, federally funded data centers (National Research Council, 2003) that must resolve terms and meanings from many highly diverse data providers. In the process, there are assumptions made in determining what knowledge is needed to sufficiently share the acquired data. The process currently involves person-to-person contact and is not only expensive and time-consuming, but it can introduce problems through error or omission, which is inherent in human interactions. This approach also requires each data provider to deal with each data center and their specific requirements. Even when accurate information is available electronically, it is offered as human-readable documents, which prohibit the ability to discover, harvest and translate it through automated processes and it precludes the ability to build upon the capabilities of the Semantic Web (W3C, 2014), such as using ontologies for term harmonization and inference mechanisms (Allemang & Hendler, 2011). Much of the advancement in implementation of standards-based technologies has resulted from discussions with data center staff who are developing interoperability standards to share their data collections to the world and not focusing on infusion of data from services into their own data stores.

Some progress has been made to develop data interoperability at its creation but the efforts have been primarily focused on automating installation and discovery of sensors for real-time observation networks (del Rio et al., 2013). The mechanisms are important and there is an added value in that they can associate sensors with unique identifiers. The ability to associate the observations with content-rich descriptions of these "born-interoperable" observations provides a significant means of capturing knowledge early and accurately. For example, the documents needed to describe the observations can reside within the sensor itself as a payload. Or, if bandwidth or storage of the content is limited, the mechanisms to associate the unique-id generated by the "born-interoperable" sensor with a web-accessible resource serving the content-rich description of provenance can be utilized to build web-based access to observational provenance.

Leadbetter (2015, see chapter 2 this book) reviews the importance of capturing provenance to build trust in collaborative environments. The focus of this chapter is on promoting methods to encourage the description of sensor provenance and process lineage in standards-based, machine-harvestable frameworks, adding value to archived data and enabling data quality assessment for newly discovered data collections. The implementation of these methods will enable providers to offer data through one standardized service, fully described for machine-to-machine harvesting by any data center or collaborative environment.

The development of communities of practice (Wenger, 1998) is also promoted here as a means to provide domain expertise in knowledge management across layered architectures. A community of practice (CoP) is by definition a self-organizing, highly creative assembly of individuals that enables knowledge to be shared through the creation of dynamic community-building infrastructures (Wenger, McDermott & Snyder, 2002). For example, a grass-roots organization called QARTOD (Quality Assurance for Real-Time Oceanographic Data) was formed to address community needs and to collect its body of knowledge with regards to assuring data quality in real-time oceanographic data offerings. The group had a small steering team but attracted about 80 participants from academia, industry and governmental agencies at its CoP meetings (http://www.ioos.noaa.gov/qartod/meetings.html). It has promoted the development of best practices and their implementation.

In contrast, the IOOS-QARTOD project (http://www.ioos.noaa.gov/qartod/welcome.html) has specific tasks and deliverables to create documents and define processes for implementation by participating organizations. A CoP does not have a specific agenda – it is dynamic and defines its own agenda within its area of expertise. The purpose of a CoP is to openly develop and share knowledge and to promote practices that move technologies or practices towards self-determined desired outcomes. The groups can form around technical or scientific interests. For example, the Group on Earth Observations (GEO) has formed numerous communities of practice (http://www.earthobservations.org/about_geo.shtml) that are focused on (but are not limited to): air quality, atmospheric chemistry, biodiversity, carbon-cycle, coastal zone, cryosphere, energy, forests, geohazards, global agricultural monitoring, health and environment and water cycle. An expert from a particular domain may have the knowledge and passion to participate in more than one related community. CoPs membership is self-subscribing and continues only as long as the interest exists. The Geo-Seas project has organized around the need for data sharing across twenty-nine institutions in the marine geophysics and geology domains (Diviacco, Lowry & Schaap, 2011).

As shown below, CoPs are paramount in addressing knowledge management at different stages of the data lifecycle. For example, a CoP with knowledge about measuring oceanic waves may form around a technology that uses wave buoys and accelerometers and a separate CoP may form to address issues relating to measuring oceanic currents and waves using Doppler current meters. There may be individuals who will bring together the knowledge about the output (oceanic waves parameters) to harmonize technical differences in each technology. These three CoPs, as described, will assure development of knowledge management technologies that can provide information that is needed to assess the data for an unknown use. Each CoP knows what is important given its own unique perspective, such as what can affect data quality. It can define procedures and requirements that assure that data are shared with sufficient information to make it valuable for long-time archival and for interdisciplinary use.

Once data are shared in a machine-harvestable, standards-based framework, a CoP that more closely relates to the consumption of knowledge and data will apply its own specialized expertise to define its specific set of requirements. They can define what is needed to know how to use and accurately understand the data. For example, a researcher investigating extreme events will definitely want to know if a range check or outlier test was applied to the data and, if so, what the parameters were for the process. Or, a research community may define the need to have data or metadata available in a particular format or standards framework. The development of end-user CoPs that define overarching capabilities and requirements will enable tools to be built to have a broader impact for the community at large.

But, this chain of creation, integration, aggregation and knowledge management cannot be implemented without the capture of provenance at each transitional point along the data lifecycle.

Where Does Knowledge Begin?

As data are generated, decisions are made that directly impact how it can be interpreted for reuse. For example, what default assumptions are made in a sensor configuration and what are the implications on the outcomes? What are the performance limits of the sensor (range, precision, accuracy)? When and how was the sensor calibrated? Were the data filtered and, if so, how? Figure 1 depicts a tiered flow of data generation, collection, sharing and aggregation, as data are generated and passed through various levels of decision-making.

Figure 1. Data lifecycle and knowledge domains. Typically, the "Data Sharing" layer is where discussions of interoperability begin. At this point, much of the knowledge relating to quality assessment is lost. Data aggregation can become a huge task when depending on human communication. The process is not only costly, but it can introduce misunderstandings, omissions and errors. To fully describe an observation, information must be captured at the point where it is best understood.

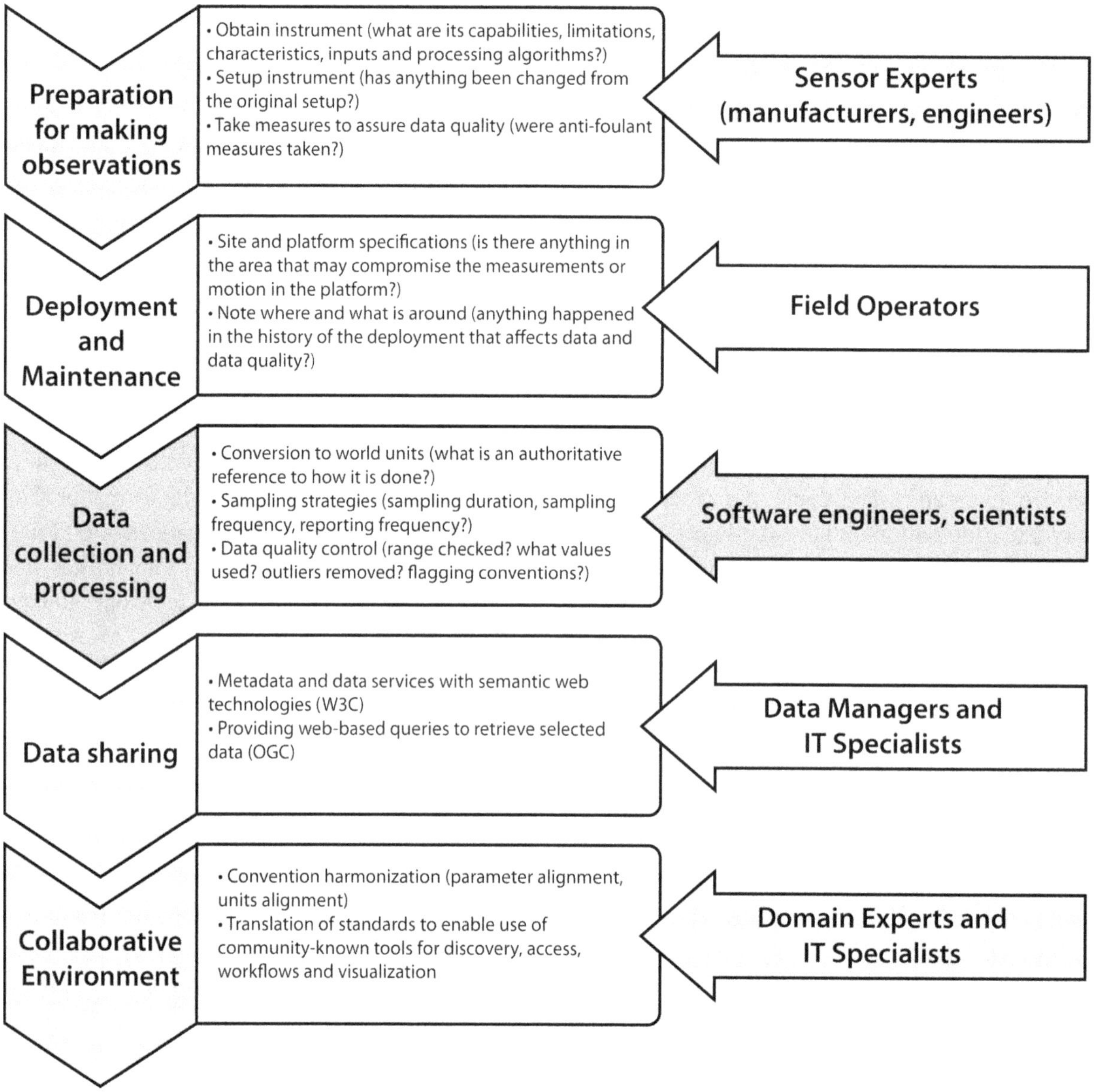

BACKGROUND

Layered Architectures Approach

Through efforts like the NSF EarthCube (http://www.earthcube.org), capabilities needed to enable cross-domain, inter-agency, international access to research data are being defined to prepare for the development of an infrastructure which will reduce the burden of data management and enable the creation of repeatable, well-documented research environments (Gomez, Pearthree & Cutcher-Gershenfeld, 2013). By providing translation mechanisms across standards-based frameworks, brokering technologies (Nativi, Craglia & Pearlman, 2013) will play an important role. Use standards! But which ones? We will no longer need to have this debate as tools are being built to translate across numerous community-adopted standards enabling data discovery and access to occur across frameworks of choice. Brokering technologies can be made available as a service or built as a component of a collaborative environment. The ability to translate between standards will not only push the edges of how we communicate but also what content will be expected and how it will be communicated. The opportunity is upon us now to promote the inclusion of content-rich provenance that can persist across the brokering technologies, as we replace our humans-in-the-loop with automated processes.

This chapter describes a practical approach to methodologies which enable the capture of knowledge at each level in the data lifecycle, by including data and provenance descriptions that are enhanced by referencing World Wide Web (W3C) Semantic Web technologies (http://www.w3.org/standards/semanticweb) within OGC® (http://www.opengeospatial.org) adopted web services, which enables services to be built that can subset data based upon specific requests. The use of standards-based frameworks enables knowledge to persist across layered architectures. The descriptions are implemented to enable semantic mediation by incorporation of references to domain generated vocabularies and ontologies. Currently, at best, metadata can be expected to contain encoded information about data. Here we propose to extend the expectation of metadata to also fully describe how an observation came to be. In all cases, the use of semantics within web-based data services is important to enable machine-to-machine harvesting of knowledge for harmonization and categorization of content. The capability is particularly important when terms are not consistently defined, which is not only the case across communities but also within communities.

INTEROPERABILITY OF THE OGC® SENSOR WEB ENABLEMENT FRAMEWORK

The Open Geospatial Consortium (OGC®) OpenGIS® standards are openly and freely available and are developed to geo-enable the web. International participation, as of January 2014, consists of 472 members from industry, government agencies and academia. The community-adopted standards of the OGC® are developed working with other standards bodies, such as the World Wide Web Consortium (W3C, http://www.w3c.org) and the International Organization for Standardization (ISO, http://www.iso.org). The adoption of standards will lead to the development of powerful tools for reducing the effort needed to conduct research across-disciplines. The tools will not only provide data access, processing and visualization but also enable brokering between standards, facilitating broader integration of global resources across disciplines and cultural boundaries.

Sensor Web Enablement (SWE) is an OGC® adopted framework that provides mechanisms to enable machine-to-machine interoperability with content-rich description of data and data provenance through web-based services (Botts,

Percivall, Reed & Davidson, 2008). The oceanographic community is building an infrastructure that utilizes the capabilities of SWE (Sorribas, Olive, Diviacco & Bermudez, 2012). As a component of the SWE framework, the Sensor Model Language (SensorML) provides standards for encoding knowledge relating to sensor provenance and process lineage. Using SensorML, a provider can supplement the OpenGIS terms, such as descriptions; coordinate reference systems; and, time encodings with SensorML terms that define process lineage and sensor characteristics, capabilities and other significant metadata about how an observation was created. The underlying design of SensorML is built upon the description of processes and uses tags that define process related terms such as input, output, and process. The SWE framework binds together basic services relating to a data resource. SWE Sensor Observation Service (SOS) has three required methods:

- *GetCapabilities* provides the ability to harvest information about what the SOS has to offer. It links the observational provenance (DescribeSensor) with the observational offerings (GetObservation). In recent Earthcube activities, the author was able to register a GetCapabilities link with three brokering systems. The GEOSS Discovery and Access Broker (http://www.eurogeoss-broker.eu) and the ESRI GeoPortal Server (http://sourceforge.net/apps/mediawiki/geoportal/index.php?title=Geoportal_Server_as_a_Broker#Features_of_a_Broker) were able to automatically harvest metadata about the services and populate their respective discovery catalogues. Using the NOAA Environmental Research Division's Data Access Program (ERDDAP) as a data and metadata service broker (http://coastwatch.pfeg.noaa.gov/erddap/index.html), a user can request the author's metadata from the GetCapabilities response and retrieve it in other standards-based encodings, such as ISO 19115 (NOAA, 2013).
- *GetObservation* harvests requested data. Using a standards-based query, the service can enable a user to specify a geospatial or temporal range; a selection of parameters that are parsed from a data collection; and/or, definition of other conditions, such as whether to harvest all data with or without quality control (QC) flags or only data that passed the QC tests. The observations are returned in Observations and Measurement (O&M) encodings providing the information about the data itself with terms such as those needed to define a data record, unit of measure, fields, etc. Depending upon the implementation, the service can enable a provider the ability to serve only the data of interest. The author registered a data service (SOS) with the NOAA ERDDAP brokering service, which provides the ability to access the SOS offering in other frameworks (netCDF, JSON, CSV) rather than its native O&M without any work by the provider or the client.
- *DescribeSensor* returns SensorML documents that describe observational provenance. Typically, if it exists at all, it is sparsely populated and does not fully utilize the content-rich capabilities of SensorML to describe sensor provenance and process lineage. This information provides the basic building block for sharing data across brokering services and within aggregation services, such as the descriptive provenance required in collaborative environments. Automated generation of these documents by sensor manufacturers, field operators and data providers will facilitate the development of content-rich

documents and thereby enable a better understanding of our data resources as they pass through layers of transformation and integration.

We are at a critical crossroads. We can now not only describe data in standards-based frameworks, but we can also describe its provenance in machine-harvestable encodings to document how an observation came to be. The information about the data, the offerings and its provenance are associated with the data itself using the standards-based SWE framework.

The following section focuses on a method of implementation and the benefit of enabling semantic mediation of terms that are defined within the web service for upstream integration, including discovery, brokering and workflow development that can be implemented within collaborative environments. A model implementation, developed by a multi-institutional team called Q2O (http://q2o.whoi.edu), for QARTOD-to-OGC, was funded by NOAA to enable dynamic quality assessment of real-time oceanographic data offerings encoded using the OGC® SWE framework. The multi-institutional Q2O team was comprised of information specialists, domain experts, sensor manufacturers, data managers and field operators and worked closely with QARTOD and another CoP called the Marine Metadata Interoperability project, or MMI (http://marinemetadata.org).

A MODEL FOR THE INTEGRATION OF SEMANTIC WEB TECHNOLOGIES IN A LAYERED ARCHITECTURE

Development of Registered Terms, Vocabularies, and Ontologies

The use of semantic web technologies is being implemented throughout the oceanographic community (Leadbetter, Arko, Chandler, Shepherd & Lowry (2015, see chapter 12 this book). The first step towards enabling semantic interoperability of observational data services is the creation of role-based vocabularies that include the definition of terms used in the description of data and its associated provenance. Here, the term "provenance" refers not only to the origin of the data but to a continuous documentation of all information that is needed to assess and describe how an observation came to be: not only who, what, when and where but also how.

The vocabularies must be encoded using standards-based technologies that enable machine-enabled semantic mediation. The development of an encoded vocabulary using the W3C mechanisms of the Semantic Web (W3C, 2014) provides a mechanism for machine-to-machine harvesting, rather than human readable documents, and realizes the benefits of formalization (access, versioning, etc.). A more complete description of the concepts that are summarized in this section can be found in Graybeal, Isenor & Rueda (2011). To create Semantic Web enabled encodings, the Q2O team used the Ontology Registry and Repository (ORR) that provides tools to create standards-based encodings of vocabularies and a registry service and can be accessed at http://mmisw.org. The ORR (Rueda, Bermudez & Fredericks, 2009) was created to meet the needs of the Marine Metadata Interoperability MMI project and provides for easy creation of version-controlled content-rich vocabularies that are encoded in Web Ontology Language (OWL) and the Resource Description Framework (RDF). The content-rich vocabularies are converted from comma-separated content (e.g., from a spreadsheet generated by Microsoft Excel) to OWL/RDF making them available for semantic mediation. Once registered, the terms can be individually and uniquely referenced as a Uniform Resource Locator (URL). Although the terms within the vocabularies are domain specific, the model can be utilized throughout the geosciences. Each vocabulary focuses on different aspects of data generation, enabling controlled vocabulary management by role-based owners.

For example, participants in the Q2O project registered several vocabularies, including: observable properties (sensible), properties (observations), QC flags, flagging conventions, QC tests, processes (including authoritative references to processing methods), event lists, and sensor and platform identifiers. Figure 2 shows the structure and content in one of the vocabularies.

Terms within the vocabulary can be referenced using the SPARQL query language (W3C, 2014) to provide specific information about a registered term. For example, Figure 3 shows the content accessed by querying the URL that describes how directional wave data are computed from pressure and velocity.

This methodology also provides manufacturers, field operators and data providers with a mental framework to guide them in how to organize, create, populate and maintain registered controlled vocabularies. For the Q2O model, the sensor manufacturer created its own vocabulary that can be referenced by any user of the manufacturer's sensors. The Q2O project also created vocabularies and SensorML documents describing standard QC tests used in the real-time oceanographic community.

In the next sections, we describe how the URLs, which reference encoded terms, can be incorporated into SWE to enable referencing and association of terms to discover relationships and hierarchies within a collaborative environment, as well as how the methodology can enable brokering of knowledge about sensor provenance and process lineage.

Figure 2. The processes that are involved in the computation of waves at the Martha's Vineyard Coastal Observatory (http://www.whoi.edu/mvco) are each described. This figure represents a collection of terms (a vocabulary). Each term in the vocabulary 'process' can have fields for a short name, authoritative reference, notes, an ID, equations and a definition. An example of one of the process ontologies is shown in Figure 3.


```
– <rdf:RDF xml:base="http://mmisw.org/ont/mvco/process">
  + <owl:Ontology rdf:about=""></owl:Ontology>
  + <owl:Class rdf:about="http://mmisw.org/ont/mvco/process/Process"></owl:Class>
  + <owl:DatatypeProperty rdf:about="http://mmisw.org/ont/mvco/process/ShortName"></owl:DatatypeProperty>
  + <owl:DatatypeProperty rdf:about="http://mmisw.org/ont/mvco/process/Reference"></owl:DatatypeProperty>
  + <owl:DatatypeProperty rdf:about="http://mmisw.org/ont/mvco/process/Notes"></owl:DatatypeProperty>
  + <owl:DatatypeProperty rdf:about="http://mmisw.org/ont/mvco/process/ID"></owl:DatatypeProperty>
  + <owl:DatatypeProperty rdf:about="http://mmisw.org/ont/mvco/process/Equations_s"></owl:DatatypeProperty>
  + <owl:DatatypeProperty rdf:about="http://mmisw.org/ont/mvco/process/Definition"></owl:DatatypeProperty>
  + <Process rdf:about="http://mmisw.org/ont/mvco/process/computeStdDev"></Process>
  + <Process rdf:about="http://mmisw.org/ont/mvco/process/pressureObsProcess"></Process>
  + <Process rdf:about="http://mmisw.org/ont/mvco/process/linearInterpolation"></Process>
  + <Process rdf:about="http://mmisw.org/ont/mvco/process/pressureQCchain"></Process>
  + <Process rdf:about="http://mmisw.org/ont/mvco/process/minThresholdTest"></Process>
  + <Process rdf:about="http://mmisw.org/ont/mvco/process/timeseriesChain"></Process>
  + <Process rdf:about="http://mmisw.org/ont/mvco/process/minThresholdSeriesTest"></Process>
  + <Process rdf:about="http://mmisw.org/ont/mvco/process/rangeSeriesTest"></Process>
  + <Process rdf:about="http://mmisw.org/ont/mvco/process/puvAnalysis"></Process>
  + <Process rdf:about="http://mmisw.org/ont/mvco/process/velocityObsProcess"></Process>
  + <Process rdf:about="http://mmisw.org/ont/mvco/process/flagOutliers"></Process>
  + <Process rdf:about="http://mmisw.org/ont/mvco/process/velocityQCchain"></Process>
  </rdf:RDF>
```

Figure 3. Integrating authoritative references. The term, puvAnalysis, in the MVCO vocabulary 'process' includes encodings of an identifier (ID), definition, authoritative reference and a short name. As seen in Figure 2, the ontology can also include equations or notes.

```
- <RDF>
  - <Description rdf:about="http://mmisw.org/ont/mvco/process/puvAnalysis">
    - <ns1:Definition>
        Triplet waves processing that converts pressure (P) output from the PQ_Chain and horizontal velocity (U)
        and vertical velocity (V) from the VQ_Chain to spectral estimates for computation of wave direction
      </ns1:Definition>
      <ns1:ID>puvAnalysis</ns1:ID>
    - <ns1:Reference>
        Dean, R.G. and R. A. Dalrymple, 1984. Water Wave Mechanics for Engineers and Scientists, Englewood
        Cliffs, N.J., Prentice-Hall, Inc., 353 pp.
      </ns1:Reference>
      <rdfs:label>puvAnalysis</rdfs:label>
      <ns1:ShortName>PUV</ns1:ShortName>
      <rdf:type rdf:resource="http://mmisw.org/ont/mvco/process/Process"/>
    </Description>
  </RDF>
```

Enabling Semantic Web Development in Sensor Observation Services

As described in the Manzella (2015, see chapter 9 this book), a knowledge management system must include open and free access to data and data products. The authors promote the need of having vocabulary discussions that "can make explicit how different participants understand the terms they use". Vocabulary discussions will result in the establishment of relationships of terms, yielding machine-harvestable ontologies. This section describes a method to include the Semantic Web technologies within the SWE framework, enabling association of the resulting ontologies with the observations. This can be accomplished by referencing the fully described, registered terms within the Sensor Observation Service (Figure 4).

The RDF/OWL encoded terms are incorporated into the SOS encodings using the XML W3C Xlink tag (DeRose, Maler, Orchard & Walsh, 2010) or by referencing the term as a definition (Figure 5). The OGC® SWE framework interoperability is not affected by the inclusion of the references. By referencing the W3C encoded terms, the meaning behind each SWE term description persists through the subsequent layers and also provides access to the SPARQL endpoint at the MMI ORR. Semantic mediation within a CoP can be conducted using the registered vocabularies separately from the SWE implementation, yet it will be associated to the semantic ontologies by referencing the URL.

A client can access the data through the SOS with the expectation of adherence to the rules of the OGC® SWE framework. The reference to specific ORR registered terms within the SWE documents, as described above, enables a broker to harvest the reference to the term and search ontologies for associations to utilize the power of the Semantic Web to discover, infer, harmonize, and categorize a particular piece of information returned in the SWE offering.

Figure 4. Integration of semantics within SWE frameworks. The SOS offerings include references to the registered terms. A client can be a broker, which can harvest the OWL/RDF encoded terms for semantic mediation.

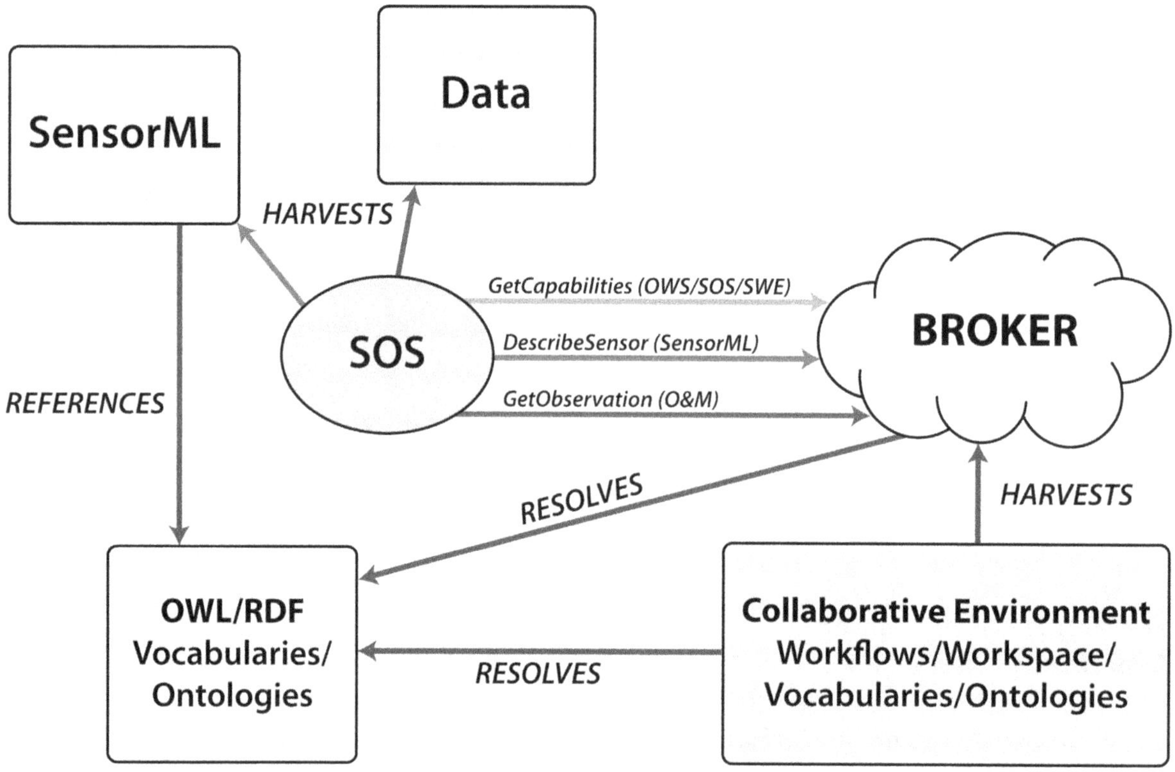

Figure 5. The OWL/RDF encoded terms are referenced within the SWE documents. This provides a connection between the Semantic Web framework of the W3C and the OGC® Sensor Observation Service.

```
- <swe:field name="waveHeightAll">
  - <swe:Quantity definition="http://mmisw.org/ont/mvco/properties/waveHeightAll">
      <swe:uom code="cm"/>
    </swe:Quantity>
  </swe:field>
```

Semantic Interoperability in Cross-Domain Brokering and Collaborative Environments

Domain experts who understand the meaning of the terms must create and register the content-rich vocabularies. Those who understand the processes and the data management systems can implement the web services enhanced with the references to the encoded terms. Domain experts also create the ontologies for relating terms and building ontologies with their particular knowledge base. As discussed in Boldrini, Craglia, Mazzetti & Nativi (2015, see chapter 14 this book), brokering technologies will be enabled to incorporate semantic mediation of the referenced terms to provide the ability to discover data and infer relationships of the data resources within a collaborative environment (Figure 6).

Figure 6. Roles of communities of practice to define requirements in knowledge generation, harmonization and collaboration. Each SOS references its respective vocabulary, associating linked terms in the data service with registered, W3C encoded terms. Domain experts access the terms to build and register ontologies for use in Semantic Web technologies. Expertise of a knowledge provider will focus on its particular technology used to create an observation with particular understanding of what is important. Brokering technologies may exist within or external to the collaborative environment and may access external or internal ontologies that relate terms to each other and/or to associated, more complex ontologies. The brokering of data services, e.g., SOS to netCDF, is also depicted above (A). Communities of Practice (CoP) from the vantage of information consumers are also needed to create ontologies and promote best practices in the creation of collaborative environments and use of brokering. A CoP within this layer will be needed to determine how to harmonize and categorize knowledge across the technologies. Knowledge consumers will define knowledge requirements and preferred frameworks to enable access to information through brokered data and catalogue services.

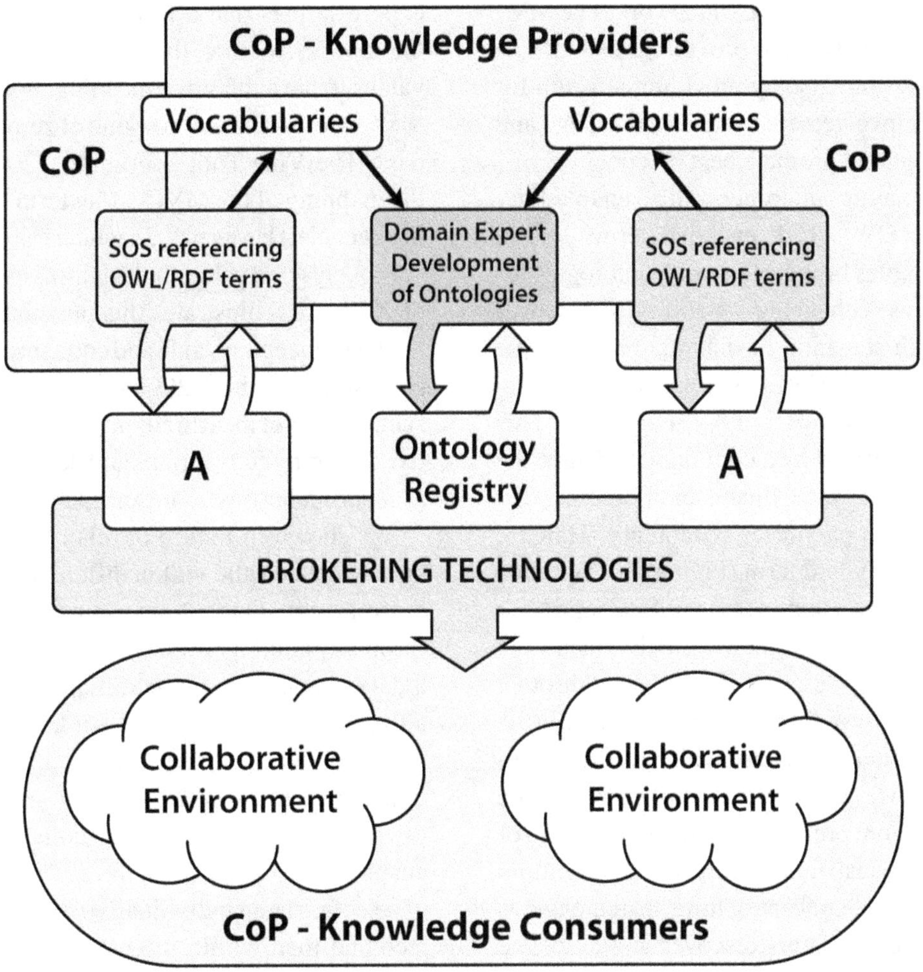

Domain experts for a particular collaboration have the understanding required to associate terms. These terms can be associated by simple concepts, such as "same as", "narrower than", etc. and they can also be defined as having more complex relationships, such as classes and categorizations. Tools are available that enable domain experts to create ontologies without needing to understand the underlying technologies. In the marine community, we can use tools such as the MMI Vine tool (Bermudez, 2007) to harmonize data and to categorize terms using the Simple Knowledge Organization System (SKOS) (Isaac & Summers, 2009) of the Semantic Web framework. The MMI serves as a CoP not only by providing needed tools and resources but also by providing a forum for community involvement to share knowledge and define, adopt and promote best practices.

The semantic interoperability enabled by referencing OWL/RDF encoded terms within the SOS enables brokers to search ontologies for related terms, enhancing its ability to discover data through semantic mediation. The example below depicts the MVCO specific term relating to wave height, waveHeightAll (Figure 7a). This term can be defined by a community to have the same meaning as the Climate and Forecast (CF) standard names parameter vocabulary (Hankin, et al., 2010) standard term (Figure 7b). Within a collaborative environment, use of Semantic Web technologies will enable the MVCO data identified with the property description referenced through the URL below (waveHeightAll) to be treated in the same way as data referenced with the URL defining sea_surface_wave_significant_height, while keeping its more precise meaning. Developers of workflows, visualization tools and model input can focus on the concept, letting search engines and semantic mediators discover and associate the details of the terminology

A hierarchical ontology is another powerful tool enabled by the Semantic Web. By forming more complex relationships, we can enable inference. For example, in a domain-expert developed ontology, a term referenced as sea_surface_wave_significant_height may be semantically defined as being in a subclassOf OceanicPhenomena. If the MVCO term and the CF term are defined as being the same (above), by inference the MVCO SOS waveHeightALL can also be discoverable as a subclassOf OceanicPhenomena.

The use of code space to associate a controlled vocabulary term with specific values is also useful in linking web services and workflows to the powers of the Semantic Web. When a domain expert defines that a term in one vocabulary is the same as another, they can associate disparate values to have the same meaning. For example, a MVCO term defines the value of zero (0) to mean pass. The Vine Tool (Bermudez, 2007) enables the mapping of the MVCO flag term 'pass' to the SeaDataNet flag term '_2', which has the meaning "probably good value" but has an associated value of 2. Figure 8 illustrates this mapping.

The concepts of code and code space were used extensively in the Q2O model, confining terms such as units of measurement (uom), history event lists, flagging conventions and defining processes as a particular type, for example QCcategory.

As shown in Figure 8, this relationship may or may not be valid within different collaborative environments. One community may determine that 'probably good' is good enough to be included as a data resource and would map the term 'probably_good' to pass rather than fail to determine how the associated data are treated. Another community may not want any data that are categorized as 'probably good'. Another collaboration may enable a particular user to have access to a means of selecting as an individual, so data can be evaluated and then within the workflow excluded or included, as determined by enabling the flexibility to dynamically assess conditional requests using the related concepts (not the values).

Figure 7. (a) Below, an RDF encoded term is shown that specifically defines a property that is output in the SOS service of the MVCO (http://q2o.whoi.edu/node/129). Note: to access the RDF rather than the html rendering include "form=rdf", as shown in this link: http://mmisw.org/ont?form=rdf&uri=http://mmisw.org/ont/mvco/properties/waveHeightAll. (b) An encoded term of a similarly defined parameter in the CF vocabulary.


```
- <RDF>
  - <Description rdf:about="http://mmisw.org/ont/mvco/properties/waveHeightAll">
    - <ns1:Definition>
        Significant wave height defined as the average of the highest one-therd of the wave heights
        from ADCP velocity spectral analysis calculated as HM0 (0.05 < f < 0.5 Hz)
      </ns1:Definition>
      <ns1:Reference>http://mmisw.org/ont/q2o/reference/IAHR_1989</ns1:Reference>
      <ns1:LongName>Wave Height</ns1:LongName>
      <ns1:ShortName>Hs_all</ns1:ShortName>
      <ns1:ID>waveHeightAll</ns1:ID>
      <rdfs:label>waveHeightAll</rdfs:label>
      <ns1:Symbol>Hmo</ns1:Symbol>
    - <ns1:Equations_s>
        4*sqrt(sum(Snn).*dF) (f > 0.05) Snn from surface velocity
      </ns1:Equations_s>
      <rdf:type rdf:resource="http://mmisw.org/ont/mvco/properties/Properties"/>
    </Description>
  </RDF>
```


(a)


```
- <RDF>
  - <Description rdf:about="http://mmisw.org/ont/cf/parameter/sea_surface_wave_significant_height">
      <ns1:canonical_units>m</ns1:canonical_units>
      <skos:definition>Height is the vertical distance above the surface.</skos:definition>
      <rdf:type rdf:resource="http://mmisw.org/ont/cf/parameter/Standard_Name"/>
    </Description>
  </RDF>
```


(b)

ROLE-BASED KNOWLEDGE: CREATION AND ACCESS

As shown in Figure 1, the agents making decisions at various points in the data lifecycle best understand observational provenance relating to their respective decisions and implementation. In this section, we describe model for the construction of a suite of SensorML documents that can rationally and fully describe how an observation came to be. Figure 9 depicts the model that identifies categories of documents that are needed to accommodate creation by each of those agents.

In earth observing systems, data are typically generated by technologies that sense observable properties. For example, an acoustic Doppler

Figure 8. The example above shows a representation of the ontology created by the MMI Vine Tool that maps terms that can be considered to have the same meaning, but utilizing code space to incorporate defined (different) values to be associated with the mapped terms.

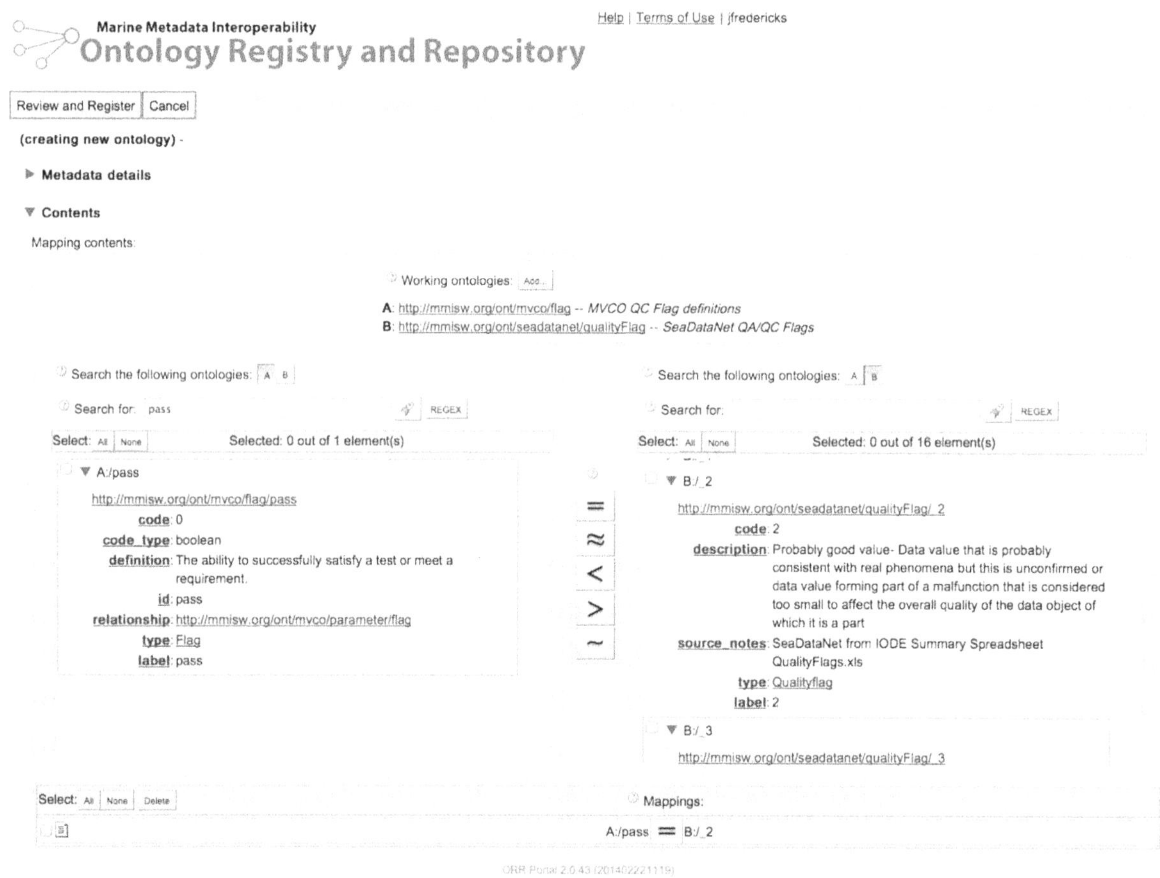

current meter computes velocity based upon the Doppler shift of sound waves reflected from a moving particle. Salinity can be computed from observing the conductivity of a known volume of seawater and its concurrently measured temperature at a known depth. Each sensor manufacturer has relevant information about how data are created. In an original equipment manufacturer (OEM) SensorML file, information will be related to these technologies and the manufacturer. For example, contact information will be about the sensor manufacturer. The document may specify a recommended calibration cycle. Model specifications will be included. Any agent who uses that particular instrument model can use the OEM SensorML document. Often a suite of sensors will contribute to the data stream (components) and they should each be described.

The next decision step in a data lifecycle is the configuration of the instrument. At this point, the contact information will be the institutional contact(s). Here a field operator will determine how the instrument is configured. A decision will often be made that changes valid instrument ranges and expected accuracies of output. These decisions can have consequences in understanding repurposed data. For example, an hourly reporting frequency may have observed an average over the

Figure 9. Role-based description of observational provenance. A SensorML system file (category 1) provides the over-arching connections of the other role-based SensorML documents. The Sensor and Configuration/Deployment files (categories 2/3) are to be created by sensor manufacturers and their applications, which will then be updated with relevant information relating to data ownership, configuration of the sensor, deployment(s) and servicing by field operators. The data management system conducts quality control (QC) and processing for derived products. Processes should be described in Process files (categories 4/5).

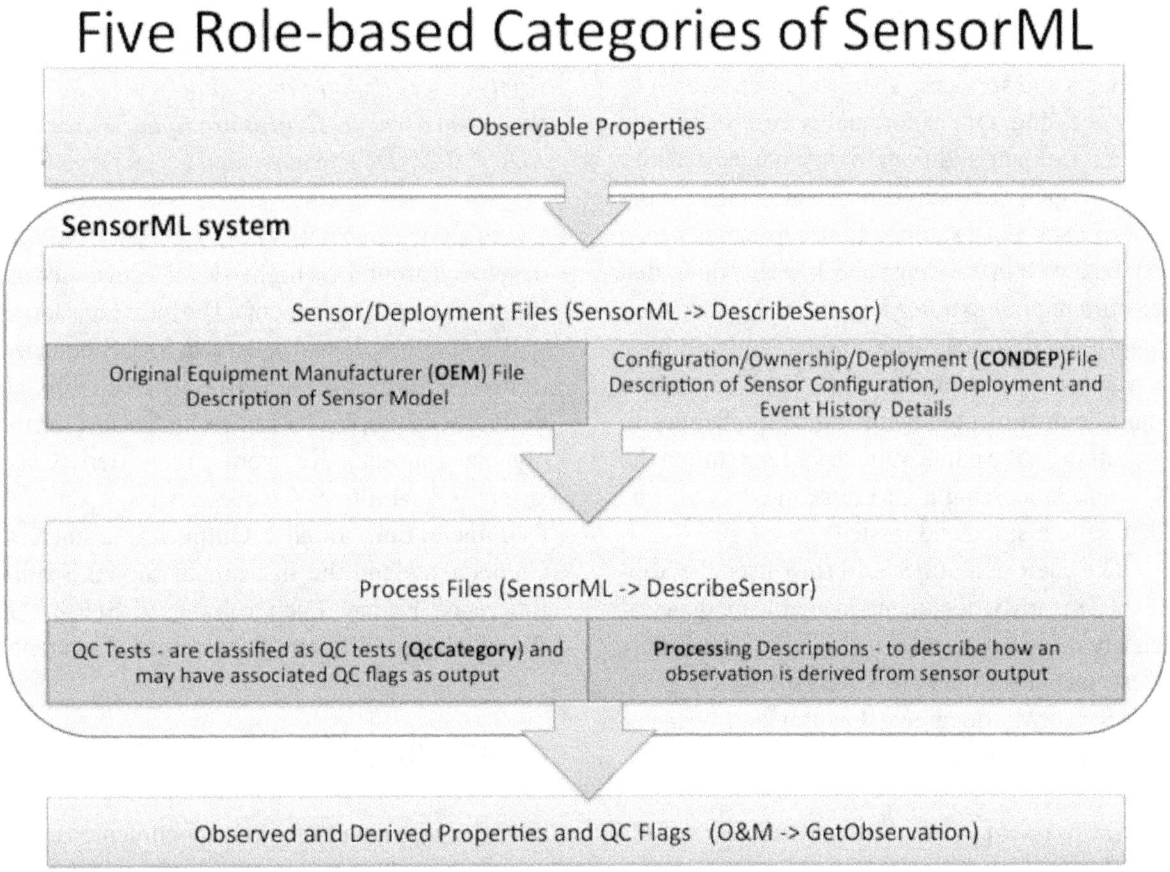

hour, or it may be a snapshot every hour. Typically, the operator uses manufacturer's software that currently locks away this knowledge in binary files or reported in human-readable reports. Access to this knowledge for machine harvesting is not available at this point, other than to be carried along for later human consumption, if specialized software is available to read it. The configuration/deployment (CONDEP) file will include the information about the sensor(s) and how each was configured (calibration dates, serial numbers or unique IDs, etc.) in standards-based encodings (SensorML).

In long-term deployments, the instances of an instrument may be swapped out. A data stream may be generated by a series of instruments, as long as the technology behind the observation does not change. In the CONDEP file, a history of events should be kept describing in SensorML the occurrence of changes that can effect data quality. For example, perhaps a diver cleaned off sensor faces or a real-time observatory conducts adaptive sampling and shifts the configuration to better capture data for particular events. Machine-harvestable, time-stamped events are critical to better understanding the data provenance.

Sensor data are normally offered through a data service. It typically is here that a data manager and the associated software processing create data products and services. There are often two types of processing. One is for quality control and the other is to create additional data products. Quality control (QC) decisions can also affect the proper use of data. For example, future researcher may not be aware that a range check was applied that has filtered out extreme events. And, processing algorithms that create data products typically have a variety of implementations. Each algorithm must be documented by including a reference to an authoritative publication. By documenting the QC and processing algorithms, the data will be better understood and trusted.

The GetCapabilities offering lists the top-level SensorML documents available for query of observational provenance. Below is an example DescribeSensor query to access the ADCP_System SensorML document that provides the list of components files and connections.

http://mvcodata.whoi.edu/downloads/sensorML/v1.0/examples/sensors/ADCP_2.2/ADCP_System.xml

SensorML files are not designed for human consumption. Note that the display of the XML documents does not work in all browsers. For example, the document cannot be viewed using the Safari browser, but it can be viewed using Mozilla Firefox. The SensorML PrettyView Uploader produces human-readable representations of the encodings and is accessible from Botts, Inc. (accessed 4/10/2014 from http://www.sensors.ws/SensorMLforms/upload.jsp). The table construct also helps developers of software packages to consume and parse the knowledge at a higher level. The Q2O waves page has example links to view the model OEM, CONDEP, QC and PROCESS files (http://q2o.whoi.edu/node/129). Below a query is shown for accessing the same ADCP_System document through the PrettyView SensorML uploader:

http://www.sensors.ws/SensorMLforms/PrettyView?url=http://mvcodata.whoi.edu/downloads/sensorML/v1.0/examples/sensors/ADCP_2.2/ADCP_System.xml

As shown in PrettyView, a parser would be developed around the higher-level terms, such as: Inputs, Output, Components. The individual terms will have references to registered, SKOS-encoded terms. Similarly, the OEM file, accessed through the PrettyView uploader depicts higher-level terms organized around: Keywords, Identifiers, Classifiers, Capabilities, Characteristics, Contacts, Documentation, Inputs, Outputs, Parameters, Components and the Instrument Model Spatial Reference Frame. Each role-based SensorML file will have a different expectation of content.

CONCLUSION

Knowledge can and should be documented at the point where it is most completely understood. Fueled by the understanding of domain experts, ontologies can be created to associate and categorize terms for discovery and harmonization. Brokers and workflows can access the knowledge and understanding through the power of the Semantic Web and the data through the syntactic and semantic interoperability of standards-based web services. Through the embedded references, the specific meaning of the terms within the data services can persist.

The use of web services for providing well-described data enables the discovery of and selective access to data, offering what you want and only what you want. And the infusion of links to registered, W3C standards-based encoded terms extends the ability to utilize domain expertise within communities of practice (CoPs) to associate and categorize information that is carried along with the web service. As we develop the ability to discover and access data across technologies (brokering), semantic mediation will provide a broader ability to incorporate knowledge and enable it to persist across the layered architecture by utilizing and harmonizing domain developed ontologies. Its implementation will enable providers to provide one service that can be harvested by multiple clients (e.g., data centers, collaborative environments).

A first step towards implementing a mechanism to capture knowledge at its source is to engage sensor manufacturers in a CoP to develop, share and promote knowledge transfer and best practices in implementing the capabilities described here. For each sensor model, manufacturers can document accurately and completely important information about the sensor and its configuration. This knowledge can be brought out of human-readable documents and binary proprietary headers for machine-to-machine capture of knowledge. Engaging sensor manufacturers has been difficult because we are not clearly and strongly defining our requirements and goals. By providing much needed tools and guidance, vendors can be guided to provide consistent, accurate and complete content. By producing content-rich SensorML documents, infused with links to their registered OWL/RDF vocabularies that they must also create, they will provide critical knowledge that can be automatically discovered and referenced as URLs. Domain experts can then harmonize the terms between vendors, while keeping more specific meanings associated within the vendor's documents. The meaning, as completely defined by the vendor, can remain associated with the data service – but the concepts can be associated to other controlled vocabularies to facilitate harmonization and conceptualization of terms in brokering services and in building tools for creating collaborative environments. SensorML is based on semantics that can define sensors (e.g., 'characteristics', 'capabilities'), events and process lineage (e.g., 'input','output','component'). These attributes can be used in any observing system and by promoting their implementation consistently and across domains, interdisciplinary collaborators can have an expectation of common content and can build upon shared tools and capabilities.

From field operators and data managers, the best knowledge of the details relating to deployment and events that affect data quality must be captured. Terms describing processes should include references to authoritative sources of algorithms and processing methods to provide understanding of limitations and qualifications of observations. By providing web-based tools for creation and validation of content-rich vocabularies and SensorML documents, each agent in the data lifecycle can be encouraged to create controlled vocabularies and SensorML documents that reference the terms. When appropriate, the documents could be built upon the vendor documents and the CoP ontologies.

These activities will require the creation of SensorML editors/generators that guide consistent content generation and a facility or capabilities within existing infrastructures where the SensorML documents can be registered, similar to what is done for the MMI vocabularies and ontologies. This will provide persistence, versioning and discovery of the documents. A cross-domain focus in the implementation of these technologies is needed to assure shared knowledge and best practices. The use of the Semantic Web and the OGC® SOS are not domain specific technologies. By building cross-domain CoPs, we will enable a shared expectation of content and capabilities.

Beneficiaries of the detailed and accurate provenance include federal and international data centers. By enabling data creators a means of easily documenting sensor specific information and process lineage, it will be more easily integrated into our systems of systems. Also, this methodology will enhance our ability to respond to extreme events. For example, when the BP Deep Horizon blow-out resulted in an urgent need to evaluate the dynamics of the Gulf of Mexico and to assess current conditions, access to historical and real-time data was required. Issues of discoverability, access and trust hindered the activity. What are these data? Who own it? How do I contact the owner? When was this sensor calibrated? Are there any data in this area? Rapid response will become more important as we are affected by climate change and our exploration for resources force us to go deeper and take on more risk. Rapid response demands well-documented and accurate data, which are easily accessible and discoverable.

The resulting vocabularies, ontologies and web services will provide the framework needed for the development of collaborative environments and brokering services that enhance the discovery of and facilitate the access to data with sufficient understanding to build trust in our global data assets.

ACKNOWLEDGMENT

Funding for the development of the Q2O model was provided under NOAA's Cooperative Agreement FY 2007 Regional Integrated Ocean Observing System Development (NOS-CSC-2007-2000875), 2008-2011. The brokering 'hack-a-thon' activities referenced in the SWE section were funded through two NSF EarthCube concept awards (EAR 1239702, 1239841), 2012-2013. Contributions to this chapter were made through funding provided by the Center for Ocean, Seafloor, and Marine Observing Systems (COSMOS) of the Woods Hole Oceanographic Institution.

REFERENCES

Allemang, D., & Hendler, J. (2011). Inferring Class Relationships. In *Semantic web for the working ontologist: Effective modeling in RDFS and OWL* (2nd ed., pp. 273–276). Waltham, MA: Morgan Kaufmann Publishers.

Bermudez, L. (2007). Vocabulary integration environment. *VINE Revista Avance en Sistemas e Informatica*, *4*(2), 53–60.

Boldrini, E., Craglia, M., Mazzetti, P., & Nativi, S. (2015). The Brokering Approach for enabling collaborative scientific research. In P. Diviacco, P. Fox, C. Pshenichny, & A. Leadbetter (Eds.), *Collaborative Knowledge in Scientific Research Networks*. Hershey, PA: IGI Global.

Botts, M., Percivall, G., Reed, C., & Davidson, J. (2008). *OGC® sensor web enablement: Overview and high level architecture*. Lecture Notes in Computer Science, 4540, 175–190. doi:10.1007/978-3-540-79996-2_10

Butterfield, M. L., Pearlman, J. S., & Vickroy, S. C. (2008). A system-of-systems engineering GEOSS: Architectural approach. *Systems Journal, IEEE*, *2*(3), 321–332. doi:10.1109/JSYST.2008.925973

Chave, A. D., Ampe, T., Arrott, M., Graybeal, J., James, M., & Meisinger, M., … Schofield O. (2011). Cyberinfrastructure for the US Ocean Observatories Initiative. Underwater Technology (UT), *2011 IEEE Symposium 2011 Workshop on Scientific Use of Submarine Cables and Related Technologies (SSC)*. doi:10.1109/UT.2011.5774109

del Rio, J., Toma, D. M., O'Reilly, T. C., Broring, A., Dana, D. R., & Bache, F. et al. (2013). Standards-based plug & work for instruments in ocean observing systems. *IEEE Journal of Oceanic Engineering*, *99*, 1–14. doi:10.1109/JOE.2013.2273277

DeRose, S., Maler, E., Orchard, D., & Walsh, N. (2010). *XML linking language (XLink) version 1.1*. Retrieved 1/28/2014 from http://www.w3.org/TR/xlink11/#concepts

Diviacco, P. (2015). Reconciling Knowledge and Collaborative e-Research. In P. Diviacco, P. Fox, C. Pshenichny, & A. Leadbetter (Eds.), *Collaborative Knowledge in Scientific Research Networks*. Hershey, PA: IGI Global.

Diviacco, P., Lowry, R., & Schaap, D. (2011). Marine seismic metadata for an integrated European scale data infrastructure: the FP7 Geo-Seas project. *Bollettino di Geofisica Teorica ed Applicata*. Retrieved 01/2013 from http://www.odip.org/documents/odip/downloads/23/bgta0051_diviacco.pdf

Gomez, B., Pearthree, G., & Cutcher-Gershenfeld, J. (2013). Cyber-infrastructure and synergistic opportunities across the geosciences. *Eos, Transactions, American Geophysical Union, 94*(49), 473. doi:10.1002/2013EO490005

Graybeal, J., Isenor, A. W., & Rueda, C. (2011). Semantic mediation of vocabularies for ocean observing systems. *Computers & Geosciences, 40*, 120–131. doi:10.1016/j.cageo.2011.08.002

Hankin, S., Blower, J., Carval, T., Casey, K., Donlon, C., & Lauret, O. ... Harlan, J. (2010). NetCDF-CF-OPeNDAP: Standards for ocean data interoperability and object lessons for community data standards processes. In *Proceedings of OceanObs'09: Sustained Ocean Observations and Information for Society* (Vol. 2). Venice, Italy: ESA Publication WPP-306. doi:10.5270/OceanObs09.cwp.41

Isaac, A., & Summers, E. (2009). *SKOS Simple Knowledge Organization System Primer*. Retrieved 1/28/2014 from http://www.w3.org/TR/2009/NOTE-skos-primer-20090818

Juan, A., Daradoumis, T., Roca, M., Grasman, S. E., & Faulin, J. (2012). *Collaborative and distributed E-Research: Innovations in Technologies, Strategies and Applications* (pp. 1–419). Hershey, PA: IGI Global. doi:10.4018/978-1-46660-125-3

Leadbetter, A. (2015). Examining trust in collaborative research networks. In *Collaborative Knowledge in Scientific Research Networks*. IGI Global Press.

Leadbetter, A., Arko, R., Chandler, C., Shepherd, A., & Lowry, R. (2015). Loose integration of local information to generate collaborative marine science knowledge. In P. Diviacco, P. Fox, C. Pshenichny, & A. Leadbetter (Eds.), *Collaborative Knowledge in Scientific Research Networks*. Hershey, PA: IGI Global.

Mazella, G. (2015). Knowledge building and computer tools. In P. Diviacco, P. Fox, C. Pshenichny, & A. Leadbetter (Eds.), *Collaborative Knowledge in Scientific Research Networks*. Hershey, PA: IGI Global.

National Research Council. (2003). *Government data centers: meeting increasing demands*. Washington, DC: The National Academies Press. Retrieved 4/10/2014 from http://www.nap.edu/catalog/10664.html

National Research Council. (2012). *The future of scientific knowledge discovery in open networked environments: Summary of a workshop*. Washington, DC: The National Academies Press. Retrieved 4/10/2014 from http://www.nap.edu/catalog/18258.html

Nativi, S., Craglia, M., & Pearlman, J. L. (2013). Earth science infrastructures interoperability: The brokering approach. *IEEE Journal of Selected Topics in Applied Earth Observations and Remote Sensing, 6*(3), 1118–1129. doi:10.1109/JSTARS.2013.2243113

NOAA, National Coastal Data Development Center. (2013). *Introduction to Geospatial Metadata – ISO 191** Metadata*. Retrieved 4/28/2014 from ftp://ftp.ncddc.noaa.gov/pub/Metadata/Online_ISO_Training/Intro_to_ISO/presentations/2_ISO_101.pptx

Rueda, C., Bermudez, L., & Fredericks, J. (2009). The MMI Ontology Registry and Repository: A portal for marine metadata interoperability. *OCEANS 2009, MTS/IEEE Biloxi – Marine Technology for Our Future: Global and Local Challenges*. Retrieved from http://ieeexplore.ieee.org/stamp/stamp.jsp?tp=&arnumber=5422206&isnumber=5422059

Sorribas, J., Olive, J., Diviacco, P., & Bermudez, L. (2012). Representing oceanographic vessels by means of sensor web enablement standards. In *Proceedings of the First ACM SIGSPATIAL Workshop on Sensor Web Enablement*. ACM. doi:10.1145/2451716.2451717

Wenger, E. (1998). *Communities of practice: Learning, meaning and identity*. Cambridge University Press. doi:10.1017/CBO9780511803932

Wenger, E., McDermott, R., & Snyder, W. M. (2002). Cultivating communities of practice: A guide to managing knowledge. Harvard Business School Press.

Willis, Z. (2012). US Integrated Ocean Observing System (IOOS®) – Delivering benefits to science and society. *OCEANS 2012, Yeosu*. Doi: 10.1109/OCEANS-Yeosu.2012.6263613

World Wide Web Consortium (W3C). (n.d.). *Semantic web*. Retrieved 1/28/2014 from http://www.w3.org/standards/semanticweb

KEY TERMS AND DEFINITIONS

Brokering: Mechanisms used to translate information for data consumers across standards-based frameworks. Brokering can be used to harmonize meaning for discovery (semantics), translate formats or provide access services.

Community of Practice (CoP): A self-organizing group that develops and promotes the use of best practices within a community.

Domain Expert: An individual with knowledge of scientific requirements and process assessment issues. A domain expert, or subject matter expert, can define what is needed to enable use of data and data services in a meaningful way.

Interoperability: The ability to share knowledge (data/metadata) based on the expectation of common content and a disciplined adherence to the use of standards. Syntactic interoperability is based upon encodings using well-defined packaging (e.g., XML, SOS). Semantic interoperability is based on the ability to have relationships and classifications of the content through the adoption of the Semantic Web technologies.

Workflow: A machine-harvestable description of processing steps that defines how observations came to be. A workflow can also be an executable document that generates output from input. Workflow can be described by defining input, algorithms, parameters and output. Defining a workflow enables well-documented, repeatable scientific research.

Chapter 14
The Brokering Approach for Enabling Collaborative Scientific Research

Enrico Boldrini
CNR - Institute of Atmospheric Pollution Research, Florence, Italy

Paolo Mazzetti
CNR - Institute of Atmospheric Pollution Research, Florence, Italy

Max Craglia
European Commission - Joint Research Centre, Ispra, Italy

Stefano Nativi
CNR - Institute of Atmospheric Pollution Research, Florence, Italy

ABSTRACT

Collaborative research poses important challenges and opportunities to scientific communities, such as the realization of effective interactions between different groups and organizations. Data sharing is growing increasingly important, as repositories are growing in size, number, and variety in the different domains. The underlying challenge is interoperability: between data, services, applications, models, and ultimately people. Abstraction and standardization (e.g. of data and services) have historically played an essential role in enabling interoperability. However, they proved not to be sufficient alone to let large-scale collaborative research thrive. For different spheres (e.g. social/economic realms), intermediation approaches were successfully applied to pursue similar goals in the so-called brokering approach. This chapter argues that brokering is useful to enable collaborative research as well, both by addressing technological interoperability and supporting socio-organizational challenges. A technological brokering framework, implemented to help the earth system science collaborative research, is finally presented, along with success stories.

COLLABORATIVE RESEARCH CHALLENGES

Collaborative studies can be characterized and categorized according to the organizations involved and their goals. Organizations span from academic to private entities and collaborations can be either tight or loose. In keeping with that, taxonomies distinguish among different types of research collaborations (e.g. disciplinary, multidisciplinary, participatory, interdisciplinary) (Balsiger, 2004) (Tress, Tress, & Fry, 2006). Research groups can belong to one or more organizations and Communities of practice (CoP). They may setup one

DOI: 10.4018/978-1-4666-6567-5.ch014

Copyright © 2015, IGI Global. Copying or distributing in print or electronic forms without written permission of IGI Global is prohibited.

or more of these collaborative research types, depending on the scale and scope of their research.

Relations among different organizations (or CoPs) have been often viewed as a high burden with unclear benefits. The prominent motivations for general inter-organization relations to emerge are an internal need for resources or the achievement of external benefits (Van de Ven, 1976). Bos et al. (Bos et al., 2007) identify three types of barriers specific to large scientific collaborations (that are to be added to the normal challenges of working across distance):

- Scientific knowledge is difficult to aggregate (e.g. information semantics and data structures heterogeneity);
- Scientists usually work independently (e.g. large meetings, hierarchical structures, resource transfers, written contracts are perceived as barriers);
- Institutional related barriers (e.g. legal, economic, cultural issues).

In addition to these organizational and human/cultural interoperability challenges, in the present Web era, issues typical of complex distributed systems are to be taken into account as well; especially, when the research gets larger such as, building global and multidisciplinary e-infrastructures. Technological and semantics interoperability challenges are mainly related to the e-infrastructure:

- **Scale:** It is increasingly difficult to make a large number of systems and organizations work together. Organization patterns need to be applied in order to guide the activities of the entities involved.
- **Heterogeneity:** It can be difficult to make systems and organizations that are different work together. Mediation between parties is often a fundamental requirement. For instance, the real challenge in complex systems such as the Internet and The World Wide Web is not the number of entities involved, but their heterogeneity (Gordon & Grace, 2012). Standards and policies issued by dedicated international organizations (such as the World Wide Web Consortium) ease such interoperability processes to take place.

Due to the growing importance of (big) data sharing, great attention and effort have been dedicated to address technological and semantic interoperability, overlooking important organizational and human/cultural barriers. However, data and interoperability scientists are not expert in data content. Therefore for example, to develop cross-domain data models or semantics, organizational and cultural barriers must be addressed in addition to data structure mapping. Data sharing is another noticeable example, where the technological challenges are not as significant as policy ones. In turn, data sharing represents a fundamental building block for subsequent achievements in collaborative research requiring both technological and cultural interoperabilities, such as:

- Creation of inter-domain models and applications making use of the datasets;
- Creation and composition of workflows orchestrating heterogeneous models and datasets;
- Creation of web-based platforms for collaborative research.

The rest of the chapter argues about the role that some technological interoperability solutions can play to facilitate domain experts lowering human and cultural interoperability barriers.

Data Sharing and Interoperability

According to the new "fourth paradigm of discovery" (Hey, Tansley, & Tolle, 2009), data-intensive science is playing an increasingly role to enable scientific breakthroughs. Therefore, scientists have looked at data sharing as a first important objective to re-use scientific research (products) developed by other colleagues and establish collaborations.

Interoperability is commonly defined as: "a property of a product or system, whose interfaces are completely understood, to work with other products or systems, present or future, without any restricted access or implementation" [IEEE] (Geraci et al., 1991).

Interoperability can be described by means of different levels, as multiple models, such as the Levels of Conceptual Interoperability Model (LCIM) (Turnitsa, 2005) (Tolk, 2006) have documented in the literature.

Data sharing between two different systems eager to collaborate is a complex interoperability task, involving different levels. LCIM shown in Figure 1 distinguishes 6 interoperability levels, with an increasing capability of interoperation from the lowest to the highest level. Data sharing interoperability can be described according to these levels:

- **Level 0 – No interoperability:** At this level no data is shared, the involved systems are just connected nodes of a network.
- **Levels 1-2 – Integretability:** Sharing of files over a network can be reached at this level. Such type of interoperability is also called Integretability of systems *(Technical-Syntactic Interoperability).* Prior to the Web era, sharing of files between different systems was a difficult task because of technical problems (e.g. different file encodings in different operating systems, such as big endian and little

Figure 1. Levels hierarchy of conceptual interoperability showing an increasing capability for interoperation from the lowest to the highest level. Different enablers operate at the different levels. Figure adapted from Turnitsa (2005).

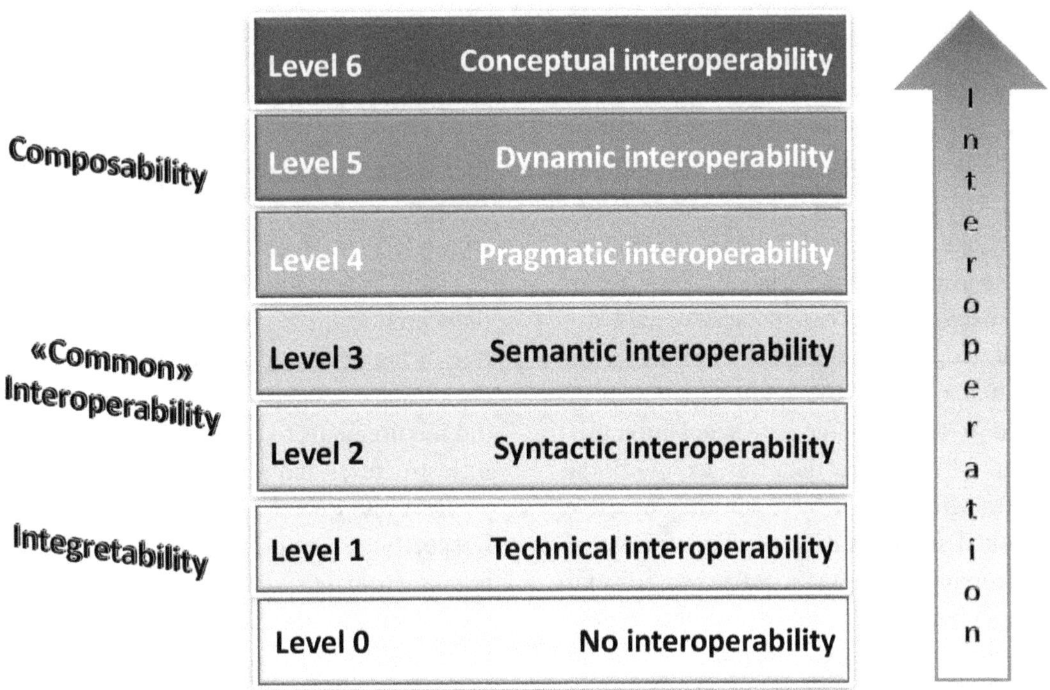

endian). Currently file sharing is easily achieved using common Internet protocols defining a specific syntax, such as HTTP and FTP.

- **Levels 2-3 – Interoperability:** The commonly defined interoperability between systems can be subdivided in three different levels *(Syntactic-Semantic Interoperability).* Data is usually being stored in files according to standard formats (defining both syntax and semantics), such as NetCDF-CF, GeoTiff and Shape files. Formats such as XML permit also the design of a well distinct data structure (semantics), described by a separated XML schema. This is the case of Geographic Markup Language (GML), a XML encoding of geographical features according to a precise data structure formalized by OGC. Brokering frameworks work at this level to make possible sharing of data between systems which are using heterogeneous data formats and structures (syntax and semantics) by taking charge of the needed conversions.
- **Levels 4-5-6 – Composability:** At the highest levels of *Pragmatic-Dynamic-Conceptual Interoperability*, data sharing is limited by increasing semantics difficulties. Different systems (and communities) associate different terms (and semantics) to the available data variables. These terms are usually collected in thesauri and can relate to others to express complex conceptual models. The models can then be used to document the data acquisition process (why and how data was acquired), the meaning of dataset variables, their possible use and so on; sharing of this information (occurring in Pragmatic interoperability) is fundamental to enable further data exploitation. Dynamic Interoperability happens when the effect of information exchange is unambiguously defined and shared as well. At present, only humans are able to act as data sharing mediators at these levels. By understanding the different models and aligning the corresponding concepts they finally realize Conceptual Interoperability. The process can be eased by adequate "mediation policies", such as the creation of working groups where experts from different domains can meet and discuss. Standardization is nowadays occurring even at these levels (e.g. INSPIRE directive (European Union, 2007) mandates specific thematic conceptual models to be implemented by the member states).

In the following we will concentrate on data sharing at syntax and semantics levels (data formats and structures): at those levels software tools such as brokering frameworks (which are the subject of the chapter) operate; they are enablers for the higher levels of data sharing.

Data sharing at such levels doesn't come anyway without difficulties.

The different data sources being deployed in the scientific community are found to be: autonomous, heterogeneous, distributed, and in need for change because of multiple drivers. In fact, data sharing represents a high burden for scientists: they feel this task to be out of their "core" activities –for example, facing data formats heterogeneity when trying to share their datasets (Meyer 2009).

To help scientists in publishing and sharing their products, data interoperability emerged as one of the most important areas: in the last two decades, it has been considered one of the essential requirements to design and develop research infrastructures and systems.

Data interoperability has been firstly pursued by specifying new and better interfaces to discover, access, evaluate, and use scientific data. Technological standards and (mandatory) guidelines have been drafted to fully describe data content and structures, as well as the internet-based protocols to share them.

Afterwards, the burden of implementing data interoperability has typically been charged to one of the two actors:

1. A data provider desiring to publish its datasets to the largest set of client applications needs to implement service interfaces compliant with multiple, heterogeneous standards.
2. A client application desiring to access datasets from the available data providers needs to implement multiple, heterogeneous standards.

In addition, it seems unrealistic that a limited number of effective standards –or a single data policy– will emerge as the solution for the data sharing problem, although several international efforts are going in this sense (e.g. international standardization bodies and international directives, such as EU INSPIRE). There is evidence for a big contrasting trend: the autonomous and heterogeneous nature of scientific organizations and communities; the heterogeneity of scientific domains; Internet and data technologies are fast moving areas and require a quick and flexible process to be fruitfully adopted. Traditional data integration approaches, such as federated schema systems and data warehouses, have failed to meet the requirements of a constantly changing organizational and technical environment (Zhu, et al., 2004). Organizational and multidisciplinary complexity, as well as sustainability, seem to be the main impediments.

It seems more realistic that the ecosystem of standards, (best) practices and data policies will constantly continue to flourish and evolve according to the new requirements posed by the different scientific domains and organizations. Each community can be thought to form a large island, in which common standards, practices and data policies are in use, hence scientific data can easily be shared within.

Not surprisingly, social and organizational challenges will play a key role to achieve data interoperability, as outlined by an investigation on the factors influencing Spatial Data Sharing (SDS) –conducted by Omran, Bregt and Crompvoets (Omran, Bregt, & Crompvoets, 2007). The research considered 23 hypotheses, such as: the positive relationship between attitude and the intention for SDS is stronger in collectivist cultures than in individualist cultures; trust positively influences egalitarian organizations to share spatial data; data policies and legal barriers, etc. Nevertheless, traditionally, the main focus of the information system literature on globally distributed teams has been on technical aspects rather than social aspects such as formal and informal communications, trust, motivation and social ties (Kotlarsky & Oshri, 2005).

Recently, the European Commission has considered organizational interoperability as important as technological and semantics ones for data infrastructure interoperability (Kubicek & Cimander, 2009). Organizational interoperability challenges include:

- Modeling organizational processes;
- Aligning information architectures with organizational goals;
- Helping organizational processes to co-operate;
- Recognizing and aligning data policies;
- Legal, cultural, and people harmonization.

Social and organizational aspects proved to be the main ones to establish research collaborations in complex and uncontrolled environments –e.g. global, multi-disciplinary, and voluntary-based programs.

Therefore, in addition to the standardization (and implementation) process managed by the different CoPs, there is a clear need to introduce a complementary method which is capable to support technological interoperability leaving organizations autonomous and interconnecting their standards. In other words, there is the need to implement a mediation process. Mediation processes are carried out by actors called Brokers.

Achieving interoperability through a Broker approach includes many benefits, such as better and dynamic maintainability of intermediation in a specialized layer and relief of implementation burden from both data provider and consumer organizations.

MEDIATION PROCESSES AND BROKERS

Interoperability between different organizations, communities, and ultimately between people is a challenge common to different contexts and realms. The mediation approach and brokers have been long used to address interoperability problems occurring in the context of socio-organizational, technological and scientific disciplines.

Brokers are distinct entities which apply the mediation paradigm to let different parties work together. They stand between the parties without belonging to a specific one and use mediation to enable technological, semantics and organizational interoperability.

The figure of brokers (or mediators) is long standing in the broad socio-organizational and technological context. A common definition of broker is "a person who buys and sells goods or assets for others". (Oxford University Press, 2014).

The next sections briefly describes the key role played by the figure of brokers (or mediators) in diverse contexts –from sociological to technological realms– and in its salient features.

Finally, gathering from these experiences, the brokering approach is applied to solve technological problems in the collaborative research context and to support as well organizational challenges.

Sociological and Organizational Context

The figure of brokers often occupy important roles in the societies; it is sufficient to think about the role of cultural mediators to resolve conflicts or simply the role of financial brokers (or mediators) and retailers in economy. A review of social mediators is provided by Meyer (Meyer 2010) covering spheres such as: economy (an agent, promoter, dealer, fixer, trader); politics (a diplomat, mediator, negotiator); information (someone who knows how to access or acquire information); education and dissemination (a proactive facilitator who connects people, and establishes the conditions to create something new or add value to something that already exists (Jackson, 2003)).

A broker is an intellectual worker enabling, catalyzing and getting things moving (Osborne, 2004). Brokering activities require a variety of tools, such as organizing seminars or meetings (Sverrison, 2001), developing databases, and producing plain-language booklets (Kramer & Wells, 2005). Mediators must often avoid two opposite tendencies: being pulled in to become full members and being rejected as intruders. Indeed, their contributions lie precisely in being neither in nor out. Mediating requires an ability to manage carefully the coexistence of membership and non-membership, yielding enough distance to bring a different perspective, but also enough legitimacy to be listened to (Wenger E., 1998). Brokers represent indeed a third party entity that could however include members of the different groups: these members can especially be good at mediation, leveraging their original domain expertise.

Brokers operate at the interface between institutions, their job is to move knowledge around and create connections between researchers and their various audiences (Meyer 2010). Brokering involves processes of translation, coordination, and alignment between different models and perspectives. It also requires the ability to link practices by facilitating transactions between them (Meyer 2010). Brokers receive specialized information from the diverse research groups they connect, so they can serve as translators and resolve conflicts occurring because of different scientific assumptions and jargon (Gray, 2008).

In summary, important features characterize a broker:

1. Brokers make possible exchange between two or more parties.
2. Brokers translate between a common *language (model)* and heterogeneous specific *languages (models)*. The creation of a common language and the mapping process are fundamental steps, they must be carried on in conjunction with experts of the different domains.
3. Brokers have a special expertise, which pose them between different groups (or systems) without being part of them –i.e. they constitute a third-party entity (that could still be composed by members of the different groups).

Brokers hold key structural positions in networks (Long, Cunningham, & Braithwaite, 2013) playing an important connectivity function[1]. A broker acts as an intermediary between two unlinked actors/clusters, aiming to facilitate some transaction, resolve conflict or increase power or social capital; a knowledge broker keeps separate groups in a network coordinated or informed, aiming to improve network information flows and prevent fragmentation (Long, Cunningham, & Braithwaite, 2013). Broker nodes link groups of actors who are not otherwise connected to each other (Gray, 2008). Brokers occupy "structural holes" at the crossroads between groups of actors. Thus, broker nodes intervene by building linkages and increasing information flow among previously unrelated parties.

Technological Context

Learning from the successful mediation approaches and intuitions found in social networks, brokers have been used for large technological systems, too.

In the infrastructures domain, mediators are often referred as technological gateways (Knobel, 2007). They are commonly introduced during the consolidation phase of a large technological system (LTS) formation (Knobel, 2007). They are used to resolve competition among viable technological systems and make different networks interoperate. Examples of current LTS are the AC/DC power converters for consumer electronics and telephone adapters for international travel. Another successful example comes from the healthcare where data integration between several institutes (e.g. primary care practice, social services, hospitals) is achieved thanks to a software architecture leveraging mediation (Budgen, Rigby, Brereton, & Turner, 2007).

Also software engineering adopted mediation as a good practice to solve complex interaction problems –dealing with software entities, in this case. In software engineering, a good practice (i.e. a general reusable solution to a commonly occurring problem) is called a design pattern. The *Mediator* pattern is used when different objects (called colleagues) need to interact between each other. Without this pattern, all the software objects are burdened by the load of implementing the different interfaces of all the other colleagues (Vlissides, Helm, Johnson, & Gamma, 1995).

The Mediator pattern has the following characteristics (Vlissides, Helm, Johnson, & Gamma, 1995):

Figure 2. Structure of the Mediator Pattern. Figure adapted from Vlissides, Helm, Johnson, & Gamma (1995).

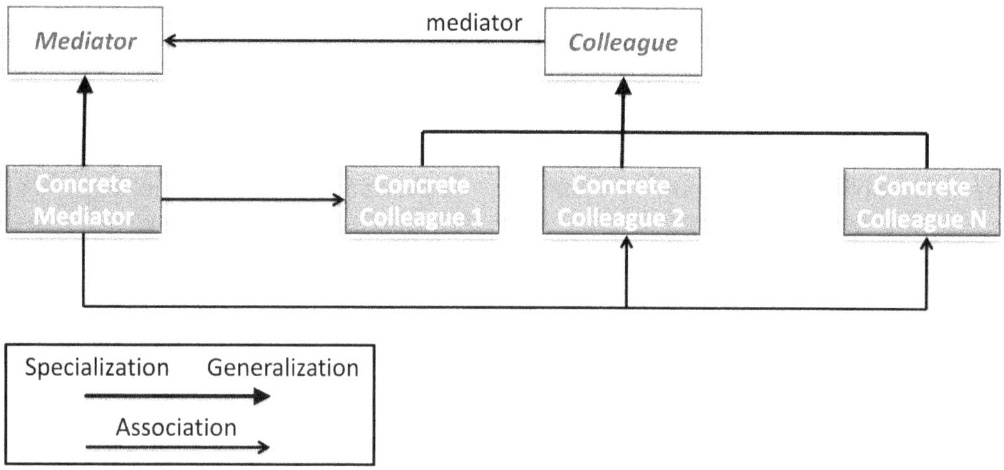

1. A mediator localizes behavior that otherwise would be distributed among several objects.
2. A mediator promotes loose coupling between colleagues.
3. A mediator replaces many-to-many interactions with one-to-many interactions between the mediator and its colleagues. These are easier to understand, maintain, and extend.
4. It highlights interaction between objects, helping to understand better the system.
5. It centralizes control.

The mediation pattern can be applied to data systems integration–i.e. specific software objects. In a naïve solution, each system implements mediators from and to each other system, so that data entries coming from a sender system are normalized according to the schema of the recipient system. This solution shows evident scalability problems. To solve this issue a brokering pattern was introduced (Clark, Krumm, & Bieleski, 1992) making use of an external mediator system, as depicted in Figure 3.

The brokering pattern is characterized by the following principles:

- Each system only interfaces to the broker and does not exchange data directly with other systems. The Broker takes care of forwarding relevant data to all other systems.
- Each system can interface to the broker in whatever format is most easily implemented –this is important to leverage legacy systems.
- Data received by the broker is examined, translated into a neutral format, and forwarded to the appropriate systems.
- Each legacy system still maintains control (e.g. over its own database).

A broker solution offers the following advantages with respect to the naïve fully interconnected approach, see Figure 3:

- **Maintainability:** Maintenance is primarily required on only one system: the Broker.

Figure 3. Complexity comparison between an interfaced and a brokered approach. As the number of systems to be connected increases, the naïve fully interfaced solution is quickly outperformed by the broker solution. The number of connections grows quadratically with the number of nodes composing the system in the first case, linearly in the latter (Clark, Krumm, & Bieleski, 1992).

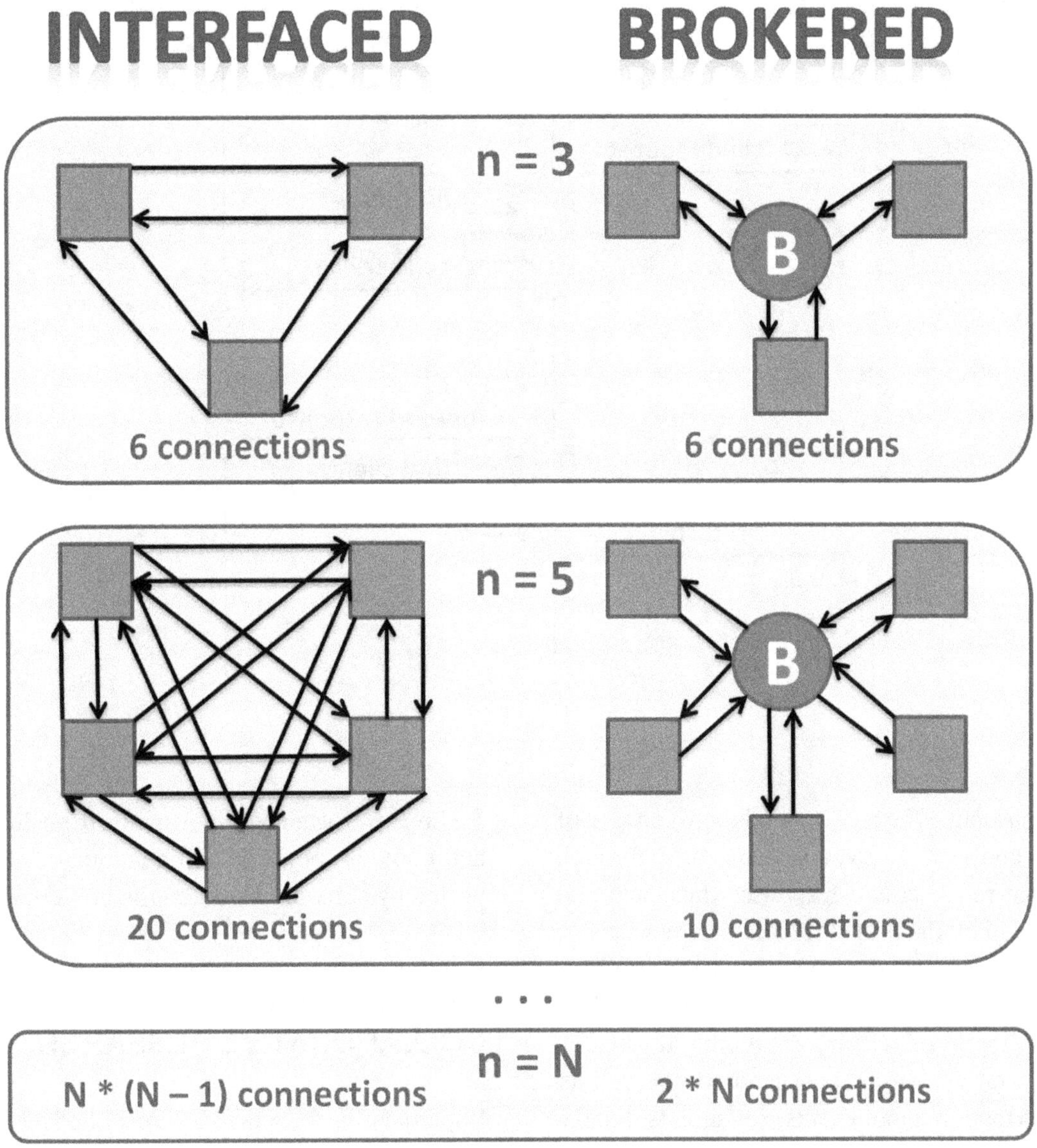

The number of interfaces grows linearly with the number of systems being interfaced: 2n instead of n*(n-1) interfaces required by a fully connected system.

- **Extensibility:** Adding another system to the existing network require the addition of only 2 interfaces.

The Broker Pattern for distributed software systems represents a further elaboration (Schmidt, Stal, Rohnert, & Buschmann, 2013) (see Figure 4). In this pattern a broker component is used to coordinate communication and transmission of requests, results and exceptions among distributed and decoupled components.

Figure 4. Sequence diagram of the Broker Pattern. The broker component coordinates communication between distributed and decoupled clients and servers. Figure adapted from: Schmidt, Stal, Rohnert, & Buschmann (2013).

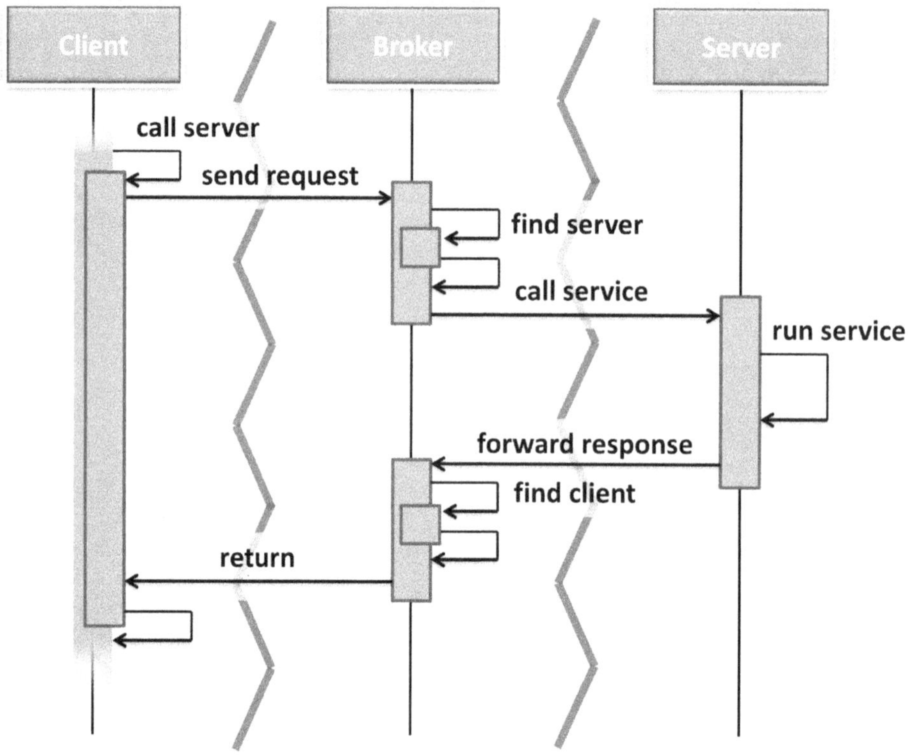

In particular the broker stands between a set of clients and a set of servers which interact through message requests and responses translated and forwarded by the broker middleware service.

A sample application of the broker pattern from (Schmidt, Stal, Rohnert, & Buschmann, 2013) is the developing of a city information system (CIS). In this scenario, a user client needs to retrieve a set of information about events, restaurants, hotels, monuments and transportation. These information is typically distributed on a set of heterogeneous servers, which makes complex its access: a broker service is hence used in this case to provide a uniform and convenient interface to the client.

This example shows as well how typical human activities requiring intermediation (such as finding heterogeneous information from different, autonomous sources) can be automatically supported by software based on the broker pattern.

A BROKERING APPROACH FOR COLLABORATIVE RESEARCH

Implementing the previously described mediation principles and extending them –to consider System of Systems (SoS) and Network of Networks (NoN) requirements– an advanced brokering approach was introduced to achieve large (i.e. multi-organizational) and multidisciplinary research infrastructures (Nativi, Craglia, & Pearlman, The brokering approach for multidisciplinary interoperability: a position paper, 2012). Naturally the final goal of these complex infrastructures is to enable collaborative research (Vaccari, Craglia, Fugazza, Nativi, & Santoro, 2012).

To put into action SoS and NoN, the main principles of this advanced brokering approach must be considered (Nativi, Craglia, & Pearlman, 2013):

1. **Autonomy:** Keep the existing disciplinary infrastructures as autonomous as possible, not asking them to implement any "more general" service bus.
2. **Subsidiarity:** Supplement but not supplant disciplinary infrastructure mandates and governance arrangements by interconnecting and mediating their service buses.
3. **Interconnection:** Build incrementally on existing infrastructures and introduce distribution and mediation functionalities to interconnect the heterogeneous service buses characterizing any domain specific or other infrastructure.
4. **Low Entry Barrier:** Assure a low entry barrier for both users and resource providers of any disciplinary infrastructure.
5. **Flexibility:** Be flexible enough to accommodate existing and future information systems and information technologies that will augment the service bus implemented by any discipline.
6. **Effectiveness:** Address the full range of information exchange needs (discovery, access, semantics, workflow, etc.).

Two additional principles have been identified and proved to be important:

- **Transparency:** In spite of technologies and communities' changes, a brokering framework must remain transparent to both resource providers and end users. This is the reason to introduce an actual framework. The brokering framework takes care of shifting requirements relieving the participating organizations to take actions.
- **Separation of Concerns:** The brokering framework is maintained by a brokering group acting as an inter-mediator between provider and consumer organizations and systems (i.e. a three tier approach not only for the technological aspects but also for the organizational ones). Brokers are facilitators for information and knowledge communication and exchange, with the final aim of thriving collaboration –in keeping with that, a brokered SoS/NoN is an example of multidisciplinary research.

The advanced brokering based on these extended set of principles has been successfully applied in the context of Earth Sciences to achieve large collaborative infrastructures and research.

SUCCESSFUL BROKERAGE IN THE GEOSCIENCES

The GEO Discovery and Access Broker

The advanced brokering approach was first applied in the context of the European FP7 project EuroGEOSS (a European contribution to GEOSS) (Vaccari, Craglia, Fugazza, Nativi, & Santoro, 2012). The project has investigated in depth the brokering approach for building multi-disciplinary SoS. It implemented a brokering framework consisting of a Discovery Broker, a Semantic Discovery Broker, and an Access Broker –taking care of disparate data discovery, semantic retrieval, and access, respectively.

The framework builds on the GI-cat (Nativi & Bigagli, Discovery, mediation, and access services for Earth Observation data, 2009) (Bigagli, Nativi, Mazzetti, & Villoresi, 2004), GI-sem, and

GI-axe technologies (Nativi, Craglia, & Pearlman, 2013), developed by the Earth and Space Science Informatics Laboratory (ESSI-Lab) of the National Research Council of Italy (CNR).

The added value of the EuroGEOSS discovery broker was successfully demonstrated at the 2011 GEO Plenary in Istanbul (GEO, 2011). By adopting the brokering approach and integrating the EuroGEOSS broker as a key component of the GEOSS Common Infrastructure (GCI), it was possible to significantly extend the pool of scientific data products (such as satellite and sensor acquisitions) available through GEOSS –from a few thousand to about 2 million (GEO, 2011). This was possible because the broker overcomes the need of different communities to make changes to their standards and specifications in order to participate in GEOSS.

The Global Earth Observation System of Systems (GEOSS) has been envisioned by the group of the eight most industrialized countries in 2003 (Group on Earth Observations, 2005) and provides a framework to integrate the Earth Observation efforts of the 90 governments, plus the European Commission, which are GEO-members and the 77 participating organizations.

A major role of GEOSS is to promote scientific connections and interactions between the observation systems that constitute the system of systems and to address some of the scientific challenges identified by the International Council for Science (ICSU) with a particular focus on nine societal benefit areas (Disasters, Health, Energy, Climate, Agriculture, Ecosystems, Biodiversity, Water, and Weather).

To pursue such ambitious goals, GEO has promoted a GEOSS Common Infrastructure (GCI) enabling collaborative research.

Through the GEOSS Common Infrastructure (GCI), GEOSS resources, including Earth observation data (satellite, in situ, models), information services, standards and best practices, can be searched, discovered and accessed by scientists, policy leaders, decision makers, and those who develop and provide information services across the entire spectrum of users (Group on Earth Observations, 2011-2013).

The current operational GCI configuration includes the evolution of the EuroGEOSS Brokering framework: the GEOSS Discovery & Access Broker (GEO-DAB) (Nativi, Craglia, & Pearlman, 2013) –see Figure 5.

The recent "GEO Report on Progress" prepared for the Geneva Ministerial Summit, held in January 2014, recognizes the key role played by GEO-DAB in GEOSS:

In 2013, the number of discoverable resources available via the GCI increased from 14 million to 65 million. Of these discoverable resources, more than 50 million (products, files, images) are tagged as GEOSS DataCORE, meaning that they can be accessed without restriction and free of charge or for the cost of reproduction. The GEOSS Portal is the single Internet gateway to the comprehensive data produced by the GEOSS community (see Figure 5). A new version of the portal was launched in December 2013. Developed by the European Space Agency (ESA), this new Portal makes it easier and faster to integrate diverse data sets; identify relevant data and portals of contributing systems; access models and other decision-support tools; and preview data before downloading. The Portal is powered by the Discovery & Access Broker (DAB) developed by CNR. The DAB connects users to an ever-increasing number of databases and information systems around the world.[2] *(Group on Earth Observations, 2011-2013).*

GEO-DAB allows GEOSS to read and mediate between the diverse standards and specifications already in use by the heterogeneous scientific Communities. By building transparent bridges

Figure 5. The GEOSS Portal is the single Internet gateway to the comprehensive data produced by the GEOSS community. GEO-DAB is the discovery and access broker enabling this framework. A recent report on the GCI achievements is included in the GEO Report on Progress at the Geneva Ministerial Summit (Group on Earth Observations, 2011-2013).

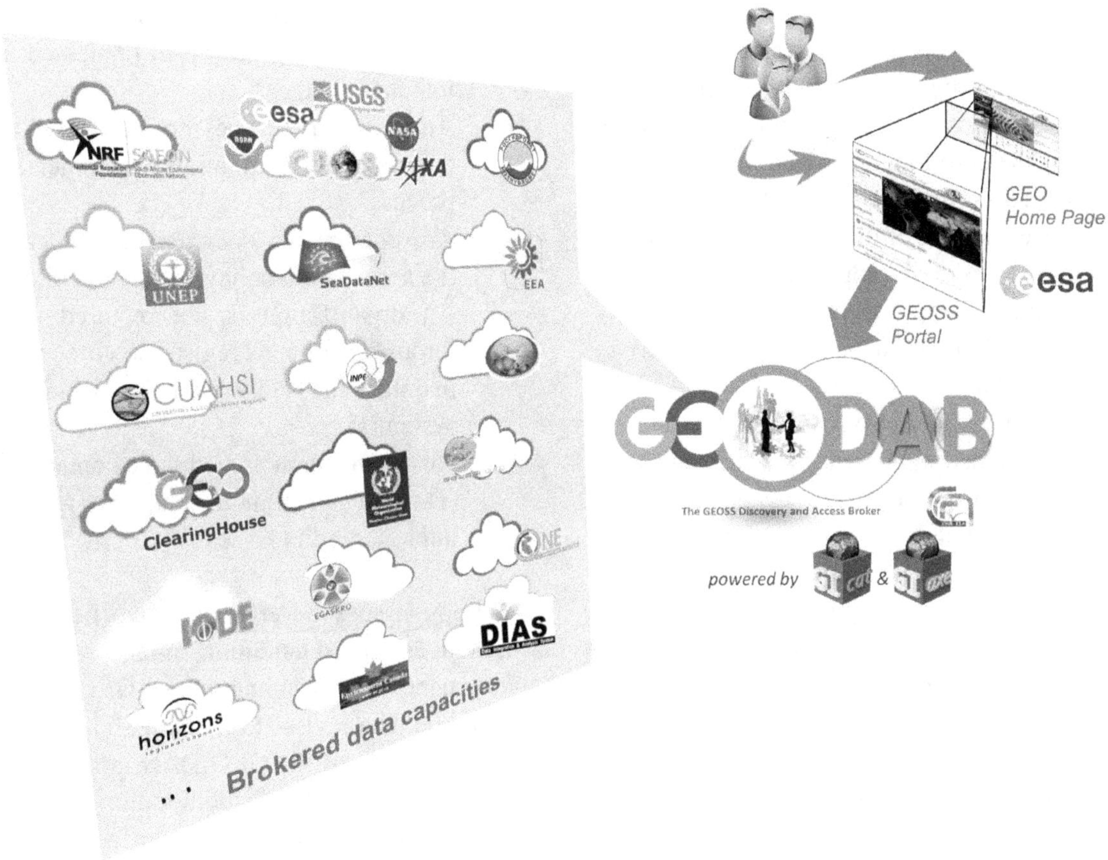

among Communities' practices, GEO-DAB makes possible to find, access, and combine the resources made available through, attaining to different thematic areas and helping to generate new knowledge.

This approach supersedes the usual centralized system approach, where each organization is burdened by the task of registering its own datasets to a central repository (using specific metadata and data models) in order to make them discoverable and accessible from a central portal. This approach supersedes as well the distributed federated approach, where each organization is burdened by the task of publishing their resources according to specific (and often proprietary) service protocols to the aim of making them discoverable and accessible from a central portal.

GEO-DAB is operatively maintained by a team whose members belong to: the GEO secretariat, ESSI-Lab of CNR-IIA, and the JRC in Ispra (Italy). This team is the third-party organization in charge of guaranteeing the efficiency and effectiveness of research and access of those resources published by the GEO members. The team takes care of brokering new capacities, as well. A set of main activities characterizing the work of the DAB operation and maintenance team has been identified:

1. Present and promote the GEO-DAB (e.g. through networking during international conferences, participating in communities of practices, and virtual or face to face meetings).
2. Individual organizations contact the DAB team to register a new system. For the registration, little effort must be required for the enrolling organization: information requested must be minimal, leaving to the DAB the task of discovering system metadata. As to interoperability information characterizing a given system, commonly, only its Internet endpoint address and service type(s) are requested:
    a. If the service type is already supported, registration is immediately achieved and the organization system can immediately be searched and accessed through GEO-DAB.
    b. If the service type is new (e.g. a profile of an existing one), the broker needs to create a new mediation module –often this is achieved by tailoring an existing one.
3. Individual organizations assess their system brokerage by discovering and accessing published data through GEOSS.
    a. If improvements in the search and access functionalities are needed, organization (or users, more generally) can contact the brokering team and point 2b) is reiterated.

These kind of activities, enabled by the GEO-DAB mediation capabilities, address most of the organizational interoperability challenges, described in the first part of the chapter. In fact, they largely relieve the GEOSS organizations from the burden required to share resources in a multi-organizational environment:

- An organization is not required to change the scope and objective of its system to comply with the GEOSS goals.
- An organization does not have to introduce any new (and federal) interoperability technology, training people to implement and maintain that.
- An organization does not have to change the data acquisition and documentation process.
- An organization does not have to change the agreed data policy.
- An organization is not required to get through a complex registration process and maintain the registered information on its system(s).
- An organization has a unique contact point (i.e. the DAB team) to contact for resolving interoperability issues.

Some issues nevertheless have been highlighted and need attention, most prominently governance and sustainability of GEO-DAB. A plan should be devised to guarantee both duration and efficiency of the GEO-DAB infrastructure. The plan should identify both funding strategies and mechanisms for an active management, such as an open board of member organizations.

Crossing the Digital Divide of Atmospheric and Hydrological Sciences

The advanced brokering framework demonstrated its utility for regional and interdisciplinary collaborative research, too. This was the case of crossing the digital divide characterizing two traditionally separated disciplines: hydrology and atmospheric sciences.

Traditionally, the two different disciplines of hydrology and atmospheric science have created large data repositories from observed time series at discrete geographical features in hydrologic science and modeled or remotely sensed coverages or grids over continuous space and time domains in atmospheric science. A clear divide existed preventing integration of information across the two domains which could lead to interesting collaborative research (see Figure 6).

Also this time the GI-cat-based discovery broker (Nativi & Bigagli, 2009) has been tailored and used to fill this gap (Salas, Boldrini, Maidment, Nativi, & Domenico, 2012) (Nativi, Caron, Domenico, & Bigagli, 2008). GI-cat is a technological framework allowing a client to search for science datasets across heterogeneous data providers by mean of standard query interfaces.

From an information technology point of view, hydrology and atmospheric science communities utilize, within their respective scientific disciplines, a single or several service buses –i.e. the middleware glue that underpins a Client-Server architecture (Nativi, Craglia, & Pearlman, 2012)– enabling the different capabilities for interoperations and build disciplinary data exchange infrastructures and domain applications. GI-cat lies in a separate brokering layer and it is able to interconnect system components (clients and servers) across different scientific disciplines, by means of functional and data model mediation, as the diagram depicted in Figure 7.

In the experimental system GI-cat was used to harvest approximately 400,000 metadata records describing gridded resources from the Motherlode THREDDS server hosted at Unidata. These resources include gridded data from the National Center for Environmental Prediction (NCEP), the Unidata Real-time Regional Model, Next Generation Radar (NEXRAD), satellites and others. At the same time GI-cat was able to perform distributed search over the millions of hydrology time series

Figure 6. The digital divide between hydrology and atmospheric science. A brokering framework can be used to bridge the gap. Figure adapted from Salas, Boldrini, Maidment, Nativi, & Domenico, (2012).

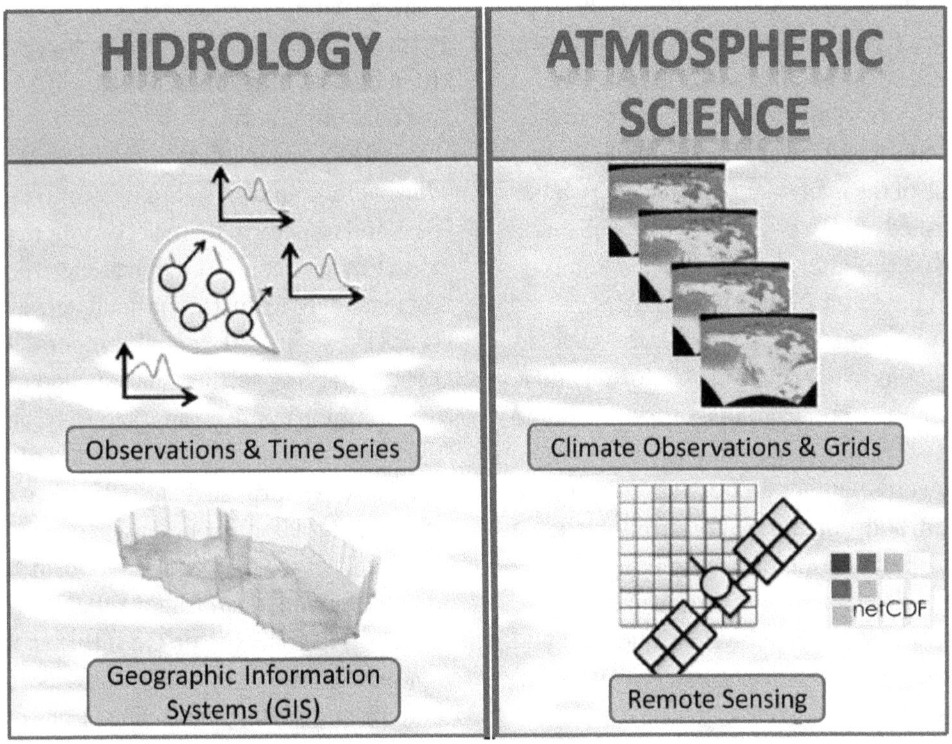

Figure 7. Disciplinary Layered Client-Server architectures and the Brokering oriented architecture for multi-disciplinary environments. Figure adapted from Nativi, Craglia, & Pearlman (The brokering approach for multidisciplinary interoperability: a position paper, 2012).

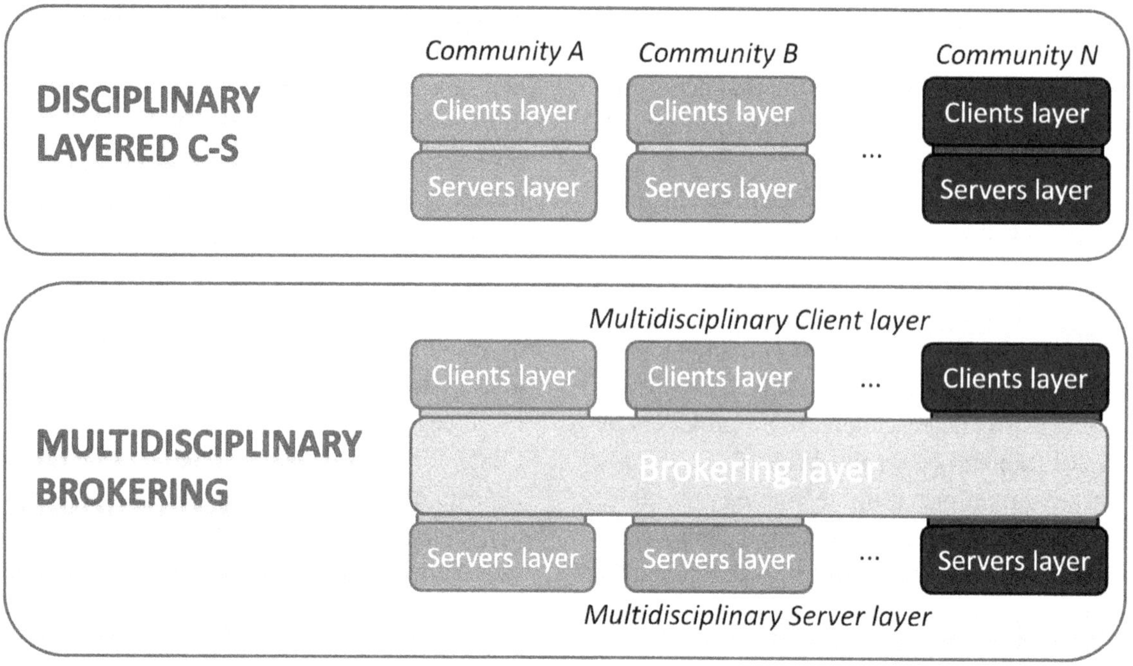

acquisitions made available by the Consortium of Universities for the Advancement of Hydrologic Science, Inc. (CUAHSI) Hydrology Information System (HIS).

The metadata of all those resources were then made available for discovery (and possibly access) through one of the many standard discovery interfaces published by GI-cat (e.g. CSW, OpenSearch, OAI-PMH, etc.), after mediation to a common data model based on ISO 19115.

This deployment makes it possible for scientists of both domains to query the broker for scientific data by using the preferred discovery client and specifying a set of constraints (e.g. geographic extent, keywords, temporal extent etc.). Resources coming from both domain are uniformly available as results for evaluation, access and use. In this experiment, different clients were used to retrieve metadata from the described system, queries could also be issued through a user friendly Web portal within GI-cat.

A further step in this research experiments is the use of the GI-axe access broker, in conjunction with GI-cat. If the GI-cat can be successfully used for metadata integration, GI-axe is its counterpart for data integration.

Indeed, once interesting datasets have been discovered, an access broker can be used to access and transform them according to a Common Grid Environment (CGE) characterized by a given data standard format, coordinate reference system (CRS), extent and spatial-temporal resolution; hence enabling further analysis and processing.

Common applications (such as QuantumGIS) exist that can leverage the results of the access broker to perform such activities (e.g. overlay of raster and vector resources).

In this case GI-axe can transform Motherlode datasets available through OPeNDAP protocol (e.g. to NetCDF-CF or GeoTIFF at the desired resolution and CRS) and CUAHSI data available through HIS (e.g. to O&M or GML). Once in this case, transformation is achieved through mediation to the internal broker data model, this time based on NetCDF for coverage data, GML for vector data and O&M for sensor data.

Integrative applications are hence eased by the use of a brokering framework; one example is validation of precipitation models against in-situ sensor acquisitions.

CONCLUSION

Collaborative research poses interesting challenges and opportunities to the scientific community. Due to the importance of data-intensive scientific discovery, data sharing emerged as a fundamental requirement for establishing collaborative research. The real challenge is implementing interoperability among different systems, networks, organizations and ultimately people, working together to share information. Interoperability has been commonly pursued addressing technological and semantic aspects, but organizational aspects are equally important.

The mediation approach has been successfully applied in different contexts. Brokers, implement mediation capabilities interconnecting heterogeneous parties in organizational contexts (i.e. people, organizations and communities). Effective applications of the mediation approach are also present in technological contexts. In these contexts, brokers have been applied to make heterogeneous technological systems work together, such as in the case of electricity adapters or the mediator and broker patterns in software engineering.

Gathering from these successful use cases, a brokering approach to enable collaborative research has been introduced. In addition to the classic intermediation principles, this new brokering application introduced additional capabilities to support the Network of Networks (NoN) and System of Systems (SoS) approaches. Brokering was conceived to address technological interoperability problems and support organizational challenges.

As any intermediation approach, the brokering approach introduces an overhead which must be governed and minimized as far its effects on performance and scalability. The main challenges identified are:

- **Governance:** In a brokering based approach, the communities and organizations involved delegate a third party, the brokering framework, to solve interoperability issues between them. This relieves participating organizations from dealing with an increasing –and often unmanageable complexity– however, a sustainability plan for the brokering framework must be considered. Examples from other contexts could be considered (e.g. the role played by organizations which maintain electrical infrastructures, standing between electricity producers and consumers).
- **Complexity and Efficiency:** Most of the complexity of interconnecting heterogeneous systems and organizations is shifted from those parties to the brokering framework. Therefore, a brokering framework must assure the necessary capabilities to face such complexity in terms of scalability, extensibility, flexibility, etc. The recent development of cloud computing technology offered a valuable instrument –as demonstrated by the GEO-DAB operational instance.
- **Flexibility:** To be successful, large brokering frameworks must take into account the constant shift and development of standards, systems, policies and organizations. This is pursued by designing brokers which are modular, easily extensible, and extremely flexible –software engineering

patterns help in that. The GEO-DAB architecture applied these patterns, proving their success. Organizationally, a brokering team must consider different expertise, including: computer scientists, data scientists, application domain/organization experts, and social scientists.

Successful organizational and technological implementations of the brokering approach are worth the overhead given the value they provide: collaboration. The brokering approach was fruitfully used to work out collaborative research environments in the Earth Sciences realm. Two of them were presented and discussed in the chapter: a global multidisciplinary environment, the GEO-DAB, and GI-cat, used as an intercommunity collaboration environment for atmospheric and hydrological sciences.

REFERENCES

Azad, B., & Wiggins, L. L. (1995). Dynamics of Inter-Organizational Geographic Data Sharing. In H. J. Onsrud, & G. Rushton (Eds.), Sharing Geographic Information (pp. 22-43). Rutgers.

Balsiger, P. W. (2004). Supradisciplinary research practices: History, objectives and rationale. *Futures*, *36*(4), 407–421. doi:10.1016/j.futures.2003.10.002

Bansler, J. P., & Havn, E. (2003). Technology-use mediation: Making sense of electronic communication in an organizational context. In *Proceedings of the 2003 international ACM SIGGROUP conference on Supporting group work* (pp. 135-143). ACM. doi:10.1145/958160.958182

Bigagli, L., Nativi, S., Mazzetti, P., & Villoresi, G. (2004). GI-Cat: a Web service for dataset cataloguing based on ISO 19115. In *Proceedings of Database and Expert Systems Applications,* (pp. 846-850). Munich, Germany: IEEE.

Bos, N., Zimmerman, A., Olson, J., Yew, J., Yerkie, J., Dahl, E., & Olson, G. (2007). From shared databases to communities of practice: A taxonomy of collaboratories. *Journal of Computer-Mediated Communication*, *12*(2), 652–672. doi:10.1111/j.1083-6101.2007.00343.x

Budgen, D., Rigby, M., Brereton, P., & Turner, M. (2007). A data integration broker for healthcare systems. *Computer*, *40*(4), 34–41. doi:10.1109/MC.2007.112

Clark, D., Krumm, J. M., & Bieleski, J. (1992). Broker: A system integration approach. In *Proceedings of the Second International Conference on System Integration (ICSI'92)* (pp. 162-170). IEEE. doi:10.1109/ICSI.1992.217273

Cluet, S., Delobel, C., Siméon, J., & Smaga, K. (1998). Your mediators need data conversion! *SIGMOD Record*, *27*(2), 177–188. doi:10.1145/276305.276321

European Union. (2007, April 25). Directive 2007/2/EC of the European Parliament and of the Council of 14 March 2007 establishing an Infrastructure for Spatial Information in the European Community (INSPIRE). *Official Journal of the European Union, 50*.

GEO. (2011). *Report of the Architecture and Data Committee (ADC), Document 19, GEO-VIII Plenary – 16-17 November 2011*. Retrieved from http://www.earthobservations.org/documents/geo-viii/19_Report%20of%20the%20Architecture%20and%20Data%20Committee%20(ADC).pdf

Geraci, A., Katiki, F., McMonegal, L., Meyer, B., Lane, J., Wilson, P., & Springsteel, F. (1991). *IEEE standard computer dictionary: Compilation of IEEE standard computer glossaries*. IEEE.

Gordon, B., & Grace, P. (2012). Emergent middleware: Tackling the interoperability problem. *Internet Computing*, *16*(1), 78–82. doi:10.1109/MIC.2012.7

Gray, B. (2008). Enhancing transdisciplinary research through collaborative leadership. *American Journal of Preventive Medicine, 35*(2), S124–S132. doi:10.1016/j.amepre.2008.03.037 PMID:18619392

Group on Earth Observations. (2005). *Global Earth Observation System of Systems GEOSS*. Noordwijk: ESA Publications Division.

Group on Earth Observations. (2014). *About GEO*. Retrieved from Group On Earth Observations: https://www.earthobservations.org/about_geo.shtml

Group on Earth Observations. (2013). *Report on Progress*. Geneva: Author.

Hey, T., Tansley, S., & Tolle, K. (2009). *The Fourth Paradigm: Data-Intensive Scientific Discovery*. Microsoft Research.

ISO. (2003). *ISO 19115:2003 Geographic information -- Metadata*. Geneve: ISO.

ISO. (2005). *ISO 19119:2005 Geographic information -- Services*. Geneve: ISO.

Jackson, N. (2003). Introduction to brokering in higher education. In *Engaging and changing higher education through brokerage* (pp. 3–20). Ashgate.

Knobel, C. (2007). *Understanding infrastructure: Dynamics, tensions, and design*. Academic Press.

Kotlarsky, J., & Oshri, I. (2005). Social ties, knowledge sharing and successful collaboration in globally distributed system development projects. *European Journal of Information Systems, 14*(1), 37–48. doi:10.1057/palgrave.ejis.3000520

Kowalczyk, S., & Shankar, K. (2011). Data sharing in the sciences. *Annual Review of Information Science & Technology, 45*(1), 247–294. doi:10.1002/aris.2011.1440450113

Kramer, D. M., & Wells, R. P. (2005). Achieving Buy-In Building Networks to Facilitate Knowledge Transfer. *Science Communication, 26*(4), 428–444. doi:10.1177/1075547005275427

Kubicek, H., & Cimander, R. (2009). Three dimensions of organizational interoperability. *European Journal of ePractice, 6*.

Long, J. C., Cunningham, F. C., & Braithwaite, J. (2013). Bridges, brokers and boundary spanners in collaborative networks: A systematic review. *BMC Health Services Research, 13*(1), 158. doi:10.1186/1472-6963-13-158 PMID:23631517

Mazzetti, P., Nativi, S., & Caron, J. (2009). RESTful implementation of geospatial services for Earth and Space Science applications. *International Journal of Digital Earth, 2*(S1), 40–61. doi:10.1080/17538940902866153

Meyer, E. (2009). Moving from small science to big science: Social and organizational impediments to large scale data sharing. *e-Research. Transformation in Scholarly Practice, 3*(2), 147–159.

Meyer, M. (2010). The rise of the knowledge broker. *Science Communication, 32*(1), 118–127. doi:10.1177/1075547009359797

Nativi, S., & Bigagli, L. (2009). Discovery, mediation, and access services for Earth Observation data. *IEEE Journal of Selected Topics in Applied Earth Observations and Remote Sensing*, 233-240.

Nativi, S., Caron, J., Domenico, B., & Bigagli, L. (2008). 09 01). Unidata's Common Data Model mapping to the ISO 19123 Data Model. *Earth Science Informatics, 1*(2), 59–78. doi:10.1007/s12145-008-0011-6

Nativi, S., Craglia, M., & Pearlman, J. (2012). The brokering approach for multidisciplinary interoperability: A position paper. *International Journal of Spatial Data Infrastructures Research, 7*, 1–15.

Nativi, S., Craglia, M., & Pearlman, J. (2013). Earth Science Infrastructures Interoperability: The Brokering Approach. *IEEE Journal of Selected Topics in Applied Earth Observations and Remote Sensing*, *6*(3), 1118–1129. doi:10.1109/JSTARS.2013.2243113

Nativi, S., Khalsa, S., Domenico, B., Craglia, M., Pearlman, J., Mazzetti, P., & Rew, R. (2011). *The brokering approach for earth science cyberinfrastructure*. NSF EarthCube White Paper.

Omran, E., Bregt, A., & Crompvoets, J. W. (2007). Spatial data sharing: A cross-cultural conceptual model. *Research and Theory in Advancing Spatial Data Infrastructure Concepts*, 79-97.

Online Etimology Dictionary. (n.d.). *Etimology for broker*. Retrieved from Online Etimology Dictionary: http://www.etymonline.com/index.php?allowed_in_frame=0&search=brocour&searchmode=none

Osborne, T. (2004). On mediators: Intellectuals and ideas trade in the knowledge society. *Economy and Society*, *33*(4), 430–447. doi:10.1080/0308514042000285224

Oxford University Press. (2014). *Definition of broker in Oxford dictionary (British & World English)*. (M. Proffitt, Ed.). Retrieved from Oxford Dictionaries: http://www.oxforddictionaries.com/us/definition/english/broker

Papazoglou, M. P., Traverso, P., Dustar, S., & Leymann, F. (2007). Service-oriented computing: State of the art and research challenges. *Computer*, *40*(11), 38–45. doi:10.1109/MC.2007.400

Pawlowski, S. D., Robey, D., & Raven, A. (2000). Supporting shared information systems: boundary objects, communities, and brokering. In *Proceedings of the twenty first international conference on Information systems*. Association for Information Systems.

Salas, F. R., Boldrini, E., Maidment, D. R., Nativi, S., & Domenico, B. (2012). Crossing the digital divide: An interoperable solution for sharing time series and coverages in Earth sciences. *Natural Hazards and Earth System Sciences*, *12*(10), 3013–3029. doi:10.5194/nhess-12-3013-2012

Schmidt, D. C., Stal, M., Rohnert, H., & Buschmann, F. (2013). *Pattern-Oriented Software Architecture, Patterns for Concurrent and Networked Objects* (Vol. 2). John Wiley & Sons.

Sheth, A. P., & Larson, J. A. (1990). Federated database systems for managing distributed, heterogeneous, and autonomous databases. *ACM Computing Surveys*, *22*(3), 183–236. doi:10.1145/96602.96604

Sverrison, Á. (2001). Translation networks, knowledge brokers and novelty construction: Pragmatic environmentalism in Sweden. *Acta Sociologica*, *44*(4), 313–327. doi:10.1080/000169901526964 02

Tolk, A. (2006). What comes after the Semantic Web - PADS Implications for the Dynamic Web. In *Proceedings of the 20th Workshop on Principles of Advanced and Distributed Simulation (PADS'06)*. IEEE. doi:10.1109/PADS.2006.39

Tress, B., Tress, G., & Fry, G. (2006). Defining concepts and the process of knowledge production in integrative research. In B. Tress, G. Tress, G. Fry, & P. Opdam (Eds.), Landscape research to landscape planning (pp. 13-26). Academic Press. doi:10.1007/978-1-4020-5363-4_2

Trigg, R. H., & Bødker, S. (1994). From implementation to design: tailoring and the emergence of systematization in CSCW. In *Proceedings of the 1994 ACM conference on Computer supported cooperative work* (pp. 45-54). ACM. doi:10.1145/192844.192869

Turnitsa, C. (2005). *Extending the Levels of Conceptual Interoperability Model. In Proceedings, Summer Computer Simulation Conference 2005*. Philadelphia: Curran Associates, Inc.

Vaccari, L., Craglia, M., Fugazza, C., Nativi, S., & Santoro, M. (2012). Integrative research: The EuroGEOSS experience. Academic Press.

Van de Ven, A. H. (1976). On the nature, formation, and maintenance of relations among organizations. *Academy of Management Review*, *1*(4), 24–36. doi:10.2307/257722

Vlissides, J., Helm, R., Johnson, R., & Gamma, E. (1995). *Design patterns: Elements of reusable object-oriented software* (Vol. 49). Addison-Wesley.

Wenger, E. (1998). *Communities of practice: Learning, meaning, and identity*. Cambridge University Press. doi:10.1017/CBO9780511803932

Wenger, E. C., & Snyder, W. M. (2000). Communities of practice: The organizational frontier. *Harvard Business Review*, *78*(1), 139–146. PMID:11184968

Zhu, F., Turner, M., Kotsiopoulos, I., Bennett, K., Russell, M., & Budgena, D. et al. (2004). Dynamic data integration using web services. In *Proceedings of the IEEE International Conference on Web Services (ICWS'04)* (pp. 262-269). IEEE.

KEY TERMS AND DEFINITIONS

Broker: A distinct entity which applies the mediation paradigm to let different parties work together. It stands between the parties without belonging to a specific one and use mediation to enable technological, semantics and organizational interoperability.

Data Access: The activity of retrieving selected datasets from a given data repository. The repository may support access options to enable access functionalities such as data subset, CRS reprojection, format conversion. (ISO, 2005).

Data Discovery: The activity of searching for interesting datasets across one or more data repositories. (ISO, 2005).

Data Set: An identifiable collection of data (e.g. a result of instrumental acquisition or a model output). (ISO, 2003).

Data Sharing: Activities related to making existing datasets available to other investigators (e.g. publication of datasets in public data repositories).

Geographic Metadata: Data about data concerning phenomena associated with a location relative to the Earth. (ISO, 2003).

GEOSS: The Global Earth Observation System of Systems provides a framework to integrate Earth Observation efforts across the world. It is led by the Group on Earth Observations, a voluntary partnership of governments and international organizations. (Group on Earth Observations, 2014).

Interoperability: The ability of a system to interoperate by means of well-defined interfaces with other (present or future) systems, without any restricted access or implementation. [IEEE] (Geraci et al., 1991).

Mediation: A process conducted between two or more conflicting parties to reach a shared settlement or position. Typically a third party, the mediator, cooperates in the process. (Meyer M., 2010).

ENDNOTES

1. Social network analysis views social relationships in terms of network theory, consisting of nodes (representing individual actors

within the network) and ties (which represent relationships between the individuals, such as friendship, kinship, organizations, etc.). These networks are often depicted in a social network diagram, where nodes are represented as points and ties are represented as lines.

2 http://www.geoportal.org The GEO portal provides a user friendly entry point to GEO-DAB. User queries are processed by GEO-DAB and transparently executed against heterogeneous and distributed sources. GEO-DAB translates back the results according to a common model, which makes possible for the GEO portal to show search results in a uniform way. Interoperability between the portal and the different publishing services is thus achieved.

Chapter 15
Building Collaborative Ontologies:
A Human Factors Approach

Tatiana A. Gavrilova
Saint Petersburg University, Russia

Irina A. Leshcheva
Saint Petersburg University, Russia

ABSTRACT

The chapter describes the research performed within the KOMET (Knowledge and cOntent structuring via METhods of collaborative ontology design) project, which was aimed at developing a new paradigm for knowledge structuring. By knowledge structure, the authors define the main domain concepts and relations between them in a form of graph, map, or diagram. The approach considers the specifics of individual cognitive style. Two stages of research have been completed: research into correlations between the expert's individual cognitive style and the peculiarities of expert's subject domain ontology development; and study of correlations between the expert's individual cognitive style and the group ontology design (including the design performed in groups consisting of experts either of similar or of different cognitive styles). The results of this work can be applied to organizing collaborative ontology design (especially for research and learning purposes), data structuring, and other group analytical work. Implications for practice are briefly delineated.

INTRODUCTION

One of the main objectives of the research process is achieving maximal effectiveness from the creation, transfer and dissemination of new knowledge. This effectiveness can be measured by the quality and speed of memorization of the principal concepts of a particular domain and of the relationship between these concepts. Wide evidence exists that the use of visual thinking to address the subject of study is positively connected with the quality and speed of memorization, and thus with the effectiveness of knowledge dissemination. Visualization working as a cognitive tool also facilitates communication within research communities.

DOI: 10.4018/978-1-4666-6567-5.ch015

Copyright © 2015, IGI Global. Copying or distributing in print or electronic forms without written permission of IGI Global is prohibited.

Special interest in such forms of knowledge codification can be observed in education science, especially within learning where the students are engaged in group knowledge sharing and co-creation processes with continuous feedback.

Mutual understanding and mentalization in research is of special interest in collective study or discovery. One of the most productive methods of research collaboration promises to be group ontology design. The specific problem being addressed in this work deals with the problem of improving the quality of group or collaborated ontologies. We are also interested in filling the gaps in understanding the group ontology design process specifics, such as the causes of differentiations between the form and the content of individual ontologies.

This paper presents the main results of the KOMET (Knowledge and cOntent structuring via METhods of collaborative ontology design)) project which was devoted to developing methods of using group visual ontology design in research and education with regards to the respondents' individual cognitive styles. The group ontology design was tested in the medical domain (smaller group) (Gavrilova, Ravodin, Bolotnikova & Kotko, 2012) and computer science (informatics) domain (larger group of participants). All the 79 respondents were graduate students of the School of Computer Science of Saint Petersburg Polytechnic University. Almost all the students had 1-2 years' experience of research in computer science, and were in their fifth year of study on the Masters programme. The preliminary results of this study were partly discussed in (Gavrilova, et al., 2013).

During the last decade, visual knowledge representation has become one of the key considerations in knowledge engineering methodology, and it is strongly associated with ontology design and development. These ontologies, which form a conceptual skeleton of the modelled domain, might serve various purposes such as better understanding, knowledge creation, knowledge sharing, and collaborative learning, problem solving, seeking advice, or developing competences by learning from peers. Recently, the ontological engineering perspective has gained interest in many research domains, such as medicine, business and computer science (Schnotz & Kurschner, 2008; Pfister & Eppler, 2012; Oltramari & Ferrario, 2009; Brochhausen, Spear, Cocos, Weiler, Martin, Anguita & Tsiknakis, 2011). These studies rely heavily on theory and tools from knowledge engineering analysis that already has a long-standing tradition in the knowledge-based systems domain (Mizoguchi & Bordeau, 2000). The largest number of knowledge engineering research articles was generated around the theme of descriptive logics and formal foundations of ontology design (Baader, Horrocks & Sattler, 2005; Kuznetsov, Obiedkov & Roth, 2007).

The tools and techniques developed in this domain can be applied fruitfully in the field of research knowledge structuring and design (Schreiber, 2000; Dicheva & Aroyo, 2004; Dicheva & Dichev, 2005, 2007; Knight, Gašević & Richards, 2006) and semantic web applications (Davies, van Harmelen & Fensel, 2002). The idea of using ontologies and visual structuring in research description and introduction has been discussed in many works (Sherlock, 2000; Fonesca, Davis & Camara, 2003; Brusilovsky, Yudelson & Gavrilova, 2005; Tansley & Tolle, 2009; Diviacco, 2015 (see chapter 19 this book)) and is now being implemented in several research projects and software tools (Bard, & Rhee, 2004; Hevner, 2007).

BACKGROUND OF ONTOLOGY ENGINEERING AND VISUAL KNOWLEDGE MAPPING

This project was targeted at developing a paradigm of data and knowledge structuring with regard to individual cognitive styles, using recent advances in knowledge engineering and conceptual structuring, aimed at creating structurally holistic knowledge bases for various areas of science and technology.

This aim was decomposed into such objectives as:

- Research of correlations between the expert's individual cognitive style and the peculiarities of the expert's subject domain ontology development,
- Research of correlations between the expert's individual cognitive style and the group ontology design (including design performed in groups consisting of experts with either similar or different cognitive styles), and
- Research of formal ontology evaluation methods from the cognitive ergonomics point of view and informal evaluation by peers.

The idea of using visual structuring of information to improve the quality of understanding and mentalization among research colleagues is not new (Shneiderman, 1996). For more than twenty years, concept mapping (Grosslight, Unger, Jay & Smith, 1991; Sowa, 1984; Jonassen, 1999; Conlon, 1997) has been used to compile maps and mental models that support the process of knowledge sharing.

Ancient scholars used diagrams and figures to describe taxonomies and other complex sets of terms. As such, the visual representation of general domain concepts facilitates and supports understanding of both substantive and syntactic knowledge. Many scholars, especially those who also teach sciences courses, operate as knowledge analysts or knowledge engineers by making visible the skeleton of the studied discipline and showing the domain's conceptual structure (Kinchin, 2005). This structure is frequently represented by a so-called "ontology".

From a philosophical viewpoint, "ontology" is the branch of philosophy which deals with the nature and organization of reality. Ontologies aim at capturing domain knowledge in a generic way and providing a commonly agreed understanding of a domain, which may be reused and shared across applications and groups (Chandrasekaran, Josephson & Benjamins, 1999). Neches and colleagues (Neches, Fikes, Finin, Gruber, Patil, Senator & Swartout, 1991) gave the classical definition as follows "An ontology defines the basic terms and relations comprising the vocabulary of a topic area as well as the rules for combining terms and relations to define extensions to the vocabulary".

An ontology is a set of definitions we make in understanding and viewing the world. The visual approach to presenting an ontology is not only compact but also comprehensive. It makes ontology a powerful mind tool (Johanssen, 1998; Gavrilova & Voinov, 1996).

By definition, ontology is a declarative representation of a certain precise domain specification, including the glossary of the domain terms and the logical expressions describing the meanings and the relationships of these terms, thus allowing structured sharing of knowledge related to the domain (Gruber, 1993). The relationships between the concepts in ontologies can be of different types, e.g. "is", "has part", "has a property of", etc. The concepts and relationships are universal for a certain class of objects in a subject area. Conceptual model visualization methods such as ontologies are also widely and effectively used in education, and many learning ontologies have been developed for a number of disciplines (Barros, Verdejo, Read & Mizoguchi, 2002; Gaeta, Orciuoli & Ritrovato, 2009; Gavrilova, Leshcheva & Bolotnikova, 2012).

However, the ontology-based approach to knowledge representation in research and pedagogy is a relatively new development. There are numerous definitions of this milestone term (Neches, Fikes, Finin, Gruber, Patil, Senator & Swartout, 1991; Gruber, 1993; Guarino & Giaretta, 1998; Gómez-Pérez, Fernández-López & Corcho, 2004). Together, these definitions clarify the ontological approach to knowledge structuring while giving enough freedom for open-ended, creative thinking. Many researchers and practitioners have argued about the distinctions between ontology and a conceptual model. We propose that ontology corresponds to the analyst's view of the conceptual model, but is not de facto the formal model itself. There are more than a hundred techniques and notations that help to define and visualize conceptual models. Ontologies are now considered as the most universal and shareable forms of such modelling.

Ontologies are useful structuring tools, in that they provide an organizing axis along which every researcher (or student) can mentally mark his/her vision in the information hyper-space of domain knowledge. Frequently, it is impossible to express all the information as a single ontology. Accordingly, subject knowledge storage consists of a set of related ontologies. Some problems may occur when moving from one ontological space to another, but constructing group meta-ontologies may help to resolve these problems.

For both formative and summative assessment purposes, creation of ontologies and explanation of the processes involved can clearly indicate the extent and nature of the knowledge and understanding. Knowledge entities that represent the static knowledge of the domain are stored in hierarchical order in the knowledge repository and can be reused by others. At the same time, those knowledge entities can be reused in description of the properties or methodological approach as applied in the context of another related knowledge entity.

Of course, the ontologies are inevitably subjective to a certain extent, as knowledge by definition includes a component of personal subjective perception; however, using the ontologies developed by others is a convenient and compact means of acquiring new knowledge. At the same time, collective ontology development experience allows the participants in the process to gain the fullest possible understanding of the subject area.

Meta-ontology provides a more general description dealing with higher-level abstractions. Figure 1 illustrates different ontology classifications in the form of the mind map. This representation may be called the knowledge map. Such maps are graphical tools for organizing and representing knowledge. Later in this paper and in our study we will use two particularly appropriate forms of knowledge maps: mind maps (Buzan, 2005) and concept maps (Novak, 1998; Novak & Canas, 2006).

Knowledge maps are now widely used for visualizing ontologies at the design stage, while ontology editors (like Protégé) facilitate the development stage.

Research on knowledge mapping in the last 12 years has produced a number of consistent findings (O'Donnell, Dansereau & Hall, 2002). People recall more central ideas when they learn from a concept map than when they learn from text, and those with low verbal ability or low prior knowledge often benefit the most. It seems that knowledge maps reduce cognitive load. The use of knowledge maps also appears to amplify the benefits associated with scripted cooperation (Stephens, 2012). Learning from maps and communication via maps are enhanced by active processing strategies such as summarization or annotation and by designing maps according to gestalt principles (Wertheimer, 1938; Gavrilova, 2011) of organization.

INDIVIDUAL COGNITIVE STYLE AND GROUP ONTOLOGY DESIGN

Cognitive Styles Features

The cognitive-styles view has acquired great influence within the education field, and is frequently encountered at levels ranging from kindergarten to graduate school. There is a thriving industry devoted to publishing cognitive-style tests and guidebooks for teachers, and many organizations

Figure 1. Summarizing the ontology classifications in a mind-map

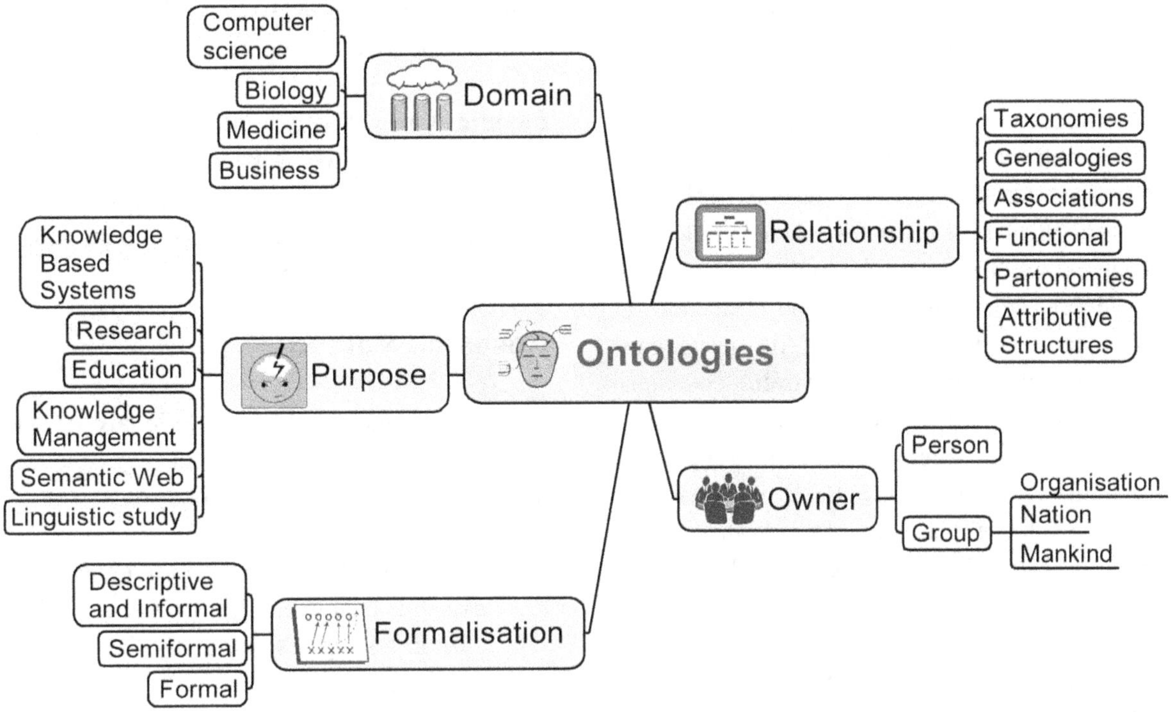

offer professional development workshops for teachers and educators built around the concept of learning styles (Peterson, Rayner, & Armstrong, 2009).

However, we will use the concept of cognitive style only for the predefined aim. As the aim of the KOMET project was to develop a paradigm of structuring data and knowledge with regard to individual cognitive styles, we had to choose the appropriate parameters or features of cognitive style.

The cognitive styles explain and describe how an individual acquires knowledge and how an individual processes information. The cognitive styles are related to problem solving, and generally to the way that information is acquired, structured and used.

Among the main features of cognitive style (Hayes & Allinson, 1998) we can name:

- Field dependence versus field independence,
- Impulsivity versus reflection,
- Narrowness versus width of the categories,
- Rigidity versus flexibility,
- Levelling versus sharpening,
- Scope of cognitive equivalence,
- Visual/audio/kinesthetic preferences, etc.

Three characteristics have been chosen from the plethora of cognitive style characteristics described in the literature (Kholodnaya, 2004): field dependence/field independence (FD/FID), impulsivity/reflection, and narrowness/width of the category.

According to the definition by Witkin (Witkin, Moore, Goodenough & Cox, 1977), FD/FID is "a structuring ability of perception". The field-independent style is defined by a tendency to separate details from the surrounding context. It can be compared to the field-dependent style, which is defined as a relative inability to distinguish detail from other information around it. The FD/FID characteristic can be interpreted as a proxy of the structuring capability of an individual mind. The characteristic of this style does influence the structuring process as a whole (e.g. ontology development "from scratch" in the research setting described in this paper), and even more it affects the restructuring process (the merging of individual ontologies). FD/FID exerts considerable influence on the collective problem-solving process. In dyads where members have cognitive styles differentiated by the FD/FID characteristic, the final solution is usually closer to the variant suggested by the FID participant. The FID dyads experience difficulty in developing common decisions on arguable points, while the FD dyads are more successful in coming to agreement in collective problem solving.

Psychologists in our research group were used to working with on-line text based on the popular modification of Witkin's method of "embedded figures" which is aimed at the search for a simple figure hidden within the complicated one (Witkin, Oltman, Raskin & Karp, 1971).

The impulsivity/reflection characteristic considers the amount of information collected prior to making a decision: impulsive individuals are able to make decisions on a considerably bounded information basis, while the reflective individuals are more inclined to make decisions considering completely full information on the respective situation. For assessing the respondents' impulsivity/reflection features, the "Matching Familiar Figures Test" (Kagan, 1966) method has been used.

As for the narrowness/width of the category, the main difference between the extreme poles of this characteristic is that narrowly categorizing individuals are inclined to restrict the area of application of a certain category, while the wide categorizers are, conversely, inclined to include a plethora of more-or-less related examples into a single category. The psychologists that help us with the experimental part advised us to use a modification by Pettigrew (1958) and Fillenbaum (1959): their method of so-called "range width task". The procedure is based on respondents' opinion on the minimum, average and maximum evaluations of a concept or category.

Research Method

The research into an expert's individual cognitive style and ontology development objective was divided into four consecutive phases:

1. Identifying the significant individual cognitive style characteristics on the basis of the on-line testing results, using the software developed by Elena Kotova and Andrew Pisarev.
2. Creating the "informatics" research domain ontologies using the Protégé (Protégé) tool and estimating the ontology metrics using the COAT software environment (Gavrilova, Bolotnikova & Gorovoy, 2012).
3. Merging individual ontologies in face-to-face group ontology design.
4. Performing statistical analysis in order to find out significant relationships between the young researchers' (experts') individual cognitive style characteristics and the ontology metrics.

The second part of the research was performed using the same test sample and included ontology development in/with the use of the Protégé tool. All the tested students were given the task of developing an ontology for the informatics domain. They did it by using visual mapping approach.

The quality of the developed ontologies was assessed by two methods:

- An expert method, where the ontology analyst and domain experts (both professors in computer science) assessed the quality by such criteria as completeness, imbalance, relevance, etc.
- A formalized method, where any ontology was assessed by a set of quantitative metrics using COAT software.

The formalized method was preferable as it was free from experts' and analysts' subjective interpretations and had the potential to be automated.

For step 3 the KOMET-DILIGENT collective ontology development was designed (Gavrilova, Leshcheva, Bolotnikova, Blagov & Yanson, 2013). This methodology enriches the findings of the Neon Project (Neon Project).

All the individual and collective ontologies (both in pairs and in groups of 3-5 members) were analyzed. In the KOMET project, specificity of the collective ontology development has been researched. The test sample consisted of the students of the Saint Petersburg State Polytechnic University, Institute of Information Technology and Control Systems (IITCS), attending the intelligent systems development course. The experiments were also aimed at figuring out how the collective categorization style is being developed.

Cognitive Ergonomic Metrics

In our research the developed ontologies were assessed by an augmented set of metrics (such as minimal depth, absolute width, etc.) suggested in Bolotnikova, Gavrilova and Gorovoy (2011). In evaluating the quality of the designed ontologies, the following two aspects are most important: (1) correctness and depth of reflection of the subject domain, and (2) ergonomic aspect of the ontology representation from the point of view of quality and human speed of perception.

The notation used to describe the metrics is the following:

- "g", a graph representing an ontology; the concepts (classes and exemplars) of the ontology are the graph nodes, the relationships between the concepts are the graph edges;
- "G", a set of all the nodes g;
- "E", a set of all the edges g.

A minimal depth:

$$m = N_{j\in P},\ \forall i(N_{j\in P} \le N_{i\in P})$$

where $N_{j\in P}$ and $N_{i\in P}$ are the path lengths j and i from the set of paths P of the graph g.

An absolute width:

$$m = \sum_{j}^{L} N_{j\in L}$$

where $N_{j\in L}$ is a number of nodes of degree j from the set of nodes L of the graph g.

An average width:

$$m = \frac{1}{n_{L\subseteq g}} \sum_{j}^{L} N_{j\in L}$$

where $N_{j\in L}$ is a number of nodes of degree j from the set of degrees L of the graph g, $n_{L\subseteq g}$, a number of all degrees of the graph (a maximal graph depth augmented by 1, if considering only a chosen dominant relationship).

90% line depth:

$$m = P_{90}(N_{j\in P})$$

where $P_{90}(N_{j\in P})$ is a 90% percentile of the graph depth (possible value of the graph path length, not exceeding the length of 90% of the graph paths).

Root-mean-square deviation of neighboring levels/degrees width ratio:

$$m = \frac{\sum_{i=2}^{n_{L\subseteq g}} \left(\frac{N_{l_i \in L}}{N_{l_{i-1} \in L}} - \frac{1}{n_{L\subseteq g} - 1} \sum_{i=2}^{n_{L\subseteq g}} \frac{N_{l_i \in L}}{N_{l_{i-1} \in L}} \right)^2}{n_{L\subseteq g} - 1}$$

Complexity metric:

A number of nodes with multiple inheritance to the set of all the graph nodes:

$$m = \frac{N_{v\in MI}}{n_G}$$

where

$$MI = \{v \in G | \exists a_1, a_2 (isa(v, a_1) \wedge isa(v, a_2)\}$$

is a set of all the graph nodes with more than one "is-a" relationship arc, $N_{v\in MI}$ is a number of all the elements of this set, n_G is a number of the graph nodes.

Average number of the parent nodes of a graph node:

$$m = \frac{1}{n_G} \sum_{v}^{G} N_{S_{v\in G}}$$

where

$$S_{v\in G} = \{a \in G | isa(v, a)\}$$

is a set of all the parents of the node v, $N_{S_{v\in G}}$ is a number of all the parent nodes of the node v, n_G is a number of the graph nodes.

Many aspects affect the quality of an ontology from the cognitive point of view. The COAT software environment provides calculation of more than 20 metrics. Metrics of this kind were first proposed by the research group of Aldo Gangemi (Gangemi, Catennaci, Ciaramita & Lehman, 2006). The ontology evaluation based on these metrics is formal but it helps to assess the ontology quality. The complete list of metrics was presented in detail in two works (Bolotnikova, Gavrilova & Gorovoy, 2011; Gavrilova, Bolotnikova & Gorovoy, 2012).

These metrics can help to understand what should be corrected in the description of the subject domain in order to improve it from the point of view of cognitive ergonomics or better perception. Thus it is supposed that each next version of the ontology will be better and it can be perceived faster by users.

The metrics can also be used in evaluating ontologies of the same subject domain produced by different people/teams. The calculated metrics help to estimate which of them is better from the point of view of cognitive ergonomics and to choose the best of them if the evaluations of other important criteria differ insignificantly.

Analysis of the Individual Construction Affected By Cognitive Style

As mentioned in the introduction, the research sample consisted of 79 students, enrolled in the intelligent systems development course. All the tested students were given the task of developing an ontology for the informatics research domain. Due to the professional specificity of the sample, a bias toward narrow, reflective and field-independent test persons was found in the sample. However, a statistically significant Spearman's negative correlation between the FID score and the time of the first answer in the Kagan was calculated, showing that the sample was dominated by the fast FID and slow FD respondents.

The two methods mentioned above were used for assessment of the result.

On the basis of the literature review and the practical ontology development experience, the following hypotheses are suggested:

Hypothesis 1: Individuals belonging to the FID extreme point of the FD/FID cognitive style characteristic tend to have highly developed cognitive structuring capabilities; thus, the quality of ontologies developed by the FID individuals would be higher.

Hypothesis 2: Impulsive individuals tend to develop superficial ontologies lacking sufficient categorization in the upper level, while the reflective individuals tend to develop deeper ontologies.

Hypothesis 3: Ontologies developed by the individuals described as "imprecise" in the Kagan impulsivity/reflectivity test results tend to be more complex.

Hypothesis 4: The "narrowness/ width of the category" cognitive style characteristic exerts significant influence on the ontology width: the "wide categorizers" tend to develop broader ontologies.

Table 1 presents a part of the two series of testing results. It describes the correlation coefficients for several metrics and main parameters of cognitive style:

- **I/R:** Impulsiveness/reflexivity; and
- **NC/WC:** Narrowness/width of category.

The correlation between the cognitive style and ontology metrics values was assessed by Spearman's coefficient (rank correlation). The significant correlation between the metrics and such features as field dependence/ field independence was not found, that is why it is not presented in the table. Empty cells in the table mean that no correlation was found.

Table 1. Correlation matrix illustrating the correspondence between ontology metrics and the cognitive styles' indices

Metrics	Test Results		
	I/R		NC/WC
	The Time of the First Answer	The Number of Mistakes	The Size of the Category
Number of classes			0,44
Number of leaves			0,46
Absolute depth			0,39
Minimum depth		0,54	
90th percentile depth	0,34		
The average width			0,48
The standard deviation of the relative width			0,48
Average number of parents of a graph node		0,47	
The absolute cardinality of families			0,44
Branching factor		0,50	
The absolute cardinality of leaves			0,46
The weighting factor branching leaves		-0,39	

Hypothesis 1 was not confirmed, as no significant correlation between the FD/FID metric and the quality of the ontologies was found; this result gave rise to optimistic feelings about the whole project, as it shows that it is possible to teach any individual to develop ontologies of a high quality.

Hypothesis 2 was partially confirmed: the "90% line depth" metric demonstrated significant positive correlations with the time of the first answer in the Kagan test, thus showing that reflective test persons tend to develop deeper ontologies; however, no significant negative correlation between the time of the first answer and the ontology width was found.

Hypothesis 3 was confirmed, as the number of mistakes in the Kagan test demonstrated a significant positive correlation with the values of the "Average number of parents of a graph node" metric that characterizes the ontology complexity.

Furthermore, the number of mistakes in the Kagan test demonstrated significant positive correlations with the metrics of the "Minimal depth of the ontology" and the "Families branching coefficient" and significant negative correlation with the weighted leaves branching coefficient.

Hypothesis 4 was fully supported: the wide categorizers developed bigger ontologies in terms of the number of concepts, achieved mainly by the greater number of "children" of each parent concept.

Respectively, the results of the "range width task" test demonstrated significant correlations with such metrics as the "Average ontology width", "Number of leaves", "Absolute cardinality of families", etc. These results also demonstrated significant correlation with the root-mean-square deviation of the average ontology width. This result shows that the number of concepts belonging to the neighbouring levels and to different branches is significantly different, indicating imbalance in the ontologies developed by the wide categorizers.

Despite the objectivity of the quantitative metrics-based method of ontology assessment, this method has the significant drawback of being too formalized and lacking semantic analysis elements. Having augmented the quantitative metrics-based analysis by a semantic analysis performed manually, we found that the ontologies developed by the field-independent test persons tend to have simpler and clearer structure. However, this simplicity and clarity tends to be achieved by truncating the concepts that do not fit into the developed ontology, thus sacrificing the ontology's completeness and integrity for formal logical consistency.

As for the collective ontology development, including wide categorizers into a group together with a FID individual can be useful, with the wide categorizers generating a plethora of sub-classes and the FID participant restructuring these. This hypothesis has been tested on the stage of research dedicated to the collective ontology development.

So, the following relationships between the respondent' individual cognitive styles and the peculiarities of respondents' subject domain ontology development have been identified as a result of the research:

- Considering the "impulsivity/reflection" scale, the reflective individuals tend to develop deeper ontologies;
- The ontologies developed by the imprecise individuals (as defined in the Kagan test) tend to be more complex;
- The "narrowness/ width of the category" cognitive style affects the ontology branching coefficient, i.e. the ontology width.

Collective Ontology Design

Specificity of the collective ontology development has also been researched, both in dyads as well as in the groups of 3-5 persons. An objective of this study was to establish how the collective categorization style was developed.

The KOMET-DILIGENT collective ontology face-to-face design methodology proposed within the KOMET project uses the following algorithm (Gavrilova, Leshcheva, Bolotnikova, Blagov & Yanson, 2013):

1. Preliminary individual ontology development by the participants and consequent mutual ontology matching;
2. Ontologies analysis, merging and alignment;
3. Ontology revision and redesign.

Students first developed individual ontologies and then were asked to develop a collective common ontology on the same topic, "computer science". The time allowed was one hour.

The experiments aimed at figuring out how a collective categorization style was developed. Two strategies were identified:

S1: A strategy of collective ontology development "from scratch", and
S2: A strategy of common ontology development on the basis of two or more individual drafts.

These strategies were affected by the peculiarities of analyzing and merging individual ontologies in the collective ontology development methodology suggested and previously explained by the authors.

The second strategy, S2, is of greater practical interest. In this case the respondents effectively applied all three basic ontology engineering operations (matching, merging and alignment).

The experiments have shown that merging usually follows either of the two scenarios:

1. Absorption scenario (60-70% of all the tested groups);
2. Compromise (synthesis) scenario (30-40% of the tested groups).

The absorption scenario has been researched by us in more detail, as this scenario was used more often than the others. The implications of this scenario have led to the design of two models:

A disjunctive model, in which the higher power ontology absorbs the lower power one, with further merging of the same-degree nodes in the resulting ontology.

A conjunctive model in which the reduction of nodes leads to the resulting ontology including only the disjunction or intersection of the same-degree nodes.

Comparison of these scenarios with the cognitive styles of the test participants revealed the following relationships:

- Field-independent (FID) test persons tend to prefer the conjunctive scenario;
- Field-dependent (FD) test persons tend to prefer the disjunctive scenario.

Figures 2, 3 and 4 illustrate the experiment and demonstrate the variation from the synthesizing scenarios. This variation is the best alternative, with collective effort showing the synergetic effect of "jumping" to a higher level of abstraction. The higher level always demonstrates deeper knowledge and better understanding of the domain specifics. These ontologies can be criticized as they were developed by young researchers. However, our experiment was targeted at the study of collaborative ontology design, not the production of a serious domain ontology.

Solutions and Recommendations

This paper addresses the conceptual limitations of traditional research communication and proposes using a visual metaphor for illustration and presentation of the the research state-of-the art and main findings.

The first most obvious possible application of the results can be in the supervising Masters and PhD research and thesis writing at different

Figure 2. Example of individual A ontology of computer science

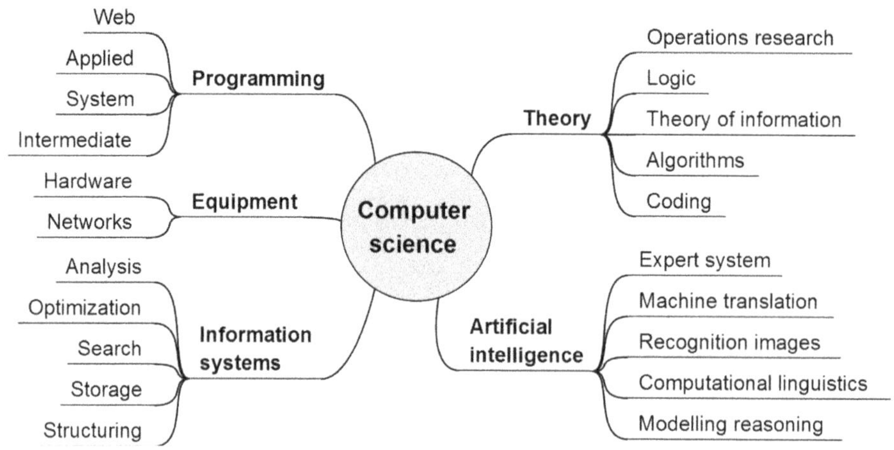

Figure 3. Example of individual B ontology of computer science

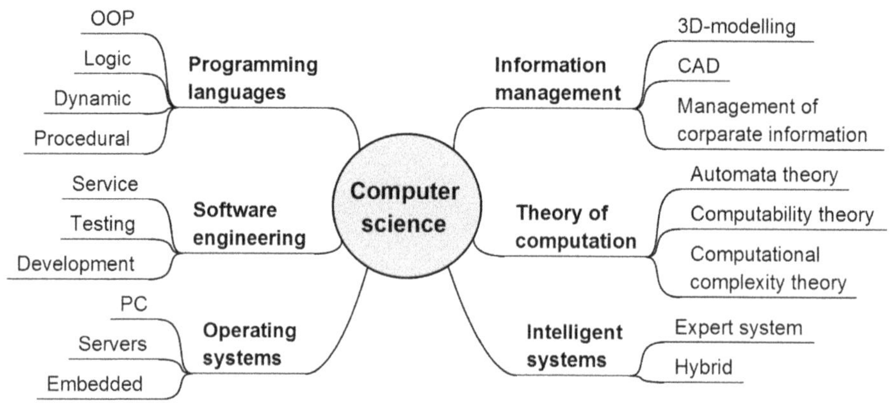

levels of education. Indeed, team ontology development by the students can lead these students to a deeper and more holistic understanding of their respective subject domains, especially when the instructor takes the students' cognitive styles into consideration when organizing individuals into groups for team ontology development.

The results may also help in organizing brain storming and design thinking sessions (Eppler & Bresciani, 2013; Eppler & Burkhard, 2008).

Finally, the impact of individual cognitive styles on team ontology development can be taken into consideration in all the possible areas where team intellectual work with ontology development as a possible method can be used, e.g. fundamental science, R&D or management consulting.

Using visual inspection of the ontology it is possible to detect gaps and misunderstandings in the state-of-the-art knowledge level and cognitive model of the domain knowledge. However, there is as yet little consensus on the useful design and orchestration of such structures. Furthermore, in many cases it is not known what the structure of socially legitimate knowledge patterns looks

Figure 4. Collaborative AB ontology of computer science

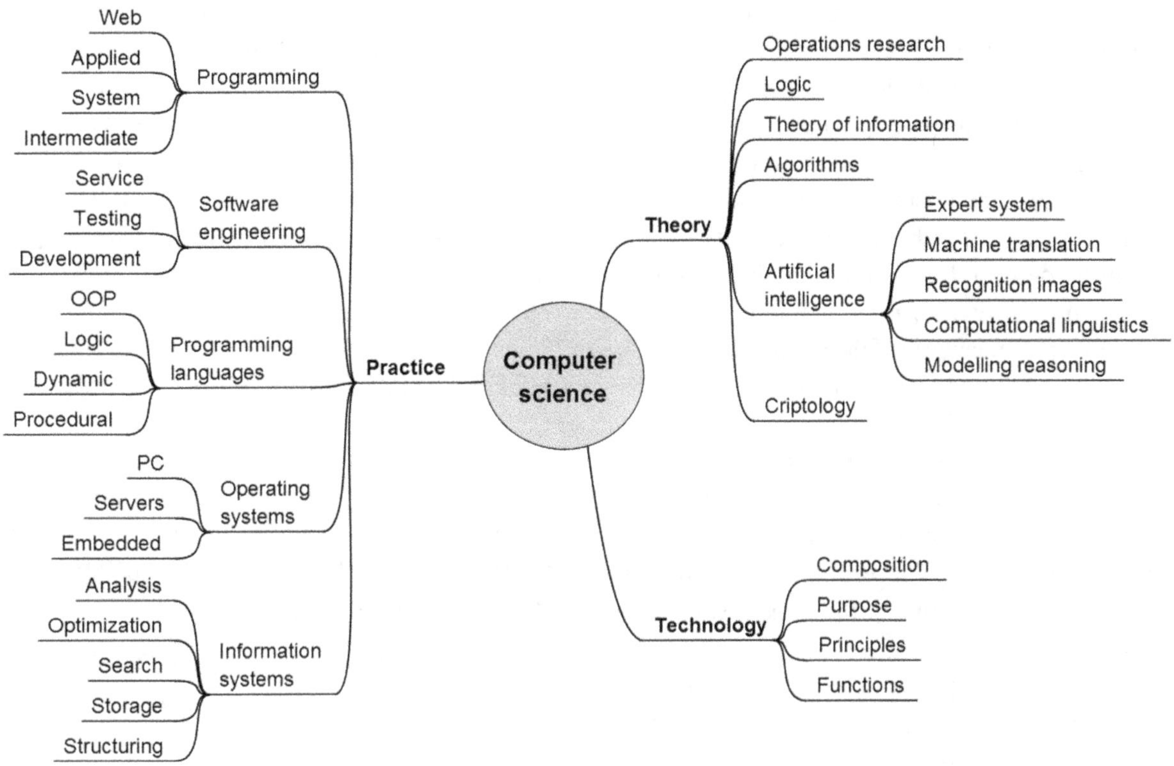

like, or how a particular instance of a knowledge model deviates from that "ideal" state (e.g. guru's view) (Cross, Parker, Prusak & Borgatti, 2001). However, researchers are individuals, and they may disagree among themselves.

FUTURE RESEARCH DIRECTIONS

The main focus of future research is concentrated around the deeper understanding of the cognitive issues of collaborative work. This discourse on the concept of the 'cognitive style' encompasses a number of interconnected topics, ranging from the impacts of the psychological theories to current perceptions about relationships between learning, understanding and teaching. As the book focuses on the research issues it may be interesting to aim further investigation at the collaborative interdisciplinary ontology design for mapping the emerging research areas.

It may be interesting to discover the new possibilities of such design and to investigate how can ontologies can be arranged to support a process of generic, collaborative learning. The future research relates such issues to the specific features of interdisciplinary collaboration.

CONCLUSION

The study described here is only a first step in the interdisciplinary research field dedicated to the inquiry into the affect of the expert's individual cognitive style parameters on the group structuring design activity. Our results are therefore of a preliminary nature.

Group ontology design, both in dyads and groups of more than two participants, is performed either "from scratch" or on the basis of the drafts prepared by the members of the group individually.

Merging of the individual drafts into a single group ontology follows either an absorption or compromise scenario, or a synthesis scenario. The absorption scenario of the merging of ontologies can be implied from either a conjunctive or a disjunctive model.

From the different cognitive styles, the field-dependent participants tended to prefer the disjunctive model of merging ontologies, with the higher power ontology absorbing the lower power one and further merging of the same-level nodes in the resulting ontology.

The field-independent participants tended to prefer the conjunctive model, with the node reduction leading to the resulting ontology, with the conjunction or interception only of the same-level nodes.

Despite the preliminary character of the research results, all the findings can be used in organizing collective ontology development, data structuring and other group analytical work. Taking into consideration their students' cognitive styles when organizing them into groups for collective ontology development, the instructor enhances her or his ability of guiding the ontology development process to reach deeper levels of students' understanding of the subject domain. Our work presents a novel approach on collaborative ontology development from the psychological point of view. Knowing of the cognitive personal peculiarities helps to leverage the subjectivity of the individually-designed ontologies.

Using recent advances in knowledge engineering and a human factors approach, we aim at creating new consistent knowledge bases for various areas of science and education.

ACKNOWLEDGMENT

The authors are very thankful to Elena Kotova (psychologist), Andrew Pisarev (programmer), Eugene Blagov and Ekaterina Bolotnikova who helped both with the experimental part and at the stage of paper preparation. The KOMET project was funded by the Russian Foundation for Basic Research (grant 11-07-00140).

REFERENCES

Baader, F., Horrocks, I., & Sattler, U. (2005). Description Logics as Ontology Languages for the Semantic Web. In Mechanizing Mathematical Reasoning (LNAI), (vol. 2605, pp. 228-248). Berlin: Springer-Verlag.

Bard, J. B., & Rhee, S. Y. (2004). Ontologies in biology: Design, applications and future challenges. *Nature Reviews. Genetics*, *5*(3), 213–222. doi:10.1038/nrg1295 PMID:14970823

Barros, B., Verdejo, M. F., Read, T., & Mizoguchi, R. (2002). Applications of a collaborative learning ontology. In C. A. Coello Coello et al (Eds.), MICAI 2002 (LNAI), (vol. 2313, pp. 301-310). Berlin: Springer-Verlag.

Bolotnikova, E. S., Gavrilova, T. A., & Gorovoy, V. A. (2011). To a Method of Evaluating Ontologies. *Journal of Computer and Systems Sciences International*, *50*(3), 448–461. doi:10.1134/S1064230711010072

Brochhausen, M. M., Spear, A. D., Cocos, C. C., Weiler, G. G., Marti'n, L. L., Anguita, A. A., & Tsiknakis, M. M. (2011). The ACGT Master Ontology and its applications — Towards an ontology-driven cancer research and management system. *Journal of Biomedical Informatics*, *44*(1), 8–25. doi:10.1016/j.jbi.2010.04.008 PMID:20438862

Buzan, T. (2005). *Mind Map handbook*. London: Thorsons.

Chandrasekaran, B., Josephson, J. R., & Benjamins, V. R. (1999). What are ontologies, and why do we need them? *Intelligent Systems and their Applications, 14*(1), 20-26.

Conlon, T. (1997). Towards Diversity: Advancing Knowledge-based Modelling with Knowledge Acquisition. In *Proceedings of 8th International PEG Conference* (pp. 379-386). Sozopol.

Cross, R., Parker, A., Prusak, L., & Borgatti, S. (2001). Knowing what we know: Supporting knowledge creation and sharing in social networks. *Organizational Dynamics, 30*(2), 100–120. doi:10.1016/S0090-2616(01)00046-8

Davies, J., van Harmelen, F., & Fensel, D. (Eds.). (2002). *Towards the Semantic Web: Ontology-Driven Knowledge Management*. New York, NY: John Wiley & Sons, Inc. doi:10.1002/0470858060

Dicheva, D., & Aroyo, L. (2004). Concepts and Ontologies in Web-based Educational Systems. *Int. J. Cont. Engineering Education and Life-long Learning, 14*(3), 187–190.

Dicheva, D., & Dichev, C. (2005). Authoring Educational Topic Maps: Can We Make It Easier? In *Proceedings of 5th IEEE International Conference on Advanced Learning Technologies, ICALT 2005*, (pp.216-219). Kaohsiung, Taiwan: IEEE. doi:10.1109/ICALT.2005.73

Dicheva, D., & Dichev, C. (2007). Helping Courseware Authors to Build Ontologies: the Case of TM4L. In *Proceedings of 13th Int. Conference on Artificial Intelligence in Education* (pp. 77-84). Academic Press.

Diviacco, P. (2015). Maps, graphs and annotations as Boundary Objects in Knowledge Networks, Distributed Cognition and Collaborative e-Research. In P. Diviacco, P. Fox, A. Leadbetter, & C. Pshenichny (Eds.), *Collaborative Knowledge in Scientific Research Networks*. IGI Global Press.

Eppler, M. J., & Bresciani, S. (2013). Visualization in management: From communication to collaboration: A response to Zhang. *Journal of Visual Languages and Computing, 24*(2), 146–149. doi:10.1016/j.jvlc.2012.11.003

Eppler, M. J., & Burkhard, R. A. (2008). Knowledge Visualization. In M. Jennex (Ed.), *Knowledge Management: Concepts, Methodologies, Tools, and Applications* (pp. 781–793). Hershey, PA: Information Science Reference.

Fillenbaum, S. (1959). Some stylistic aspects of categorizing behavior. *Journal of Personality, 27*(2), 187–195. doi:10.1111/j.1467-6494.1959.tb01828.x PMID:13665498

Fonesca, F., Davis, C., & Camara, G. (2003). Bridging ontologies and conceptual schemas in geographic information integration. *GeoInformatica, 7*(4), 355–378. doi:10.1023/A:1025573406389

Gaeta, M., Orciuoli, F., & Ritrovato, P. (2009). Advanced ontology management system for personalised e-Learning. *Knowledge-Based Systems, 22*(4), 292–301. doi:10.1016/j.knosys.2009.01.006

Gangemi, A., Catennaci, C., Ciaramita, M., & Lehman, J. (2006). Modelling ontology evaluation and validation. In Y. Sure & J. Domingue (Eds.), *The Semantic Web: Research and Applications, LNCS 4011* (pp. 140–154). Berlin: Springer-Verlag.

Gavrilova, T. (2011). Knowledge Engineering: Structural View. In *New Trends in Software Methodologies, Tools, and Techniques Frontiers: Proceedings of the Tenth SoMet_11* (pp. 219 – 227). IOS Press.

Gavrilova, T., Bolotnikova, E., & Gorovoy, V. (2012). New Ergonomic Metrics for Educational Ontology Design and Evaluation. In *New Trends in Software Methodologies, Tools, and Techniques Frontiers» in Artificial Intelligence and Applications series: Proceedings of the Tenth SoMet_12* (vol. 246, pp. 361-378). IOS Press.

Gavrilova, T., Leshcheva, I., & Bolotnikova, E. (2012). Using visual conceptual models in teaching. In *Proceedings of the 8th International Conference on Education ICE-2012* (pp. 191–197). Research and Training Institute of East Aegean.

Gavrilova, T., Leshcheva, I., Bolotnikova, E., Blagov, E., & Yanson, A. (2013). Measuring Psychological Impact on Group Ontology Design and Development: Empirical Approach. In *Proceedings of 4th International Conference "Knowledge Engineering and the Semantic Web"* (vol. 394, pp. 29-43). Springer.

Gavrilova, T. A., Ravodin, R. A., Bolotnikova, E. S., & Kotko, A. V. (2012). Collaborative development of Dermatovenerology Ontology. In *Proceedings of Int. conf. Med-e-Tel 2012: Global Telemedicine and eHealth Updates: Knowledge Resources,* (vol. 5, pp. 39-43). ISfTeH, G.D. of Luxembourg.

Gavrilova, T. A., & Voinov, A. (1996). Visualized Conceptual Structuring for Heterogeneous Knowledge Acquisition. In *Proceedings of International Conference on Educational Multimedia and Hypermedia, EDMEDIA'96* (pp. 258-264). Boston: MIT.

Gómez-Pérez, A., Fernández-López, M., & Corcho, O. (2004). *Ontological Engineering with examples from the areas of Knowledge Management, e-Commerce and the Semantic Web*. London: Springer-Verlag.

Grosslight, L., Unger, C., Jay, E., & Smith, C. L. (1991). Understanding models and their use in science: Conceptions of middle and high school students and experts. *Journal of Research in Science Teaching*, *28*(9), 799–822. doi:10.1002/tea.3660280907

Gruber, T. (1993). A translation approach to portable ontology specifications. *Knowledge Acquisition*, *5*(2), 199–220. doi:10.1006/knac.1993.1008

Guarino, N., & Giaretta, P. (1998). Ontologies and Knowledge Bases: Towards a Terminological Clarification. In Towards Very Large Knowledge Bases: Knowledge Building & Knowledge Sharing (pp. 25-32). IOS Press.

Hayes, J., & Allinson, C. W. (1998). Cognitive style and the theory and practice of individual and collective learning in organizations. *Human Relations*, *51*(7), 847–871. doi:10.1177/001872679805100701

Hevner, A. R. (2007). The three cycle view of design science research. *Scandinavian Journal of Information Systems*, *19*(2), 87.

Jonassen, D. H. (1999). Designing constructivist learning environments. In C. M. Reigeluth (Ed.), Instructional design theories and models (2nd ed.), (Vol. 2, pp. 215-239). Mahwah, NJ: Lawrence Erlbaum.

Kagan, J. (1966). Reflection-impulsivity: The generality and dynamics of conceptual tempo. *Journal of Abnormal Psychology*, *71*(1), 17–24. doi:10.1037/h0022886 PMID:5902550

Kholodnaya, M. (2004). *Kognitivnye stili. O prirode individual'nogo uma* (2nd ed.). Saint Petersburg: Piter. (in Russian)

Kinchin, I. M., De-Leij, F. A. A. M., & Hay, D. B. (2005). The evolution of a collaborative concept mapping activity for undergraduate microbiology. *Journal of Further and Higher Education*, *29*(1), 1–14. doi:10.1080/03098770500037655

Knight, C., Gašević, D., & Richards, G. (2006). An Ontology-Based Framework for Bridging Learning Design and Learning Content. *Journal of Educational Technology & Society*, *9*(1), 23–37.

Kuznetsov, S. O., Obiedkov, S., & Roth, C. (2007) Reducing the Representation Complexity of Lattice-Based Taxonomies. In PrissU. PolovinaS.HillR. (Eds.), *Proc. 15th International Conference on Conceptual Structures (ICCS 2007) (LNAI), (vol. 4604*, pp. 241-254). Springer. doi:10.1007/978-3-540-73681-3_18

Mizoguchi, R., & Bourdeau, J. (2000). Using Ontological Engineering to Overcome Common AI-ED Problems. *International Journal of Artificial Intelligence in Education, 11*(2), 107–121.

Neches, R., Fikes, R. E., Finin, T., Gruber, T., Patil, R., Senator, T., & Swartout, W. R. (1991). Enabling Technology for Knowledge Sharing. *AI Magazine*, *12*(3), 36–56.

Neon Project. (n.d.). Retrieved from http://www.neon-project.org/

Novak, J. D. (1998). *Learning, Creating, and Using Knowledge: Concept Maps as Facilitative Tools in Schools and Corporations*. Mahwah, NJ: Lawrence Erlbaum & Assoc.

Novak, J. D., & Cañas, A. J. (2006). The Theory Underlying Concept Maps and How to Construct Them. *Technical Report IHMC CmapTools 2006-01*. Florida Institute for Human and Machine Cognition. Retrieved online December 12, 2006, from http://cmap.ihmc.us/Publications/ResearchPapers/TheoryUnderlyingConceptMaps.pdf

O'Donnell, A. M., Dansereau, D. F., & Hall, R. H. (2002). Knowledge Maps as Scaffolds for Cognitive Processing. *Educational Psychology Review*, *14*(1), 71–86. doi:10.1023/A:1013132527007

Oltramari, A., & Ferrario, R. (2009). *Formal Ontologies Meet Industry*. Amsterdam: IOS Press.

Peterson, E. R., Rayner, S. G., & Armstrong, S. J. (2009). Researching the psychology of cognitive style and learning style: Is there really a future? *Learning and Individual Differences*, *19*(4), 518–523. doi:10.1016/j.lindif.2009.06.003

Pettigrew, T. F. (1958). The measurement and correlates of category width as a cognitive variable. *Journal of Personality*, *26*(4), 532–544. doi:10.1111/j.1467-6494.1958.tb02350.x

Pfister, R. A., & Eppler, M. J. (2012). The benefits of sketching for knowledge management. *Journal of Knowledge Management*, *16*(2), 372–382. doi:10.1108/13673271211218924

Protégé. (n.d.). Retrieved from http://protege.stanford.edu

Schnotz, W., & Kurschner, C. (2008). External and Internal Representations in the Acquisition and Use of Knowledge: Visualization Effects on Mental Model Construction. *Instructional Science: An International Journal Of The Learning Sciences*, *36*(3), 175–190. doi:10.1007/s11251-007-9029-2

Schreiber, G. (2000). *Knowledge Engineering and Management: The CommonKADS Methodology*. Cambridge, MA: MIT Press.

Sherlock, G. (2000). Analysis of large-scale gene expression data. *Current Opinion in Immunology*, *12*(2), 201–205. doi:10.1016/S0952-7915(99)00074-6 PMID:10712947

Shneiderman, B. (1996). The eyes have it: A task by data type taxonomy for information visualizations. In *Proceedings of IEEE Symposium on Visual Languages* (pp. 336-343). IEEE Computer Society Press. doi:10.1109/VL.1996.545307

Sowa, J. F. (1984). *Conceptual Structures: Information Processing in Mind and Machine*. Reading, MA: Addison-Wesley.

Stephens, S. (2012). From Tree to Map: Using Cognitive Learning Theory to Suggest Alternative Ways to Visualize Macroevolution Evolution. *Education and Outreach*, *5*(4), 603–618. doi:10.1007/s12052-012-0457-3

Tansley, S., & Tolle, K. M. (Eds.). (2009). *The fourth paradigm: data-intensive scientific discovery*. Redmond, WA: Microsoft Research.

Wertheimer, M. (1938). *Gestalt theory*. Hayes Barton Press.

Witkin, H. A., Oltman, P.K., Raskin, E. & Karp, S.A. (1971). A manual for the embedded figures tests. Consulting Psychologists Press.

Witkin, H. A., Moore, C. A., Goodenough, D. R., & Cox, P. W. (1977). Field-dependent and field-independent cognitive styles and their educational implications. *Review of Educational Research*, *47*(1), 1–64. doi:10.3102/00346543047001001

Yudelson, M., Gavrilova, T., & Brusilovsky, P. (2005). Towards User Modeling Meta Ontology. In L. Ardissono, P. Brna, & A. Mitrovic (Eds.), *UM 2005 (LNAI), (vol. 3538*, pp. 448–452). Berlin: Springer-Verlag.

ADDITIONAL READING

Adeli, H. (1994). *Knowledge Engineering*. New York: McGraw-Hill.

Alitek Consulting. What is strategic mapping? Retrieved online August 1, 2003, from http://www.alitek.com.au/strategic-mapping.htm

Allinson, C. W., & Hayes, J. (1996). The cognitive style index, a measure of intuition-analysis for organization research. *Journal of Management Studies*, *33*(1), 119–135. doi:10.1111/j.1467-6486.1996.tb00801.x

Ashburner, M., Ball, C. A., Blake, J. A., Botstein, D., Butler, H., Cherry, J. M., & Sherlock, G. (2000). Gene Ontology: Tool for the unification of biology. *Nature Genetics*, *25*(1), 25–29. doi:10.1038/75556 PMID:10802651

Boose, J. H. (1990). Knowledge acquisition tools, methods and mediating representations. In H. Motoda et al. (Eds.), *Knowledge Acquisition for Knowledge-Based Systems* (pp. 123–168). Tokyo: IOS Press, Ohinsha Ltd.

Brusilovsky, P., & Rizzo, R. (2002). Map-Based Horizontal Navigation in Educational Hypertext. *Proceedings of Hypertext*, University Of Maryland, College Park, USA.

Card, S. K., Mackinlay, J. D., & Shneiderman, B. (1999). *Readings in Information Visualization: Using Vision to Think*. Morgan Kaufmann.

Eisenstadt, M., Domingue, J., Rajan, T., & Motta, E. (1990). Visual Knowledge Engineering. *IEEE Transactions on Software Engineering*, *16*(10), 1164–1177. doi:10.1109/32.60296

Fensel, D. (2001). *Ontologies: A Silver Bullet for Knowledge Management and Electronic Commerce*. New York: Springer. doi:10.1007/978-3-662-04396-7

Gavrilova, T., Brusilovsky, P., Yudelson, M., & Puuronen, S. (2006). Creating Ontology for User Modelling Research. In: *Workshop Ubiquitus User Modeling on European Conference on Artificial Intelligence ECAI 2006*, (pp. 11-15). Italy, Riva del Garda.

Gavrilova, T., Dicheva, D., Sosnovsky, S., & Brusilovsky, P. (2005). Ontological Web Portal for Educational Ontologies. In *Proc. Of Applications of Semantic Web Technologies for E-Learning Workshop (SW-EL'05) in conjunction with 12th Int. Conf. on Artificial Intelligence in Education (AI-ED'05)* (pp. 19-29). Amsterdam.

Gavrilova, T., Farzan, R., & Brusilovsky, P. (2005). One Practical Algorithm of Creating Teaching Ontologies. In: http://www.alitek.com.au/strategic-mapping.htm, *12th International Network-Based Education Conference NBE 2005* (pp. 29-37). Finland, Rovaniemi.

Gavrilova, T., & Laird, D. (2005). Practical Design of Business Enterprise Ontologie. M. Bramer and V. Terzyan (Eds.), Industrial Applications of Semantic Web (pp. 61-81). New York: Springer.

Noy, N., & McGuinness, D. (2001). Ontology Development 101: A Guide to Creating Your First Ontology. *Stanford Knowledge Systems Laboratory Technical Report KSL-01-05 and Stanford Medical Informatics Technical Report SMI-2001-0880.*

Sosnovsky, S., & Gavrilova, T. (2006). One Approach to Development of Educational Ontology for C-Programming. *Int. Journal of Information Theories and Applications, 13*(4), 303–308.

Sternberg, R. (1997). *Thinking Styles.* Boston: Cambridge University Press. doi:10.1017/CBO9780511584152

Tu, S., Eriksson, H., Gennari, J., Shahar, Y., & Musen, M. (1995). Ontology-Based Configuration of Problem-Solving Methods and Generation of Knowledge-Acquisition Tools. *Artificial Intelligence in Medicine, 7*(3), 257–289. doi:10.1016/0933-3657(95)00006-R PMID:7581625

Vestal, W. (2005). *Knowledge Mapping: The Essentials for Success.* APQC.

Wielinga, B., Schreiber, G., & Breuker, J. (1992). A Modelling Approach to Knowledge Engineering. *Knowledge Acquisition, 4*(1), 23–39. doi:10.1016/1042-8143(92)90013-Q

Witkin, H. A., & Goodenough, D. R. (1981). *Cognitive Styles: Essence and Origins.* NY: International Universities Press.

KEY TERMS AND DEFINITIONS

Cognitive Styles: Explain and describe how the individual acquires knowledge and how an individual processes information. The cognitive styles are related to problem solving, and generally to the way that information is acquired, structured and used.

Concept Map: Directed graph whose nodes represent concepts and links present relations.

Field Independence: The tendency of an individual to separate details from the surrounding context. It can be compared with field-dependence, relative inability to distinguish detail from other information around it.

Impulsivity/Reflection Characteristic: Considers the amount of information collected prior to making a decision: impulsive individuals are able to make decisions on a considerably bounded information basis, while reflective individuals are more inclined to make decisions considering complete information on the situation.

Knowledge Engineering: Science that studies knowledge acquisition, structuring and formalization.

Laddering: Creating hierarchical structures of concepts.

Mind-Maps: Visual diagrams that represent ideas and information using a combination of words and images with lines to draw relationships between topics.

Narrowness/Width of the Category: A cognitive style parameter describing the ability of an individual to restrict (or to broaden) the area of application of a certain category.

Ontology: Hierarchically structured concept vocabulary.

Section 4

Formalization of Syntax and Semantics of Dynamic Knowledge

Referring to the schema described in the description of Section 3, this section addresses issues related to dynamic knowledge formalization methods such as mind mapping, event bushes, and Petri nets.

Chapter 16
Grammar of Dynamic Knowledge for Collaborative Knowledge Engineering and Representation

Cyril Pshenichny
Geognosis Project, ITMO University, Russia

Dmitry Mouromtsev
ITMO University, Russia

ABSTRACT

Constructive discussion must lead to a shared understanding. This understanding is commonly expressed as text; however, for the purposes of collaborative research, the tools of knowledge engineering/knowledge representation look more appropriate. The problem with them is that, to the present day, they are developed largely for the tasks that imply fixed relations between things and their properties, termed here as static. However, collaborative research often deals with fields of knowledge that represent changing environments where these relations cannot be considered fixed, and the tools to capture scenarios of evolution (i.e. the dynamic tools of knowledge engineering) are far from that evolved as static ones, mainly due to the lack of strict logical or mathematical foundation for representation of dynamic knowledge. This chapter presents an attempt to formulate a unified grammar to encode the knowledge of changing environments in any field of science.

INTRODUCTION

Constructive discussion must lead to a shared understanding – at least to understanding of points of disagreement. In modern literature, these "milestones of understanding" are often termed as boundary objects defined as "artifacts that aim to bridge concurrent cognitive models through abstraction from all the domains of the partners" (Diviacco, 2012, p. 17). These artifacts range from purely formal (e.g., file structure, blank layers of a geographic information system, keys of

DOI: 10.4018/978-1-4666-6567-5.ch016

Copyright © 2015, IGI Global. Copying or distributing in print or electronic forms without written permission of IGI Global is prohibited.

a database), which add little to understanding, to some shared meanings. So far, common practice of human discussion has been mainly to come up with a text, be it a manifesto or research paper, which aims to fix and present these meaningful boundary objects (usually in the form of generic terms).

Nevertheless, text has a few properties that undermine its ability to serve this purpose – i.e., unconstrained grammar of a natural language always leaves a room for uncertainty. Verbal narration is unlikely to present all possible consequences from a considered point (that is, to make sure that understanding is really achieved); the way the consequences are made is often unclear and non-verifiable; moreover, to be readable, text should preserve intonations, style and emphases, which may stress some points and obscure others. Finally, text has a beginning and end and thus channels the thought in one given direction, while the vision it strives to communicate may – and commonly does – require not a single-path and one-way, but a network-like, or polytree, pattern of presentation, with an opportunity to travel both ways on each path. Even the hypertext markup language is not free from this disadvantage because it rarely allows to follow several links between the documents simultaneously.

An alternative approach is offered by knowledge engineering/knowledge representation. That is, to bind verbal boundary objects not in a text but in a graphic conceptualization of the considered field of knowledge (e.g., Martin, 2002; Gavrilova et al., 2013).

However, the tools suggested so far (ontologies, conceptual graphs and others) cover only a part of the tasks arising at conceptualization of fields of knowledge – namely, the representation of static knowledge. Static is the knowledge in which the relations between objects are considered fixed and not changing (Pshenichny and Mouromtsev, 2013). This has proven to be efficient for a wide range of tasks like building data models and search algorithms or description of those domains in which things, their properties and relations are considered stable – e.g., production units of a factory or biological species.

The other part of existing tasks, which relates to the knowledge on changing events and implies consideration of processes and scenarios, is regarded as *dynamic knowledge* (Pshenichny and Mouromtsev, 2013). It remains largely uncovered by existing tools of knowledge engineering and representation. Though, there is quite a research done in the field of dynamic ontologies and ontology evolution (Stojanovic, 2004; Zablith, 2008; Murdock et al., 2010, and others) including that in collaborative environments (Noy et al., 2006), but the approach in this field is quite different and focuses on how to adapt ontology to changing phenomena, while the issue is not how to capture the changes in the phenomena themselves. Speaking about "evolution", "processes", "scenarios" and related issues, the researchers in this field mean the change of ontology proper, not of the world it describes, and look for solutions to keep ontology self-consistent despite the changing world rather than to model the changes of the latter.

Meanwhile, the tasks of representation and engineering of dynamic knowledge are especially crucial for collaborative studies where the issue of mutual understanding acquires a key importance. Ignoring the dynamic properties of many environments (e.g., a man is sleeping – a man is vigil *and not sleeping*), one would not reach understanding but imitate it – obscuring, but not resolving, the contradictions. Therefore, a formal account of dynamic nature of many environments addressed by various disciplines, from geosciences to psychology, is of vital importance for efficient collaborative studies and research networking.

The objective of this chapter is to draft a general approach to collective reasoning and representation of knowledge of changing environments. To do this, we will

- Consider the concept of static versus dynamic knowledge,

- Constrain the semantics and syntax of dynamic knowledge representation,
- Probe the pragmatic aspects of the suggested semantics and syntax.

STATIC VERSUS DYNAMIC KNOWLEDGE

Classification of knowledge into static and dynamic was suggested for the knowledge engineering and knowledge representation by Pshenichny and Kanzheleva (2011). The key concept of this approach is *environment*, understood as a situation being modeled. Environment is modeled as a set of statements of general form $S - P_1, P_2, \ldots, P_n$, where S is subject, and $P_1, P_2, \ldots, P_n$ are predicates, with or without negation. We put three restrictions herewith, (i) subject may be only one in a statement, (ii) subject may not be negated, (iii) all predicates may not be negated in a statement, at least one must be without negation (e.g., a statement $S - \neg P$, $\neg$ being the negation, is not allowed). The reason to introduce the limitation (iii) is the necessity to indicate at least one property for any subject.

These statements are called *events*. Events may be organized by different rules of inference or, informally speaking, just rules of composition. Events having the same subject and predicates denying one another:

$S' - P_1, P_2, P_3, \ldots, P_{n-1}, P_n,$

$S' - \neg P_1, P_2, P_3, \ldots, P_{n-1}, P_n,$

$S' - P_1, \neg P_2, P_3, \ldots, P_{n-1}, P_n,$

$\ldots,$

$S' - P_1, P_2, P_3, \ldots, \neg P_{n-1}, P_n,$

$S' - P_1, P_2, P_3, \ldots, P_{n-1}, \neg P_n,$

$S' - \neg P_1, \neg P_2, P_3, \ldots, P_{n-1}, P_n,$

$S' - P_1, \neg P_2, \neg P_3, \ldots, P_{n-1}, P_n,$

$\ldots,$

$S' - P_1, P_2, P_3, \ldots, \neg P_{n-1}, \neg P_n,$

$\ldots,$

$S' - \neg P_1, P_2, \neg P_3, \ldots, P_{n-1}, P_n,$

$\ldots$

$S' - \neg P_1, \neg P_2, \neg P_3, \ldots, \neg P_{n-1}, P_n,$

are considered mutually exclusive. Maximum number of mutually exclusive events having the same predicates is determined by the amount of these predicates and can be calculated by the Newton binomial. Presence of a predicate and its negation in one event is not admitted. *Dynamic* is the environment, in which mutually exclusive events are present; otherwise the environment is *static* (Figure 1).

As shown in Figure 1, static environment can be reduced to consideration of relations between classes and, possibly, individuals; hence, there is no need to identify events in it. On the contrary, dynamic environments imply, in fact, a passage from one static environment to another, incompatible with the former, and thus needs consideration of events. The passage from one event to another or, broadly speaking, from one set of events to another set is termed *change*, or *process* (Pshenichny and Kanzheleva, 2011). Dynamic environments are classified further as non-directed, directed singular path and directed alternative path environments (Figure 2).

Figure 1. The difference between static and dynamic environments. Static environments imply fixed relations between objects. Dynamic environment can be represented as a succession of static, possibly including mutually exclusive events ("Man is healthy, strong and sleeping" - "Man is healthy, strong and not sleeping"). In the simplest case shown here, each static environment consists of one event.

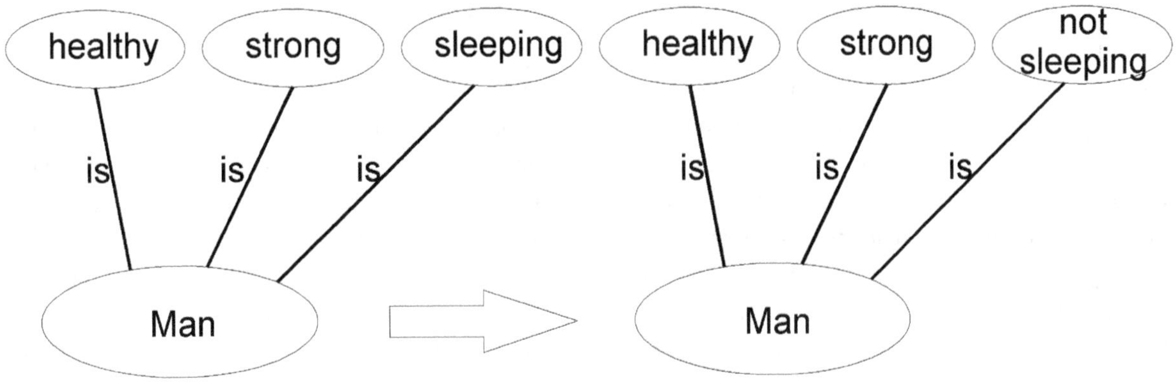

Figure 2. Classification of modeling environments. Directed changes are those having clear beginning and end. If changes are not directed, there must be either a cycle (a loop), or oscillations expressed qualitatively, or chaotic changes, which cannot be analyzed qualitatively (but not complete quiescence which would bring us to the no-change environment addressed by object-based methods).

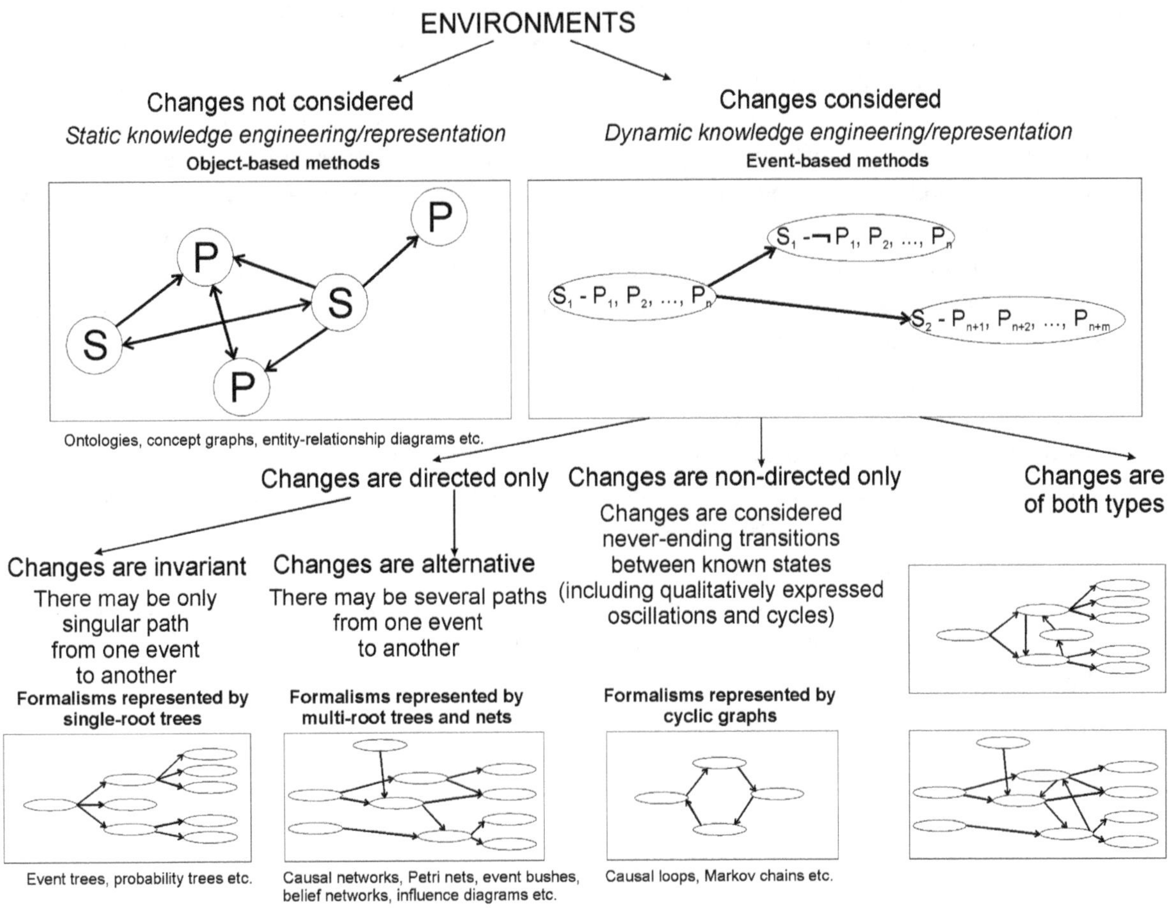

METHODS FOR ENGINEERING AND REPRESENTATION OF DYNAMIC KNOWLEDGE

Static environments are addressed by traditional knowledge representation tools (ontologies, conceptual graphs, event-driven process chains, ontology web language, knowledge interchange format, resource description framework, entity-relationship diagrams and many others), which operate with objects considered either as classes (i.e., types: concept types or relation types) or as individuals (Martin, 2002). The grammar for their representation is well-developed and rooted in classical logic and set theory.

The armory for dynamic environments is much looser and poorly organized. Different methods exist in various disciplines, from industrial engineering to geosciences, they were designed for particular tasks, and no unified grammar has been suggested so far. Non-directed dynamic environments can be captured, e.g., by causal loop diagrams, directed singular path ones, by event trees, directed alternative path, by event bushes, Petri nets, Bayesian networks and influence diagrams and some other methods. A large group of methods can capture any type of dynamic environment; however, it is unlikely that methods can be found that are equally good for static and dynamic ones. Though, some of the methods considered here as static – e.g., special types of ontologies (time, process, or so-called dynamic ones) or conceptual graphs – do describe processes in environments, but their approach is essentially static, as they do not allow a subject to change a predicate to its negation. Therefore, what they produce is either a classification or declaration of processes or events, but not scenarios of admissible changes in an environment.

FRAMEWORK FOR UNIFIED GRAMMAR OF DYNAMIC KNOWLEDGE

General Approach

To describe scenarios of a changing environment, it looks reasonable to examine in detail what changes may take place in it and attempt to formally classify these changes based on the general rule that changing environment may include pairs of mutually exclusive events. Putting a mild condition that mutually exclusive events may be, but not necessarily are, present in an environment, we are going to develop a universal framework both for dynamic and static environments. However, the cases in which said events are not present are expected to be reducible to traditional grammar of ontologies, conceptual graphs and the like.

Below we will develop a three-stage approach to classification of changes in modeling environments. The entire classification is quite big and its detailed consideration departs far from the issues of collaborative research, so it will be presented elsewhere. Another point deliberately omitted here is interrelation between the specified types of change and the above classification of environments (Figure 1). Aside from the "no-change" particular case that corresponds to the static environment, the rest cases are believed to refer each to all three types of changing environment; however, there may arise a specificity in their application to each of these.

Henceforth, an informal term "change", or "process", will be substituted by its formal equivalent, *connective*, by analogy with the mathematical logic (Kleene, 1952; Hilbert and Bernyce, 1934; Takeuti, 1975, and others).

Types and Subtypes of Connective of the Unified Grammar of Dynamic Knowledge

Each connective has its causal (left) and effect (right) sets of events, and these three together make up a *formula*:

$$\{E_1, E_2, \ldots, E_i\}\ Connective\ \{E_{i+1}, E_{i+2}, \ldots, E_j\}$$

where $i, j=1, 2, \ldots, n$, n being natural number, E is event, $\{E_1, E_2, \ldots, E_i\}$ is the left set, and $\{E_{i+1}, E_{i+2}, \ldots, E_j\}$, the right set of events.

None of these sets may be empty, all of them must be finite; within one set, none of the events may be the same (i.e., have the same subject, the same set of predicates and the same predicates negated), but the same event may occur once in the left, and once in the right set.

It should be noted herewith that while the events are analyzed in terms of their subjects and predicates (as is the case in this work), they can be regarded either as the same or as different. If some additional information is associated with events (e.g., time intervals), then perhaps it would be more reasonable to say not "the same" but "similar". However, this is a matter of the future work.

As the first step, connectives are classified by the number of events in left and right sets. Four cases are defined as follows and denoted by capital letters of the Old Slavonic alphabet zhivot, shta, yon and izhen (Figure 3):

These letters were chosen because they well represent visually the relation between the causal and effect sets, meaning causal set above, and effect, below in each letter: "many lead to many" (*zhivot*), "many lead to one" (*shta*), "one leads to many" (*yon*) and "one leads to one" (*izhen*). Old Slavonic (or Early Cyrillic) alphabet can be easily found on the Internet as pixel graphics (see, e.g., Kodeks webpage at the Otto-Friedriech-Universitaet Bamberg website (http://kodeks.uni-bamberg.de/aksl/Schrift/AkslAlphabet.htm), the Omniglot page on Old Slavonic alphabet (http://www.omniglot.com/writing/ocslavonic.htm) or Wikipedia entry for Early Cyrillic alphabet, http://en.wikipedia.org/wiki/Early_Cyrillic_alphabet), though pronunciation of letters may slightly differ depending on the source. The easiest way of using these symbols (followed by the authors of this chapter) is saving a graphic file and copying the symbols from there as graphic objects.

At the second step, each of four types of connective is examined for similarity of its left and right sets. The following subtypes can be defined,

zhivot1: The same events are not present in the left and right sets;

Figure 3. Main types of connectives of the unified grammar of dynamic knowledge: zhivot (a), shta (b), yon (c) and izhen (d)

$$\{E_1, E_2, \ldots, E_i\}\ \text{Ж}\ \{E_{i+1}, E_{i+2}, \ldots, E_j\} \qquad \text{(a)}$$

$$\{E_1, E_2, \ldots, E_i\}\ \text{Щ}\ E_{i+1} \qquad \text{(b)}$$

$$E_1\ \text{Ѩ}\ \{E_2, E_3, \ldots, E_j\} \qquad \text{(c)}$$

$$E_1\ \text{І}\ E_2 \qquad \text{(d)}$$

zhivot2: Some events are the same in the left and right sets;

zhivot3: All events are the same in the left and right sets;

shta1: The only event of the right set is present on the left;

shta2: The event of the right set is not present on the left;

yon1: The only event of the left set is present on the right;

yon2: The only event of the left set is not present on the right;

izhen1: The events at the left and at the right are the same;

izhen2: The events at the left and at the right are different.

Further classification has been made for the first type of connective, *zhivot*. Its subtype *zhivot2* is classified into three sub-subtypes,

zhivot2a: All events of the left set are present on the right (but not everything that is on the right is also present on the left);

zhivot2b: Some events of the left set are present on the right, some are not, and some events are present in the right set and are absent on the left;

zhivot2c: All events of the right set are present on the left (but not everything that is on the left is also present on the right).

This gives ground to define the *zhivot* (and other) connective subtypes in a way that resembles the definition of connectives in the propositional logic, with a few important distinctions:

- Events may change their truth value from left to right in a formula;
- Not individual events but their sets or subsets are listed in truth tables;
- It is assumed that all events in a particular (sub)set that is listed in a truth table must have the same truth value, otherwise the truth value of the formula cannot be defined (the latter, however, resembles the requirement of classical logical inference that all the premises must be true for the inference to be true).

The latter condition can be understood broadly – if there is a subset of mutually exclusive events in given set, naturally, only one event of this subset may be true, others are necessarily wrong. Then the truth value of this subset is assumed truth if one of the mutually exclusive events is true, and false, if all of them are false. (Importantly, while the presence of a predicate and its negation is forbidden in one event, the presence of mutually exclusive events in a set is allowed.)

The dynamic knowledge, being the knowledge of changing events, implies that, if only different events are considered, some of them are causes, and some, effects. Herewith, it is not necessary that all causes are placed on the left, and effects, on the right. The connectives are meant to express not only this but also much more complicated cases when causes produce effects but are inherited (and, hence, may produce other effects later), or effects pre-exist but are enhanced by additional causes, and so forth. To express these cases, both left and right sets may include subsets of causal and effect events and it is admitted that some or all of these events may be also inherited from left to right changing or preserving their truth value.

Classification of Subtypes of Zhivot

The definition of *zhivot1* is shown in Table 1, where the columns left of the connective mean the $\{E_{causal}\}$ and $\{E_{effect}\}$ subsets of the left set, and columns right of the connective, the subsets of the right set of events. In the future they will be probably generalized to formally designated sets analogous to lists of formulae in the calculi of classical logic (see, e.g., Gentzen, 1934; Takeuti, 1975).

As is seen from the table, only one combination of values, namely that in the first string, is

Table 1. Truth table for the zhivot1 subtype connectives

$\{E_{causal}\}$	$\{E_{effect}\}$	Zhivot1	$\{E_{causal}\}$	$\{E_{effect}\}$
T	F	T	F	T
T	F	F	T	F
T	F	F	T	T
T	F	F	F	F
F	T	F	F	T
F	T	F	T	F
F	T	F	T	T
F	T	F	F	F
T	T	F	F	T
T	T	F	T	F
T	T	F	T	T
T	T	F	F	F
F	F	F	F	T
F	F	F	T	F
F	F	F	T	T
F	F	F	F	F

defined to return the value "true" for the connective. Therefore, it is not necessary to list all the variants in the definition tables for the connective types and subtypes, and henceforth this will be used to minimize the size of the tables.

For the sub-subtype *zhivot2a*, the following definition is suggested (Table 2).

For the sub-subtype *zhivot2b*, the definition is in Table 3.

The *zhivot2c* sub-subtype is defined in Table 4. This illustrates the case when some events cease as a result of presence of other events.

Finally, the *zhivot3* subtype is defined as follows (Table 5).

Table 2. Truth table for the zhivot2a subtype connectives

$\{E_{causal}\}$	$\{E_{effect}\}$	Zhivot2a	$\{E_{causal}\}$	$\{E_{effect}\}$
T	F	T	T	T
Other variants		F	Other variants	

These definitions, along with definitions of the subtypes of *shta*, *yon* and *izhen* connectives that will be given elsewhere, are expected to become the basis for the formulation of event calculus by analogy with sequent or natural-sequent calculi of the classical logic (Gentzen, 1934). Indeed, scenarios, understood as successions of changes, can be parallelized with sequents of formulae, and rules of adding/removal of formulae (i.e., connectives and sets of events they connect) can be suggested.

At the third step of the considered classification, the internal structure of events of the left and right sets is addressed. To do this, one should recall the definition of event as an expression of a general form $S - P_1, P_2, \ldots, P_i$, where S is subject, and $P_1, P_2, \ldots, P_i$ are predicates, with or without negation. The options that arise herewith will be referred to as *modi (moduses) of connectives*. By analogy, consideration of the internal structure of propositions also marks the transition from the propositional to predicate logic in the classical logic framework.

To avoid controversy, an assumption should be introduced for any superset of events connected by a connective (a superset that includes left and right sets of events or, alternatively, any subsets of the said sets): if any two events within this superset have the same subject, they must have the same predicates too and may differ only by negation signs – i.e., may be either mutually exclusive or the same. If they are the same, (i) they may be only two in the superset and (ii) must belong one to the left, one, to the right sets – because, first, if to divide the superset into left and right sets, no other sets are possible within it, and then, the same event may not enter either left or right set twice. If the events in the superset are mutually exclusive, they may be either two or more in number and belong to both sets in any proportion, including zero in one set and all in the other. Armed with this assumption, we proceed to classification of modi.

At present this is done only for the *zhivot* type of connective. The same for the entire kit of con-

Table 3. Truth table for the zhivot2b subtype connectives

$\{E_{causal\text{-}non\text{-}inherited}\}$	$\{E_{causal\text{-}inherited}\}$	$\{E_{effect}\}$	*Zhivot2b*	$\{E_{causal\text{-}non\text{-}inherited}\}$	$\{E_{causal\text{-}inherited}\}$	$\{E_{effect}\}$
T	T	F	T	F	T	T
Other variants			F	Other variants		

Table 4. Truth table for the zhivot2c subtype connectives

$\{E_{causal}\}$	$\{E_{effect}\}$	*Zhivot2c*	$\{E_{causal}\}$	$\{E_{effect}\}$
T	T	T	F	T
Other variants		F	Other variants	

Table 5. Truth table for the zhivot3 subtype connectives

$\{E_{inherited}\}$	*Zhivot3*	$\{E_{inherited}\}$
T	T	T
Other variants	F	Other variants

nectives including *shta*, *yon* and *izhen* takes too much space to be included in one publication and will be reported elsewhere. In this work, we aim to demonstrate the approach that will be used.

Consider a pair of events, one belonging to the left, and one, to the right set of the type of connective. Let us call the event from the left set the left event, and that from the right, the right event. The following options are possible passing from the left event to the right one.

1. Nothing changes, the subject and the predicates (including negation) of both events are the same:

Left: $(S - P_1, P_2, \ldots, P_n)$; Right: $(S - P_1, P_2, \ldots, P_n)$.

2. Subject is the same, predicates are the same but the negation changes:

Left: $(S - P_1, P_2, \ldots, P_n)$; Right: $(S - \neg P_1, P_2, \ldots, P_n)$

or

Left: $(S - \neg P_1, P_2, \ldots, P_n)$; Right: $(S - P_1, P_2, \ldots, P_n)$.

or cases including negation of more than one (and not necessarily the first) predicate.

3. Subject changes, while predicates (including negation) remain the same:

Left: $(S_m - P_1, P_2, \ldots, P_n)$; Right: $(S_k - P_1, P_2, \ldots, P_n)$.

4. Subject changes, and the list of predicates does so too (necessarily not only by negation), so that at least one predicate is replaced or added/excluded:

Left: $(S_m - P_1, P_2, \ldots, P_{n-1}, P_n)$; Right: $(S_k - P_1, P_2, \ldots, P_{n-1}, P_{n+1})$

or

Left: $(S_m - P_1, P_2, \ldots, P_{n-1}, P_n)$; Right: $(S_k - P_1, P_2, \ldots, P_{n-1}, P_n, P_{n+1})$.

or cases including both replacement and adding/exclusion, with or without negation; common predicates may be present or not.

Let us now examine whether each of these options can apply to *all* couples of a blindly selected left event and an accidentally selected right event in a given formula with a *zhivot* type connective.

Consider the set of all possible couplings of an accidentally selected left event and an accidentally selected right event. For example, if there is a formula

$$\{E_{1,\,left}, E_{2,\,left}\}\ Zhivot\ \{E_{3,\,right}, E_{4,\,right}\},$$

there will be couples $\{E_{1,\,left}, E_{3,\,right}\}$, $\{E_{1,\,left}, E_{4,\,right}\}$, $\{E_{2,\,left}, E_{3,\,right}\}$ and $\{E_{2,\,left}, E_{4,\,right}\}$.

In a formula

$$\{E_{1,\,left}, E_{2,\,left}, E_{3,\,left}\}\ Zhivot\ \{E_{4,\,right}, E_{5,\,right}\},$$

the couples will be $\{E_{1,\,left}, E_{4,\,right}\}$, $\{E_{1,\,left}, E_{5,\,right}\}$, $\{E_{2,\,left}, E_{4,\,right}\}$, $\{E_{2,\,left}, E_{5,\,right}\}$, $\{E_{3,\,left}, E_{4,\,right}\}$ and $\{E_{3,\,left}, E_{5,\,right}\}$, and so on.

Taking into the account that in case of the *zhivot* type connective the number of events both at the left and at the right is necessarily more than one, option 1 is not possible because, admitting the opposite, we obtain, e.g., for the case of two left and two right events that if event $E_{1,\,left}$ is the same as event $E_{3,\,right}$, but there must be another couple, in which $E_{3,\,right}$ participates and is the same as its left counterpart, the $E_{2,\,left}$. But this means that $E_{1,\,left}$ is the same as $E_{2,\,left}$, which is not possible. Therefore, option 1 can be represented only by some, but not all, possible couplings in a formula with *zhivot* type connective.

Similarly, option 2 may not be present in the *zhivot* type of connectives because it implies that only one of the mutually exclusive events exists on the left, and only one, on the right, that is in contradiction with the definition of the *zhivot* type.

Likewise, option 3 can, and necessarily will, apply ubiquitously if the superset includes events with different subjects but all having the same list of predicates and their negations.

Finally, option 4 can entirely cover a wide range of said supersets.

Let us denote *extreme modi* those that reflect the case when the entire superset corresponds to one particular option, and *transitional modi*, those that require two or three options to cover the superset. Saying "transitional" one indicates what extreme modi are involved, e.g., "transitional between 1 and 2", where 1 and 2 are extreme modi.

Then, for the *zhivot* type connective, there will be extreme modi 3 and 4. The same time, as seen from the above, transitional modi can be suggested between the options 1 and 2, 1 and 3, 1 and 4, 2 and 3, 2 and 4, 3 and 4, as well as between options 1, 2 and 3, options 1, 2 and 4, options 2, 3 and 4, options 1, 2, 3 and 4. Still, at least for the *zhivot* type of connectives, transitions between 1 and 2, between 1 and 3 and between 1, 2 and 3 are not possible because either they would require more than one event with the same subject to be true in one (left or right) subset simultaneously, which is forbidden, or require an event with a different subject that would lead to the case 1-2-4.

These will be denoted as transitional modi 1-2, 1-3, 1-4 and so forth, up to 1-2-3-4. These modi, extreme and transitional, can be related to subtypes and sub-subtypes of the *zhivot* type connective.

Zhivot1 includes the extreme modi 3, 4 and transitional modus 2-4. It does not include the transitional modi 1-4, 1-2-4 and 1-3-4, because they imply inheritance of one event, which is not the case for *Zhivot1*. Hence, the following *connectives* can be finally specified in this subtype: *Zhivot1,3*, *Zhivot1,4*, *Zhivot1,2-4*, *Zhivot1,3-4* and *Zhivot1,2-3-4*, each corresponding to a particular modus.

Zhivot2a allows only the transitional cases 1-3 and 1-4 and thus gives three connectives, *Zhivot2a,1-3*, *Zhivot2a,1-4* and *Zhivot2a,1-3-4*.

Zhivot2b implies that something is inherited from left to right that means the presence of solely transitional modi between option 1 and others, but not two, and produces five connectives, *Zhivot2b,1-3*, *Zhivot2b,1-4*, *Zhivot2b,1-2-4*, *Zhivot2b,1-3-4* and *Zhivot2b,1-2-3-4*.

Zhivot2c, symmetrically to *Zhivot2a*, allows only the transitional cases 1-3, 1-4 and 1-3-4: *Zhivot2c,1-3*, *Zhivot2c,1-4* and *Zhivot2c,1-3-4*.

*Zhivot*3 gives ground solely for three transitional modi, all of which are characterized by presence of transition from option 1, and, hence, produces three connectives: *Zhivot3,1-3*, *Zhivot3,1-4* and *Zhivot3,1-3-4*.

Full definitions of the connectives of the *zhivot* type are listed in Table 6.

Table 6. Definition of the zhivot type connectives

Connective	Truth Values (Reference)	Modus of Subject and Predicate Change
Zhivot1,3	Table 1	3 (extreme)
Zhivot1,4	Table 1	4 (extreme)
Zhivot1,2-4	Table 1	2-4 (transitional)
Zhivot1,3-4	Table 1	3-4 (transitional)
Zhivot1,2-3-4	Table 1	2-3-4 (transitional)
Zhivot2a,1-3	Table 2	1-3 (transitional)
Zhivot2a,1-4	Table 2	1-4 (transitional)
Zhivot2a,1-3-4	Table 2	1-3-4 (transitional)
Zhivot2b,1-3	Table 3	1-3 (transitional)
Zhivot2b,1-4	Table 3	1-4 (transitional)
Zhivot2b,1-2-4	Table 3	1-2-4 (transitional)
Zhivot2b,1-3-4	Table 3	1-2-4 (transitional)
Zhivot2b,1-2-3-4	Table 3	1-2-3-4 (transitional)
Zhivot2c,1-3	Table 4	1-3 (transitional)
Zhivot2c,1-4	Table 4	1-4 (transitional)
Zhivot2c,1-3-4	Table 4	1-3-4 (transitional)
Zhivot3,1-3	Table 5	1-3 (transitional)
Zhivot3,1-4	Table 5	1-4 (transitional)
Zhivot3,1-3-4	Table 5	1-3-4 (transitional)

One may reasonably argue whether all or some of these connectives can be reduced to other types (*shta*, *yon* or *izhen*). Indeed, as is being done in the relational database theory, relation "many-to-many" can be split into "one-to-many", "many-to-one" and "one-to-one". However, the task of these connectives is to give expressive tools for description and modeling of knowledge, and depending on the meaning, it must be up to a human to decide whether to use the *zhivot* or another type connective. Again, an analogy with the predicate logic is clearly seen in that there is no formal instruction what formula to use to convey the meaning.

Modi of the Zhivot Type Connectives

Whatever be the formalism underlying the dynamic knowledge grammar, it should not be directly seen by end users – the scientists maintaining a collaboration in a domain of knowledge. Still, this formalism would enable to create an efficient networking tool and, desirably, a user-friendly interface that would help the collaborators choose the most appropriate tool for presenting their shared understanding and guarantee the correctness of its use regardless of the properties of the domain.

How this can work is shown below by trivial examples from chemistry and physics interpreted as the *zhivot* type connectives. The formula that illustrates each connective is visualized as a simple directed graph readily usable in collaboration activities. The graphs more or less of this kind have been widely used, e.g., in various COLLA projects (Diviacco, 2012). Vertices of the graph are events and the connective, and arcs have no specific meaning and simply unite each event with the connective. Demonstration needs triviality; however, imagine that in early days of science the formulation of these facts also could result from collaboration and mutual agreement between the scientists. For simplicity, we assumed that in most cases the list of predicates includes only one predicate (maximum two where necessary).

The *Zhivot1,3* connective (Figure 4) is exemplified by a well-known type of chemical reaction between an alkali and acid that leads to formation of salt and water.

In terms of the dynamic knowledge grammar, the contents of Figure 4 can be recorded as follows:

$\{(S_1 - P), (S_2 - P)\}$ *Zhivot1,3* $\{(S_3 - P), (S_4 - P)\}$,

where S_1 is "acid", S_2 is "alkali", S_3 is "salt", S_4 is "water", and P is "in reaction".

The *Zhivot1,4* connective is shown in Figure 5.

The formula for this example is given below:

$\{(S_1 - P_1), (S_2 - P_2)\}$ *Zhivot1,4* $\{(S_3 - P_3), (S_4 - P_4)\}$,

where S_1 is "acid", S_2 is "alkali", S_3 is "salt", S_4 is "water", P_1 is "inorganic", P_2 is "potassic", P_3 is "bitter", and P_4 is "clean".

One may argue that bringing so diverse predicates in one system is not efficient; however, the efficiency of reasoning and knowledge representation is not an issue discussed in this work, here we are concerned merely on opportunities to encode *any* kind of knowledge and reasoning about changing environments. In practice of research and communication, there do arise cases in which bringing such diverse information together is necessary.

The *Zhivot1,2-4* connective is shown in Figure 6.

The formula for this connective is:

$\{(S_1 - P_1), (S_2 - P_2, P_3), (S_3 - P_4)\}$ *Zhivot1,2-4* $\{(S_4 - P_5), (S_2 - P_2 \neg P_3, S_5 - P_6)\}$,

where S_1 is "acid", S_2 is "catalyzer", S_3 is "alkali", S_4 is "salt", S_5 is "water", P_1 is "inorganic", P_2 is "added", P_3 is "acts", P_4 is "potassic", P_5 is "bitter", and P_6 is "clean".

The formal reason for putting two predicates to the subject "catalyzer" is the requirement that if the predicate is only one in the event, it may not

Figure 4. The Zhivot1,3 connective exemplified by a simple chemical reaction (see comments in the text)

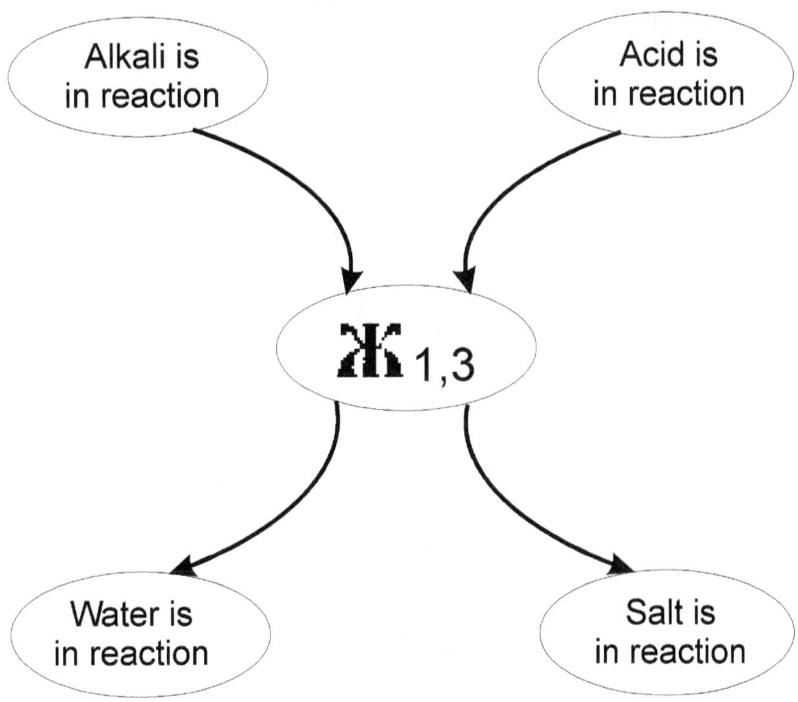

Figure 5. The Zhivot1,4 connective exemplified by a simple chemical reaction (see comments in the text)

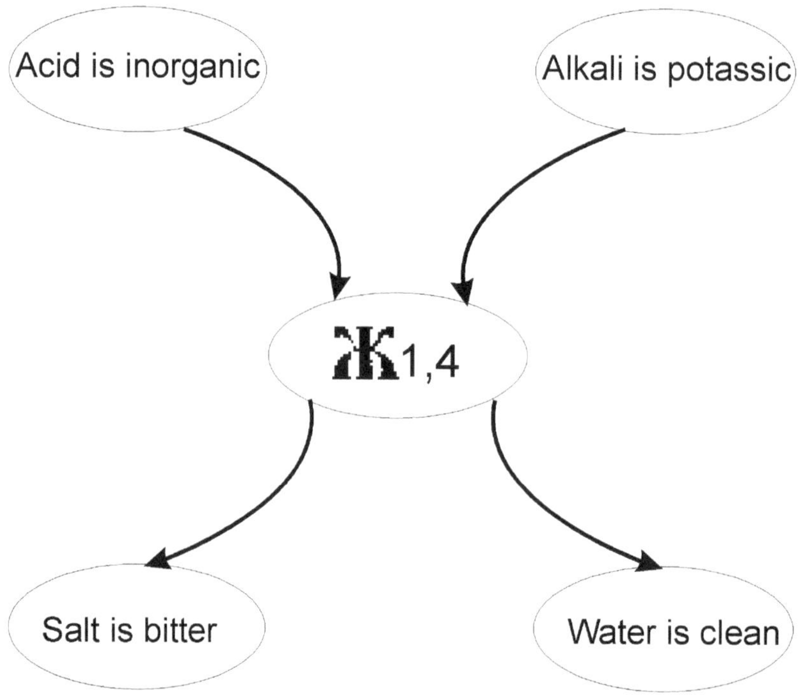

Figure 6. The Zhivot1,2-4 connective exemplified by a slightly more complicated chemical reaction assuming the presence of catalyzer (see comments in the text)

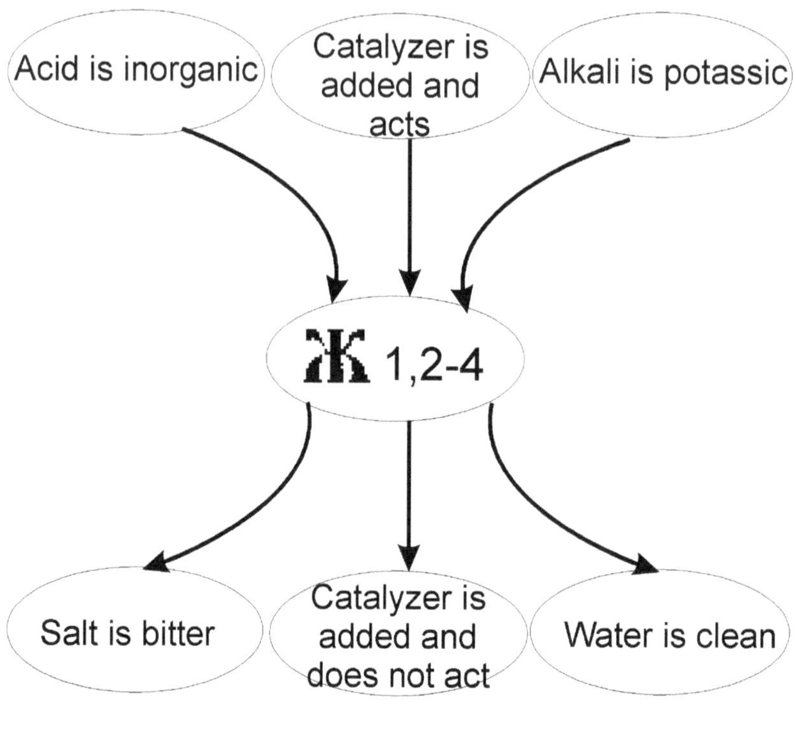

be negative, while the definition of the connective requires the negation of this predicate in either set.

Of course, in practice things would look vice versa – not the predicates should be selected to fit the formal pattern but formal pattern should be selected to fit the existing knowledge recorded in the form of events. The key issue is, therefore, to have enough patterns to be able to encode any knowledge desired.

The next connective, *Zhivot1,3-4*, is shown in Figure 7.

This example is represented by the following formula:

$\{(S_1 - P_1), (S_2 - P_1)\}$ *Zhivot1,3-4* $\{(S_3 - P_1), (S_4 - P_2)\}$,

where S_1 is "acid", S_2 is "alkali", S_3 is "salt", S_4 is "water", P_1 is "inorganic", P_2 is "clean".

The next connective, *Zhivot1,2-3-4*, is illustrated by Figure 8.

This example is represented by the following formula:

$\{(S_1 - P_1), (S_2 - P_2, P_3), (S_3 - P_1)\}$ *Zhivot1,2-3-4* $\{(S_4 - P_1), (S_2 - P_2\ \neg P_3), (S_5 - P_4)\}$,

where S_1 is "acid", S_2 is "catalyzer", S_3 is "alkali", S_4 is "salt", S_5 is "water", P_1 is "inorganic", P_2 is "added", P_3 is "acts", P_4 is "clean".

The group of *Zhivot2a* sub-subtype includes three connectives. It is illustrated by a somewhat different matter, the behavior of solution in a tank. The first one, *Zhivot2a,1-3*, is shown in Figure 9.

This example is represented by the following formula:

$\{(S_1 - P_1), (S_2 - P_1)\}$ *Zhivot2a,1-3* $\{(S_1 - P_1), (S_2 - P_1), (S_3 - P_1)\}$,

where S_1 is "tank", S_2 is "solution", S_3 is "residue", P_1 is "present".

Figure 7. The Zhivot1,3-4 connective exemplified by a trivial chemical reaction

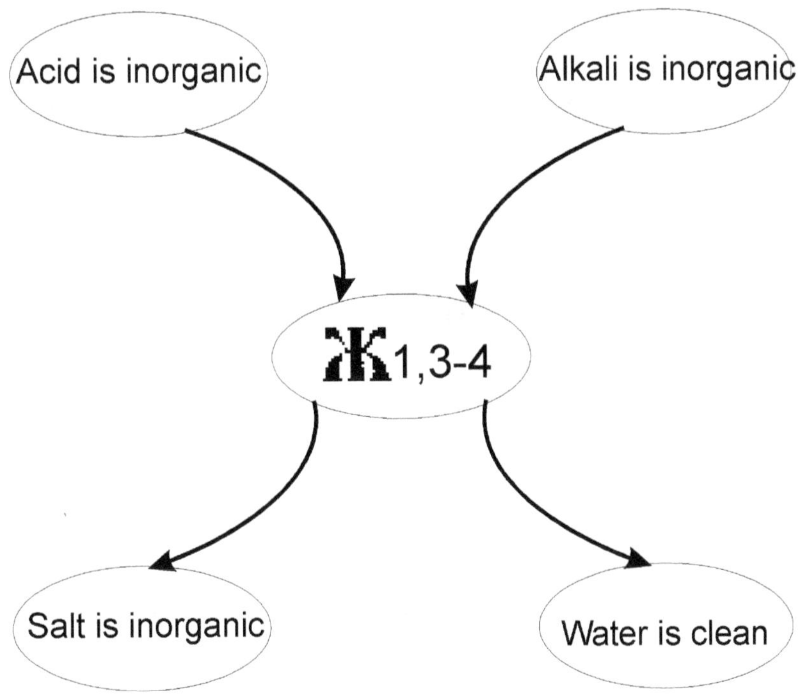

Figure 8. The Zhivot1,2-3-4 connective exemplified by a chemical reaction in presence of catalyzer

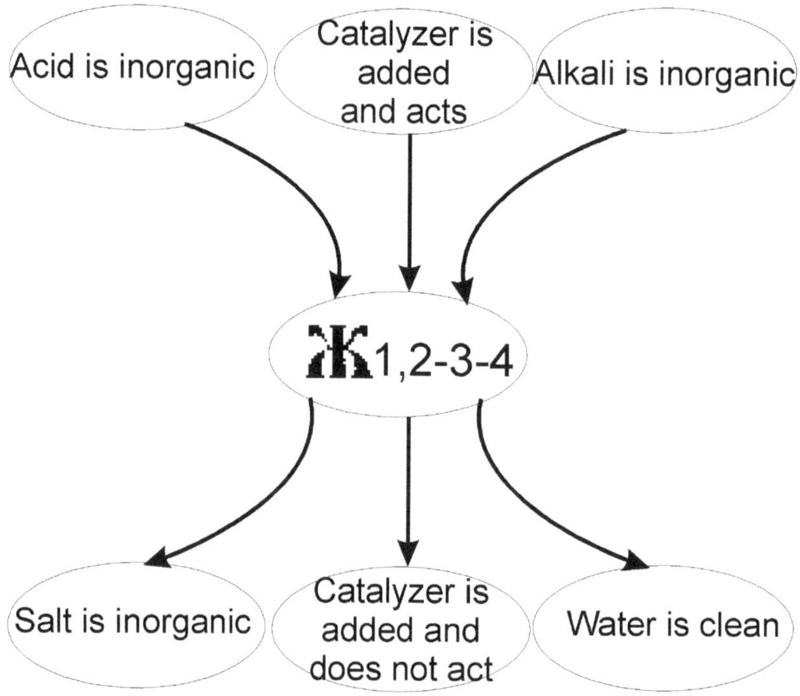

Figure 9. The Zhivot2a,1-3 connective exemplified by a physical process of precipitation of residue from a solution

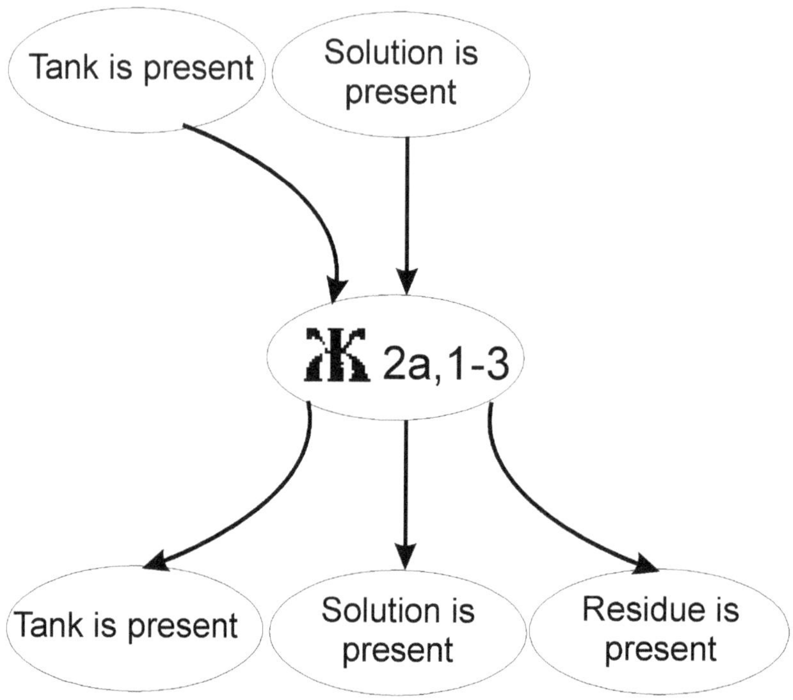

The next connective, *Zhivot2a,1-4*, is exemplified by Figure 10.

This example is represented by the following formula:

$\{(S_1 - P_1), (S_2 - P_1)\}$ *Zhivot2a,1-4* $\{(S_1 - P_1), (S_2 - P_1), (S_3 - P_2)\}$,

where S_1 is "tank", S_2 is "solution", S_3 is "residue", P_1 is "present", P_2 is "settles down".

At first sight, it looks merely a word play, a replacement of one word ("present") by another ("settles down"). Nonetheless, this change may tag to different properties associated with "settling down". This is why such slight difference in wording may appear important for representation of the knowledge and should be expressed as two formally different structures.

The last connective of the *Zhivot2a* group, *Zhivot2a,1-3-4*, is demonstrated in Figure 11.

This example is represented by the following formula:

$\{(S_1 - P_1), (S_2 - P_2)\}$ *Zhivot2a,1-3-4* $\{(S_1 - P_1), (S_2 - P_2), (S_3 - P_2), (S_4 - P_3)\}$,

where S_1 is "tank", S_2 is "solution", S_3 is "residue", S_4 is "temperature", P_1 is "installed", P_2 is "present", P_3 is "settles down", P_4 is "drops".

The next and the most abundant subtype of connectives, *Zhivot2b*, is illustrated again by an acid-alkali reaction. The first one, *Zhivot2b,1-3*, is represented by an example in Figure 12.

The formula for the *Zhivot2b,1-3* connective is written below:

$\{(S_1 - P_1), (S_2 - P_1), (S_3 - P_1)\}$ *Zhivot2b,1-3* $\{(S_4 - P_1), (S_2 - P_1), (S_5 - P_1)\}$,

where S_1 is "acid", S_2 is "catalyzer", S_3 is "alkali", S_4 is "salt", S_5 is "water", and P_1 is "in reaction".

Figure 13 illustrates the connective *Zhivot2b,1-4*.

This plot is mirrored by the formula:

Figure 10. The Zhivot2a,1-4 connective exemplified by the same example as the previous case

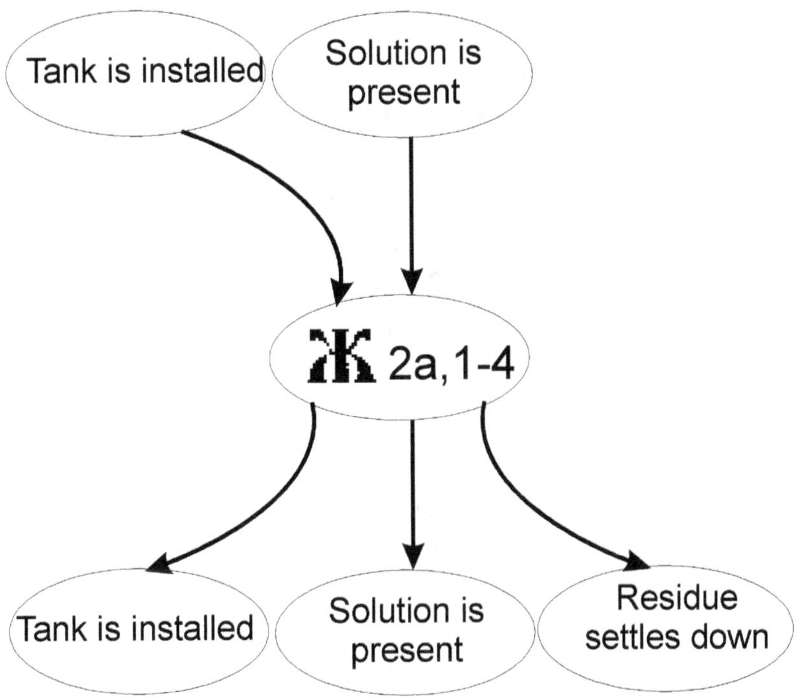

Figure 11. The Zhivot2a,1-3-4 connective exemplified by behavior of a solution in a tank with regard to environment (i.e., dropping temperature)

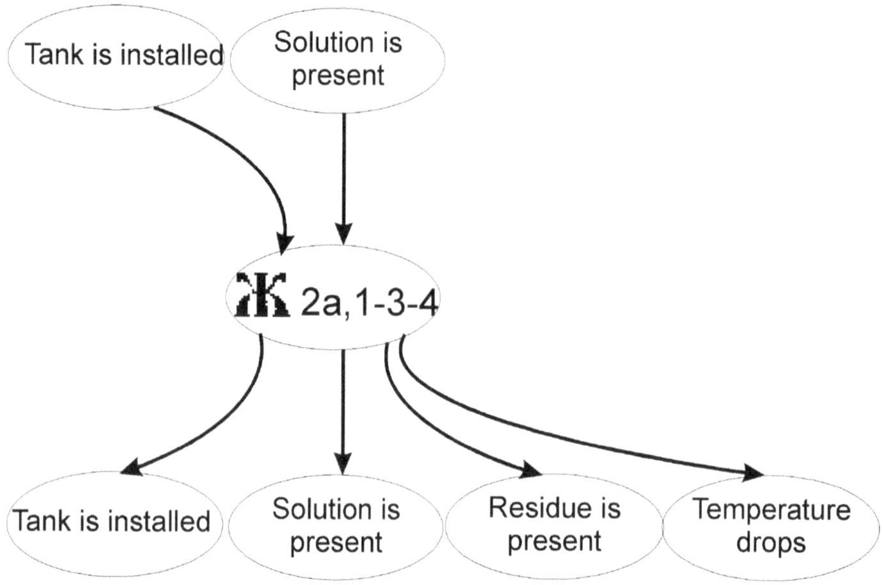

Figure 12. The Zhivot2b,1-3 connective exemplified by a chemical reaction assuming the presence of catalyzer (see comments in the text)

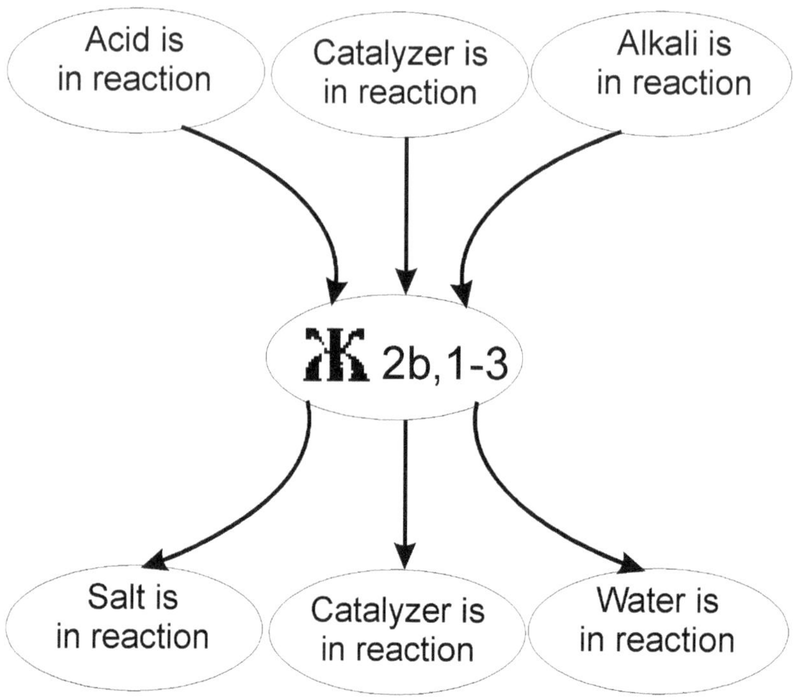

Figure 13. The Zhivot2b,1-4 connective exemplified by a chemical reaction

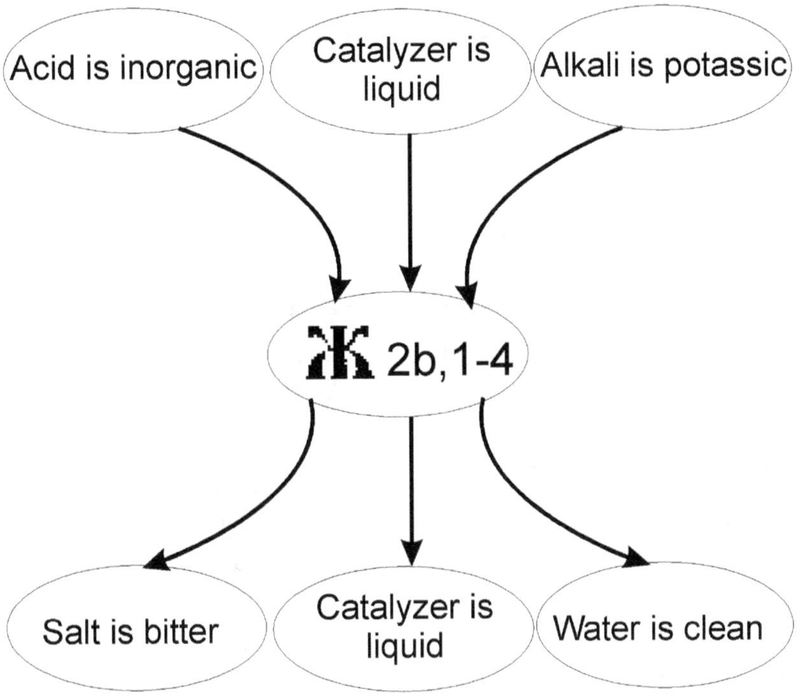

{$(S_1 - P_1)$, $(S_2 - P_2)$, $(S_3 - P_3)$} Zhivot2b,1-4 {$(S_4 - P_4)$, $(S_2 - P_2)$, $(S_5 - P_5)$},

where S_1 is "acid", S_2 is "catalyzer", S_3 is "alkali", S_4 is "salt", S_5 is "water", P_1 is "inorganic", P_2 is "liquid", P_1 is "potassic", P_1 is "bitter", and P_1 is "clean".

The next connective is *Zhivot2b,1-2-4*. It is visualized by Figure 14.

This plot is mirrored by the following formula:

{$(S_1 - P_1)$, $(S_2 - P_2)$, $(S_3 - P_3, P_4)$, $(S_4 - P_5)$} Zhivot2b,1-2-4 {$(S_5 - P_6)$, $(S_2 - P_2)$, $(S_3 - P_3, \neg P_4)$, $(S_6 - P_7)$},

where S_1 is "acid", S_2 is "catalyzer 1", S_3 is "catalyzer 2", S_4 is "alkali", S_5 is "salt", S_6 is "water", P_1 is "inorganic", P_2 is "liquid", P_3 is "added", P_4 is "acts", P_5 is "potassic", P_6 is "bitter", P_7 is "clean".

The passage from event $S_3 - P_3, P_4$ ("Catalyzer 2 is added and acts") to event $S_3 - P_3, \neg P_4$ ("Catalyzer 2 is added and does not act") reflects the treatment of time in the proposed grammar. Time is teated purely qualitatively; it is estimated solely by change of negation of a predicate – e.g., from "acts" to "does not act" (anymore). Meanwhile, the predicate "added" remains in place despite that this action (addition of catalyzer) happened well ago – it *did* happen and has not been (in fact, cannot be) negated, so the subject (catalyzer) still can be marked as "added".

Another connective of the *Zhivot2b* sub-subtype, *Zhivot2b,1-3-4*, is addressed in Figure 15.

In terms of the proposed grammar, this is:

{$(S_1 - P_1)$, $(S_2 - P_2)$, $(S_3 - P_3)$} Zhivot2b,1-3-4 {$(S_4 - P_1)$, $(S_2 - P_2)$, $(S_5 - P_4)$},

where S_1 is "acid", S_2 is "catalyzer", S_3 is "alkali", S_4 is "salt", S_5 is "water", P_1 is "Cl-bearing", P_2 is "liquid", P_3 is "potassic", P_4 is "clean".

The last connective of the *Zhivot2b* sub-subtype is *Zhivot2b,1-2-3-4* (Figure 16).

Figure 14. The Zhivot2b,1-2-4 connective exemplified by a chemical reaction

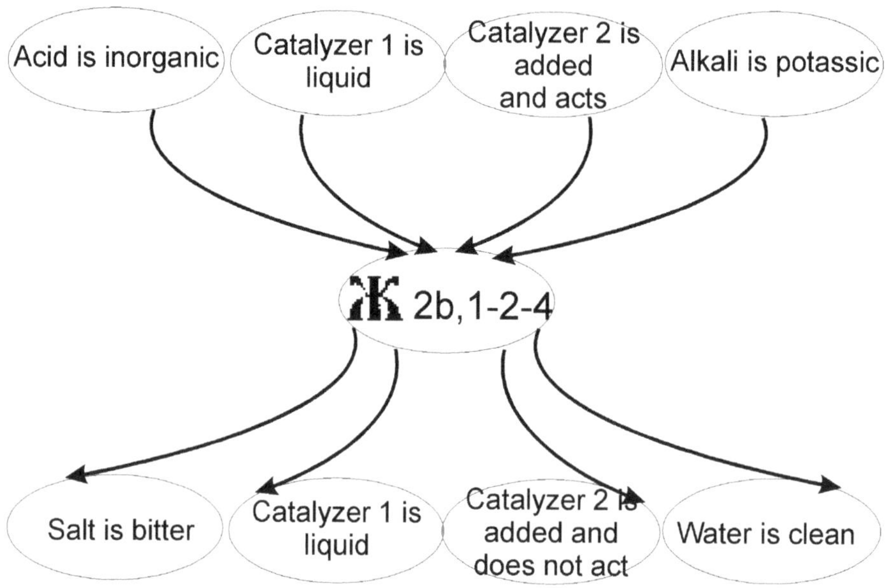

Figure 15. The Zhivot2b,1-3-4 connective exemplified by a chemical reaction in presence of catalyzer

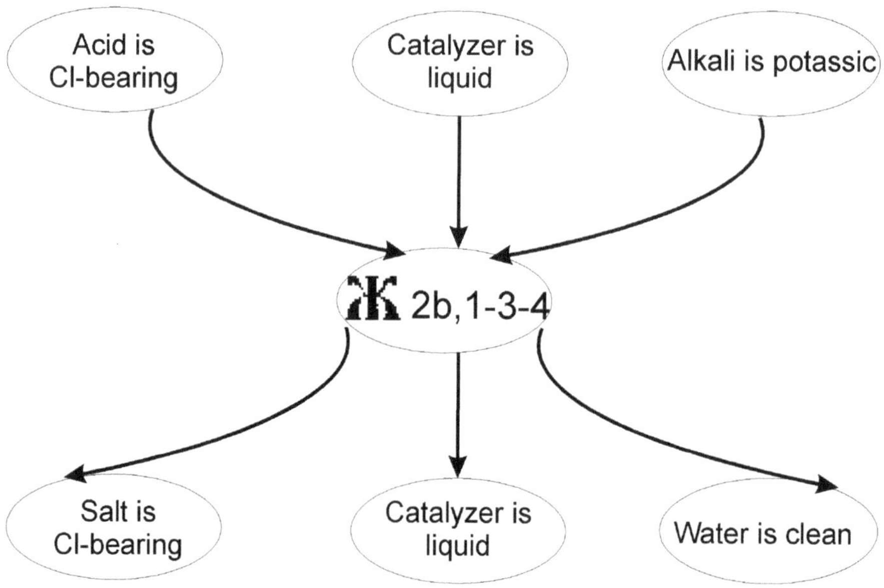

Figure 16. The Zhivot2b,1-2-3-4 connective exemplified by a chemical reaction in presence of two catalyzers

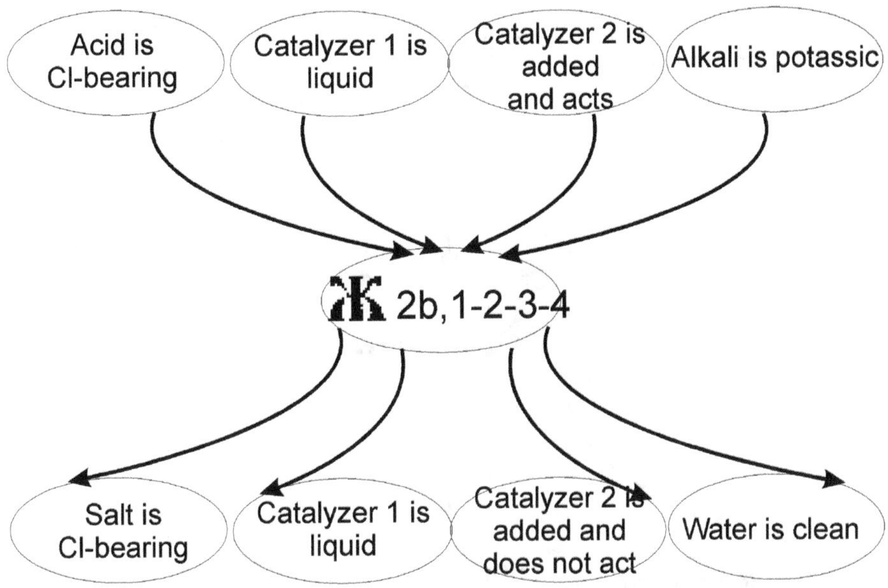

The following formula corresponds to this plot:

$\{(S_1 - P_1), (S_2 - P_2), (S_3 - P_3, P_4), (S_4 - P_5)\}$ *Zhivot2b,1-2-3-4* $\{(S_5 - P_1), (S_2 - P_2), (S_3 - P_3, \neg P_4), (S_6 - P_7)\}$,

where S_1 is "acid", S_2 is "catalyzer 1", S_3 is "catalyzer 2", S_4 is "alkali", S_5 is "salt", S_6 is "water", P_1 is "Cl-bearing", P_2 is "liquid", P_3 is "added", P_4 is "acts", P_5 is "potassic", P_6 is "clean".

Consideration of the next sub-subtype of *zhivot* connectives, *Zhivot2c*, brings us back to the example of solution in a tank. Its first connective to be considered, *Zhivot2c,1-3*, is illustrated as follows (Figure 17).

The formula for the *Zhivot2c,1-3* connective is written below:

$\{(S_1 - P_1), (S_2 - P_1)\}$ *Zhivot2c,1-3* $\{(S_1 - P_1), (S_2 - P_1), (S_3 - P_1)\}$,

where S_1 is "tank", S_2 is "solution", S_3 is "powder", P_1 is "present".

The next connective of this sub-subtype is *Zhivot2c,1-4*. It is shown in Figure 18.

The formula for the *Zhivot2c,1-4* connective is given below:

$\{(S_1 - P_1), (S_2 - P_2), (S_3 - P_3)\}$ *Zhivot2c,1-4* $\{(S_1 - P_1), (S_2 - P_1)\}$,

where S_1 is "tank", S_2 is "solution", S_3 is "powder", P_1 is "present".

Like in some previous cases, this example shows, together with the previous one, that slight rephrasing in natural language may lead to different formal patterns, e.g., in automatic encoding of dynamic knowledge from a text. Nonetheless, these different patterns seem to be equally good in expressing the sense.

The last connective of the *Zhivot2c* sub-subtype is *Zhivot2c,1-3-4*. It adds little to the representation power of this kind of connectives but is needed for formal completeness of the set (Figure 19).

The formula for the *Zhivot2c,1-3-4* connective is given below:

Figure 17. The Zhivot2c,1-3 connective shown by the example of a powder dissolving in solution in a tank

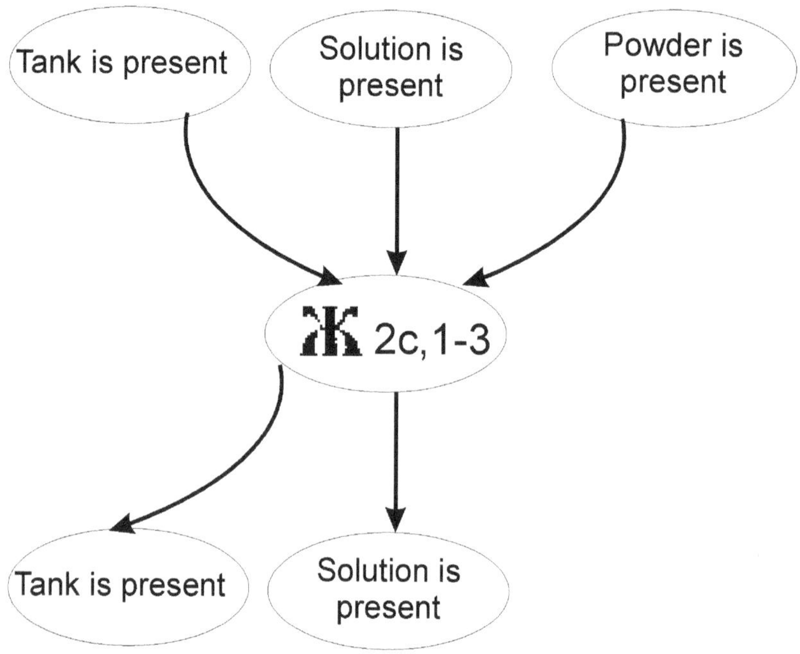

Figure 18. The Zhivot2c,1-4 connective shown by the example of a powder dissolving in solution in a tank

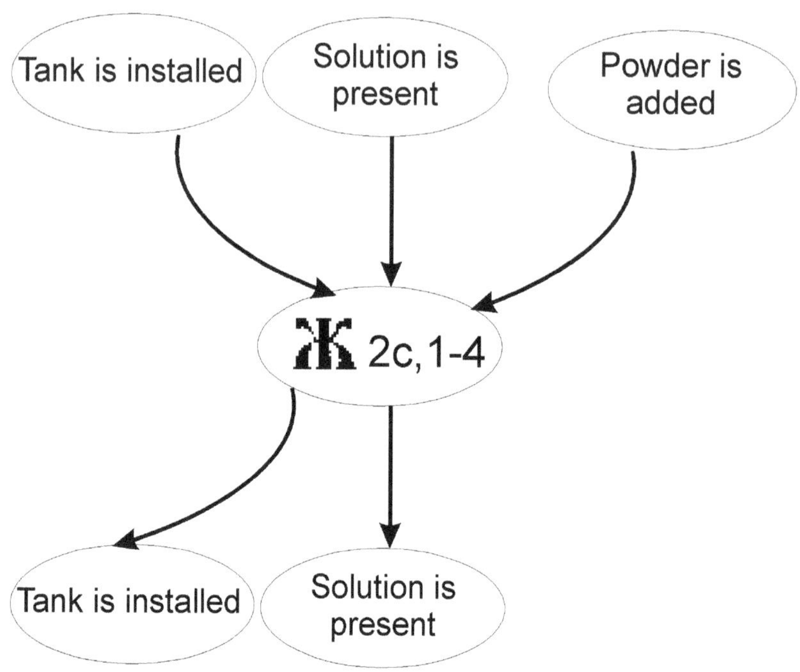

Figure 19. The Zhivot2c,1-3-4 connective shown by the example of a powder dissolving in solution in a tank

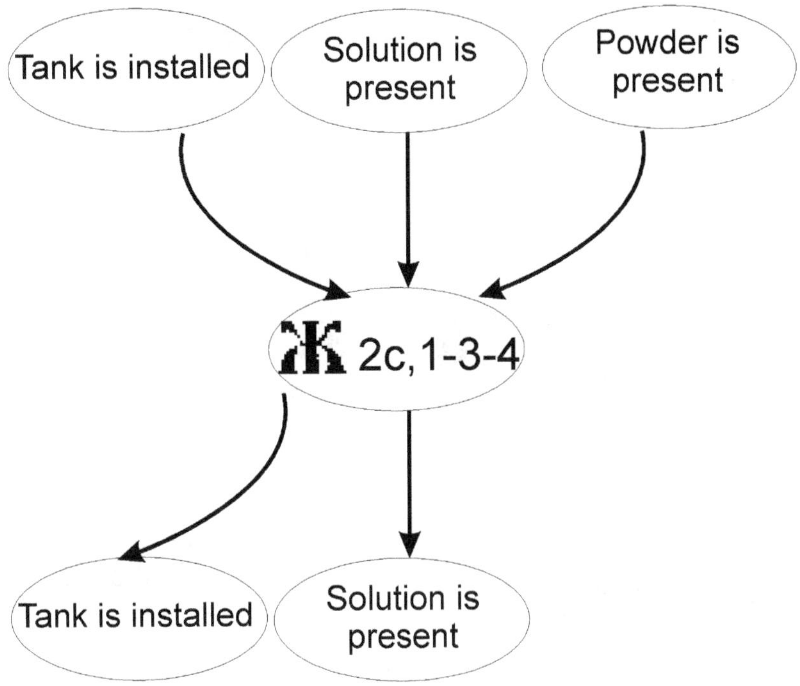

{(S_1 – P_1), (S_2 – P_2), (S_3 – P_3)} Zhivot2c,1-3-4 {(S_1 – P_1), (S_2 – P_2)},

where S_1 is "tank", S_2 is "solution", S_3 is "powder", P_1 is "present".

The last subtype of zhivot connectives, *Zhivot3*, is exemplified by the same modeling environment as the previous sub-subtype. The first connective herewith is *Zhivot3,1-3* (Figure 20).

The formula for the *Zhivot3,1-3* connective used in Figure 20 is given below:

{(S_1 – P_1), (S_2 – P_1)} Zhivot3,1-3 {(S_1 – P_1), (S_2 – P_1)},

where S_1 is "tank", S_2 is "solution", P_1 is "present".

Connective *Zhivot3,1-4* is visualized in Figure 21.

Its formula is:

S_1 – P_1, S_2 – P_2 Zhivot3,1-4 S_1 – P_1, S_2 – P_2

where S_1 is "tank", S_2 is "solution", P_1 is "present".

Finally, the connective *Zhivot3,1-3-4* is plotted in Figure 22.

Herewith, the formula is:

{(S_1 – P_1), (S_2 – P_1), (S_3 – P_2)} Zhivot3,1-3-4 {(S_1 – P_1), (S_2 – P_1), (S_3 – P_2)},

where S_1 is "tank", S_2 is "solution", S_3 is "powder", P_1 is "present", P_2 is "added".

Other types of connectives – *shta*, *yon* and *izhen*, will be also classified in this way based on the same general principles and assumption formulated above, and the results will be published elsewhere. Nonetheless, even now the *zhivot* type connectives may have direct application, at least as a reference, in the studies that aim to develop collaboration methodologies and the computer-based tools for their support. This is shown by Rogalchuk and Solomin (2015, see chapter 18 this book) by the example of Petri nets and, to a

Figure 20. The Zhivot3,1-3 connective shown by the example of a solution in a tank

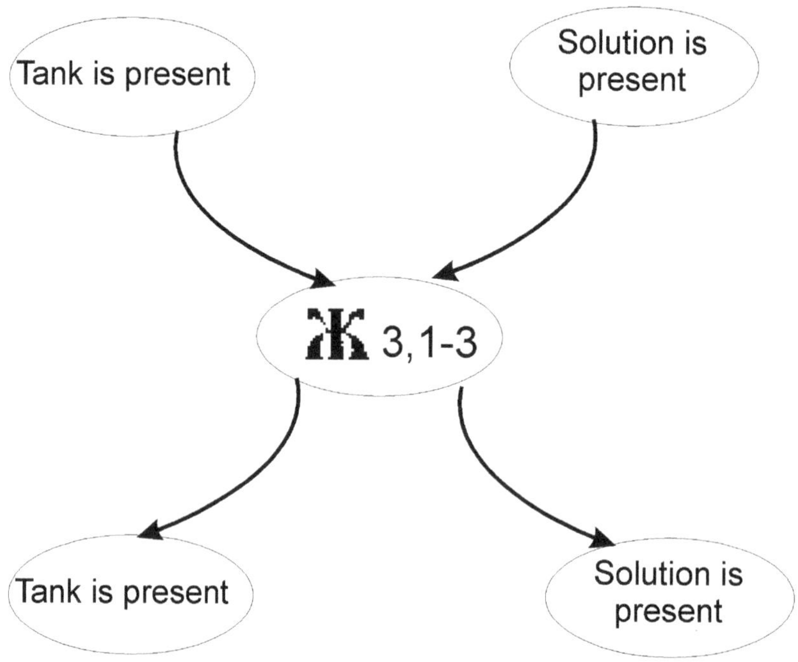

Figure 21. The Zhivot3,1-4 connective shown by the example of a solution in a tank

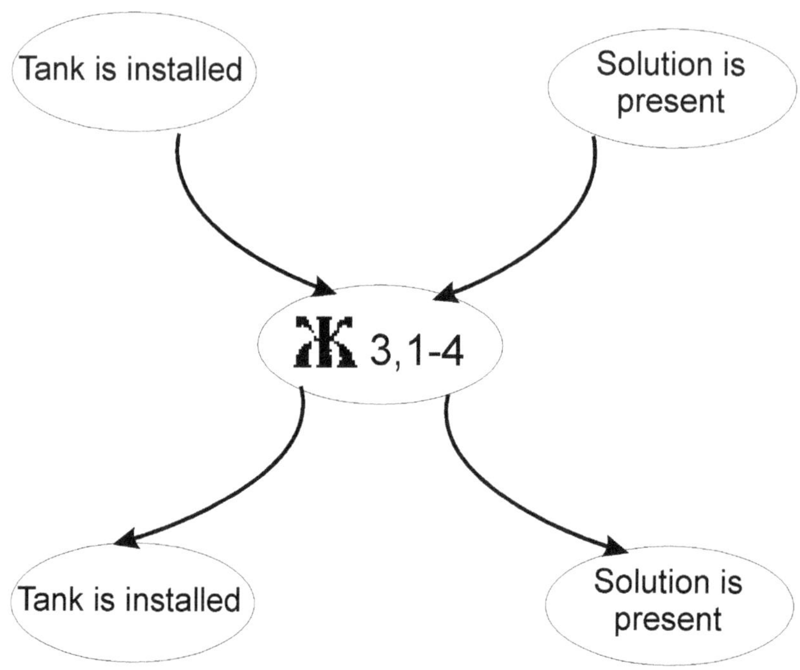

Figure 22. The Zhivot3,1-3-4 connective shown by the example of a non-dissolvable powder added to solution in a tank

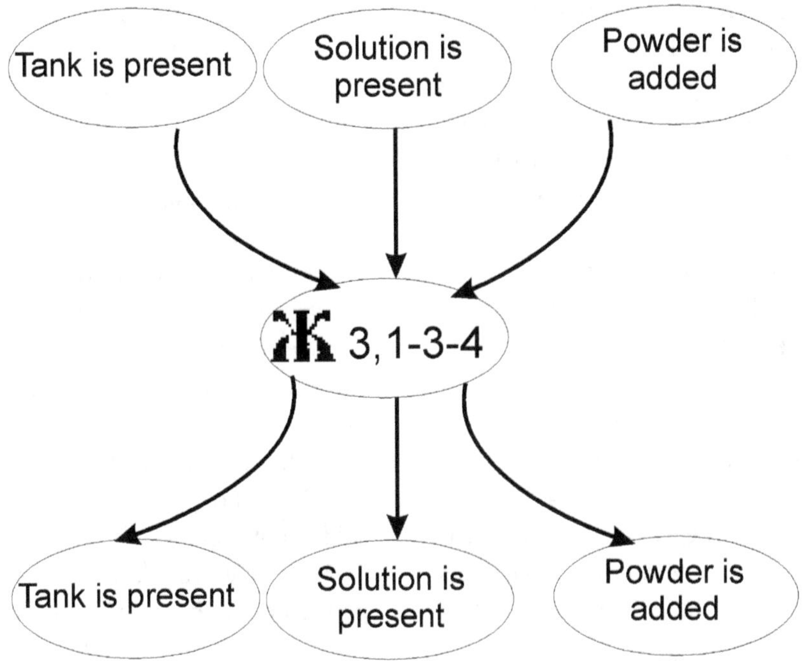

lesser degree, by Solomin and Mandrik (2015, see chapter 17 this book) for the event bush. The latter, however, uses connectives of other types than *zhivot* and therefore cannot be fully interpreted in terms of the proposed grammar so far.

DISCUSSION

Insights into Logical and Mathematical Formalization of Dynamic Knowledge Grammar

Unlike the static knowledge engineering that is well-rooted in the classical logic and sets theory, there is no consensus on that what logical or mathematical formalism to choose for formal description of dynamic methods. Still, what should be definitely said that classical logic is not applicable for operation with mutually exclusive events defined above, and sets theory would need a revision to become extendable onto the dynamic paradigm. Formal record of the event bush as a directed labeled hypergraph performed by Pshenichny and Mouromtsev (2013) was not completely successful and anyway is not enough to enable inference. The same time, it should be stressed that an opportunity of inference, though certainly desirable, is not a must at this stage. The grammar proposed here aims not to become an "inference machine" but to serve as a language to formulate particular methods of dynamic knowledge engineering, which, in turn, would perform inference. As such, it includes a semantic part (internal subject-predicate structure of events, i.e., the third step of classification of connectives) and syntactic part (definition of connectives, or the first two steps of the same classification). Still, deep formalization of this language is necessary.

There are mathematical, non-classical logical and computational approaches that should be explored, e.g., type theory (Farmer, 2008), coalgebra (Block and Leroix, 1985), graph transformation (Rozenberg and Grzegorz, 1997), process algebra, process calculi (Baeten, 2004) and some others. As a result, a purely formal theory must be obtained that grounds the definitions of events and connectives in a known framework, provides theorems for those assertions given above, which have been substantiated quite informally, and allows to infer new scenarios from existing ones in a particular method of dynamic knowledge engineering.

Use of Dynamic Knowledge Grammar in Collaborative Knowledge Networks

Graphic representation of knowledge helps scientists to build understanding. Dynamic knowledge is the matter of any scientific discussion about changing environments, be it a motion of elementary particles in physics or world's financial crisis in economics. Some of the existing methods of graphic representation of dynamic knowledge (e.g., event trees, influence diagrams or Bayesian belief networks) are being used in panel discussions relating to environmental crises or other issues of great interest and importance (Neri et al., 2008; Martí et al., 2008). Nonetheless, as is evidenced by informal communications of many scientists practically involved in these activities, both as experts and as moderators ("knowledge engineers", the persons who compose the graph that must summarize the collective view), there commonly arise problems with understanding – the experts do not grasp what they are supposed to say and in what form to fit the procedure of expert assessment. Part of the reason may be the lack of common terminology, but in most cases the expert panels include specialists who have been working in one field for years (e.g., in volcanic or seismic crisis assessment (Diviacco et al., submitted) or in medicine) and have quite a consensus on using terms. Still, the misunderstanding with each other and, the most importantly, with the moderator of the process is rooted, to our mind, in different patterns of thinking the scientists (and humans in general) possess. Where one expert has in mind a cyclic pattern, another may think of branching scenarios, and yet another, on a "web" of interrelated paths with several starting and end points, and so forth. Combinations of different patterns are innumerable, and unlikely a knowledge engineer may luckily guess which one is on the mind of an expert.

Likewise, such patterns serve as the basis for virtually all theories in the natural sciences and humanities. Therefore, the quantitative models are also explicitly or implicitly based on these. Arguing about the adequacy and applicability of the models, scientists, in fact, often dispute about the patterns of thinking these models are based on, though not a single word is explicitly said about them.

Similarly, when involved as experts in expert panels or collaborators in joint research projects, scientists encounter problems embedding their vision into the suggested framework, be it a tree, a loop, a bush, a net (polytree) or something else. Hence, this framework must be flexible and able, first, to account for any personal or group pattern of thinking and suggest (or perhaps even to develop anew) a formal method that would best fit the personal or group vision, and then, to look for an opportunity to impartially reconcile the visions or prove their incompatibility. Exactly this is the mission of the suggested unified grammar of dynamic knowledge. Its connectives, even those that seem to bear no meaningful information, must enable to encode *any* kind of reasoning in terms of events. All the scientists should do is to put their points in the form of events defined above, $S - P_1, P_2, \ldots, P_n$, where S is subject, and $P_1, P_2, \ldots, P_n$ are predicates, with or without negation, with the corresponding restrictions. This task, however, seems to be much easier than to adapt one's mind to some pre-existing formal pattern of thinking. The task of knowledge engineer would be to select the appropriate connectives and build the plot that exactly reflects the expert's standpoint. Then these plots should be compared and analyzed for compatibility with each other and with various methods (diagrams) of dynamic knowledge representation.

At best, this strategy should be enhanced with the traditional, static knowledge representation framework (ontologies, conceptual graphs and other methods) to analyze the relationship between individual subjects and predicates used by experts to formulate the events.

With further development, the universal dynamic knowledge grammar must receive a software tool that would bring it to a computer-based environment for support of collaborative studies – COLLA or the like. Desirably, this software also should be compatible with existing tools for plotting conceptual structures applicable to the dynamic knowledge, e.g., activity diagrams in the Unified Modeling Framework – or, alternatively, form its own modeling framework to host particular tools of visualization of the dynamic knowledge.

CONCLUDING REMARKS

The dynamic knowledge, describing evolving environments, is virtually ubiquitous and unavoidable in collaborative research. Meanwhile, it cannot be efficiently handled by existing tools of knowledge engineering and knowledge representation, because it implies a change of subject-predicate relation including, possibly, change of predicate to its negation. This makes it different from the static knowledge, which describes fixed, non-changing environments and is well formalized by classical logic. This claims for a different approach to is formalization. As the first step of this approach, a unified grammar for dynamic knowledge engineering and representation is being proposed.

The key concept of dynamic knowledge is event defined as a statement of general form $S - P_1, P_2, \ldots, P_i$, where S is subject, and $P_1, P_2, \ldots, P_i$ are predicates or their negations (the amount of negations being smaller than total amount of predicates). This defines the semantics of dynamic knowledge. In general, it can be represented as a succession of events. The structure of this succession may be different and determines the type of environment to which the field of knowledge belongs.

A change, or a passage, from one non-empty set of events to another is considered a formula, which is the main unit of syntax of dynamic knowledge grammar. Analyzing the passage from events to events and interrelationships between the causal and the effect (resulting) sets of events, one may recognize four possible types of change, denoted by capital letters of Old Slavonic alphabet, *zhivot* (many causes lead to many effects), *shta* (many causes – one effect), *yon* (one cause – many effects) and *izhen* (one cause – one effect). These types are specified further, analyzed in terms of possible existence of causes when the effects take place and finally classified in regard to the subject-predicate structure of individual events. As a result, a three-stage classification of possible transitions is obtained, the end members of which are called connectives of dynamic knowledge and govern the formulae.

The first type of connectives, *zhivot,* was classified thoroughly and studied in concern of its direct applicability to organization of knowledge at collaborative studies. All proposed connectives serve well for this purpose, though some redundancy may be observed. Nevertheless, the main pragmatic value of suggested grammar is that it should provide common ground for various existing and yet inexistent methods of dynamic knowledge engineering (event trees, event bushes, Petri nets, belief networks, influence diagrams and others), optimize their use and give semantic support to inference in each of them.

The development of the proposed grammar of dynamic knowledge may give a powerful tool for elicitation and reconciliation of patterns of thinking that is essential for efficient support of collaborative research.

REFERENCES

Baeten, J.C.M. (2004). *A brief history of process algebra*. Vakgroep Informatica, Technische Universiteit Eindhoven: Rapport CSR 04-02.

Block, R. E., & Leroux, P. (1985). Generalized dual coalgebras of algebras, with applications to cofree coalgebras. *Journal of Pure and Applied Algebra*, *36*(1), 15–21. doi:10.1016/0022-4049(85)90060-X

Diviacco, P. (2012), Addressing Conflicting Cognitive Models in Collaborative E-Research: A Case Study in Exploration Geophysics in Collaborative and Distributed E-Research. In A.A. Juan, T. Daradoumis, M. Roca, S.E. Grasman, & J. Faulin (Eds.), Collaborative and Distributed E-Research: Innovations in Technologies, Strategies and Applications (pp. 247-275). Hershey, PA: IGI Global.

Diviacco, P., Pshenichny, C., Carniel, R., Khrabrykh, Z., Shterkhun, V., & Mouromtsev, D. et al. (Submitted to). Organization of a geophysical information space by using an event-bush-based collaborative tool. *Earth Science Informatics*.

Farmer, W. (2008). The seven virtues of simple type theory. *Journal of Applied Logic*, *6*(3), 267–286. doi:10.1016/j.jal.2007.11.001

Gavrilova, T., Bolotnikova, E., Leshcheva, I., Blagov, E., & Yanson, A. (2013). Measuring Psychological Impact on Group Ontology Design and Development: An Empirical Approach. In P. Klinov & D. Mouromtsev (Eds.), Knowledge Engineering and the Semantic Web (pp. 29-43). Heidelberg, Germany: Springer.

Gentzen, G. (1969). Investigations into logical deduction. In M., Szabo (Ed.), The Collected Papers of Gerhard Gentzen (pp. 68-128). Amsterdam, The Netherlands: North-Holland (original work published 1934).

Hilbert, D., & Bernays, P. (1934). *Crundlagen der Mathematik (Fundamentals of Mathematics)* (Vol. 1). Heidelberg, Germany: Springer-Verlag.

Kleene, S. C. (1952). *Introduction to Mathematics*. New York: D. Van Nostrand Co., Inc.

Martí, J., Aspinall, W. P., Sobradelo, R., Felpeto, A., Geyer, A., & Ortiz, R. et al. (2008). A long-term volcanic hazard event tree for Teide-Pico Viejo stratovolcanoes (Tenerife, Canary Islands). *Journal of Volcanology and Geothermal Research*, *178*(3), 543–552. doi:10.1016/j.jvolgeores.2008.09.023

Martin, P. (2002). Knowledge representation in CGLF, CGIF, KIF, Frame-CG and Formalized-English. In *Proceedings of ICCS'02, 10th International Conference on Conceptual Structures* (pp. 77-91). Borovets, Bulgaria: Springer. doi:10.1007/3-540-45483-7_7

Murdock, J., Buckner, C., & Allen, C. (2010). Evaluating Dynamic Ontologies. In A. Fred, J.L.G. Dietz, K. Liu, & J. Filipe (Eds.), *Knowledge Discovery, Knowledge Engineering and Knowledge Management. Second International Joint Conference* (pp. 258-275). Valencia, Spain: Springer.

Neri, A., Aspinall, W. P., Cioni, R., Bertagnini, A., Baxter, P. J., & Zuccaro, G. et al. (2008). Developing an Event Tree for probabilistic hazard and risk assessment at Vesuvius. *Journal of Volcanology and Geothermal Research*, *178*(3), 397–415. doi:10.1016/j.jvolgeores.2008.05.014

Noy, N., Chugh, A., Liu, W., & Musen, M. (2006). A Framework for Ontology Evolution in Collaborative Environments. In CruzI.DeckerS.AllemangD.PreistC.SchwabeD.MikaP.UscholdM.AroyoL. (Eds.), *Proceedings of the 5th International semantic Web Conference* (pp. 544-558). Athens, GA: Springer. doi:10.1007/11926078_39

Pshenichny, C. A., & Kanzheleva, O. M. (2011). Theoretical foundations of the event bush method. In K. Sinha, L. Gundersen, J. Jackson, & D. Arctur (Eds.), Societal Challenges and Geoinformatics (pp. 139-165). Boulder, CO: GSA Special Paper 482. doi:10.1130/2011.2482(12)

Pshenichny, C. A., & Mouromtsev, D. I. (2013). Representation of the Event Bush Approach in Terms of Directed Hypergraphs. In H.D. Pfeiffer, D.I. Ignatov, J. Poelmans, & N. Gadiraju (Eds.), Conceptual Structures for STEM Research and Education (pp. 287–298). Mumbai, India: Springer. doi:10.1007/978-3-642-35786-2_22

Rogalchuk, V., & Solomin, K. (2015). Perspectives of Use of Petri Nets in Collaborative Research. In P. Diviacco, P. Fox, C. Pshenichny, & A. Leadbetter (Eds.), *Collaborative Knowledge in Scientific Research Networks*. Hershey, PA: IGI Global.

Rozenberg, G., & Grzegorz. (1997). *Handbook of Graph Grammars and Computing by Graph Transformations* (vols. 1–3). Singapore: World Scientific Publishing.

Solomin, K., & Mandrik, M. (2015). Constructing historical knowledge through graphic boundary objects. In P. Diviacco, P. Fox, C. Pshenichny, & A. Leadbetter (Eds.), *Collaborative Knowledge in Scientific Research Networks*. Hershey, PA: IGI Global.

Stojanovic, L. (2004). *Methods and Tools for Ontology Evolution* (Doctoral dissertation, University of Karlsruhe). Retrieved 10 August 2014 from d-nb.info/1001606787/34?origin=publication_detail

Takeuti, G. (1975). Proof Theory. Studies in Logic and the Foundations of Mathematics (vol. 81). New York: American Elsevier Publishing Co.

Zablith, F. (2008). Dynamic ontology evolution. In *Proceedings of the International Semantic Web Conference (ISWC) Doctoral Consortium*. Karlsruhe, Germany: Springer.

KEY TERMS AND DEFINITIONS

Change, or Formula, of Unified Dynamic Knowledge Grammar: A succession of two sets of events, called left and right, which are linked by connective. In representation of dynamic knowledge, change naturally corresponds to a process (natural, social or other), while a succession of changes, in which the right set of the previous change is the left set of the next one, describes scenarios (or paths of evolution) of the described world.

Connective of Unified Dynamic Knowledge Grammar: A symbol to connect sets of events in the way determined by its definition.

Dynamic Knowledge: Knowledge about dynamic environments, in which the relations between objects may change.

Environment: A modeled situation.

Event: A set of statements of general form S – P1, P2, …, Pn, where S is subject, and P1, P2, …, Pn are predicates, with or without negation, complying the following conditions: (i) subject may be only one in a statement, (ii) subject may not be negated, (iii) all predicates may not be negated in a statement, at least one must be without negation. Events represent the relations between objects of two types: one subject and whatever number of predicates.

Static Knowledge: Knowledge represented by static environments, in which the relations between objects are considered fixed and not changing.

Chapter 17
Constructing Historical Knowledge through Graphic Boundary Objects

Konstantin Solomin
ITMO University, Russia

Maria Mandrik
Archive of the Russian Academy of Sciences – St. Petersburg, Russia

ABSTRACT

Ironically, collaborative studies are often most needed exactly in those fields in which they are least feasible. This fully refers to history. Descriptive knowledge requires comparison, incorporation, and even merging of views, concepts, and viewpoints that either do not intersect or are in apparent contradiction. A solution could be found if one manages to suggest some kind of common reference points, the so-called boundary objects, but the way these objects have been created precludes their usefulness in history. This chapter presents an experiment of creating graphic boundary objects for historical studies by means of a dynamic tool of knowledge engineering/knowledge representation, the event bush.

INTRODUCTION

It is a well known fact that the methodology of reasoning in the descriptive fields of knowledge including humanities and, in particular, history, differs much from that in the sciences based on experiment or solid theory. By their nature, humanities are highly descriptive and context-dependent, and this undermines the value of any data; in most cases there exist equally profound but incompatible viewpoints on the same issue and, importantly, the reliability of these viewpoints cannot be proven or assessed formally. When analyzing any event in the history, scientists have to rely on documents or other pieces of material evidence, which were created by people guided by their subjective esteem. Hence, a scientist's analysis has to be a compilation of subjective representations of historical data.

In the discourse of methodology of historical science the issues of reasoning are addressed repeatedly and in various ways. To name at least few of the debated in the international literature, the following should be mentioned:

DOI: 10.4018/978-1-4666-6567-5.ch017

Copyright © 2015, IGI Global. Copying or distributing in print or electronic forms without written permission of IGI Global is prohibited.

- Definition of the central terms of theory of history, such as "historiography," "philosophy of history," "theory of history," and "memory" (Klein, 2011);
- "Historical distance" (Hollander et al., 2011) between the study and the object of study and the position of a historian as an "ideal" or biased "observer" (Kuklick, 1969);
- The syntax of objects and the representation of history (Carbonell, 2009);
- Relevance of past occurrences (whether they are meaningful per se or only in relation to the modern world – Smith, 2009);
- Reasonable grounds for comparison of cultures (Rüsen, 1996);
- Replacement of static concepts with dynamic ones in historical discourse (Reill, 1986), and many others.

Most of these issues "wrap" the problem of reasoning in the specific context of historical thought; however, to our mind, all of them are rooted in the poor tractability of reasoning in history (comparing, say, to engineering). Exactly this methodological weakness of history is its real Achilles' heel.

To cope with it, one should look at the experience of other fields where this hindrance is known (perhaps not to such high degree as in history) – medicine, management, politics, geosciences and others. The solution found there is application of the tools of knowledge engineering and representation to clarify reasoning and explicate the structure of knowledge. Though the emphasis of knowledge engineering and representation seems to be quite applied, these studies raise fundamental theoretical issues and lead, in fact, to complex rethink of the entire field of knowledge.

Herewith, the general question to be answered is: can one ever formally describe the methodology of reasoning if the latter is based on the natural language? This question is especially important for collaborative studies, which are, on the one hand needed, and sometimes even a must for research in history, and the other, are almost impossible because of difference of backgrounds in the same fields of study.

A few scientists have addressed this problem in many fields, Diviacco (2012) to be mentioned among the latest and the most comprehensive surveys. To formally describe the reasoning, he suggests an approach of boundary objects based on the work of Star and Griesemer (1989). Diviacco defines them as follows, "Boundary objects are artifacts that aim to bridge concurrent cognitive models through abstraction from all the domains of the partners" (Diviacco, 2012, p. 17). Because of this abstraction, he notes further, one can foresee their application mainly in the coordination rather than in the actual working activities. As an example, Diviacco quotes layers and polygons in geographic information systems which can be attributed any semantics and be filled with any spatial data. However, in history such "blanks" would make little sense because historical data require a context. Without being contextually rooted they have little value. This claims for some "meaningful artifacts", or specifically historical boundary objects loaded with some pieces of historical knowledge put in universal formulations in natural language. Pshenichny et al. (2009) suggest to use (i) as general as possible, (ii) purely descriptive, (iii) simple and (iv) unique, non-overlapping terms.

To make boundary objects work for representation of reasoning, they must be grouped and co-managed. One straightforward way to do that is to introduce *graphic boundary objects* representing some labeled graphs (Tutte, 1998) or alike structures (Pshenichny & Mouromtsev, 2013) corresponding to notations of knowledge engineering and knowledge representation, with verbal formulations linked to nodes of graphs, and arcs, defining some kind of inference. Thus, Diviacco (2012) suggests, as possible candidates, the following types of graph-based (or graph-looking) boundary object, mind/concept maps, workflows, and event bushes. Among these, concept maps (and another, quite similar tool, cognitive maps) have been used in history and political science (e.g., Popova, 2011)

Still, despite the obvious advantages of concept maps, such as transparency and representation power, they have a limited ability to capture the changes of objects and their properties, which is essential, in our opinion, for the representation of reasoning in history. Workflows were designed mainly for technical and administrative applications. Both cognitive maps and workflows in their present state represent essentially static means of knowledge engineering (Pshenichny & Kanzheleva, 2011; Pshenichny & Mouromtsev, 2013) and possess a grammar that does not fit the rules of dynamic event description grammar as suggested by Pshenichny et al. (2014; see this volume). The event bush, however, seems to be a good candidate for generating and managing boundary objects in history, since it is a dynamic approach with a relatively well-developed grammar representing, in fact, a kind of a controlled natural language.

This chapter aims to test the hypothesis that at least for some fields of historical studies the event bush may be used to generate graphic boundary objects, which would be usable for construction of theories in history more or less in the way this is being done in mathematics or engineering. For this test then, (i) the selected historical context will be presented below, (ii) the method of event bush will be briefly described, (iii) an event bush model will be built and (iv) the results discussed.

MODEL DEVELOPMENT

Historical Context

The period from February to November, 1917, in Russia is considered one of the most tragic episodes of its history. The decrowning of its last Emperor, Nikolai (Nicholas) the 2nd, put the army to a morally unbearable choice, either to keep senseless faith to the dethroned monarch, or to violate the oath and serve to the new government, or to take another, but similarly undesired, path. Monarchism, in turn, could be expressed in a number of ways, ranging from peaceful retirement to a riot. Other ways also proved to be very diverse, from defection from the army (practiced mostly by soldiers, especially by those who were not at the frontline of the World War First or were influenced by Bolshevik propaganda) to forming national troops of Ukrainians, Caucasian and Turkic nationals, Latvians, Estonians, Finns, Poles and so forth, as Russia is a highly multinational country. This choice, both at the personal and collective level, was sometimes very hard and often spelled death. Among the main reasons that determined it at the personal level were social stratum and social category of the military men (noblemen, philistines (a class of Russian urban inhabitants, usually craftsmen or merchants), peasants and workers), the cultural context the person had been brought up in (roughly speaking, conservative versus liberal), mother tongue (Russian or other) and so forth. Of course, along with personal moods, there were group tendencies and preferences determined by the will and respect of the upper-level commander (say, general or admiral), situation at the front, activity of Bolshevik, nationalist or other propaganda, supply of food, fuel and ammunition and other innumerable circumstances. Still, we tend to think that these circumstances, possibly making the group choice prevail over personal one in short term, gave way to personal decision in long term. This is why we focused in this work on personal preferences of Russian military men and the ways we can explain them. It should be stressed that the period addressed here ends up on the 7 November, 1917 (25 October Julian style) when the second phase of the revolution, the Bolshevik coup, brought the military men to yet harder and more complicated choice.

There is a multitude of historical studies that address this period but, as to be suspected, great many of them are in Russian, and virtually all are naturally biased depending on the political and cultural background of the author. Still, in some of these studies the reader finds statements of former Russian officers who witnessed the process of transformation of their army and published their notes in Russian abroad in 1920s-1960s (Tsvetkov, 2010; Spiridovich, 1962; Krasnov, 1922; Eliseev, republished in 2003; Kutepov, 1934) as well as the statements of former low ranks evidencing rapid desertion (e.g., Zoschenko, 1933) and nationals directly involved in formation of ethnic troops (e.g., Petliura, 1918).

When a consistent model of the discussed historical context is built, it will be certainly open for update and involvement of much greater amount of existing sources of information including abundant continental European and Anglo-American scholarship on the topic.

The Method of Event Bush

The method of event bush was designed to address a particular kind of modeling environment, which, in our opinion, is the most appropriate for historical contexts – that of directed alternative changes (Pshenichny & Kanzheleva, 2011). This type of environment can be metaphorized as an arena for various, intertangling processes, which are thought to originate from a limited number of causes or sources (but not less than two), tend to leave a definable result and can be followed as more or less distinct scenarios.

Formalisms describing such environments are expressed graphically as constructions having many "roots" or net-shaped, allowing various paths at least between some of the nodes, and permitting local cycles, i.e., notations strongly resembling graphs and likely interpretable as such. Many of the abovementioned event-based notations apply to these environments, excluding tree-shaped (event/probability trees, Bayesian trees) and completely cyclic ones (e.g., causal loops). The environment of directed alternative changes is described by the event bush as a set of events of four types arranged in certain order (defined as multiflow structure – see Pshenichny and Mouromtsev, 2013) and connected by special connectives (Pshenichny & Kanzheleva, 2011).

The Purpose

The knowledge engineering method of event bush intends to give a strict and finite but extendable display of an area of reality and corresponding domain of knowledge. Importantly, it does not intend to paint an objective and true picture of reality based solely on formal grounds. The purpose of the event bush is to more or less impartially structure and shape the subjective information gathered by an expert and communicated by expert to decision-makers. Like any other formalization, it does not state what is true or false, what is relevant or not, for the studied context. Nevertheless, it helps organize what expert (scientist) considers relevant, allowing to formulate everything that can happen given a list of premises. It imposes some requirements on these to ensure the most complete and objective inference, but the choice of premises is totally at the scientist's discretion. In a discourse or polemics, the tool of event bush may help us to formally express and compare contrasting standpoints. At the same time, formalization itself often urges an expert to revise knowledge, find gaps and strands of "wooly" reasoning, terminological and conceptual intricacy and the like. Such imperfections can be tackled and to some degree cured by the event bush. But like any other tool, in principle it can be used alternatively, e.g., to produce a beautiful, formally perfect nonsense.

The Technique

The method of event bush rests on the assumption that some areas of reality (e.g., behavior of various groups in the Russian army in our example) can be represented as shown in Figure 1. In the considered area of reality, the following events are identified.

- (*ia*) Primary internal events. These are primary, not overlapping and non-unique inputs (basic objects, processes or tendencies) – e.g., the Russian army, without any external influencing conditions). Such inputs, according to the concept of event bush, would determine any further course of events ("happenings").
- (*ib*) Primary external events stemming from environmental circumstances. They indicate the way an environment may affect basic inputs or influence them further. These are essentially indirect manifestations which "shape" different "happenings"; for instance, the decrowning of the Russian Emperor.
- (*ii*) Secondary events (processes or objects) that result from primary inputs with or without the contribution of incoming circumstances – the "happenings" proper ("the Russian army responds to the decrowning of the Emperor") formulated in a strict and concise way indicating the core features determined by causes, following the principle "one more cause – one more property".
- (*iii*) Tertiary events that denote end results, or products, generated either by primary internal or by secondary events, with or without primary external ones. Tertiary

Figure 1. Basic syntax of Event Bush

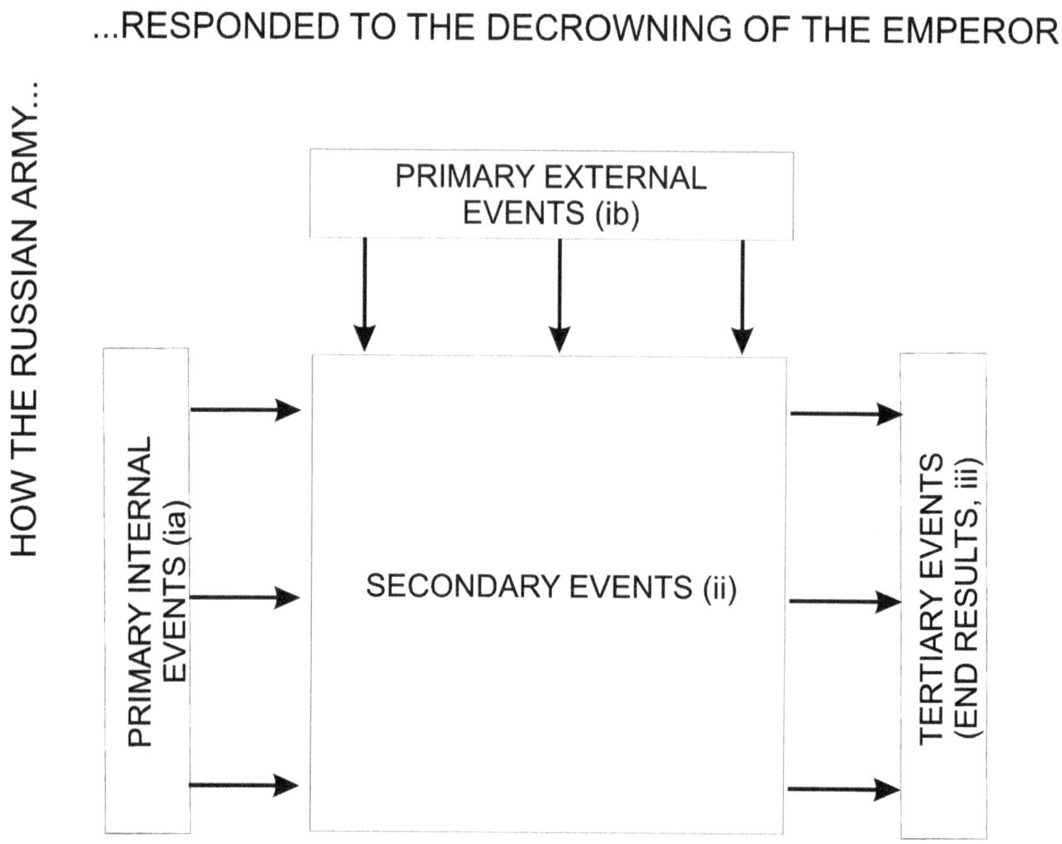

events document the completed "happenings". In the case considered here, it is reasonable to take as end results the various group choices of Russian military men.

Other types of events (e.g., those describing some other type of results – say, quaternary ones) can be added, but the presence of the (*i*) - (*iii*) is mandatory. The general interrelation of these types is shown by an example in Figure 2.

In accordance with the syntax of event bush presented in Figure 1, the relations between the (*i*) - (*iii*) is as follows:

- Events (*ia*) and (*ib*) must not lead to other (*ia*) or (*ib*);
- Event (*ib*) may lead to an event only together with (*ia*) or (*ii*);
- Event (*ii*) must not lead to (*ia*) or (*ib*);
- Event (*iii*) must not lead to (*ia*), (*ib*), (*ii*), and another (*iii*).

The connectives of the event bush, each having a graphic designation, enforce these relations.

Pshenichny and Kanzheleva (2011) define four connectives, flux, influx, conflux, and furcation, of which the first two, flux and influx, are mandatory for an event bush.

If one event leads to another, this is defined as flux:

E_1 *Flux* E_2

and denoted by simple arrow. For instance, in Figure 2,

The Russian army does not respond to the decrowning of the Emperor (E_1)

FLUX

The Russian army does not respond to the decrowning of the Emperor (E_2)

If two or more events lead to one, this refers either to influx (denoted by a rounded crossing, or "right turn" graphic sign) or to conflux (denoted by "double right turn") connectives.

Putting the right turn, we express an opinion (or an obvious possibility) that two events together may lead to another event, so that one, above the right turn sign (E_1), "flows in", or modifies, another event E_2 (to the left of it), resulting in the event below the said sign (E_3):

E_1, E_2 *Influx* E_3.

For example, in Figure 2,

The Russian army does not respond to the decrowning of the Emperor (E_1), *Emperor is decrowned* (E_2)

INFLUX

The Russian army responds to the decrowning of the Emperor (E_3).

Observably, all events (*ib*) act as modifiers. The subject of the resulting event is inherited from the main premise and the predicate is formed by the modifying one. For intuitive clarity, we adopt that the "right turn" points always down, from the modifying event to the resulting one, which denotes the influx. The main and the modifying events may change places, only if this makes sense in the considered domain of knowledge and only involves events of type (*ii*).

Once an event (*ia* or *ib*) is introduced or event (*ii*) appears in a bush and one subsequently coincides with all previously introduced events, one has to find out whether any new event results from each pair.

Along with the flux, influx is mandatory for the event bush, for the influx solely involves the events *ib* in the inference.

The connective of conflux implies that two or more events (E_1, E_2, …, E_{n-1}) "conflow" into one (E_n) playing similar (symmetric) role:

Figure 2. Examples of possible relations in an Event Bush: flux, influx and furcation (above); conflux (below)

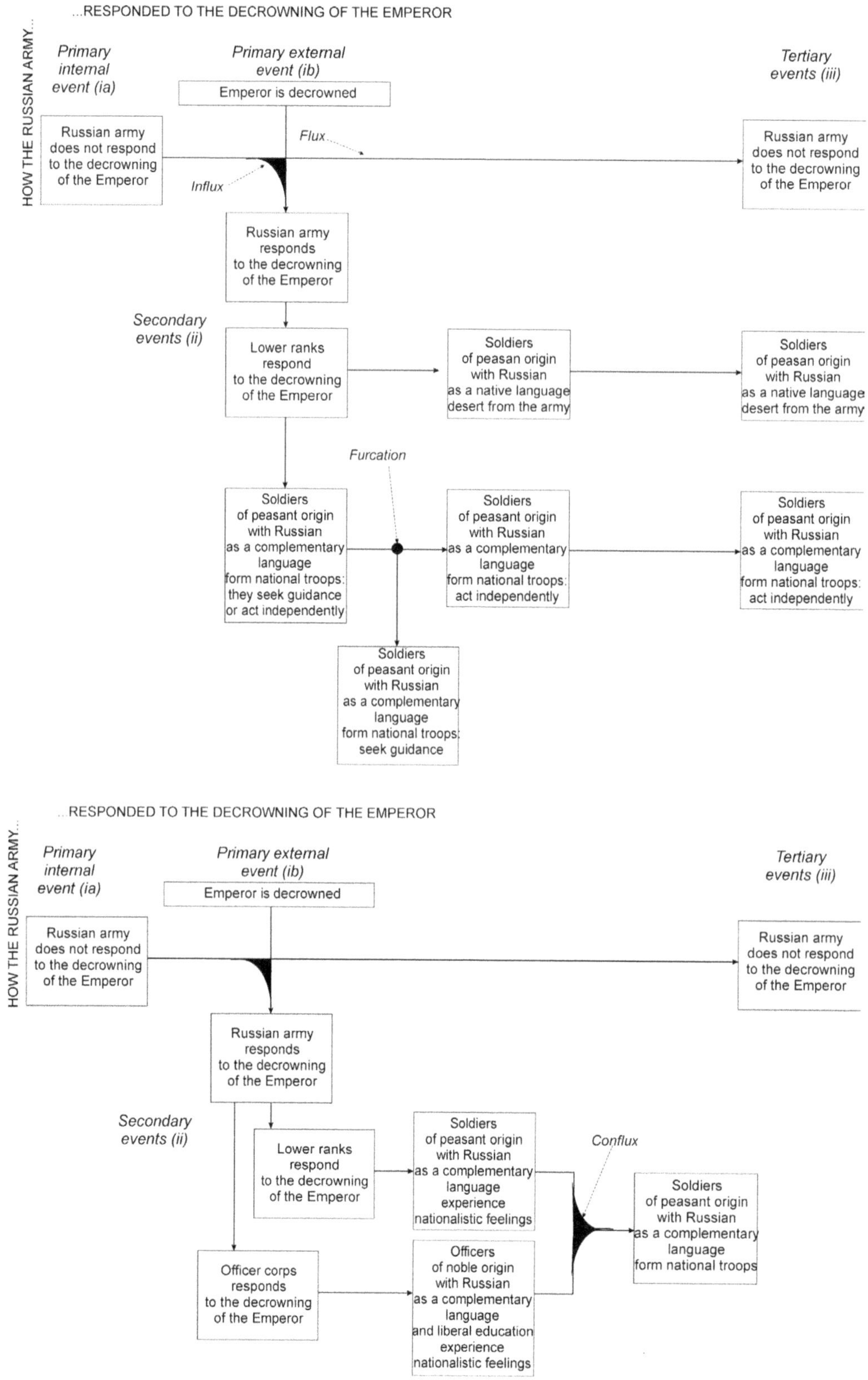

$E_1, E_2, \ldots, E_{n-1}$ *Conflux* E_n

(Pshenichny & Kanzheleva, 2011). This is not used in the present study (see below). The event bush does not admit that more than two events lead to one and play non-symmetric role herewith.

This connective can be seen in Figure 2:

Officers of noble origin with Russian as a complementary language and with liberal education experience nationalistic feelings (E_1)

Soldiers of peasant origin with Russian as a complementary language experience nationalistic feelings (E_2)

CONFLOW

Soldiers and officers with Russian as a complementary language form national troops (E_3)

Furcation (graphically denoted as circle) represents the case that one event (E_1, left of the circle) necessarily leads to one of incompatible alternatives ($E_2, E_3, \ldots, E_n$, right of it):

E_1 *Furcation* $E_2, E_3, \ldots, E_n$.

This can be exemplified by the following inference shown in Figure 2:

Soldiers of peasant origin with Russian as a complementary language form national troops: they seek guidance or act spontaneously (E_1)

FURCATION

Soldiers of peasant origin with Russian as a complementary language form national troops: they act spontaneously (E_2),

Soldiers of peasant origin with Russian as a complementary language form national troops: they seek guidance (E_3).

Wishing to express a case, in which different crisply defined options are compatible, one may put several arcs going out of one node (Figure 2), i.e., use several flux connectives. This means that an event can lead to any number of other events, in influx/conflux with others or by itself. However, this is not the case in the study here.

The rules of composition of and inference in the event bush do not envisage an opportunity of producing several events from a number of events in one single step. Thus, the listed inference options expressed by the connectives of the event bush exhaust the entire variety of possibilities.

Change of events is a correctly built expression including one connective of any type (Pshenichny & Kanzheleva, 2011). Change has one causal part (that includes the events, one or more, left of the connective) and one effect part (the events to the right of the connective).

Flow, or *scenario*, is a finite totally ordered set of changes, in which the left part of the first change is an ia event, the right part of the last change is a iii event, and the right part of every preceding change is the left part of the following change. If there is a change with influx within, the modifier of this influx (ib or ii) is included, but not the events that caused it, unless there are successive influx changes.

If there is a scenario, in which at least one event participates in causal parts of two changes (e.g., one, being flux, and another, furcation), there is another scenario, which overlaps with the former one in the interval from the beginning (*ia*) to this very event, and then ramifies. Furthermore, if there is a scenario without the ii events, first, it may only consist of one change, namely the flux from *ia* to *iii*, and then, there also must be a scenario that overlaps with this one in the beginning (i.e., at the *ia* event) and includes at least one influx. This influx of *ia* and another event, *ib* or *ii*, will produce an *ii* event, which will then route this scenario further.

There cannot be an event in the scenario not included in a bush.

For more detail on the methodology of event bush the reader is referred to the papers by Pshenichny et al. (2009) and Pshenichny and Kanzheleva (2011).

Modeling

The historical context described in Section 2 was considered as an alternative change environment, and the modeling task was formulated as "How the Russian army responded to the decrowning of the Emperor". This, in turn, is the title of the event bush (Figure 3; verbal formulation of events see in Table 1).

The above model reflects solely the vision of the authors but, in contrast to the majority of historical considerations, this subjectivity is put forth explicitly due to the rules of event bush composition. In case of disagreement, one can oppose particular subject, particular predicate or negation, particular subject-predicate link (i.e., event) or particular change between events – or argue whether the proposed events are supported by data. If they are not, either the data are incomplete that often is the case in historical studies, or the scenarios suggested in the event bush exist only in the mind of an expert (historian). However, even in this case they should not be discarded because, first, they should be stored as an unused opportunity (and perhaps the fact that such opportunity was kept in mind by those who participated in the modeled events made them behave in the way they did), and besides, it may be useful for prediction of future evolution.

Alternative vision of the same historical context may readily result in another event bush. If such bushes involve the same subjects and predicates (e.g., considering the native language, education and so forth), they would be easily (and fruitfully) comparable to each other.

The conditions that affect the behavior of the actors include military service and the decrowning of the Emperor. With the existential predicates "exists" or "takes place", these are the primary external events of the bush. One more crucial condition, the war, was not considered for this model because of space restraints. Its inclusion would have made the bush larger and more intricate to fit a publishable publication format. Still, we understand that for the modeled context its presence is essential, so the event bush presented here should not be considered complete. We will provide a full example of the model on the internet at a later point.

At the present, we will limit ourselves to the two above mentioned primary external events. The former, "Military service exists", influences all primary internals and produces new subjects, "officers" (in case of noblemen and philistines), and "soldiers" (if workers or peasants). Volunteering by noblemen or philistines as soldiers was so rare that we did not consider it here. Again, such an inclusion would account for a much larger bush.

Taken into account, however, was the Russian army, especially the one of the early 20th century, since it was quite possible for a common soldier (i.e., a former peasant or worker) to achieve the rank of officer.

This option was put into the bush based on the reason that such officers may be sympathetic to common soldiers in some cases and their decisions may differ from the choice of the rest of the officer corps.

A comment has to be made here. Plotting scenarios in the bush, we consider (or assume that) the subjects of all events behave "crisply", i.e. the whole set behaves as a single being (person); either all Russian-speaking conservative noblemen went into military service and became officers, or no one did, and so forth. This assumption allows us to put the connective of furcation in some cases (see Figure 3). This action is very convenient computationally as it divides the outcome into incompatible options of which one must have taken place. However, it is also clear that some noblemen served in the military, as some did not, and that

Figure 3. How the Russian army responded to the decrowning of the Emperor

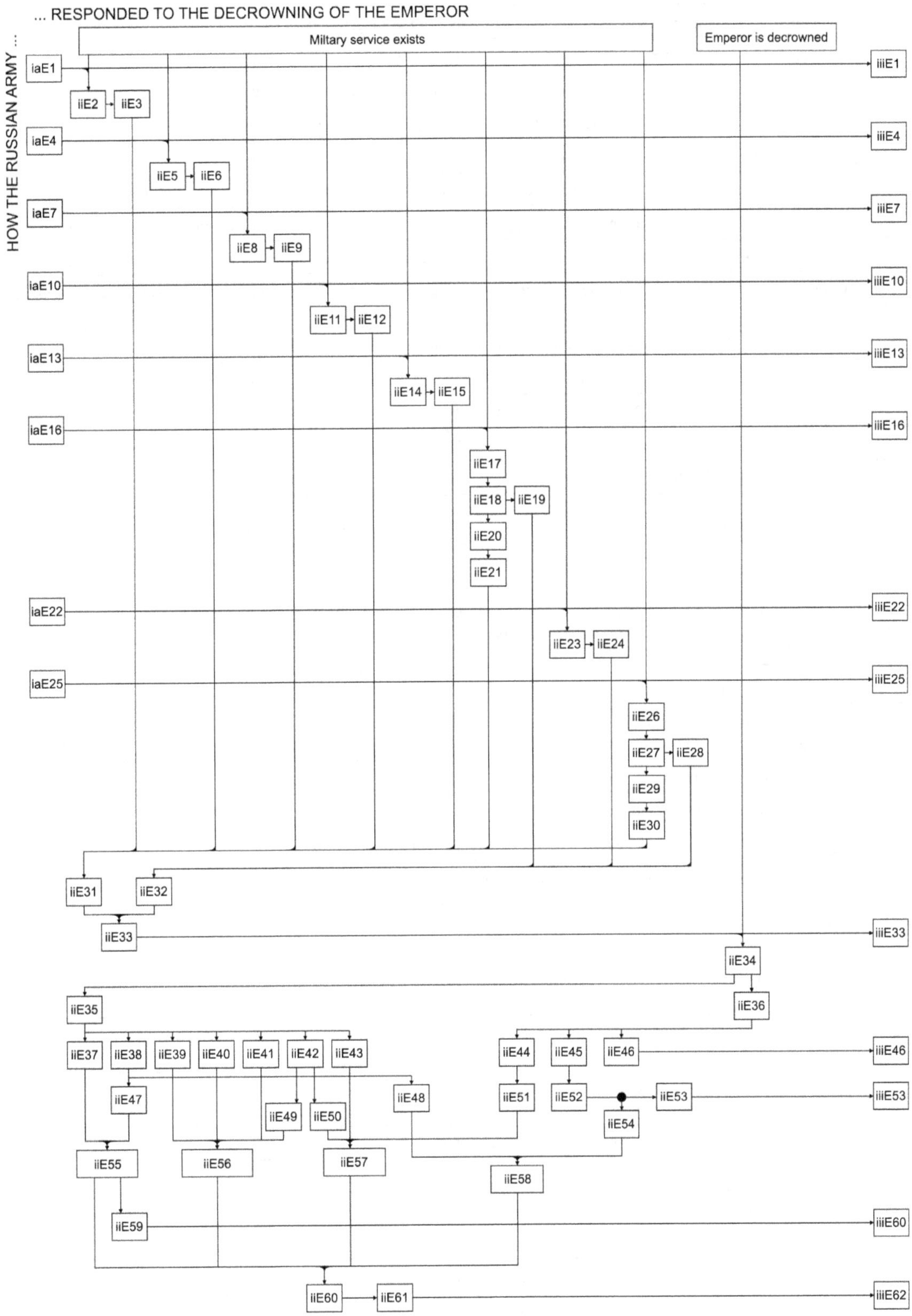

Table 1. Explanation of events from Figure 3

E_1	Noblemen with Russian as native language and with conservative education do not serve in the Russian army.
E_2	Noblemen with Russian as native language and with conservative education serve in the Russian army.
E_3	Officers of noble origin with Russian as native language and with conservative education serve in the Russian army.
E_4	Noblemen with Russian as native language and with liberal education do not serve in the Russian army.
E_5	Noblemen with Russian as native language and with liberal education serve in the Russian army.
E_6	Officers of noble origin with Russian as native language and with liberal education serve in the Russian army.
E_7	Noblemen with Russian as a complementary language and with conservative education do not serve in the Russian army.
E_8	Noblemen with Russian as a complementary language and with conservative education serve in the Russian army.
E_9	Officers of noble origin with Russian as a complementary language and with conservative education serve in the Russian army.
E_{10}	Noblemen with Russian as a complementary language and with liberal education do not serve in the Russian army.
E_{11}	Noblemen with Russian as a complementary language and with liberal education serve in the Russian army.
E_{12}	Officers of noble origin with Russian as a complementary language and with liberal education serve in the Russian army.
E_{13}	Philistines do not serve in the Russian army.
E_{14}	Philistines serve in the Russian army.
E_{15}	Officers of philistine origin serve in the Russian army.
E_{16}	Peasants with Russian as native language do not serve in the Russian army.
E_{17}	Peasants with Russian as native language serve in the Russian army.
E_{18}	Soldiers of peasant origin with Russian as a native language serve in the Russian army: they gain or do not gain promotions to become officers.
E_{19}	Soldiers of peasant origin with Russian as a native language serve in the Russian army: they do not gain promotions to become officers.
E_{20}	Soldiers of peasant origin with Russian as a native language serve in the Russian army: they gain promotions to become officers.
E_{21}	Officers of peasant origin with Russian as a native language serve in the Russian army.
E_{22}	Peasants with Russian as a complementary language do not serve in the Russian army.
E_{23}	Peasants with Russian as a complementary language serve in the Russian army.
E_{24}	Soldiers of peasant origin with Russian as a complementary language serve in the Russian army.
E_{25}	Workers do not serve in the Russian army.
E_{26}	Workers serve in the Russian army.
E_{27}	Soldiers with worker background serve in the Russian army: they gain or do not gain promotions to become officers.
E_{28}	Soldiers with worker background serve in the Russian army: they do not gain promotions to become officers.
E_{29}	Soldiers with worker background serve in the Russian army: they gain promotions to become officers.
E_{30}	Officers with worker background serve in the Russian army.
E_{31}	Officer corps are at service in the Russian army.
E_{32}	Lower ranks at service in the Russian army.
E_{33}	Russian army does not respond to the decrowning of the Emperor.
E_{34}	Russian army responds to the decrowning of the Emperor.
E_{35}	Officer corps respond to the decrowning of the Emperor.

continued on following page

Table 1. Continued

E_{36}	Lower ranks respond to the decrowning of the Emperor.
E_{37}	Officers of noble origin with Russian as a native language and with conservative education deny new government and do not seize power in their troops.
E_{38}	Officers of noble origin with Russian as a complementary language and with conservative education deny new government and do not seize power in their troops. They experience nationalistic feelings: yes or no.
E_{39}	Officers of noble origin with Russian as a native language and with liberal education support new government.
E_{40}	Officers of noble origin with Russian as a complementary language and with liberal education support new government.
E_{41}	Officers of philistine origin support new government.
E_{42}	Officers of peasant origin with Russian as a native language deny or support new government.
E_{43}	Officers with worker background deny new government and seize power in their troops.
E_{44}	Soldiers with worker background seize power in their troops.
E_{45}	Soldiers of peasant origin with Russian as a complementary language form national troops: they act spontaneously or seek guidance.
E_{46}	Soldiers of peasant origin with Russian as a native language desert from the army and go home.
E_{47}	Officers of noble origin with Russian as a native language and with conservative education deny new government and do not seize power in their troops. They do not experience nationalistic feelings.
E_{48}	Officers of noble origin with Russian as a native language and with conservative education deny new government and do not seize power in their troops. They experience nationalistic feelings.
E_{49}	Officers of peasant origin with Russian as a native language support new government.
E_{50}	Officers of peasant origin with Russian as a native language deny new government and seize power in their troops.
E_{51}	Soldiers with worker background seize power in their troops.
E_{52}	Soldiers of peasant origin with Russian as a complementary language form national troops: they act spontaneously or seek guidance.
E_{53}	Soldiers of peasant origin with Russian as a complementary language form national troops: they act spontaneously.
E_{54}	Soldiers of peasant origin with Russian as a complementary language form national troops: they seek guidance.
E_{55}	Officer corps oppositional to new government carry on their service.
E_{56}	Officer corps supporting new government carry on their service.
E_{57}	Workers' and peasants' troops carry on their service.
E_{58}	Nationalistic troops carry on their service.
E_{59}	Officers oppositional to new government leave their service.
E_{60}	Service is carried out by officer corps denying new government, officer corps supporting new government, nationalistic troops and workers' and peasants' troops.
E_{61}	Conflict occurs between officer corps denying new government, officer corps supporting new government, nationalistic troops and workers' and peasants' troops.

there were consequences from these coexistences. This then calls for a fuzzy approach to the event bush method. The possibility of such an approach was advocated theoretically (Pshenichny et al., 2009), but the corresponding formalism has not been developed. We suspect though that the needs of humanities would encourage the fuzzification of the event bush model.

Together the officers and soldiers from all social strata and categories form one more entity, or oneore subject, the Russian army, which functioned as an integral body during the World War First and, therefore, has to be considered in the bush unless influenced by a primary external event such as the "Decrowning of the Emperor". When influenced, this entity begins to disintegrate into the parts it was initially made of, the officers and soldiers, which, in turn, disintegrate further into subparts, depending on the predicates inherited from primary internal events.

One of the key ideas of modeling was to demonstrate that predicates, or to put it correctly, the properties of subjects involved at the very beginning determine different paths of evolution after the influence of similar primary external event. Importantly, the scenarios of the event bush are based not on data but on the considerations we have in mind as experts (the quality and reliability of experts is a different issue addressed in special studies). For instance, we considered that conservative officers of noble origin did not accept the new government, while liberally-minded did as did the officers coming from the group of philistines. Non-Russian officers under some conditions may have looked for their co-nationals and even led the newly formed national troops (which continued, at least nominally, to be part of the Russian army), and so forth. The result was a new type of army which rid itself of archaically-minded (and most apt?) officers who preferred to retire rather than serve without the czar. This army, however, was no longer a unified corps. Rather it was haphazardly organized based on social origins and other qualifiers, or traits, of military men which inherently results in internal conflicts. These internal conflicts could have led to factions even without a Bolshevik coup.

Discussion

Our event bush models the opportunities of transformation of the Russian army after the decrowning of the Emperor. It represents events represented by statements in boxes (i.e., graphic objects loaded with pieces of historical knowledge formulated by strict rules) and connectives ("arrows", "arcs"), which carry formal meaning for any event bush regardless of its contents. Events are formulated following the guidelines of Pshenichny et al. (2009), i.e., being as general as possible, purely descriptive, primitive by meaning and unique and non-overlapping in the modeled environment. This means that different researchers studying, e.g., the destiny of different corps, say, the Navy or the Cossacks, would find the terms of these generic for their studies with some specificity, which can be informally regarded as "specificity of absence" – e.g., there likely were no or very few officers from the workers or peasants in the Navy or liberally-minded officers among the Cossacks. Still, every subject-predicate relation presents particular case studies. Particular cases, furthermore, can serve as general cases for further specification of research and be plotted as event bushes themselves. Event bushes for general cases will be used as rules of composition of event bushes describing particular cases, as was demonstrated earlier by the example of chemistry (Pshenichny, 2013).

Moreover, alternative views can also be shown by similar events grouped into another event bush with a different set of connectives. Such bushes are termed as *family* of event bushes (Pshenichny et al., 2009). Descriptiveness and generality of terms allow the event bushes to move away from the issue of whether the entity in question is termed correctly. The approach of the event bush modeling is, "once a scientist thinks so and uses (or considers possible to use) such terms, let us see what happens next".

As seen from the above, not only events of an event bush can serve as boundary objects in a domain of historical knowledge, but also allow formalized operation with boundary objects based on the event bush's syntax and semantics. This actually means that the "historical mechanisms" that formally describe the historical processes and scenarios can be modelled using the event bush method. This action, in turn, will produce a body of impartial historical knowledge. Researchers and their schools could then employ the individual and or collective contexts for their scholarly work.

In such "purified" form, historical knowledge can serve as a potential data model for the creation of databases of historical events (that is, "filling" the event bush with certain names, dates and other facts and materialized in a computer-based tool– an example of the COLLA project embedding the event bush and related data was provided by Diviacco et al., submitted). It may be also be used for an additional task – the adjustment of the above discussed basic historical mechanisms (with necessary transformation) to other contexts (e.g., the faith of national armies in German, Turkish or Austro-Hungarian empires after decrowning of the last monarch, or behavior of army when a military dictatorship is over, and so forth) and, importantly, to some modern occasions or events.

However, as was pointed above, individual bush represents only a *vision* of a particular scientist and his/her scientific school. This approach does help to identify inconsistencies and gaps within a particular vision but cannot determine what is actually true or false. Nevertheless, if specialists agree on the semantics of events, a *family* of event bushes could be built (Pshenichny et al. 2009) in which some common scenarios can be found. These scenarios may be understood as "tautologies" in the field of knowledge modeled by the event bush. Moreover, projected onto a similar dataset, different bushes of a particular family can be evaluated as better or poorer when supported by data. Finding new evidence may change the "distribution of proof" between the bushes. This is seen as one of the main perspectives of application of the event bush method in history.

This perspective, being quite promising, nevertheless seems to be in contradiction with one of the main traditions of this science, which seems to be implicitly based on a reflection of unidirectional and invariant time by a unidirectional and invariant form of its representation, the text (or coherent story). The role of text in history was discussed in many fundamental works on historiography, White (1973) to be mentioned first of all. While text suggests a flow of information, the event bush represents "multiflow" that includes never-happened but plausible scenarios. Implementation of such vision of history would substantially enrich this science and, inter alia, notably increase its forecasting ability.

CONCLUSION

Modeling the historical context of the transformation of the Russian army after decrowning of Russia's last Emperor, Nikolai II, by means of the event bush shows that at least for the addressed domain of historical knowledge this method serves to provide graphic boundary objects and, moreover, allows to build an impartial knowledge base in what seems to be a descriptive and interpretative field from the vantage point of the hard sciences, having, nonetheless, substantial academic and practical importance to us all. Our study intends to open an opportunity for wide cross-cultural, cross-contextual and cross-presuppositional collaboration in history and may hopefully lead to a wider and more efficient computer-assisted collaboration in historical research.

REFERENCES

Carbonell, B. (2009). The Syntax of Objects and the Representation of History. *History and Theory*, *47*(2), 122–137. doi:10.1111/j.1468-2303.2009.00501.x

Diviacco, P. (2012). Addressing Conflicting Cognitive Models in Collaborative E-Research: A Case Study in Exploration Geophysics in Collaborative and Distributed E-Research. In A. A. Juan, T. Daradoumis, M. Roca, S. E. Grasman, & J. Faulin (Eds.), *Collaborative and Distributed E-Research: Innovations in Technologies, Strategies and Applications*. Hershey, PA: IGI Global. doi:10.4018/978-1-4666-0125-3.ch012

Diviacco, P., Pshenichny, C., Carniel, R., Khrabrykh, Z., Shterkhun, V., & Mouromtsev, D. et al. (Submitted). Organization of a geophysical information space by using an event-bush-based collaborative tool. *Earth Science Informatics*.

Eliseev, F. I. (2003). *S Kornilovskim konnim*. Moscow: Astrel Press. Retrieved 18 March 2014 from http://www.dk1868.ru/history/s_korn_konn1.htm (in Russian).

Feigenbaum, E. A. (1984). Knowledge engineering: the applied side of artificial intelligence. In *Proceedings of a Symposium on Computer Culture: The Scientific, Intellectual, and Social Impact of the Computer.* New York, NY: New York Academy of Sciences. doi:10.1111/j.1749-6632.1984.tb16513.x

Hollander, J. D., Paul, H., & Peters, R. (2011). Introduction: The Metaphor of Historical Distance. *History and Theory*, *50*(4), 1–10. doi:10.1111/j.1468-2303.2011.00599.x

Klein, K. L. (2011). *From History to Theory*. Berkeley, CA: University of California Press.

Krasnov, P.N. (1922). *Na vnutrennem fronte* (vol. 1). Berlin, Germany: Archive of Russian Revolution (in Russian).

Kuklick, B. (1969). The Mind of the Historian. *History and Theory*, *8*(3), 313–331. doi:10.2307/2504139

Kutepov, A. P. (1934). *General Kutepov.* Sbornik statei. Retrieved 18 March 2014 from http://www.dk1868.ru/history/Kutep2.htm (in Russian).

Petliura, S. V. (1918). Potreba viiskovoi literature. *Knigar*, *7*, 375–376.

Popova, O. V. (2011). *Political analysis and prediction*. St. Petersburg, Russian Federation: Aspect Press. (in Russian)

Pshenichny, C. A. (2013). Adjustment of the Event Bush Method to Chemical and Related Technological Tasks. In D. Mouromtsev, C. Pchenichniy, & D. Ignatov (Eds.), *Modeling States, Events, Processes and Scenarios* (pp. 74-81). Retrieved 18 March 2014 from http://iccs2013.hbcse.tifr.res.in/workshops/copy2_of_text.pdf

Pshenichny, C. A., Carniel, R., & Akimova, V. L. (2005). Decreasing the uncertainty of BBN technique by means of complex formal approach tovolcanological information treatment. *Geophysical Research Abstracts*, *7*, EGU05–A-01016.

Pshenichny, C. A., & Kanzheleva, O. M. (2011). Theoretical foundations of the event bush method. In K. Sinha, L. Gundersen, J. Jackson, & D. Arctur (Eds.), *Societal Challenges and Geoinformatics*. Boulder, CO: Geological Society of America. doi:10.1130/2011.2482(12)

Pshenichny, C. A., & Khrabrykh, Z. V. (2002). Knowledge Base of Formation of Subaerial Eruption Unit. In S. Leroy & I. Stuart (Eds.), *Environmental Catastrophes and Recovery in the Holocene (Abstracts)*. London: Brunel University. Retrieved 18 March 2014 from: http://atlas-conferences.com/cgi-bin/abstract/caiq-22

Pshenichny, C. A., & Mouromtsev, D. I. (2013). Representation of the Event Bush Approach in Terms of Directed Hypergraphs. In H. D. Pfeiffer, D. I. Ignatov, J. Poelmans, & N. Gadiraju (Eds.), *Conceptual Structures for STEM Research and Education*. Mumbai, India: Springer. doi:10.1007/978-3-642-35786-2_22

Pshenichny, C. A., Nikolenko, S. I., Carniel, R., Sobissevitch, A. L., Vaganov, P. A., & Khrabrykh, Z. V. et al. 2008). The event bush as a potential complex methodology of conceptual modelling in the geosciences. In Sanchez-MarreJ.BejarJ. ComasJ.RizzoliA.GuarisoG. (Eds.), *International Congress on Environmental Modelling and Software*, (vol. 2, pp. 900-912). Barcelona, Spain: International Environmental Modelling and Software Society.

Pshenichny, C. A., Nikolenko, S. I., Carniel, R., Vaganov, P. A., Khrabrykh, Z. V., & Moukhachov, V. P. et al. (2009). The Event Bush as a Semantic-based Numerical Approach to Natural Hazard Assessment (Exemplified by Volcanology). *Computers & Geosciences*, *35*(5), 1017–1034. doi:10.1016/j.cageo.2008.01.009

Reill, P. H. (1986). Narration and Structure in Late Eighteenth-Century Historical Thought. *History and Theory*, *25*(3), 286–298. doi:10.2307/2505192

Rüsen, J. (1996). Some Theoretical Approaches to Intercultural Comparative Historiography. *History and Theory*, *35*(4), 5–22. doi:10.2307/2505441

Smith, S. G. (2009). Historical Meaningfulness in Shared Action. *History and Theory*, *48*(1), 1–19. doi:10.1111/j.1468-2303.2009.00482.x

Spiridovich, A. I. (1962). *Velikaya voina i fevralskaya revolucia. 1917 – 1922*. Retrieved 18 March 2018 from http://www.rummuseum.ru/lib_s/spirid000.php (in Russian).

Star, S. L., & Griesemer, J. R. (1989). Institutional ecology, translations and boundary objects: Amateurs and professionals in Berkeley s museum of vertebrate zoology. *Social Studies of Science*, *19*(3), 387–420. doi:10.1177/030631289019003001

Tsvetkov, V. J. (2010). *Selected essays: Tsarskaya pamiatka, Zaveschanie imperatora, Proschalniy prikaz*. Retrieved 18 March 2014 from http://rusk.ru/st.php?idar=424636 (in Russian).

Tutte, W. T. (1998). *Graph theory as I have known it*. New York: Oxford University Press.

White, H. (1973). *Metahistory: The Historical Imagination in 19th-century Europe*. Baltimore, MD: Johns Hopkins University.

Zoschenko, M.M. (1933, July 3). *Kakie u menya byli professii*. Literaturnyi Leningrad (in Russian).

KEY TERMS AND DEFINITIONS

Boundary Object: An artifact that aims to bridge concurrent cognitive models through abstraction from all the domains of the partners.

Environment: A modeled situation. Dynamic environments were classified by Pshenichny and Kanzheleva (2011).

Event Bush: Method of dynamic knowledge engineering that represents alternative directed change environments as a set of intertangling scenarios using four types of events (primary internal, primary external, secondary and tertiary) and four connectives (flux, influx, furcation and conflux).

Event: In history is understood broadly and semi-intuitively; dynamic knowledge engineering – a set of statements of general form $S - P_1, P_2, \ldots, P_n$, where S is subject, and $P_1, P_2, \ldots, P_n$ are predicates, with or without negation, complying the following conditions: (i) subject may be only one in a statement, (ii) subject may not be negated, (iii) all predicates may not be negated in a statement, at least one must be without negation. Events represent the relations between objects of two types: one subject and whatever number of predicates.

Knowledge Engineering: A selection of methods of various origins (from statistics to psychology, from linguistics to physiology) to look at how qualitative (commonly, though not necessarily, verbal) expressions are treated by humans and thus try to minimize the human subjectivity in information modeling (Pshenichny after Feigenbaum, 1984).

Chapter 18
Perspectives of Use of Petri Nets in Collaborative Research

Vyacheslav Rogalchuk
St. Petersburg State Transport University, Russia

Konstantin Solomin
ITMO University, Russia

ABSTRACT

Petri Nets are efficient and well-formalized tools for modeling the dynamic systems. With some reasonable constraints, they can be readily applied to knowledge representation and engineering, and thus assist in collaborative research. This chapter proposes a way to do this based on the unified grammar of dynamic knowledge. Petri Nets appear interpretable in terms of the said grammar and may be considerably enhanced by the introduction of semantic rules for formulation of expressions in their places. In such form, they are quite apt to be used as a knowledge representation and visualization method in computer-based tools for support of collaborative research, and their quantification potential (i.e., marking of the net and attributing weights to its arcs) may appear very useful for this task, which is usually accomplished by purely qualitative tools.

INTRODUCTION

Petri nets are a family of mathematical models designed for modeling, analysis and development of dynamic systems, from jurisprudence to biology. The original theory was presented in 1962 by Carl Adam Petri in his Doctoral Dissertation «Communications with Automata», where he described both mathematic and graphic approaches to modeling (Petri, 1962). This approach was developed in his later works (e.g., Petri, 1973; 1977). Classical Petri nets are now also called Low Level Petri Nets; many extended models have been built on top of them, linking this method to various mathematical formalisms and thus widening its modeling capabilities. Three classes of Petri Nets are now in use – discrete nets, continuous time nets and stochastic nets.

Mathematically, Petri nets are directed labeled bipartite multigraphs (Petri, 1962). At the same time, they bear semantic information in the nodes and therefore can be used for representation of

DOI: 10.4018/978-1-4666-6567-5.ch018

Copyright © 2015, IGI Global. Copying or distributing in print or electronic forms without written permission of IGI Global is prohibited.

qualitative knowledge about the modeled systems. In doing this, they may well cover the alternative change environments (Pshenichny & Kanzheleva, 2011) and therefore should be regarded as a dynamic knowledge representation and engineering method. As such, Petri nets can be applied to use the unified dynamic knowledge grammar suggested by Pshenichny and Mouromtsev (2015, see chapter 16 this book).

The usefulness of Petri nets for organization of knowledge "both inside and outside computer" was emphasized yet by Petri (1977, p. 131). However, in last twenty years, discussing the application of Petri nets to knowledge representation, researchers were inspired mainly by an ability of one particular type of Petri net, the fuzzy one, to represent one particular, however important, type of knowledge – the fuzzy knowledge (Li & Lara-Rosano, 2000; Liu et al., 2013; Xiaozhong, 2003; Yeung & Tsang, 1994; Wang et al., 2014; Ribarić et al., 2009). Interestingly, most of these authors state that Petri nets are particularly efficient to represent the dynamic knowledge. However, virtually all of them consider knowledge "dynamic", in the sense that this knowledge is imprecise, incomplete and may change (regardless of whether it describes fixed or changing matters), while the term "dynamic knowledge" sensu Pshenichny and Mouromtsev (2015, see chapter 16 this book) means the knowledge about dynamic environments, or, in other words, the knowledge about changing events. Henceforth the term "dynamic knowledge" will be used in the latter sense. Desirably, such knowledge is precise and complete but includes incompatible statements corresponding to changing events. However, it certainly can be fuzzified and serve as the basis for any fuzzy knowledge. Therefore, consideration of "crisp dynamic knowledge" performed below is relevant to the existing studies of fuzzy knowledge representation by means of Petri nets.

Still, what is essential in this consideration is analysis of subject-predicate structure of dynamic knowledge. This issue has not been addressed by existing studies of knowledge representation by means of Petri nets. The present work intends to start filling this gap, as it is especially relevant in concern of collaborative research and formulation of shared knowledge. An important, though indirect, hint that Petri nets may appear exclusively appropriate in collaborative studies in the least formalized and highly descriptive fields of knowledge is the work of Hussein and El-Sattar (2012) on application of Petri nets to interactive storytelling.

Considering a possible use of Petri nets as a collaborative knowledge representation/engineering tool, this chapter presents a short introduction into Petri net models, brings them into the dynamic knowledge engineering context and suggests the guidelines to apply these models in collaborative studies.

DEFINITION AND PROPERTIES OF PETRI NETS

A Petri net is a directed bipartite multigraph. Directed means that for each arc of this graph the beginning and end are defined (or, in other words, the arcs go from parent nodes to child nodes). It has two types of nodes called *places* and *transitions*; directed arcs link places to transitions and transitions to places but not the nodes of similar type to each other – this is the meaning of bipartite. In some cases a place and transition may be connected by two oppositely directed arcs (thus forming a multigraph); however, this can always be substituted by a normal directed graph structure. Places represent a set of variables that define the model state. Transitions are abstractions for rules of state changes. Behavior of a system is modeled by execution (or *firing*) of transitions in a net. Not a single node may appear in the net more than one time; in other words, all names of the nodes are unique for the net.

Thus the structure of Petri Net N is defined by a tuple $\{P, T, I, O\}$, where $P = \{p_1, p_2, ..., p_n\}$ is a set of places, denoted as p_i, $T = \{t_1, t_2, ..., t_m\}$ is a set of transitions, denoted as t_j, $I: P \times T \rightarrow N$, an application of precedence, and $O: P \times T \rightarrow N$, an application of incidence, N being the set of natural numbers. $I(p_i,t_j)$ contains the number (or weights) of arcs going from p_i to t_j, and $O(p_i,t_j)$ contains the number of arcs going from t_i to p_j. Places with arcs leaving them and ending in t_i are called *input places* for t_i; Places with arcs connected to them from transition t_i are *output places* for t_i.

Each place can have a discrete number of 'tokens' representing some condition or resource. The token distribution in all places is called *marking* of a net. Mathematically, a marked net N is defined by a couple $\{N, M\}$, where N is a Petri net structure as stated above, and $M: P \rightarrow N$ is marking. $\mu(N)$ denotes the marking of net N, and $\mu(p_i)$ denotes the marking of a place p_i (or, a number of tokens within place p_i).

The initial distribution of tokens in the net describes the initial state of the system, and it is denoted as marking μ_0. From there, the state changes of system are modeled by executing transitions.

For each marking μ_i there is a set of transitions enabled to be fired, or executed. The transition t_i is enabled if for every input place p_j there is a greater number of tokens than a weight of arc leading to t_i ($\forall p_j \in P$, $\mu(p_j) \geq I(p_j, t_i)$). Figure 1 presents a marked net with t_1 disabled and t_2 enabled, since p_2 contains two tokens, which is less than $I(p_2, t_1) = 3$, and equal to $I(p_2,t_2) = 2$. Places are depicted as circles with dots within representing tokens, and transitions are depicted as rectangles.

Execution of transition creates a new marking μ` from the marking μ, as described by the *firing rule*: first subtract the weight of corresponding outgoing arcs (in Figure 1: 2 and 3 for Transition 2) from the amount of tokens for every input place (*p2* and *p3* for Transition 2) of fired transition, then for every output place (*p5* and *p6* for Transition 2) add the weight of the arc incoming from the corresponding transition:

$$\forall p, \mu`(p) = \mu(p) - I(p,t) + O(p, t),$$

where p is place, $\forall$ is universal quantifier, the rest as stated above.

Figure 1. Marked Petri net with places 1-6 (p1 – p6), transition t1 disabled and transition t2 enabled; unmarked arc is meant to have weight 1 (by default). See comments in the text.

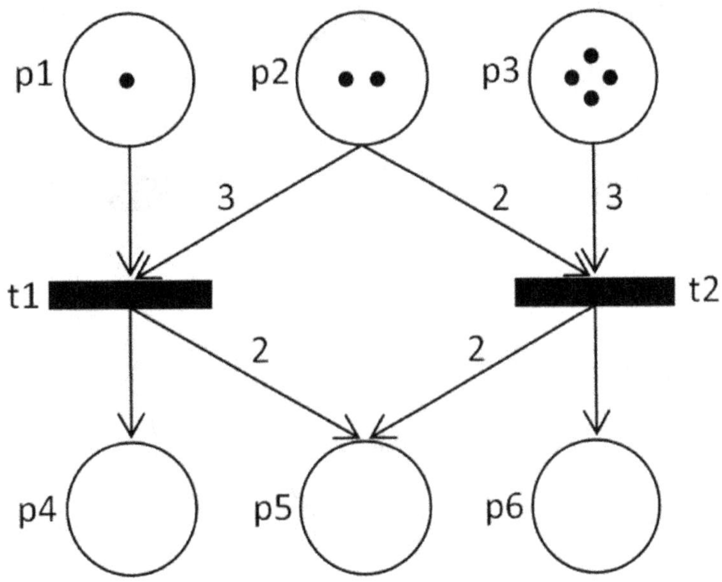

Figure 2 depicts the state of the same Petri net after t_2 being fired.

Given the marking, several transitions can be enabled, depending on the distribution of tokens. For example, if *p2* of Figure 1 had three tokens, both *t1* and *t2* would be enabled. In such event, only one transition can be fired at a time, generating the new marking. So, to find all possible next states for some system state, all enabled transitions are fired one after another, starting from the same marking. The sequence of firing is undetermined, since all transitions are considered having equal priority.

Some higher-level Petri nets, which can be of potential use for knowledge representation and engineering, are considered below.

When modeling a large system, place-transition net graphs can become almost unreadable. Several types of high-level nets were developed to avoid this. These nets add information to nodes, arcs and tokens, allowing to 'fold' original net into a more compact and readable model. Jensen (1981) introduced colored nets as an extension to basic Petri net formalism allowing to build more concise models in cases where there is a lot of similar elements or data structures.

A Colored Petri net (CPN) has a classical net structure including places, transitions and directed arcs between them. After that, a set of *colors* is defined, and a set of inscriptions (or labels) is added to elements of a structure. For a place it is a set of token colors that the tokens in that place are allowed to have. For an arc it is arc expressions, which are built from variables, constants, operators and functions. These expressions are used to find all enabled transitions. For a transition to be enabled, it must be possible to bind all the variables that appear in associated arc expressions; firing of a transition means evaluation of input arc expressions, removal of consumed token

Figure 2. Marked Petri net after execution of Transition t2

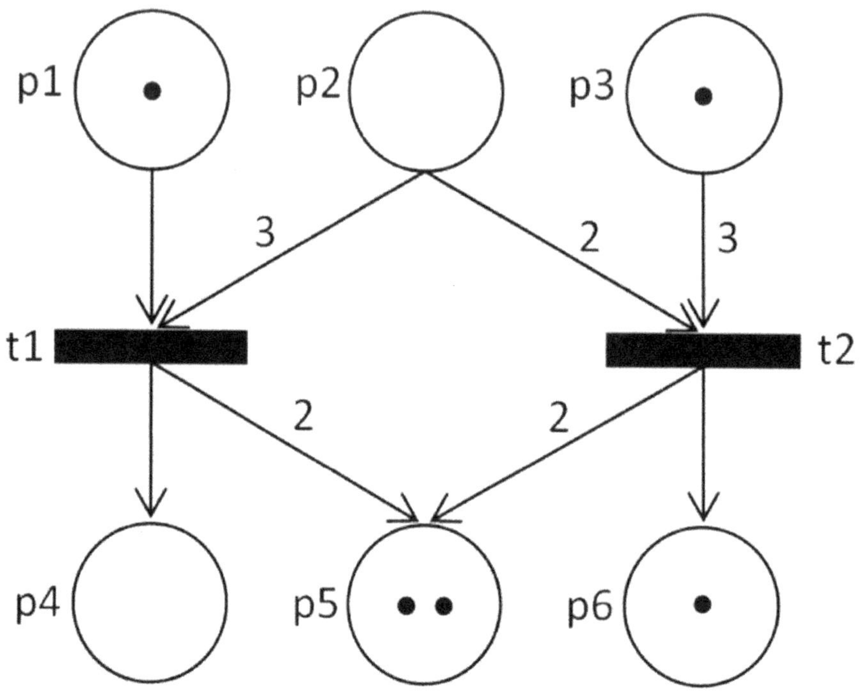

multisets from input places, evaluation of output arc expressions with bound values and addition of resulting color multisets to output places.

A colored Petri net can be seen as an abbreviation of the classical Petri net. To unfold a colored net, there must be additional places and transitions generated for every color in the net, and arcs added.

Well-formed nets are a subclass of colored Petri nets, defined to simplify their analysis without losing expression power (Girault & Pradat-Peyre, 2009). There are three types of constraints applied to nets: color domain constraint, color function constraints and guard functions usage restriction. Color domains must be Cartesian products of elementary color classes – classes that contain only colors independent of each other. Arc expressions can contain only a subset of color functions - elementary color functions – the identity function, used to select a color from color set, successor function and broadcast function.

Since original place-transition nets have no notion of time, several enhancements have been proposed to allow modeling of systems in which time is important. Among these are *timed Petri nets* (Ramchandani, 1973). Timed Petri nets extend Petri nets by associating a firing duration to transitions and adding a rule that a transition should be fired as soon as it is enabled.

Another approach to timing problem is *time Petri nets* (TPNs) suggested by Merlin (1974). TPNs associate two parameters with each transition - *min* and *max* -, which relate to the moment in which the transition becomes enabled. If a transition t is enabled at some moment τ, then t should be fired at a moment between $\tau+min$ and $\tau+max$, unless it became disabled in another marking.

A firing schedule is a pair of sequence of transition names *ts* and a sequence of relative firing dates *ds*. A schedule (*ts, ds*) is achievable from a state *S* if and only if transitions in the ts sequence are successively fireable from *S* at relative firing dates in the ds sequence. Operation of a net can thus be described by the set of its schedules achievable from its initial state. It is impossible to build reachability graph of states of time Petri net as time is continuous and there will be infinite number of nodes, so instead state classes are defined and analyzed.

Figure 3. Example of time Petri net

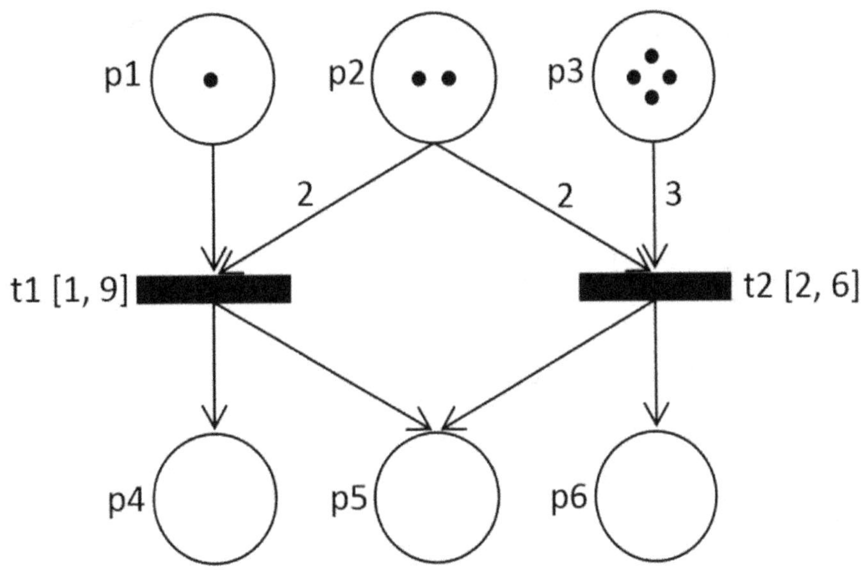

For further detail, one should refer to Merlin (1974). This type of Petri net has been considered here to illustrate one of the most probable next steps the Petri net-based knowledge representation may make, as time is of doubtless relevance to the knowledge of changing environments.

DISCUSSION: PETRI NETS AS A DYNAMIC KNOWLEDGE ENGINEERING TOOL

Instructive Example

Modeling by means of Petri nets can be illustrated by a simple example of railway traffic lights (Figure 4). Suppose that it indicates the state of two rail sections behind it. When train approaches the traffic lights, rail section 1 is free (place "free 1"). When train passes the traffic lights and, hence, enters section 1 (place "occ 1", "occ" standing for "occupied"), the section 1 is busy and the red light is on (place "red"). When train enters section 2 (place "occ 2"), the yellow light is on (place "yellow"). Finally, when train leaves the section 2 (place "free 2"), the green light is on (place "green").

The formal model presented in Figure 4 may receive many interpretations (Table 1). This is another virtue of formal models widely used in knowledge representation and engineering – see, e.g., Mouromtsev et al. (2014). For instance, "Some states help Syrian rebels" (place "green"), transition 1, "Chemical weapons are being transported to seaports" (place "occ1"), "Some states do not help Syrian rebels" (place "red"), transition 2, "Chemical weapons are being transported from seaports away from Syria" (place "occ2"), "Some states give a signal to Syrian rebels" (place "yellow"), transition 3, "Some states help Syrian rebels" (place "green").

Figure 4. Petri net that can be interpreted in a number of ways and serve, in some interpretations, the purpose of collaborative knowledge representation. See comments in the text.

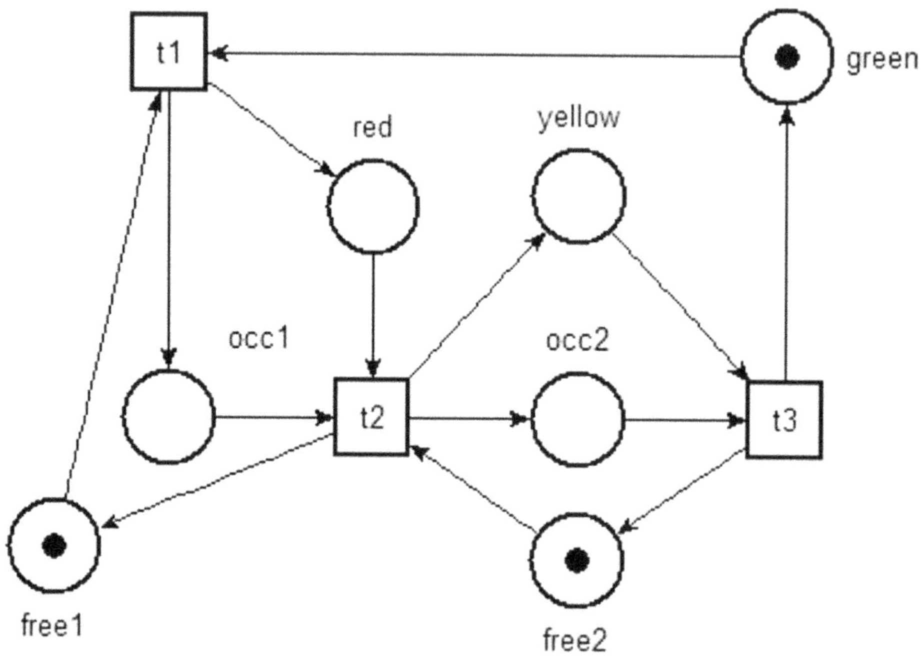

Table 1. Place interpretations in Figures 4 and 5

Place	Meaning in Figure 4 ("Railway")	Meaning in Figure 5 ("Syria")
FREE 1	Railway section 1 is unoccupied	Weapons are ready to be transported
FREE 2	Railway section 2 is unoccupied	Seaports are ready to send out ships
OCC 1	Railway section 1 is occupied by train	Chemical weapons are being transported to seaports
OCC 2	Railway section 2 is occupied by train	Chemical weapons are being transported from seaports away from Syria
GREEN	Green light is on	Some states help Syrian rebels
YELLOW	Yellow light is on	Some states give a signal to Syrian rebels
RED	Red light is on	Some states do not help Syrian rebels

The latter interpretation may well serve the purposes of collaborative research or even investigation issues. For instance, each of the places may be commented by collaborators and/or supplied with pieces of data to see how well each point of reasoning is supported. Technically this can be done by existing computer-based tools, e.g., COLLA (Diviacco, 2012; Diviacco et al., submitted). The considered example may be the case for political scientists, historians, military experts, journalists and many others. Removing "Syria" from the names of the places, we obtain a generic case applicable to a number of situations including those that have not happened yet; removing "chemical" or substituting "chemical weapons" by another subject, we arrive into a yet more generic case, thus constructing a general knowledge more or less similar to the way it was done by Solomin and Mandrik (2015, see chapter 17 this book) by means of the event bush method described by Pshenichny et al. (2009) and Pshenichny and Kanzheleva (2011).

Furthermore, if some collaborators do not agree with this model, they may formulate their disagreement in similar terms. First, they may argue the amount of tokens necessary to fire this or that transition or associated with this or that place. Then, they may claim to modify the structure of the model, e.g., deny or rearrange some of the places (Figure 5).

Thus a *family* of models can be obtained that operate similar terms but are organized differently, similarly, again, to families of event bushes alternatively describing the same problem. However, it should be stressed herewith that event bushes and other methods known to the authors do not provide a quantitative means to express disagreement. Here, the marking of places and weights associated with arcs of the net may indicate a quantitative disagreement (e.g., one accepts the model in general but argues the firing thresholds for some transitions, like that how many weapons should be transported to stop giving help to rebels) and the arcs themselves may express qualitative disagreement (e.g., even when the weapons are away the help is not resumed – see Figure 5).

GRAMMAR OF PETRI NETS

To bring Petri nets into the dynamic knowledge engineering framework and thus make them a tool of collaborative research, it is vitally important to ensure a strict and correct way of encoding knowl-

Figure 5. Alternative model of the military conflict in Syria, representing a model modified from Figure 4.

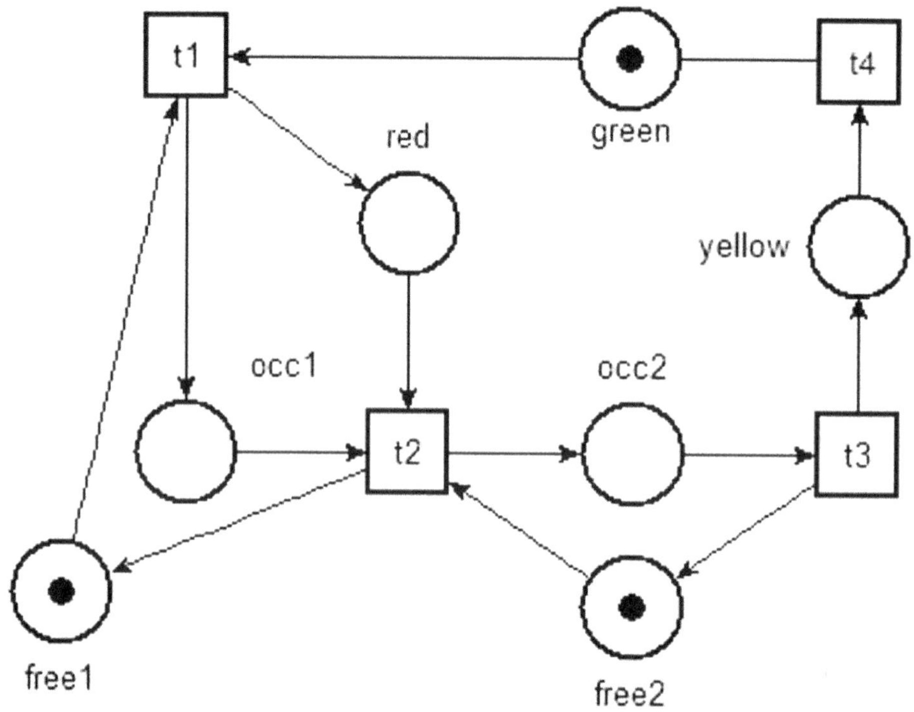

edge. This is the purpose of the unified grammar of dynamic knowledge proposed by Pshenichny and Mouromtsev (2015, see chapter 16 this book). Below the basic rules of composition of Petri nets will be compared to the semantics and syntax of this grammar to see whether and how the former complies with the latter.

This grammar includes two distinct types of entity, events and connectives. Connectives connect events into so-called changes expressed in this grammar as formulae, by analogy with the classical logic. Petri nets also possess two different types of entity, places and transitions, and syntactically their position is pretty similar to events and connectives, correspondingly – transitions necessarily connect places, which cannot be connected otherwise. Hence, it seems straightforward to compare places with events and transitions with connectives.

Events are defined by Pshenichny and Mouromtsev (2015, see chapter 16 this book) as statements of general form $S - P_1, P_2, \ldots, P_n$, where S is subject, and $P_1, P_2, \ldots, P_n$ are predicates, with or without negation. We put three restrictions herewith, (i) subject may be only one in a statement, (ii) subject may not be negated, (iii) all predicates may not be negated in a statement, at least one must be without negation (e.g., a statement $S - \neg P$, $\neg$ being the negation, is not allowed).

No special rules have been proposed for formalization of places in Petri nets, and this seems to be perhaps the most serious hindrance to use Petri nets as a knowledge engineering tool. So far, the names are given to places quite freely, as was done in Figure 1 ("red", "green", "occ1", "free2" and so forth).

Still, nothing precludes us from constraining the names of places in line with the definition of events and concomitant rules suggested in the event description grammar. Then, the places in the "railway" example will be formulated as follows (Table 2):

Table 2. Event definitions for places in Figures 4 and 5

Place Name	Place Definition as Event in Figure 4	Place Definition as Event in Figure 5
FREE 1	$S_1 - P_1, \neg P_2, \neg P_3, \neg P_4$	$S_1 - P_1, \neg P_2, \neg P_3, \neg P_4$
FREE 2	$S_1 - P_1, \neg P_2, \neg P_3, P_4$	$S_1 - \neg P_1, \neg P_2, \neg P_3, P_4$
OCC 1	$S_1 - P_1, P_2, \neg P_3, \neg P_4$	$S_1 - \neg P_1, P_2, \neg P_3, \neg P_4$
OCC 2	$S_1 - P_1, \neg P_2, P_3, \neg P_4$	$S_1 - \neg P_1, \neg P_2, P_3, \neg P_4$
GREEN	$S_2 - P_5$	$S_2 - P_5, \neg P_6, \neg P_7$
YELLOW	$S_3 - P_5$	$S_2 - P_5, \neg P_6, P_7$
RED	$S_4 - P_5$	$S_2 - P_5, P_6, \neg P_7$

- **Place "free1":** Train (S_1) is on the way (P_1), is not at the first rail section (P_2), is not at the second rail section ($\neg P_3$), is not behind the second rail section ($\neg P_4$).
- **Place "occ1":** Train (S_1) is on the way (P_1), is at the first rail section (P_2), is not at the second rail section ($\neg P_3$), is not behind the second rail section ($\neg P_4$).
- **Place "occ2":** Train (S_1) is on the way (P_1), is not at the first rail section ($\neg P_2$), is at the second rail section (P_3), is not behind the second rail section ($\neg P_4$).
- **Place "free2":** Train (S_1) is on the way (P_1), is not at the first rail section ($\neg P_2$), is not at the second rail section ($\neg P_3$), is behind the second rail section (P_4).
- **Place "red":** Red light (S_2) is turned on (P_5).
- **Place "yellow":** Yellow light (S_3) is turned on (P_5).
- **Place "green":** Green light (S_4) is turned on (P_5).

Similarly, in the chemical weapon example, places can be formulated in this way.

- **Place "free1":** Chemical weapons (S_1) are stored at military bases (P_1), are not transported from military bases to seaports ($\neg P_2$), are not stored in seaports ($\neg P_3$), are not transported away from seaports by sea ($\neg P_4$).
- **Place "occ1":** Chemical weapons (S_1) are not stored at military bases ($\neg P_1$), are transported from military bases to seaports (P_2), are not stored in seaports ($\neg P_3$), are not transported away from seaports by sea ($\neg P_4$).
- **Place "occ2":** Chemical weapons (S_1) are not stored at military bases ($\neg P_1$), are not transported from military bases to seaports ($\neg P_2$), are stored in seaports (P_3), are not transported away from seaports by sea ($\neg P_4$).
- **Place "free2":** Chemical weapons (S_1) are not stored at military bases ($\neg P_1$), are not transported from military bases to seaports ($\neg P_2$), are not stored in seaports ($\neg P_3$), are transported away from seaports by sea (P_4).
- **Place "red":** Some countries (S_2) take part in the conflict (P_5), do not help rebels ($\neg P_6$) and do not notify them to be ready to receive help ($\neg P_7$).
- **Place "yellow":** Some countries (S_2) take part in the conflict (P_5), do not help rebels ($\neg P_6$) and notify them to be ready to receive help (P_7).
- **Place "green":** Some countries (S_2) take part in the conflict (P_5), help rebels (P_6) and do not notify them to be ready to receive help ($\neg P_7$).

Comparing these two interpretations of similar formal model, one may notice that formulations of events look rather unusual for the common sense. Indeed, where common sense tells us just to note, "train is at second section" or just "second section busy", we are supposed to add that train is on the way (that seems obvious) and besides is neither at the first one nor behind the second one. Regarding the chemical weapons, we have to note that they cannot be simultaneously at military bases and on the way to seaports, or on the way and stored and seaports, and so forth – though, both for the train and for weapons it is true that they can be in two states simultaneously (e.g., train cannot jump form one section to the other; while the head is there, the tail is still here; the more so this refers to chemical weapons that may be partly sent off by sea, and partly still in place) – but this is rather an issue of quantitative fuzzy approach that should be imposed over the initial crisp formulations that are the issue in this work. These formulations, however, should be exhaustingly thorough to avoid ambiguity and, ideally, allow automated parsing.

Still we should admit that at present this goal is far from achieved, as is evidenced by rather wordy and "wooly" formulation of some of them ("notify rebels to be ready to receive help"). These predicates obviously have complex structure which at present is not addressed by the proposed grammar – e.g., predicates "to help rebels" and "to notify rebels" have common name that refers to similar object but this is in no way pronounced in formal expressions (P_5 and P_6). Hopefully this will be done in the future, with development of logical and mathematical grounds of the said grammar.

Interestingly, despite the similar structure of the model (which is supposed to correspond to syntax of the unified grammar considered below), the semantics, or internal structure of the events, in the two interpretations is notably different. In the "railway" case, there are four subjects, of which three have similar predicate. In the "weapons" case, subjects are only two, and each has a list of predicates, necessarily at least with one which is not negated.

In general, the above analysis shows that places of Petri nets look quite compliable with the requirements to events of unified grammar of dynamic knowledge (Pshenichny and Mouromtsev, 2015, see chapter 16 this book) and can be regarded as such and termed "events".

Transitions of Petri nets, in turn, should be compared with connectives of this grammar. Four types of connective are defined by Pshenichny and Mouromtsev (2015), denoted by letters of Old Slavonic alphabet: *zhivot*, *shta*, *yon* and *izhen*:

$\{E_1, E_2, \ldots, E_i\}$ *Zhivot* $\{E_{i+1}, E_{i+2}, \ldots, E_i\}$;

$\{E_1, E_2, \ldots, E_i\}$ *Shta* E_{i+1};

E_1 *Yon* $\{E_2, E_3, \ldots, E_i\}$;

E_1 *Izhen* E_2.

where E_i is an individual event, $i=1, 2, \ldots, n$.

These letters were chosen because they well represent visually the relation between the causal and effect sets, meaning causal set above, and effect, below in each letter: "many lead to many" (*zhivot*), "many lead to one" (*shta*), "one leads to many" (*yon*) and "one leads to one" (*izhen*).

Only one type of connective, *zhivot*, has been classified down to particular connectives. Still, even at the existing level of generality, applicability of these connective types to transitions of Petri nets can be examined.

Starting from the simplest, the *izhen* type can definitely be found in transitions of Petri nets (Figure 6). This can be exemplified as, say,

{Electric power (S_1) is turned on (P_1)} Izhen {Lamp (S_2) is shining (P_2)}

corresponding to Figure 6a, and

{Organism (S_1) is growing (P_1)} Izhen {Organism (S_1) is growing (P_1)}

Figure 6. An example of an unmarked Petri net transition that corresponds to the izhen type of connective: (a), when the event is not inherited, (b), when the event is inherited – this is the case of Petri nets represented by a multigraph. In case (a), place p1 corresponds to event "Electric power (S_1) is turned on (P_1)", p2, to "Lamp (S_2) is shining (P_2)", transition t1 is the connective of izhen type. In case (b), place p1 corresponds to event "Organism (S_1) is growing (P_1)", and transition t1, to a connective of izhen type. Mind that the inheritance of an event means inheritance of both its subject and the full list of predicates with negations, if any.

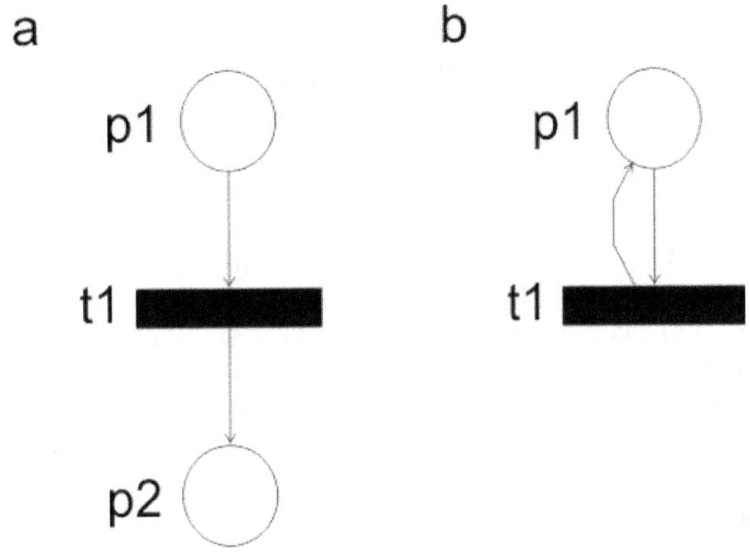

corresponding to Figure 6b.

In Figure 5, the *izhen* type connective similar to that shown in Figure 6a may also be exemplified by the following change:

{Yellow light (S_3) is turned on (P_5)} Izhen {Green light (S_4) is turned on (P_5)}.

As for the *yon* and *zhivot* type connectives, they have been fairly well exemplified by Figures 1-5. For instance, in Figure 4,

{Train (S_1) is on the way (P_1), is not at the first rail section ($\neg P_2$), is at the second rail section (P_3), is not behind the second rail section ($\neg P_4$)} Yon {Train (S_1) is on the way (P_1), is not at the first rail section ($\neg P_2$), is not at the second rail section ($\neg P_3$), is behind the second rail section (P_4); Yellow light (S_3) is turned on (P_5)},

{Train (S_1) is on the way (P_1), is not at the first rail section (P_2), is not at the second rail section ($\neg P_3$), is not behind the second rail section ($\neg P_4$); Green light (S_4) is turned on (P_5)} Zhivot{Train (S_1) is on the way (P_1), is at the first rail section (P_2), is not at the second rail section ($\neg P_3$), is not behind the second rail section ($\neg P_4$), Red light (S_2) is turned on (P_5)},

or just

{(S_1 - P_1, $\neg P_2$, P_3, $\neg P_4$)} Yon {(S_1 - P_1, $\neg P_2$, $\neg P_3$, P_4), (S_3 - P_5)},

{(S_1 - P_1, P_2, $\neg P_3$, $\neg P_4$), (S_4 - P_5)} Zhivot {(S_1 - P_1, P_2, $\neg P_3$, $\neg P_4$), (S_2 - P_5)}.

The *shta* type connectives are not demonstrated here but can be easily imagined in the Petri net framework based on the above examples, e.g.,

{(Mud (S_1) flows in a mudflow (P_1)), (Stones (S_2) flow in a mudflow (P_1))} Shta {(Mudflow (S_3) includes mud (P_2) and stones (P_3))}, or

{(S_1 - P_1), (S_2 - P_1)} Shta {(S_3 – P_2, P_3)}.

Considering the truth values of places interpreted as events, one may put forth that the set of places that enables transition is true, and the one that does not, is false. Importantly, in truth tables of connectives of the dynamic knowledge grammar not individual events but sets of events are attributed the values of truth and false, and this appears very useful in regard to Petri nets – e.g., in Figure 1, the set of *p1* and *p2* is false, and the set of *p2* and *p3*, true, though *p2* is present in both.

The marking of Petri net puts additional constraint to enforceability of the connectives of dynamic knowledge grammar. To allow an inference, not only an event must be present in left part of a change but also have enough tokens to be fired. Moreover, the amount of tokens indicates how many times an inference can be performed. This is a very important opportunity, which, possibly, could be later introduced into the dynamic knowledge grammar to be borrowed by other methods based on it.

The examples of interpretation of Petri nets in terms of dynamic knowledge grammar show, though do not prove formally yet, that this grammar can be used to represent the Petri nets for the purpose of knowledge representation given the formulation of places as events. This opens a wide opportunity of using this formalism in collaborative research and combining it with other methods of dynamic knowledge representation and engineering.

PETRI NETS COMPUTER-BASED TOOLS

As Petri Nets have a very wide use, lots of tools are available to automate editing, analysis and visualization tasks for every type of net. A tools database maintained by Frank Heitmann from University of Hamburg lists more than 80 packages accessible online (http://www.informatik.uni-hamburg.de/TGI/PetriNets/tools/quick.html). Among them the CPNTools package is the most widely known, with full support for colored nets (see http://cpntools.org).

For a distributed research project a high valued feature is the support of PNML data interchange format. Petri Net Markup Language (PNML) is defined by the standard ISO/IEC 15909 (part 2, transfer format). It defines the XML-based format for place-transition, high-level and symmetric Petri Nets. Other types of networks are supported via extension mechanism of PN Type Definitions, which is yet to be defined in ISO/IEC 15909-3. There are already several packages supporting PNML format including CPNTools, TiNA (Time Petri Net Analyzer, http://projects.laas.fr/tina) and PNML Framework (http://pnml.lip6.fr).

CONCLUDING REMARKS

Applicability of Petri nets as a tool for representation and engineering of dynamic knowledge has been illustrated by the example of simple place-transition marked nets. However, the more sophisticated types of Petri net are based on these ones and can, except the temporal nets, be reduced to these.

Simple Petri nets can be successfully represented by the unified dynamic knowledge grammar, in which the places correspond to events, and transitions to connectives. The semantic requirements imposed by this grammar on formulation of places in no way decreases the presentation power of Petri nets but urges to rethink and explicate the subject-predicate relations underlying the places of the net.

Marking of the nets and attributing weights to arcs bring an opportunity of specification of disagreement that is especially valuable for reconciliation of points in collaborative research. If some collaborators do not agree with a model, they may formulate their disagreement in similar terms. First, they may argue the amount of tokens necessary to fire this or that transition or associated with this or that place. Then, they may claim to modify the structure of the model, e.g., deny or rearrange some of the places. Thus qualitative disagreement, being a more fundamental one, can be discerned from quantitative disagreement, and an adequate action can be taken. Then, special methods for treating qualitative (e.g., logic) and quantitative (e.g., probability theory) information and their interplay may be considered in relation to Petri nets. Also, quantification of the nets theoretically should allow combination of Petri nets with other quantitative models (e.g., physical models), an interesting option that should be explored in the context of collaborative studies.

Temporal nets are based on the same theoretical grounds but offer an opportunity of adding time intervals to transitions. Whether this feature, potentially very useful for collaborative studies, can be expressed directly using the dynamic knowledge grammar, will be inquired in the future. However, there were attempts, at least theoretical, to add time intervals to some of the dynamic knowledge engineering tools (see, e.g., Pshenichny et al., 2009).

A straightforward task of future research is the comparison of Petri nets with other tools of dynamic knowledge engineering and representation (e.g., the event bush) to find out their relative capabilities in solving similar tasks and look for a strategy of their joint application in collaborative research.

However, right now one can definitely say that, provided the formulation of places and transitions in terms of the dynamic knowledge grammar, the Petri nets are quite usable in collaborative studies. A computer-based tool for it can be, e.g., COLLA, an asynchronous system of support of collaborative projects, and another interesting avenue of technical research is potential interoperability between the COLLA and the tools available for editing Petri nets.

REFERENCES

Diviacco, P. (2012). Addressing Conflicting Cognitive Models in Collaborative E-Research: A Case Study in Exploration Geophysics. In A. Juan, T. Daradoumis, M. Roca, S. Grasman, & J. Faulin (Eds.), Collaborative and Distributed E-Research: Innovations in Technologies, Strategies and Applications. Hershey, PA: IGI Global. doi:10.4018/978-1-4666-0125-3.ch012

Diviacco, P., Pshenichny, C., Carniel, R., Khrabrykh, Z., Shterkhun, V., & Mouromtsev, D. et al. (Submitted). Organization of a geophysical information space by using an event-bush-based collaborative tool. *Earth Science Informatics.*

Girault, C., & Pradat-Peyre, J.-F. (2009). High level Petri Nets. In M. Diaz (Ed.), *Petri Nets: Fundamental models, verification and applications.* Hoboken, NJ: Wiley.

Hussein, H. K., & El-Sattar, A. (2012). State of the Art in Interactive Storytelling Technology: An Approach Based on Petri Nets. In P. Pawlewski (Ed.), *Petri Nets - Manufacturing and Computer Science.* Rijeka, Croatia: InTech.

ISO IEC 15909-2. (2011). *Systems and software engineering: High-level petri nets: Part 2: Transfer format.* Retrieved August 10, 2014 from http://www.iso.org/iso/catalogue_detail.htm?csnumber=43538

Jensen, K. (1981). Coloured Petri nets and the invariant-method. *Theoretical Computer Science, 14*(3), 317–336. doi:10.1016/0304-3975(81)90049-9

Li, X., & Lara-Rosano, F. (2000). Adaptive fuzzy petri nets for dynamic knowledge representation and inference. *Expert Systems with Applications, 19*(3), 235–241. doi:10.1016/S0957-4174(00)00036-1

Liu, H.-C., Lin, Q.-L., Mao, L.-X. L.-X., & Zhang, Z.-Y. (2013). Dynamic Adaptive Fuzzy Petri Nets for Knowledge Representation and Reasoning. *Systems, Man, and Cybernetics: Systems. IEEE Transactions*, *43*(6), 1399–1410.

Merlin, P. M. (1974). *A Study of the Recoverability of Computing Systems*. (PhD Thesis). University of California, Irvine, CA.

Mouromtsev, D. I., Pshenichny, C. A., & Yakovlev, A. V. (2014). Semantic and structural delineation of market scenarios by the event bush method. *Decision Support Systems and Electronic Commerce*, *57*, 34–41. doi:10.1016/j.dss.2013.07.008

Petri, C. A. (1962). *Kommunication mit Automaten*. (PhD Dissertation). University of Bonn, Bonn, Germany.

Petri, C. A. (1973). Concepts of Net Theory. In *Mathematical Foundations of Computer Science: Proceedings of Symposium and Summer School*. Bratislava, Slovakia: Mathematical Institute of the Slovak Academy of Sciences.

Petri, C. A. (1977). General Net Theory. In B. Shaw (Ed.), *Computing System Design: Proceedings of the Joint IBM University of Newcastle upon Tyne Seminar*. Newcastle Upon Tyne, UK: University of Newcastle upon Tyne.

Pshenichny, C. A., & Kanzheleva, O. M. (2011). Theoretical foundations of the event bush method. In K. Sinha, L. Gundersen, J. Jackson, & D. Arctur (Eds.), *Societal Challenges and Geoinformatics*. Boulder, CO: Geological Society of America. doi:10.1130/2011.2482(12)

Pshenichny, C. A., & Mouromtsev, D. I. (2015). Grammar of Dynamic Knowledge for Collaborative Knowledge Engineering and Representation. In P. Diviacco, P. Fox, C. Pshenichny, & A. Leadbetter (Eds.), *Collaborative Knowledge in Scientific Research Networks*. Hershey, PA: IGI Global.

Pshenichny, C. A., Nikolenko, S. I., Carniel, R., Vaganov, P. A., Khrabrykh, Z. V., & Moukhachov, V. P. et al. (2009). The Event Bush as a Semantic-based Numerical Approach to Natural Hazard Assessment (Exemplified by Volcanology). *Computers & Geosciences*, *35*(5), 1017–1034. doi:10.1016/j.cageo.2008.01.009

Ramchandani, C. (1973). *Analysis of Asynchronous Concurrent Systems by Timed Petri Nets*. (PhD Thesis). Massachusetts Institute of Technology, Cambridge, MA.

Ribarić, S., Pavešić, N., & Zadrija, V. (2009). Intersection Search for a Fuzzy Petri Net-Based Knowledge Representation Scheme. Knowledge-Based and Intelligent Information and Engineering Systems (LNCS), (vol. 5711, pp. 1-10). Berlin: Springer.

Solomin, K., & Mandrik, M. (2015). Constructing historical knowledge through graphic boundary objects. In *Collaborative Knowledge in Scientific Research Networks*. Hershey, PA: IGI Global.

Wang, W.-M., Peng, X., Zhu, G-N., Hu, J., & Peng, Y-H. (2014). Dynamic representation of fuzzy knowledge based on fuzzy Petri net and genetic-particle swarm optimization. Expert Systems with Applications, 41, 1369–1376.

Xiao-Zhong, L. (2003). A Kind of Fuzzy Petri Nets and its Dynamic Characteristics. *Acta Simulata Systematica Sinica,* 2003-S1.

Yeung, D. S., & Tsang, E. C. C. (1994). Fuzzy knowledge representation and reasoning using Petri nets. *Expert Systems with Applications*, *7*(2), 281–289. doi:10.1016/0957-4174(94)90044-2

KEY TERMS AND DEFINITIONS

Dynamic Knowledge Grammar: A grammar of modeling language allowing to represent knowledge about changing events in a strict and uniform way.

Environment: A modeled situation.

Event: A set of statements of general form $S - P_1, P_2, \ldots, P_n$, where S is subject, and $P_1, P_2, \ldots, P_n$ are predicates, with or without negation, complying the following conditions: (i) subject may be only one in a statement, (ii) subject may not be negated, (iii) all predicates may not be negated in a statement, at least one must be without negation. Events represent the relations between objects of two types: one subject and whatever number of predicates.

Petri Net: A graph-based mathematical language for the modeling of behavior of distributed concurrent systems.

Section 5
Knowledge Handling Avoiding Formalization

This section addresses how knowledge can be handled without formalization. As mentioned above, this need comes from the issues related to the difficulties in conceptualizing and communicating knowledge.

Chapter 19
Maps, Graphs, and Annotations as Boundary Objects in Knowledge Networks, Distributed Cognition, and Collaborative E-Research

Paolo Diviacco
Istituto Nazionale di Oceanografia e di Geofisica Sperimentale (OGS), Italy

Alessandro Busato
Istituto Nazionale di Oceanografia e di Geofisica Sperimentale (OGS), Italy

ABSTRACT

Scientific communities tend to speciate in tribes that evolve their culture so that after some time they do not have the same understanding of the same terms or concepts. Tacit knowledge, that is the knowledge that cannot be expressed or formalized, complicates this even more. Since we cannot fully know what a concept refers to, we would have limitations in developing IT systems to support collaborative and distributed cognition activities and knowledge networks trying to map knowledge between paradigms. To address this, the authors propose to switch to another perspective where knowledge is kept implicit and the referential communication function (iconic signification) is exploited instead. In this perspective, they develop a specific virtual research environment named COLLA-ANT, which, albeit being still a prototype and needing more use cases, has proved successful in addressing the above-mentioned issues.

INTRODUCTION

Chapter 1 of this book (Diviacco, 2015) addresses the mechanism that allows a single researcher to reach the outer world, the problem of subjectivity and the supposed existence of its counterpart, that is objectivity. There, it was demonstrated that no path towards ontology can be traced, while only through the epistemic we can identify states of affairs upon which reasoning can produce visions, theories and eventually knowledge. We said in Diviacco (2015, see chapter 1 this book) that it is

DOI: 10.4018/978-1-4666-6567-5.ch019

Copyright © 2015, IGI Global. Copying or distributing in print or electronic forms without written permission of IGI Global is prohibited.

possible to speak of the epistemic as judgments, in the sense that these latter represent or portray things. They are about things. Judgments are connected to representations. We compared and saw similarities in authorship in Science and other domains such as Art, for example. There, the author reaches the outer world through representation as well. The path of Diviacco (2015, see chapter 1 this book) continues with other issues, while here we intend to stem from this specific point to explore the use of representation, its limitations and how we can address them in the perspective of collaborative work and knowledge networks. Representation will be here associated with signs that bear the possibility of expressing different meaning in different contexts. We saw, in fact, how scientific reasoning, especially in domains where falsification is complex, can introduce a form of speciation in the scientific communities by which "tribes" tend to be created. These latter will develop their own culture and will have difficulties in understanding those of the others.

BACKGROUND

Communication

Communication models analyze the conditions under which representations are produced, transmitted and received. Most of these models identify an *author*, who produces a *message* which is received by a *reader*. The message refers to a certain *state of affairs* and is expressed in a *code*. The code must be *shared* as much as possible between the author and the reader. In all this, a *medium* is required that connects the author and the reader.

All these factors influence the structure of every communication.

A communicative function is the aspect under which a communication can be understood. Among all the possible communicative functions[1] we will concentrate on two of them which are mostly important, and that are the (i) referential and the (ii) meta-linguistic functions.

The referential function is mostly important because relates the representation with the referent, that is the state of affairs[2]: Following Kant (1781) there are three possible modes of presenting states of affairs, and namely (i) assertoric: when a state of affairs is affirmed to exist[3] (ii) hypothetic: that refers to the possible existence of a state of affairs. [4](iii) apodictic: refers to a necessary existence of a state of affairs upon logical inference from another state of affairs.

The meta-linguistic function was introduced by Carnap (1942) to ensure communal understanding within the scientific community. It is an abstract level where the concept is defined in order that the reader can understand it. Carnap (1947) developed this function in the view to explicate concepts in a denotative fashion. This means that the concept is defined by its *intension*. The *intension* of a concept is a sort of list of all its pertinent attributes. The *extension* of a concept is the set of all individuals that match the intension. The model is based on the assumption that it is the *intension* that determines the *extension*, which means that only once we have defined the set of properties of a concept "A" can we realize if an entity is "A". The problem is that, following this vision, a change in the intension will change the contents of the set of related entities (Putnam, 1975) and therefore it is easy to understand that denotation and the idea that representation can be "customized" by the needs of a community do not go together well.

MAIN FOCUS OF THE CHAPTER

Introducing the Context

Traditionally three levels are identified in communication:

The syntactic level: here the aim is that the message arrives exactly as it is issued. At this level we do not enter inside the content of the message, we only take care that there could be no filtering on what sent. This is similar to the case of an electronic connection where we want that the input can be received correctly. In fact the formalization of the problem made by Shannon and Weaver (1949) has been done essentially in this domain. At this level the efforts are often spent on standardization, in order that information is entered in the right way and fashion[5]. A deeper analysis of cases and problems related to syntactical communication in scientific research can be found in Diviacco (2012).

The semantic level: here we enter into the domain of meaning. Semantics is indeed defined as the study of meaning. Meaning is what is denoted, what is referred to. In linguistics meaning is what is communicated through the use of language, and semantics is primarily connected with statements and reasoning.

The semiotic (or pragmatic) level is a further abstraction where the link to meaning is extended to consider contexts and signs.

The origin of the word context refers to the binding of something with the circumstances to which it is linked, and therefore at this level multiple and contrasting interpretations or visions of the same concept or object can coexist considering different circumstances.

Signs are dyads composed by a signifier, that is the form that the sign takes, and the signified, that is the concept, the object it represents. Signification is the process of connecting the signifier with the signified.

Whereas semantics focuses on what words mean, semiotics is concerned with how signs mean (Sturrock 1986, 22). Umberto Eco (1979) instead maintained a wider definition stating that semiotics is concerned with everything that can be taken as a sign.

It is very important to note that something becomes a sign only when we invest it with meaning it did not have before. Anything can be a sign as long as someone interprets it as 'signifying' something.

In Saussure's vision (1916), signs are totally arbitrary. While the arbitrary essence of signs is not a recent discovery[6], the importance given to it in contemporary semiotics reveals its power to be a radical concept, since it states the autonomy of representation in relation to reality or even that signs do not 'reflect' reality but rather can construct it. From the point of view of Science this is overwhelming, since, as described extensively in Diviacco (2015, see chapter 1 this book), knowledge is built and transmitted through representation.

After Pierce (1931), following a criterion of increasing arbitrarity, three modes of signification exist:

- **Indexical:** A mode in which the signifier is directly connected to the signified through a causal or physical connection. This refers to observations, to measurements that pertain to a certain phenomenon. To clues that link to it, which is exactly the matter that abduction[7] can process such as, for example, sparse samples of a geologic setting, medical symptoms, historical recordings.
- **Iconic:** A mode in which the signifier is perceived as resembling or imitating the signified. This, science-wise, is the case of a map, of a model that represents a phenomenon. This is the most critical layer, where the subject reaches the outer world, and therefore possible partners, projecting his background in the means of production of representation.
- **Symbolic:** That in which the signifier does not resemble the signified. This is purely arbitrary and conventional, so that the relationship must be established and learnt. This is the case for example of language or scientific notation. An example of arbitrary symbolism is mathematics. This does

not even need to refer to an external world: here concepts are linked to signifieds via a system of relations.

It is important to note that these three modes are not mutually exclusive and that some signs can be characterized by their interference.

Saussure (1916) introduced the term *motivation* to express the strength with which a signifier is linked to its signified. In this sense, the more a signifier is bound to the signified, the more *'motivated'* the sign is, whereas, the less *motivated* the sign, the more learning of an agreed convention is required: indexical and iconic signs are highly motivated; symbolic signs are unmotivated.

Practical Application of Signification Modes

In Diviacco (2015, see chapter 1 this book) we saw that the subject, the author, reaches the outer world through representation and we also described the issues related to this process. How can we match that with the perspective we are drawing here?

We claim that this should be possible, particularly in the case of abductive sciences, mirroring the different signification modes drawn above, in the flow that accounts for Scientists progressive abstraction from observations to knowledge building and theories.

Researchers consider first *motivated* clues and symptoms that are represented as indexical signs: the base of the structure. Processing and reasoning on the particular matter of interest produce a framework that is generally translated in a model, which normally is represented using N-dimensional graphs or maps that therefore can be associated to the iconic level. Further abstraction can be obtained when researchers formalize their visions using *unmotivated,* already existent or ad-hoc symbols to obtain a proper theory.

In a collaborative experience or knowledge network, this abstraction process involves sharing signification at all levels. When arbitrary meaning is not or cannot be negotiated[8] the symbolic/coded/cultural component of representation becomes predominant. This is the moment when the background of each researcher has the major chances to bias the process of representation and eventually the results of the research project.

As explained in detail in Diviacco (2015, see chapter 1 this book) in several domains where falsification of theories is not an easy task, where reasoning itself cannot be reduced to the inductive/deductive loop of the classic scientific method, communities of researchers tend to speciate forming incommensurable tribes that tend to preserve their territories due to, not only ethical, but also political reasons. This attitude together with the difficulties in sharing representational codes hamper the possibility to converge to common visions and ultimately build knowledge together.

At the indexical level clues are shared in an assertoric referential fashion. Researchers say, "Here are the observations, here are the measurements". They generally consider this a straightforward task that does not need any concern. On contrary also at this level there might be room for interpretation and bias, in fact observations are always acquired using a strategy, which generally can the consequence of the necessity to demonstrate an idea. We will overlook the issue of rhetorics applied too often in the quest to obtain support to a theory (that is nevertheless a very important issue, see Pera (1994)) but we cannot avoid highlighting that necessarily any experiment is bound to a theory. Observation parameters setting and the observation itself is bound to what someone is expecting to see, which of course depends on the theory the scientist already has in his mind. Quoting Hanson (1958, p.19)

Observation of x is shaped by prior knowledge of x.

"Opportunistic" reuse of data in other projects and their integration with other data may introduce useful asymmetries but do not remove original shaping effects.

Iconic signification is a particularly critical phase in which each researcher, explicitly or not, consciously or not, projects his own background on the representation he shares with the others, and therefore filters and customizes the process of collaboration. The presentation of the state of affairs, here, is proclaimed and often perceived as assertoric, while mostly it is hypothetic since it is the outcome of reasoning and processing by a researcher. Each tribe tends to develop his own vision, practices, tacit knowledge, so that any representational product of one tribe potentially might be not understandable by another (Becher, & Trowler, 2001; Kuhn, 1962). Besides understandability, particularly at this level, concerns might rise also regarding the possibility to perceive the same way[9].

The symbolic level is where notations and languages are created. Here theoretically we could find the keys for bridging different communities, in fact the symbolic level is that which can be thought of as having a meta-linguistic function, and where through a proper use of logic inferences (apodictic referential function) we should be able to reference properly to the state of affairs as viewed from multiple and different perspectives. Unfortunately this is not so easy to put in practice. Concepts do not map straightforwardly because: (i) knowledge cannot be fully captured and (ii) mapping needs to be negotiated.

As Polanyi (1966, p.4) said: *"we know more that we can tell"*. Therefore, even if there is the will to do so, (which is not always the case) we cannot extract fully what is in the heads of the members of a community. Besides knowledge is embedded in practices or even tools so that some knowledge will always remain obscure and implicit.

On the other hand the activity of mapping across different cultures will have to deal also with the social and political issues discussed above. Concepts will be defined, and therefore mapped, not only upon what really needed but, in the quest for positioning, also upon convenience. Real consensus is problematic. From a political point of view "big fish" will tend to drive the discussion, "herd" opinion, in order to fulfill their personal/institutional interests, but there are also intrinsic limitations in judgment aggregation that will need to be considered[10]

Summing up, shared knowledge building involves considering signification at all levels, which opens a wide issue, since any 'fixing' of 'the chain of signifiers' is both temporary and socially determined (Coward, & Ellis, 1977). Following Barthes (1957) signs are ideological.

Theory-Ladenness

Pierre Duhem's holistic vision (Duhem, & Wiener, 1954), at the end of the 19th century, maintained that an experiment is not simply an observation, but rather an interpretation of observations from within a theoretical framework. It is therefore impossible to isolate hypotheses from observations from background assumptions and from the involved theories.

Many Authors stemming from the work of Duhem, envisioned that our experience is categorized and preconditioned by prior belief, and that the process of building knowledge always involves the use of concepts from some previous view. Studies on perception demonstrated the influence of motivation and expectation (Bruner, & Postman, 1949; Warren, 1970) in how people shape reality. They tend to interpret ambiguous figures so that they see what they want to see, and besides, social relations can further complicate this.

All this is generally referred to as the theory-ladenness argument (Bishop, 1992; Hanson, 1958).

Since experiments are based on observations, and since observations are theory-laden, then experimentation has to be theory-laden too. Experiments are based on some prior theory.

Together with "logical" theory-ladenness, "perceptive" theory-ladenness has been largely discussed in literature, both in general and in the particular case of Science. This is an argument particularly evident in the case of visual representation and is known as the conceptual penetrability of the visual system, meaning the possibility that theories condition how something is perceived visually. This holds both in the case of observations of phenomena, of states of affairs, in the case of representation, and in the exchange of them. Epistemological radicals such as for example Kuhn (1958), Churchland (1979) and Feyerabend (1975), maintain that each observation made within different contexts and therefore within competing frameworks, being inevitably theory-laden, produce necessarily, as a result, observations that are affected by theories, and will be therefore different and very likely incompatible. Radicals move the point up to a vision where scientists not "shaped" (or "trained" using a more acceptable terminology) within the specific framework will not be able to perceive the word in terms of the theory's observational concept. There are good reasons to believe that experience and training can alter our visual system so that we are capable of recognizing new things that before were not accessible. Following Fodor (1983) it should be possible to test when this happens if a new feature is identified in a fast, mandatory and non-voluntary fashion. I take as an example Exploration Seismic data interpretation that to the layman only very hardly can mean anything, while after years of training and tutoring allow interpreters to identify geological features as naturally emerging from an apparent chaotic pattern.

To demonstrate interpreters' tendency to project their cognitive framework even where no geology is involved, we conducted an informal experiment submitting to a group of interpreters a processed image of a draft of a section of the text of this chapter. Vertical and horizontal letters and spaces between words, sometimes, get lined horizontally and vertically to resemble geologic faults or layers, while this of course has nothing to do with geology.

Seismic data is made of vertical traces that record the contrast of acoustic impedance between horizontal layers. A seismic profile is made of a horizontal sequence of many seismic traces that sample a geological setting. This is generally plotted as a sequence of vertical wiggle and variable area trace such as that portrayed in Figure 1 (center and right). The text was converted to such presentation system using a specific software that we use to de-materialize old seismic sections, and that is able from a printed section to restore a full digital seismic section that can be later processed by any processing software or loaded into interpretation software. In detail, this software follows and digitize the shape of a printed seismic trace path, in our experiment we substituted real printed seismic data with "seismic-look-alike" rotated printed text. The de-materializing software made no complaints, accepted the rotated text and gave us back a seismic section representing in wiggle and variable area form the input rotated text. This was loaded into a piece of interpretation software and proposed to geologists and seismic interpreters, who, of course, were not aware of the provenance of data. After complaints on data quality, most of them were able to recognize patterns resembling layers and faults even if no geology was involved whatsoever. (Figure 1)

Apophenia is the unmotivated seeing of connections and specific experience of an abnormal meaningfulness. It has to deal with the human tendency to seek patters in random situations. This is the case from the poetic recognition of faces and shapes in the clouds to the prosaic misconception of gambling regularities. The case of the experiment above mentioned fits perfectly this, with scientists identifying (due to their trained visual system) features that could be compatible with what they can normally interpret as a seismic unit, in an absolutely out-of-context and non-geologic object.

Data, Information, Knowledge

The model devised so far can be compared with another one commonly used within IT related

Figure 1. The process and outcomes of an informal experiment on theory-ladenness, aiming to test how scientists tend to project their cognitive frameworks on data, graphs and maps even if they are not related to the physical phenomena behind. A section of text has been selected from this chapter, and has been rotated (left) converted to a seismic section (center) and submitter to seismic data interpreters. They were able to see some coherence in signals even if the graph has no relation to a geologic phenomenon (right).

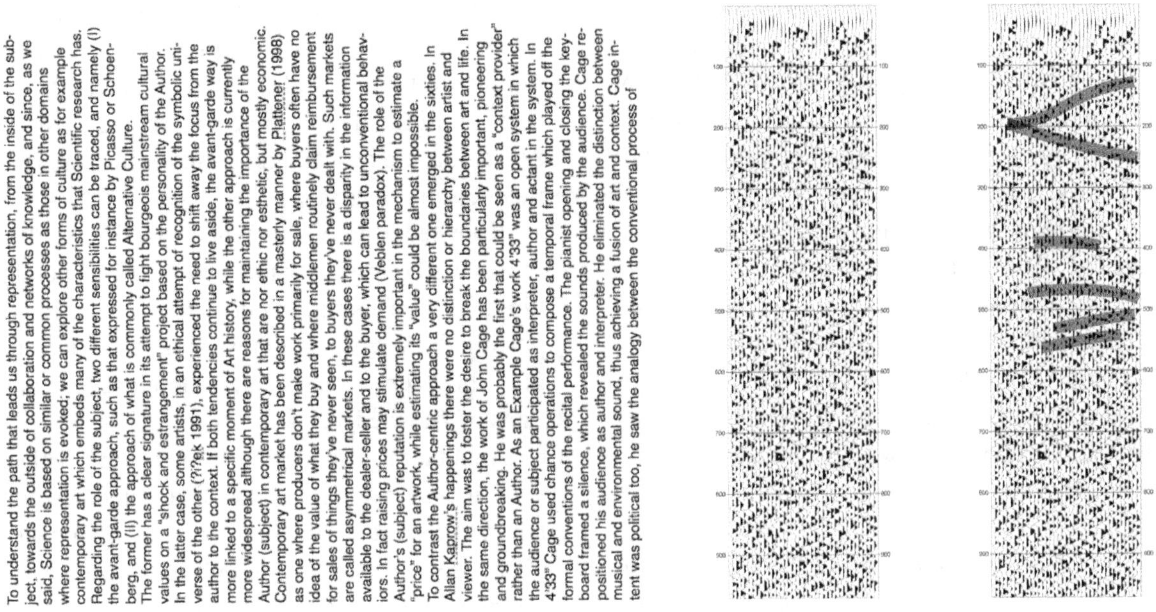

domains that is generally called the Data/Information/Knowledge (DIK) model. The two do not match straightforwardly because developed in two different domains, but through their mapping we can try to devise practical (IT based) solutions for the issues we have seen so far.

The DIK model is often represented as a layered pyramid. At the very base level we have raw data and observations: the state of affairs expressed in an assertoric referential fashion, which are understood as indexical signs.

Data do not hold the whole phenomena they observe, they are only snapshots of life, and as in the case of life, which has to have a meaning even if often not clear to us, data need meaning to be useful. In other words data is not organized and need meaning to be such.

If we add meaning to data we obtain information. Since, as mentioned above, meaning is what is denoted, what is referred to, information will therefore embed a referential function.

Meaning can be bound to data in many ways. For example, meaning can be added to processing the latter in an apodictic referential fashion such as in the case of extracting temperature trends from time-lapse soundings. The diagrams and the maps we can draw out of this can be understood as an iconic sign in its function to portrait the phenomenon.

The last layer of the DIK pyramid is knowledge, which is a very difficult concept to define in itself. Besides, often, in the literature, information and knowledge conceptualizations have been quite largely blurred and muddled, so that the need to distinguish them has been felt even by Frank Zappa who highlights, in one of his songs, that "*Information is not knowledge*" (Zappa, 1979).

Knowledge is what we know. Following Zeleny (2005, p.15) *"All knowledge is tacit"*. If this seems a complete closure, actually it is not fully so. If here tacit means indeed that some knowledge is inaccessible because pertains to the deep maps of conceptual connections we have in our brain and that cannot be fully explored, it also means that some knowledge is simply unexpressed. This knowledge is mirrored in procedures, practices, processes, actions and propositions that we can gather under the term Technologies. These can be at least partially made explicit and then formalized using several notations and very likely, then, also processed. This is the domain of knowledge engineering that will be explored more in detail in section 3 and 4 of this book.

Our brain, our background, our know-how, together with the environment in which we live and our attitude toward action - in a word, the *context* - are essential components of knowledge. Summing up: information with the addition of context is knowledge, or as in the definition of the European Committee for Standardization, knowledge is:

A set of data and information and a combination of, for example know-how, experience, emotion, believes, values, ideas, intuition, curiosity, motivation, learning styles, attitude, ability to trust, ability to deal with complexity, ability to synthesize, openness, networking skills, communication skills, attitude to risk and entrepreneurial spirit to result in a valuable asset which can be used to improve the capacity to act and support decision making. (CEN CWA 14924, 2014)

In the perspective of collaborative work and collaborative knowledge networks things can get complicated. In fact, since multiple and concurrent contexts will coexist at the same time, the possible clash of different visions will need to be always considered and addressed.

A very important aspect of the DIK model is that we can identify not only an ascending trend[11] within it, but also a descending one. In fact, it is very important to highlight that knowledge mirrors back into information and data. As maintained above, both cases of indexical and iconic signification modes are shaped by knowledge and therefore in both cases of data and information, knowledge and therefore context can influence how users access them, which is probably one of the reason why the three layers of the DIK model have been often confused.

An interesting method that allows to bind knowledge to data and information, for example on observations but also maps or diagrams derived from them, and to accommodate both the ascending and descending trends within the DIK model is annotation.

Knowledge here is somehow glued to the data in a hypothetic referential way by the scientist him/herself. This can be put in practice linking or posting drawings or/and text notes on maps and graph but even to formal and symbolic representations of knowledge.

Annotation is a particularly interesting method since it can be imagined in a distributed and collaborative sense. Following the ideas of Diviacco (2015, see chapter 1 this book) collective and distributed reasoning would need an information space to become real. This could be granted by an IT system where all data, information, annotation as sketching and messaging, will be made available to all actors, in the same place. The hinge of this chapter relates indeed to the reasons why, the possibilities and the modalities in which this can be done.

How to Address Possible Contrasting Knowledge

As mentioned before, the most striking limitations in handling knowledge are that this is often tacit, that backgrounds can only with great difficulty be isolated and that socio-political issues are very much embedded into scientists' positions, mental habits and actions as well.

We claim that, in itself, this is not negative, since it guarantees some kind of "bio-diversity" in knowledge. If the reader will allow us to continue in the similarity with biologic evolution, we could recall that bio-diversity is known to be a useful mechanism to overcome crises. Allowing different and competing knowledge to co-exists can be a good strategy to have more chances of solving complex and ever changing problems, as in the case of scientific research, and especially in the case of abductive sciences.

Of course we need ways to handle and support this bio-diversity, in the perspective of fostering collaboration and knowledge networks, otherwise the result will be the fragmentation of communities and isolation of partners in tribes and therefore the loss of any possible sharing of cognitive processes and knowledge.

Several approaches can be followed. A major distinction is on whether they make knowledge explicit or whether they keep it implicit.

In the first case the idea is to eventually exploit the symbolic meta-linguistic functionality of mapping explicit formalizations of concepts across different contexts. This can be done in a denotative semantic fashion, focusing therefore on language, such as in the case of controlled vocabularies or of the web ontologies techniques, or differently on the base of a graph as in the case of workflows, mind maps or event bushes.

In both cases, as we already mentioned, methods based on knowledge formalization could be prone to difficulties in capturing knowledge due to tacit knowledge and socio-political issues.

Paths towards explicit formalization of knowledge are explored in two specific sections of this book, we will here focus instead on implicit knowledge handling to support collaborative activities and knowledge networks. The reason for this is that the formalization-based approach, together with its limitation, is most commonly used, while we would like to highlight here the advantages of the implicit knowledge approach, and highlight that in many cases it has to be used. In fact we claim that the relevance of tacit knowledge, straightforwardly and inevitably makes the implicit approach a must, and that the ratio between the involvement of the implicit and the explicit approach in developing knowledge networks systems must follow somehow the ratio between tacit and "formalizable" knowledge of the issue to handle. In the case of abductive scientific research this ratio is very much influenced by tacit knowledge and therefore knowledge network systems in scientific research domains such as the geosciences, human sciences, medicine and so on and so forth, would need to handle knowledge in a less formalized approach.

How can this be put in practice?

The formalization approach to knowledge, as mentioned above, is mostly aiming to act as a meta-linguistic communicative channel where the signification mode is symbolic, referencing to states of affairs in a hopefully apodictic fashion[12].

The implicit approach cannot have a direct meta-linguistic functionality where concepts could be "hand-made" or mapped, since, of course, they are implicit. Here the referential communicative function has to be un-tethered with the states of affairs. This can be granted only at a semiotic level using the iconic signification mode, which allows the needed loosening of coupling of referent and reference.

Further on, in the perspective of collaborative work and knowledge networks, this loosening of the significations allows also multiple and contrasting signification to coexist so that multiple paradigms can address the same issue from different point of views.

But eventually we need something that somehow attracts paradigms, otherwise there will be no common place for shared cognition.

Diviacco (2012) analyzes two possible solutions: (i) social tagging and (ii) Boundary objects. The former support the creation of a shared cognitive space in a very flexible way, but at the same time knowledge here is also rather difficult to "capture". Ambiguity, synonymy and discrepancies in granularity (Golder, & Huberman, 2005) suggest to introduce semantics (Specia, & Motta, 2007) so that the theoretical advantages of social tagging as an implicit approach is reduced.

Boundary objects (Star, & Griesemer, 1989) are based on the studies of Susan Leigh Star and

James R. Griesemer. They found that different communities could collaborate effectively notwithstanding the fact that they could have rather different cognitive models[13]. They found that, in this sense, communities can be gathered in the same cognitive space by artifacts (the boundary objects) that aim to bridge concurrent cognitive models through abstraction from all the domains of the partners. Boundary objects are weakly structured in common use, and become strongly structured in individual use. They contain sufficient detail to be understood by one partner, although it is not necessary that he/she understands the context in which the other partners use it. They are artifacts related to sets of information, conversations, interests, rules, or plans, and are at the centre of communities of practices (Diviacco, 2012; Star, & Griesemer, 1989).

Diviacco (2012) analyzes several possible types of boundary objects: maps, graphs, workflows, mind maps, event bushes while focuses on practical implementations of the last three types because they stem from an approach aiming at formalization of knowledge. These graphs, using nodes of a network, formally represent state of affairs, workflows, relations between concepts or events. If the graph structure is rather strict, this is not the case for the node definition, that can be more relaxed. Nodes are only labels to containers. This allows the introduction also here of implicit, and therefore possibly contrasting knowledge. Event bushes are particularly interesting in this since multiple and contrasting paths through its nodes can be devised so that the same representation can fit several points of view (Diviacco, & Pshenichny, 2011). Diviacco (2012) does not consider maps and graphs that instead are the subject of this chapter.

Maps, Graphs and Annotation as Boundary Objects

As mentioned above, this chapter is about exploring solutions able to handle implicit knowledge using boundary objects. We stated above that implicit knowledge can be handled properly at a semiotic level and that in this the iconic signification mode allows the needed detachment between reference and referent. This signification mode is that to which maps, graphs and annotation can be associated to. We therefore can close the loop and propose these latter as boundary objects and extend the work of Diviacco (2012) that did not include them. The gain in this sense is considerable since employing formalized graphs, such as mind maps etc, as boundary objects allow us freedom only at the level of nodes, meaning that some rigidity from the framework of the graph itself is imposed. In fact, one of the limits we found in the use of these types of boundary objects is that once devised, it is rather complex to change them. This would need a reconfiguration not only of the graph but also of the content of the nodes. After a change in the graph structure, in fact, information associated with one node might be better associated with another one and changing its position might be a complex and time-consuming process. The use of formalized graphs as boundary objects in a collaborative work is more effective once the schema is well agreed upon or if it is a workflow that all partners are well aware of. On the contrary the use of maps and graphs allows collaborative work also outside a proper workflow or framework. Using a metaphor there is no need for a musical score. Rather, here, partners are free to play in a jam session, which of course is more in the sense of what described in Diviacco (2015, see chapter 1 this book): of shared cognition processes and of context creation instead of being strictly "standard"-science-wise and objective while at the same time producing highly personality-driven research.

Referring to the DIK schema, maps and graphs pertain to the information layer that is built on the data layer that essentially stores raw observations and measurements.

As maintained earlier in this chapter annotation has a particular role since it allows knowledge to be driven to the lower levels in an easy way, while at the same time allowing a flexible leveraging of implicit/explicit knowledge. To exist, annotation needs a frame, which is data, information and possibly also graphic representations of knowledge. From a practical point of view, implementing such system will therefore need to consider all the three levels.

At the data level, observations and measurements will need to be stored in the system, generally in a database or in a file-system with links to files stored in a database. Together with storage, specific means to access and process (at least basically) data should be provided in order to produce the graphs and maps needed for the information level. It should be possible for scientists to visualize maps and graphs in order to be enabled to reason and interpret information to develop new ideas, new visions and theories, which ultimately will constitute their contexts. Depending on the scientific domain it could be more or less important to have interactive tools to produce maps and graphs. In the case of the geosciences this is almost always a good idea. Interactive processing tools, even basic ones, allow changing "the perspective" from which information is analyzed, in order to study possible changes in the scientific paradigm. Besides, if maps and graphs can be produced interactively, there will be no need to rebuild completely the system but only to change the settings of representation for the new needs.

Once processing and visualization tools have produced maps and graphs, at this precise moment annotation can be introduced. Annotation essentially is metadata, consisting of the addition of knowledge to information. In a very simple vision, here, researchers should be enabled to identify features and post messages conveying their ideas.

Features can be identified using (i) symbolic notation or (ii) unstructured drawing.

The first case is more complex since would require the introduction of meta-linguistic functionalities, meaning essentially the possibility to refer to a shared taxonomy of symbols or a controlled vocabulary that will be then domain dependent and, even more critically, context dependent. Notwithstanding this in case of small teams with only minor differences in background[14], this could be a useful approach.

Unstructured drawing is conversely the most simple but often the most effective way when crossing domains. In this case features will be identified by a simple sketch that will be overlaid on maps and graphs. From a certain point of view, these sketches are not directly correlated to the other layers unless the will of the scientist links them. If no symbolic notation is used (but often even in that case) to express the researcher vision and knowledge something more is needed. A good candidate for this is simple text. Scientists can link or post text messages to identified features in order to append knowledge to information, closing then the loop, that we have discussed above, between the ascending trend from data to information to knowledge and the descending one from knowledge to information.[15]

As mentioned above, annotation is a particularly interesting method from a knowledge network, distributed cognition and collaborative work point of view. In fact sketches and texts can be shared among partners in a collaborative project or in a knowledge base in order that the system collecting them, in its very meaning and aim to be a shared space, would constitute the abstraction (although very physical) needed for collaboration, distributed cognition or knowledge. This recalls the identification of the shared cognitive space with the experiment itself in the case of the Hadron Collider for Knorr Cetina (1999) and the forms of distributed cognition discussed by Giere (2002).

Towards a Practical Implementation

So far we have discussed the theory behind and the functional specifications of a system able to handle distributed cognition and knowledge using boundary objects such as maps, graphs and annotations. Now we need to put all these concepts into something that can be practically used by a community of researchers. In this perspective we will introduce similar initiatives, studies and implementations. These, mirroring the complexity of the subject, have been often named differently in different contexts. Collaborative Virtual Environments, Collaboratories, Gateways, Science Gateways, Virtual Organisations are among the possibilities, while a widely used term is Virtual Research Environment (VRE). Following Fraser (2004) these are "sets of online tools, systems and processes interoperating to facilitate or enhance the research process within and without institutional boundaries". To delve into the motivations and the components of such infrastructures we suggest to refer to Carusi and Torsten R. (2010). There, it is also possible to find interesting information about the distribution of VREs among the different disciplinary fields, such as their demanded in the humanities which is very much of concern in the perspective of Diviacco (2015, see chapter 1 this book). Case studies are listed such as Max Plank's eSciDoc, CSIRO's MemRe and many others. Some VRE use collaborative tools such as for example MS Jammer, Alfresco or one of the many similar others. We can excise these form the discussion here, since they focus mainly on the process of syncing partners in a team, which, of course, is an important aspect of collaboration, but is not related with the issues of knowledge we wanted to handle in this chapter.

Instead, of specific interest for our discussion, is the JISC funded and MyGrid related Myexperiment (http://www.myexperiment.org/) initiative. This is a web-based resource supporting the sharing of workflows and related resources for scientific experiments. Workflows here are essentially graphical representations of processing scripts that can be used for "in silico" sperimentation throught the integration of myexperiment with another platform called Taverna (http://www.taverna.org.uk/). While being developed mainly in the field of the bio-sciences the approach theoretically can span also other domains. Workflows or scripts are important because represent procedures and therefore can be used to formalize tacit knowledge, but in themselves do not cover the whole spectrum of the issues described above. A complementary initiative to myexperiment is research objects (http://www.researchobject.org/), here data, metadata, results, pubblications can be gathered in order to enable reproducible and transparent research. Knowledge here is somehow "packed" and in some cases (through myexperiment workflows) the process of scientific discovery can be re-run. This is very interesting, but cannot to our knowledge account for contrasting contexts and visions. Annotation, in the above mentioned initiatives, is more oriented towards messaging, while we are interested in gathering discussion around iconic features, and besides, in myexperiment, visualization relies on external systems.

Since no existing framework would have provided us with the needed tools, to tests our vision, we decided to develop our own VRE that we named COLL-ANT.

The Parents of the Current Implementation

So far we have discussed the theory behind and the functional specifications of a system able to handle distributed cognition and knowledge using boundary objects such as maps, graphs and annotation. We will now discuss a practical implementation of such ideas in a web based piece of software we developed and are starting to use intensively in the field of geophysics. The motivations to focus on this specific domain are explained more in detail in Diviacco (2003; 2005; 2012) and Diviacco and Busato (2013) but in few words, they have to deal with the economic value of geophysical data (especially seismic data) and with the will to avoid external users to access them in an un-controlled manner. The collaborative web based software we have developed, which name is COLLaborative ANnotation based Tool (COLL-ANT) offers to external users the possibility to work on the data (depending on user privileges) without having a local copy of them, and at the same time to collaborate among them.

The developed system is an integration between two lines of web based system we were already developing and continuously improving since quite some time ago: (i) S.N.A.P[16] and (ii) COLLA[17]. The reason for having two distinct products was that they were addressing two different DIK levels and namely: S.N.A.P. was addressing the data and information level while COLLA was aiming at creating a collaborative space to handle implicit knowledge.

S.N.A.P (Diviacco, 2005; Diviacco, 2007; Diviacco, 2011; Diviacco, 2013) is a web based system aimed at managing geophysical data, focusing especially on seismic data. It allows the discovery of data using metadata parameters (see Diviacco 2012 on the metadata model used) or geographic maps. Once the data have been discovered it is possible to visualize them interactively using the available web based facilities. These latter offer also basic processing tools that are very effective in highlighting the existence of possible geologic features. S.N.A.P is a framework that has been used within several international initiatives such as the Antarctic Seismic Data Library (SDLS) (http://sdls.ogs.trieste.it/), which collects all seismic data acquired in the Antarctic sector by all research institutes worldwide, or the GeoSeas FP7 EU project (http://www.geo-seas.eu/)that federates 26 marine research institutes to create an European geologic and geophysics data space.

Colla instead is a web based collaborative tool able to handle implicit knowledge leveraging formal knowledge representation using boundary objects such as mind maps, workflows or event bushes (Diviacco, 2012; Diviacco, & Pshenichny, 2010; Diviacco, & Pshenichny, 2011; Pshenichny et al., 2013). Graphs are visualized as Scalable Vector Graphics (SVG) (http://www.w3.org/Graphics/SVG/). This XML dialect allows to identify objects, for example nodes or arcs of a graph, and associate them with functions such as for example web linking or message pop up. Web links allow to drive users to areas (tasks or nodes in COLLA jargon) where file uploading and discussion threads take place. COLLA has been tested in several both smaller and larger scale collaborative projects. It gave good results in defining within a large community the controlled vocabulary for geophysical terms within the GeoSeas project. Within the development of this book it has also been used to create an aggregation space and knowledge base for the editors and Editors Advisory Board of the book. In this case the graph used as a reference has been a tentative content map of the book, meaning a mind map representing the arguments we wanted to explore. Proposals, references, ideas have been all gathered in the corresponding node so that each user was able to find it easily taking advantage of the mind map itself.

COLL-ANT

COLL-ANT is a prototype developed to demonstrate the possibility and effectiveness of the above suggested approach to address the difficulties in handling contrasting context dependent knowledge, paradigms, visions and practices. COLLA-ANT allows for knowledge to remain implicit while at the same time shared, making use of boundary objects that handle storage, retrieval, visualization and annotation of data and information.

With several new functionalities COLL-ANT integrates the ideas behind Colla and SNAP, living on top of both of them and providing means to access one from the other and vice-versa.

Figure 2. COLL-ANT workflow schema

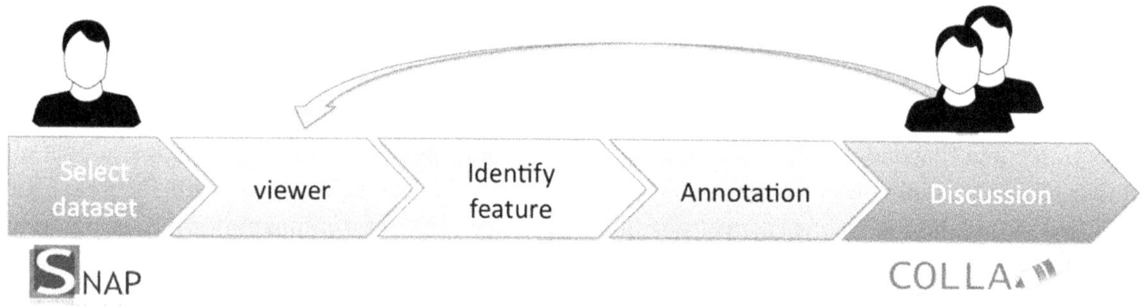

A lot of work has been done to redesign the SNAP framework, which so far was based on a server side pre-cooked SVG maps files and a client side SVG map rendering paradigm. This was changed to a server side GeoServer (http://geoserver.org) and a client side OpenLayers (http://openlayers.org/) libraries paradigm, which allowed to extend the interaction with the systems in the direction of OGC (http://www.opengeospatial.org/) web services and in particular towards WMS (http://www.opengeospatial.org/standards/wms) and WFS (http://www.opengeospatial.org/standards/wfs). Within the system data are held in a database or in files linked to from within the database. They are made available via WMS/WFS to OpenLayers, the map client side rendering facility, which shows the map in the browser, with the correct projection and scale. Each data-type is collected in a different layer which can be switch on or off to visualize it or not, in order to avoid cluttering the map.

A further map layer is made available (client side) where annotation can take place. These are saved in the database as geographic features, together with information on the user so that the collaborative work could be managed. The system allows then to recall annotations from the database and display it on the map layer, so that the annotation is automatically positioned in the correct geographic location.

Maps show position of data, such as for example, tracks of seismic lines or multibeam swath or DTM grids. Each object in the map is clickable so that a pop-up window can be open where metadata about that object is displayed. Among this information the window shows also a web link to a web based data viewing facility. This is based on a servlet that, after having identified the dataset, processes and plot it interactively upon the requests from the client.

In Figure 4 it is possible to see the case of a web based viewer for seismic data. This software is an evolution of the web based seismic viewing facility described in Diviacco and Busato (2013) with several ameliorations to integrate it even more with the new SNAP framework.

A very important feature of COLL-ANT is that not only maps but also its data and graphs visualization facilities offer also a specific layer devoted to annotations. Similarly to the case of maps these are stored in the database using as coordinate system the coordinate system of the graph displayed.

Annotations described so far correspond to the identification of an area on a map or a graph. These are supposed to trigger discussions that are held on COLLA. In fact even before the development of COLL-ANT, COLLA offered facilities to handle threads of discussion and means to upload files to be used within a discussion. The link between

Figure 3. COLL-ANT mapping and data discovery facility

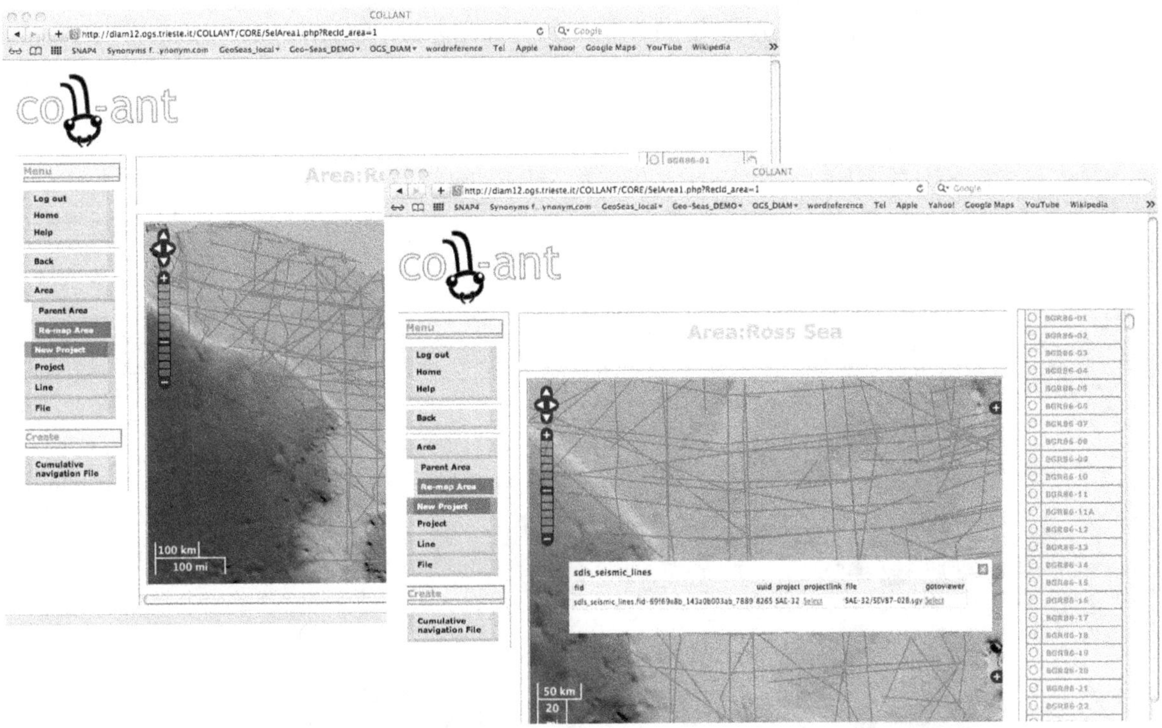

SNAP and COLLA is provided by simple URLs that are automatically generated in both systems upon the creation of a sketch in SNAP.

COLL-ANT Usage

The usage of COLL-ANT starts from the SNAP side which embeds all the functionalities of a web based data discovery and visualization tool.

Users are required to register so that they can be identified, and "enrolled" in the collaborative projects.

data can be searched upon metadata parameters, such as for example line name, data owner, or domain specific parameters such as for example sampling frequency of a seismic line, or ping rate of a multibeam survey. Data is plotted on geographic maps and each dataset is clickable, so that its selection opens a window where through web link it is possible to reach specific metadata web pages, or the viewing and annotation tools. (Figure 3)

Once the designated community has been identified and the dataset they are interested in has been isolated, collaborative work can start. Users are enabled, independently, to visualize maps and graphs, changing interactively plotting parameters and performing basic processing flows. Zooming and panning is easily achieved using mouse/trackpad gestures (Figure 4). Graphs show on the upper right corner a list of icons offering various tools to draw rectangles, circles or free hand lines that can be used to identify geologic features.

Sketches populate a specific layer of the visualization and are stored in the database in order to be recalled in the discussion. The act of drawing a sketch triggers the creation (i) of a node (in the COLLA side of COLL-ANT), (ii)

Figure 4. COLL-ANT sketching and annotation facility. The creation of a sketch triggers the creation of a COLLA note/task and the automatic link of them, so that clicking on the object sends the user to COLLA. The link between SNAP and COLLA is provided by simple URLs that are automatically generated in both systems upon the creation of a sketch in SNAP.

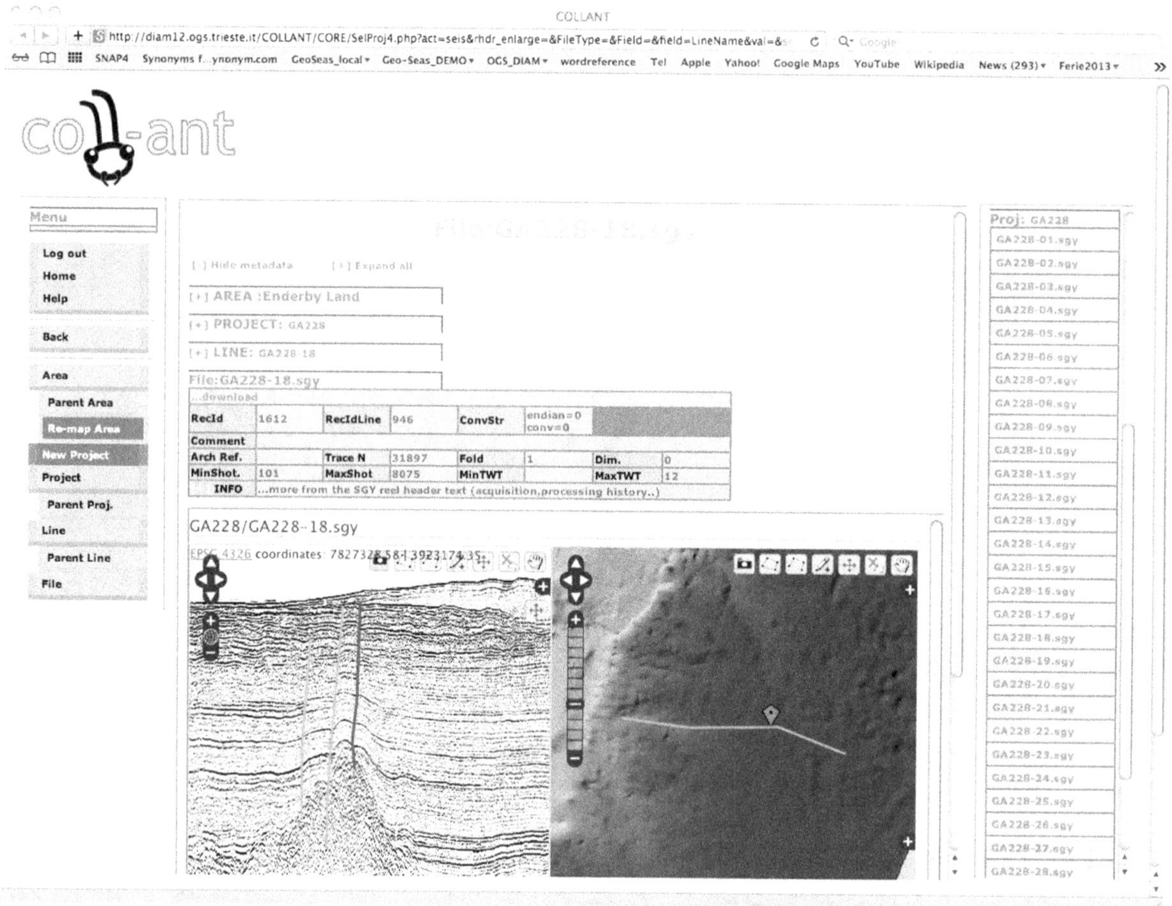

of a corresponding URL (in the SNAP side of COLLA-ANT) pointing to that node, and (iii) of a corresponding URL (in the COLLA side of COLLA-ANT) to get back to the SNAP side. COLLA nodes host discussion threads and files pertaining to them. Every change in the status of a node, such as for example the upload of a file or the submission of a text to the discussion, is notified to all users enrolled in the collaborative project through an automatic e-mail notification system. These automatic e-mails contain "magic-links" that allow to avoid the tedious process of login and allow users to enter immediately the node and the thread of the topic under discussion.

FUTURE RESEARCH DIRECTIONS

So far, COLL-ANT has been used mainly for the definition of the datasets needed within research projects. Users, collectively, after data discovery use the visualization tools to realize if the data they are looking at are interesting or not. The decision is based on a process of identification of geologic features (where existent) and link of this latter to a COLLA node discussion thread, where they can debate on the actual existence of the feature and on the importance of the data in studying it. Thereafter users prefer to load and interpret the subset data into the vertical commercial

Figure 5. Each COLL-ANT sketch links to a COLLA node where the actual discussion and file handling takes place

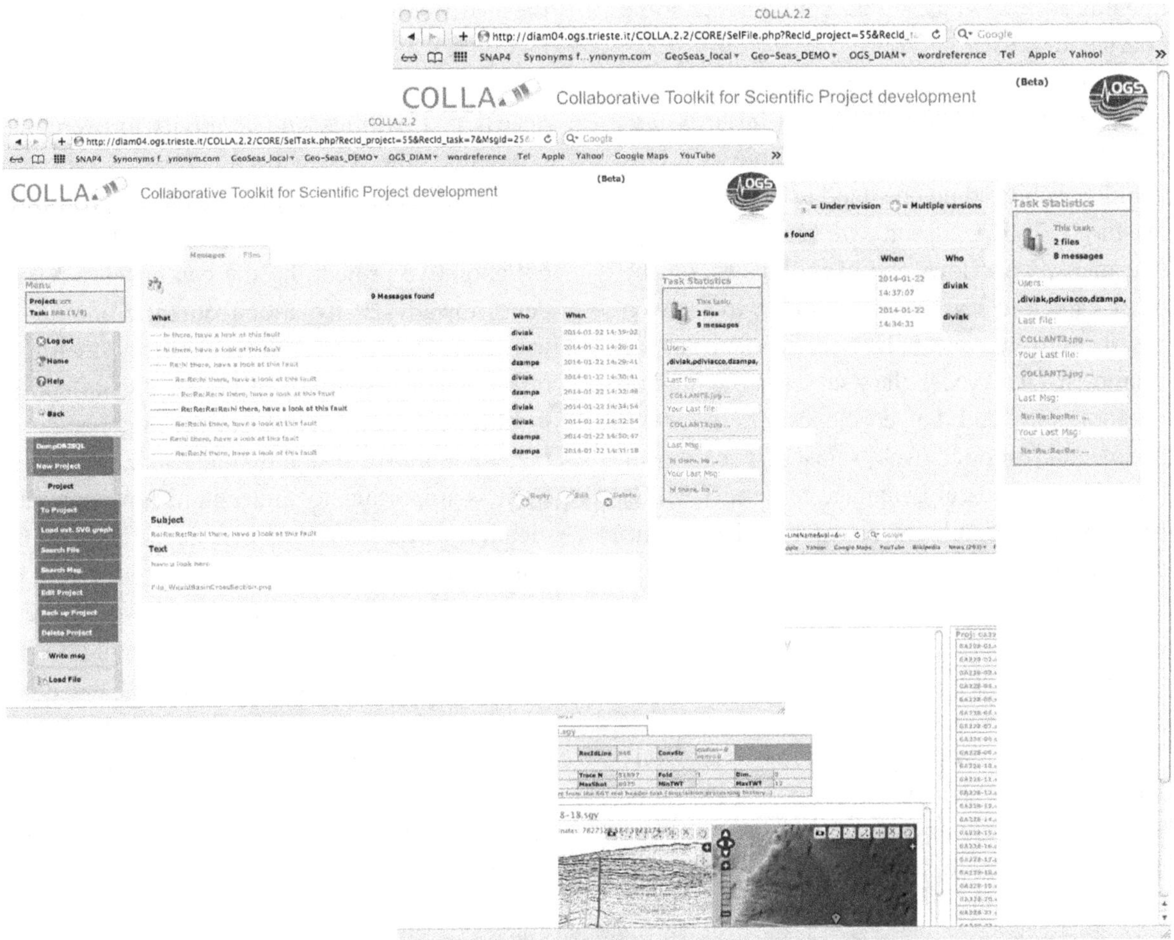

software they are more used to work with, such as for example HIS- Kingdom (http://www.ihs.com/products/oil-gas-information/analysis-software/kingdom-seismic-interpretation/index.aspx) or Schlumberger-Petrel (http://www.software.slb.com/products/platform/Pages/petrel.aspx). COLL-ANT of course does not aim to substitute this software, but we think that the future development of COLL-ANT should be in the direction of integration with them. This could be achieved devising extensions of COLL-ANT to connect and extract information from the interpretation software databases and providing interpreted features backward to COLL-ANT. This of course has to be done in a controlled way, in order that users could be enabled to access only to the data they are entitled to see. This of course is not straightforward since it means that an integration between accounts on interpretation software and on COLL-ANT will need to be achieved as well.

Further to these, we see the need to extend COLL-ANT in order to integrate even more SNAP and COLLA in a single product. So far, in fact, COLL-ANT users clicking on the sketch in SNAP are driven to an independent instance of COLLA, where the actual discussion takes place. It would be advisable, of course, to display everything directly within the same map or graph and avoid opening a new window that could create some confusion to users.

Another interesting extension of COLL-ANT would be to integrate formal and informal approaches to knowledge within the same tool. This could be achieved on one hand using the already existing COLLA capacity of handling conceptual maps, workflows and event bushes, on the other hand integrating the possibility to use web based controlled vocabulary services such as for example BODC vocabulary server in the field of oceanography and the geosciences (http://www.bodc.ac.uk/products/web_services/vocab/), or the Unified Medical Language System (UMLS) (http://www.nlm.nih.gov/research/umls/) in the case of medicine.

CONCLUSION

This chapter stems from the arguments of Diviacco (2015, see chapter 1 this book) addressing the importance of representation in the process of understanding the outer word and of knowledge creation. The aim of this chapter is to devise a system aiming to support distributed cognition and networked knowledge within scientific communities. In this perspective, we first describe the approach we took in developing the system; we then devise its functional specifications and eventually present the implementation we developed.

The approach extends the findings of Diviacco (2015, see chapter 1 this book), and the importance of representation, not only in the path from the subject towards his/her understanding of the outer world but also in that to be followed in order to reach other actors and communities in general. We highlight that the subject presents him/herself to the community bringing with him/her his/her background, context, academic and socio-political position. This cannot be overridden and inevitably bias his/her relations with the other partners. At the same time it is also important to preserve these different perspectives because: on one hand, the "bio-diversity" scientists introduce in the collaborative work, potentially, can bring the necessary flexibility and creativity to overcome problems, otherwise difficult to solve, within a "one-way" only paradigm. On the other hand if some partner cannot find his/her context in the collaborative project, he/she can develop a tendency toward self-isolation from the community, triggering what is called the community "melting" effect, which essentially is a form of disaffection towards the collaborative project in general and the collaborative system in particular.

It is important to understand how communication takes place and can be influenced by this phenomenon. Essentially, it is important to account for different contexts to have different maps of the same concept. Meaning will point to different targets. A denotative approach will force one vision only or will introduce the need for a meta-linguistic functionality that will take place as a mapping between several of them. In this latter case knowledge will need to be put explicit, but as we have demonstrated the importance of tacit knowledge and of socio-political issues only very seldom allow this.

We decided then to move from the meta-linguistic approach to another one that would grant us the necessary flexibility. We focused on the referencing functionality with an iconic signification mode, which in practice can be translated as a form of representation. In this perspective, representation becomes the boundary objects that allow to bridge between different paradigms, to handle multiple and contrasting visions, and theory-ladenness itself. We do not claim that formal and implicit approaches to handle knowledge are completely immiscible; there must be ways to leverage both. Rather we want to highlight the importance of the implicit knowledge approach and underpin the cases in which it should be used.

In this perspective we developed a web-based system that embeds the ideas we described above and that demonstrate their importance. This system, named COLL-ANT, allows users to access data and information, visualize it and, using annotation, collaboratively and in a distributed fashion, support a scientific research community.

COLLA-ANT is still in its infancy, and has not been used enough to understand its real potential, but, from the fact itself that we were able to translate the ideas behind it to a working environment, and from the first period of tests where there has been a good and wide acceptance of the tool, we are rather optimistic on its capability to address the above described issues of collaborative work, distributed cognition and knowledge network.

REFERENCES

Barthes, R. (1957). *Mythologies*. London: Ebenezer Baylis & Son Ltd.

Becher, T., & Trowler, P. R. (2001). *Academic tribes and territories intellectual enquiry and culture of disciplines*. Buckingham, UK: The Society for Research into Higher Education and Open University Press.

Bishop, M. A. (1992) Theory-Ladenness of Perception Arguments. In *Proceedings of the Biennial Meeting of the Philosophy of Science Association* (pp. 287-299). Academic Press.

Bruner, J. S., & Postman, L. (1949). On the perception of incongruity: A paradigm. *Journal of Personality, 18*, 206-223. Retrieved from http://www.jstor.org/stable/192762

Carnap, R. (1942). *Introduction to Semantics*. Harvard Uni. Press.

Carnap, R. (1947). *Meaning and Necessity: A Study in Semantics and Modal Logic*. University of Chicago Press.

Carusi, A., & Torsten, R. (2010). *Virtual Research Environment collaborative landscape study*. Bristol: JISC.

Churchland, P. M. (1979). *Scientific realism and the Plasticity of Mind*. Cambridge, UK: Cambridge University Press. doi:10.1017/CBO9780511625435

Coward, R., & Ellis, J. (1977). *Language and Materialism: Developments in Semiology and the Theory of the Subject*. London: Routledge & Kegan Paul.

de Saussure, F. (1983). *Course in General Linguistics* (R. Harris, Trans.). London: Duckworth. (Original work published 1916)

Diviacco, P. (2005). An open source, web based, simple solution for seismic data dissemination and collaborative research. *Computers & Geosciences, 31*(5), 599–605. doi:10.1016/j.cageo.2004.11.008

Diviacco, P. (2012). Addressing Conflicting Cognitive Models in Collaborative E-Research: A Case Study in Exploration Geophysics. In *Collaborative and Distributed E-Research: Innovations in Technologies, Strategies and Applications*. IGI Global Press. doi:10.4018/978-1-4666-0125-3.ch012

Diviacco, P. (2015). Reconciling Knowledge and Collaborative e-Research. In Collaborative Knowledge in Scientific Research Networks. Hershey, PA: IGI Global.

Diviacco, P., & Busato, A. (2013, September). The Geo-Seas seismic data viewer: A tool to facilitate control of data access. *BGTA, 54*(3), 257–270. doi:10.4430/bgta0078

Diviacco, P., Glaves, H., & Shaap, D. (2011). *Marine Seismic Metadata for an Integrated European Scale Data Infrastructure: The Fp7 Geo-Seas Project*. BGTA. doi:10.4430/bgta0051

Diviacco, P., & Pshenichny, C. A. (2010). Concept-referenced spaces in computer-supported collaborative work. In *Geophysical Research Abstracts* (Vol. 12). Vienna, Austria: EGU.

Diviacco, P., & Pshenichny, C. A. (2011). A case study on the use of event bushes as a formal representation for computer supported collaborative work in the geosciences. In *Geophysical Research Abstracts* (Vol. 12). Vienna, Austria: EGU.

Diviacco, P., & Wardell, N. (2003). Reprocessing and dissemination of Antarctic Seismic Data. *Terra Antarctica Reports 2003, 9*, 133-136.

Duhem, P., & Wiener, P. P. (1954). La Théorie Physique: Son Objet et sa Structure [The Aim and Structure of Physical Theory]. Princeton University Press.

Eco, U. (1979). *A Theory of Semiotics*. Indiana University Press.

European Committee for Standardization (CEN). (2014). *European Guide to good Practice in Knowledge Management* (CWA 14924 report). Retrieved January, 2014, from http://www.cen.eu/cen/Sectors/Sectors/ISSS/CWAdownload/Pages/Knowledge%20Management.aspx

Feyerablend, P. (1975). *Against Method*. London: New Left Books.

Fodor, J. (1983). *The Modularity of Mind*. Cambridge, MA: MIT Press.

Fraser, M. (2004). *Supporting Virtual Research Environments: How can we help?* Retrieved June 27, 2011 from http://users.ox.ac.uk/~mikef/rts/vre/img2.html

Giere, R. (2002, December). Discussion Note: Distributed Cognition in Epistemic Cultures. *Philosophy of Science*, 69.

Golder, S., & Huberman, B. A. (2005). *The structure of collaborative tagging systems*. HP Labs Technical Report. Retrieved august 2011 from http://www.hpl.hp.com/research/idl/papaers/tags

Hanson, N. R. (1958) Patterns of Discovery. Cambridge, UK: Cambridge University Press.

Kant, I. (1781). *The Critique of Pure Reason*. Retrieved from http://www.gutenberg.org/ebooks/4280

Knorr Cetina, K. (1999). *Epistemic Cultures: How the Science make knowledge*. Cambridge, MA: Harvard University Press.

Kuhn, T. S. (1962). *The structure of scientific revolutions*. Chicago, IL: University of Chicago Press.

List, C. (2005). Group knowledge and group rationality: A judgement aggregation perspective. *Episteme: A Journal of Social Epistemology, 2*(1), 25-38

Peirce, C. S. (1958). Collected Writings (8 Vols.). (C. Hartshorne, P. Weiss, & A. W. Burks, Eds.). Cambridge, MA: Harvard University Press.

Pera, M. (1994). Scienza e Retorica. University of Chicago Press.

Polanyi, M. (1966). The Tacit Dimension. New York: Anchor Day Books.

Pshenichny, C. A., Carniel, R., & Diviacco, P. (2013). *Engineering of Dynamic Knowledge in Exact Sciences: First Results of Application of the Event Bush Method in Physics. ICCS 2013 - MSEPS 2013 – Modeling States*. Events, Processes and Scenarios.

Putnam, H. (1975). *Mind, language and reality: Philosophical papers* (Vol. 2). Cambridge, UK: Cambridge University Press. doi:10.1017/CBO9780511625251

Shannon, C. E., & Weaver, W. (1949). *A mathematical model of communication*. Chicago, IL: University of Illinois Press.

Specia, L., & Motta, E. (2007). Integrating folksonomies with the semantic web. In *Proceedings of the European Semantic Web Conference*. ESWC Press.

Star, S. L., & Griesemer, J. R. (1989). Institutional ecology, translations and boundary objects: Amateurs and professionals in Berkeley s museum of vertebrate zoology. *Social Studies of Science*, *19*(4), 387–420. doi:10.1177/030631289019003001

Sturrock, J. (1986). *Structuralism*. London: Paladin.

Warren, R. (1970). Perceptual restoration of missing speech sounds. *Science*, *167*(3917), 392–393. doi:10.1126/science.167.3917.392 PMID:5409744

Zappa F. (1979). *Packard Goose on Joe's Garage* [LP]. Zappa Record, CBS

Zeleny, M. (2005). *Human Systems Management: Integrating Knowledge, Management and Systems*. World Scientific. doi:10.1142/4929

ADDITIONAL READING

Sheriff, R. E. (1973). *Encyclopedic dictionary of exploration geophysics*. Tulsa, OK: The Society of Exploration Geophysics.

KEY TERMS AND DEFINITIONS

Abduction: A way of reasoning where the conclusions do not follow logically form the premises, but can be considered (in relative terms) the best explanation.

Boundary Object: An artifact that allow two different and incommensurable communities to nevertheless collaborate. They are weakly structured in common use, and become strongly structured in individual use.

Context: The set of circumstances to which something is bound. In the chapter it is used to highlight the possible co-existence of multiple contexts linked to the same entity, object or a concept. This co-existence can trigger the issue of having possibly contrasting interpretations or visions of the same concept or object can coexist considering different circumstances.

Data: Is raw observations or measurements. It is not organized and does not present any meaning. Adding meaning to data we get information.

Knowledge: Is what we know. It is the output of the process of reasoning which is based on information and it is conditioned by the context in which the researcher lives. Knowledge mirrors back in all human activities and decisions so that its shadow is projected over any scientific experiment, data or information.

Information: Is data with meaning. In the perspective of this chapter information can be represented by maps and graphs, where data is framed.

Representation: A belief that one's own culture is superior to other cultures.

Tacit Knowledge: Knowledge that cannot be made explicit because obscured in socio-political phenomena or embedded in practices or even tools.

Theory Ladenness: Refers to the fact that experience is categorized and preconditioned by prior belief, and that the process of building knowledge always involves the use of concepts from some previous view. This holds both from a logical and a perceptive point of view.

ENDNOTES

1 For a complete explanation refer to Engelhardt and Zimmerman, 1982.

2 To understand why we talk about state of affairs and not facts please refer to Diviacco (2015, see chapter 1 this book).

3 Originally Kant referred to facts, but considering what said above this is very unlikely to be the case while we can consider states of affairs instead, referred to through the use of scientific data and observations.

4 A specific form of hypothetical claim is an interpretation. Interpretations always imply the possible existence of other interpretations.

5 In many cases the position itself of information can be linked to meaning, and therefore to higher levels than syntax. This is the case for example of the headers of a file, where metadata have to be entered in appropriate locations otherwise the whole sense of the representation is lost.

6 Plato (Cratylus) already highlighted that words can be exchangeable for the same object or concept.

7 See Diviacco (2015, see chapter 1 this book) for further information on abductive reasoning.

8 And even in the case negotiation takes place concerns might rise on the process of consensus achievement, see later on aggregation of judgment.

9 See later section on theory-ladenness.

10 See section on judgment aggregation in Diviacco (2015, see chapter 1 this book).

11 From data to information, from information to knowledge.

12 The apodictic mode requires the necessary existence of a state of affair, while the formalization approach, as a tool, can produce perfect non-senses.

13 Or as in the vision of Diviacco (2015, see chapter 1 this book), extending the argument outside scientific research, culture in general.

14 This is often the case when judgment aggregations strategy is based on the division of the cognitive work into subgroups of partners that can better elaborate on the premises.

15 To close the loop from knowledge to data we would need to be able to reach the phase of acquisition of the observations as mentioned in the section of this chapter regarding Theory-Ladenness.

16 SNAP is the acronym for Seismic data Network Access Point.

17 COLLA in Italian means glue, so that it is supposed to keep researchers together.

Compilation of References

Abidi, S. S. R., Hussini, S., Sriraj, W., Thienthong, S., & Finley, G. A. (2009). Knowledge sharing for pediatric pain management via a Web 2.0 framework. In K.-P. Adlassnig et al. (Eds.), *Medical Informatics in a United and Healthy Europe* (pp. 287–291). IOS Press.

Aboelela, S. W., Larson, E., Bakken, S., Carrasquillo, O., Formicola, A., Glied, S. A., ... Gebbie, K. M. (2007). Defining interdisciplinary research: Conclusions from a critical review of the literature. *Health Services Research, 42*(1p1), 329–346.

Abrams, L., Cross, R., Lesser, E., & Levin, D. (2003). Nurturing interpersonal trust in knowledge-sharing networks. *The Academy of Management Executive, 17*(4), 64–77. doi:10.5465/AME.2003.11851845

Ackoff, R. L. (1974). *Redesigning the Future: Systems approach to societal problems*. New York: Wiley.

Ackoff, R. L. (1989). From data to wisdom. *Journal of Applied Systems Analysis, 16*, 3–9.

Adams, T. L., & Demaiter, E. I. (2007). *Knowledge workers in the New Economy: Skill, flexibility and credentials*. American Sociological Association. Retrieved from http://citation.allacademic.com/meta/p178393_index.html

Akers, K. G. (2013). Looking out for the little guy: Small data curation. *Bulletin of the American Society for Information Science and Technology, 39*(3), 58–59. doi:10.1002/bult.2013.1720390317

Ala-Mutka, K., Gáspár, P., Kismihók, G., Suurna, M., & Vehovar, V. (2010). Status and Developments of eLearning in the EU10 Member States: The cases of Estonia, Hungary and Slovenia. *European Journal of Education, 45*(3), 494–513.

Alavi, M., & Leidner, D. E. (2001). Review: Knowledge management and knowledge management systems: Conceptual foundations and research issues. *Management Information Systems Quarterly, 25*(1), 107–136. doi:10.2307/3250961

Allemang, D., & Hendler, J. (2011). Inferring Class Relationships. In *Semantic web for the working ontologist: Effective modeling in RDFS and OWL* (2nd ed., pp. 273–276). Waltham, MA: Morgan Kaufmann Publishers.

Allen, R. (2012). Transforming Earthquake Detection? *Science, 335*(6066), 297–298. doi:10.1126/science.1214650 PMID:22267802

Allen, T. D. (2007). Mentoring relationships from the perspective of the mentor. In B. R. Ragins & K. Kram (Eds.), *The handbook of mentoring at work: Theory, research and practice* (pp. 123–147). Thousand Oaks, CA: Sage.

Al-Rodhan, N. R., & Stoudmann, A. G. (2006). Definitions of Globalization: A Comprehensive Overview and a Proposed Definition. Geneva: Centre for Security Policy. Retrieved from http://gcsp.ch/content/download/1267/9834/file/Definitions%20of%20Globalization%20-%20A%20Comprehensive%20Overview%20and%20a%20Proposed%20Definition.pdf

Altintas, I., Berkley, C., Jaeger, E., Jones, M., Ludäscher, B., & Mock, S. (2004). Kepler: an extensible system for design and execution of scientific workflows. In *Proceedings of 16th International Conference on Scientific and Statistical Database Management*, (pp. 423–424). doi:10.1109/SSDM.2004.1311241

Ames, D. P., Horsburgh, J. S., Cao, Y., Kadlec, J., Whiteaker, T., & Valentine, D. (2012). HydroDesktop: Web services-based software for hydrologic data discovery, download, visualization, and analysis. *Environmental Modelling & Software*, *37*(0), 146–156. doi:10.1016/j.envsoft.2012.03.013

Amidon, D. M. (1991). *Memo to Ken Olson*. Retrieved from http://www.entovation.com/gkp/origins.htm

Anderson, J. L. (1998). Embracing uncertainty: the interface of Bayesian statistics and cognitive psychology. *Conservation Ecology*, *2*, 2. Retrieved from http://www.consecol.org/vol2/iss1/art2

Andrews, K. M., & Delahaye, B. L. (2000). AT Influences on knowledge processes in organizational learning: The psychosocial filter. *Journal of Management Studies*, *37*(6), 797–810. doi:10.1111/1467-6486.00204

Arko, R., Chandler, C., Clark, P., Shepherd, A., & Moore, C. (2012). *Rolling Deck to Repository (R2R): Linking and integrating data for oceanographic research*. Paper presented at 2012 Fall Meeting, American Geophysical Union. San Francisco, CA.

Arnell, N. W., Tompkins, E. L., & Adger, W. N. (2005). Eliciting information from experts on the likelihood of rapid climate change. *Risk Analysis*, *25*(6), 1419–1431. doi:10.1111/j.1539-6924.2005.00689.x PMID:16506972

Aspinall, W. P. (2006). Structured elicitation of expert judgment for probabilistic hazard and risk assessment in volcanic eruptions. In H.M. Mader, S.G. Coles, C.B. Connor, & L.J. Connor (Eds.), Statistics in Volcanology. Geological Society of London.

Azad, B., & Wiggins, L. L. (1995). Dynamics of Inter-Organizational Geographic Data Sharing. In H. J. Onsrud, & G. Rushton (Eds.), Sharing Geographic Information (pp. 22-43). Rutgers.

Baader, F., Horrocks, I., & Sattler, U. (2005). Description Logics as Ontology Languages for the Semantic Web. In Mechanizing Mathematical Reasoning (LNAI), (vol. 2605, pp. 228-248). Berlin: Springer-Verlag.

Baddeley, M. A., Curtis, A., & Wood, R. A. (2004). An introduction to prior information derived form probabilistic judgements: elicitation of knowledge, cognitive bias and herding. In A. Curtis & R. A. Wood (Eds.), Geological Society, London Special Publications: Geological Prior Information: Informing Science and Engineering (vol. 239, pp. 15–27). Academic Press.

Baeten, J.C.M. (2004). *A brief history of process algebra*. Vakgroep Informatica, Technische Universiteit Eindhoven: Rapport CSR 04-02.

Bagley, P. (1968). *Extension of programming language concepts*. Philadelphia: University City Science Center.

Baier, V. (2006). *Forty years of acting electron-positron colliders*. arXiv:hep-ph/0611201

Baker, K., & Chandler, C. (2008). Enabling long-term oceanographic research: Changing data practices, information management strategies and informatics. *Deep-sea Research. Part II, Topical Studies in Oceanography*, *55*(18-19), 2132–2142. doi:10.1016/j.dsr2.2008.05.009

Bak, P., & Sneppen, K. (1993). Punctuated equilibrium and criticality in a simple model of evolution. *Physical Review Letters*, *71*(24), 4083–4086. doi:10.1103/PhysRevLett.71.4083 PMID:10055149

Balsiger, P. W. (2004). Supradisciplinary research practices: History, objectives and rationale. *Futures*, *36*(4), 407–421. doi:10.1016/j.futures.2003.10.002

Bansler, J. P., & Havn, E. (2003). Technology-use mediation: Making sense of electronic communication in an organizational context. In *Proceedings of the 2003 international ACM SIGGROUP conference on Supporting group work* (pp. 135-143). ACM. doi:10.1145/958160.958182

Baranik, L. E., Roling, E. A., & Eby, L. T. (2010). Why does mentoring work? The role of perceived organizational support. *Journal of Vocational Behavior*, *76*(3), 366–373. doi:10.1016/j.jvb.2009.07.004 PMID:20401322

Bard, J. B., & Rhee, S. Y. (2004). Ontologies in biology: Design, applications and future challenges. *Nature Reviews. Genetics*, *5*(3), 213–222. doi:10.1038/nrg1295 PMID:14970823

Barjak, F. (2006). *Team diversity and research collaboration in life sciences teams: Does a combination of research cultures pay off?* University of Applied Sciences Northwestern Switzerland. Retrieved from http://netreact-eu.org/documents/DPW2006-02_Team%20Diversity%20_Barjak_Franz.pdf

Barros, B., Verdejo, M. F., Read, T., & Mizoguchi, R. (2002). Applications of a collaborative learning ontology. In C. A. Coello Coello et al (Eds.), MICAI 2002 (LNAI), (vol. 2313, pp. 301-310). Berlin: Springer-Verlag.

Barthes, R. (1957). *Mythologies*. London: Ebenezer Baylis & Son Ltd.

Barthes, R. (1977). *Image, Music, Text*. New York: Hill and Wang.

Bazilian, M., Rogner, H., Howells, M., Hermann, S., Arent, D., & Gielen, D. et al. (2011). Considering the energy, water and food nexus: Towards an integrated modelling approach. *Energy Policy*, *39*(12), 7896–7906. doi:10.1016/j.enpol.2011.09.039

Beaman, R. J., & Harris, P. T. (2007). Geophysical variables as predictors of megabenthos assemblages from the northern Great Barrier Reef, Australia. In B. J. Todd & H. G. Greene (Eds.), Geological Association of Canada Special Paper: Mapping the seafloor for habitat characterization (vol. 47, pp. 247–263). Academic Press.

Beaman, R. J., Daniell, J. J., & Harris, P. T. (2005). Geology–benthos relationships on a temperate rocky bank, eastern Bass Strait, Australia. *Marine & Freshwater Research*, *56*(7), 943–958. doi:10.1071/MF04306

Becher, T., & Trowler, P. R. (2001). *Academic tribes and territories intellectual enquiry and culture of disciplines*. Buckingham, UK: The Society for Research into Higher Education and Open University Press.

Bechhofer, S., Ainsworth, J., Bhagat, J., Buchan, I., Couch, P., Cruickshank, D., et al. (2010). Why linked data is not enough for scientists. In *Proceedings ofIEEE Sixth International Conference on e-Science*, (pp. 300-307). IEEE. doi:10.1109/eScience.2010.21

Bechhofer, S., Buchan, I., De Roure, D., Missiera, P., Ainsworth, J., & Bhagata, J. et al. (2013). Why linked data is not enough for scientists. *Future Generation Computer Systems*, *29*(2), 599–611. doi:10.1016/j.future.2011.08.004

Belhajjame, K., Corcho, O., Garijo, D., Zhao, J., Missiera, P., Newman, D., et al. (2012). Workflow-Centric Research Objects: A First Class Citizen in the Scholarly Discourse. In *Proceedings of the ESWC2012 Workshop on the Future of Scholarly Communication in the Semantic Web (SePublica2012)*. ESWC.

Bell, D. (1976). The Coming of the Post-Industrial Society. *The Educational Forum*, *40*(4), 574–579. doi:10.1080/00131727609336501

Berk, R. A., Berg, J., Mortimer, R., Walton-Moss, B., & Yeo, T. P. (2005). Measuring the Effectiveness of Faculty Mentoring Relationships. *Academic Medicine*, *80*(1), 66–71. doi:10.1097/00001888-200501000-00017 PMID:15618097

Bermudez, L. (2007). Vocabulary integration environment. *VINE Revista Avance en Sistemas e Informatica*, *4*(2), 53–60.

Berners-Lee, T. (2006). *Linked Data – Design Issues*. Retrieved from http://www.w3.org/DesignIssues/LinkedData.html

Berners-Lee, T., Hendler, J., & Lassila, O. (2001). The Semantic Web. *Scientific American*, *284*(5), 34–43. doi:10.1038/scientificamerican0501-34 PMID:11396337

Bettiol, M., & De Pietro, L. (2003). Comunità di pratica e organizzazione: Un rapporto da scoprire. *T D Tecnologie Didattiche*, *30*, 53–60.

Beyer, H., & Holtzblatt, K. (1998). *Contextual design: Defining customer centered systems*. New York, NY: Academic Press.

Bhagia, J., & Tinsley, J. A. (2000). The mentoring partnership. *Mayo Clinic Proceedings*, *75*(5), 535–537. doi:10.1016/S0025-6196(11)64225-0 PMID:10807085

Bigagli, L., Nativi, S., Mazzetti, P., & Villoresi, G. (2004). GI-Cat: a Web service for dataset cataloguing based on ISO 19115. In *Proceedings of Database and Expert Systems Applications,* (pp. 846-850). Munich, Germany: IEEE.

Biloslavo, R., & Prevodnik, M. (2010). Impact of Organizational Culture on Knowledge Management in Higher Education. In D. Harorimana (Ed.), *Cultural Implications of Knowledge Sharing, Management and Transfer: Identifying Competitive Advantage* (pp. 152–179). Hershey, PA: Information Science Reference. doi:10.4018/978-1-61520-965-1.ch418

Bird, A. (2000). *Thomas Kuhn*. Bloomington, IN: Acumen Publishing.

Birnholtz, J. (2006). What does it mean to be an author? The intersection of credit, contribution, and collaboration in science. *Journal of the American Society for Information Science and Technology*, *57*(13), 1758–1770. doi:10.1002/asi.20380

Birnholtz, J. P. (2007). When do researchers collaborate? Toward a model of collaboration propensity. *Journal of the American Society for Information Science and Technology*, *58*(14), 2226–2239. doi:10.1002/asi.20684

Bishop, M. A. (1992) Theory-Ladenness of Perception Arguments. In *Proceedings of the Biennial Meeting of the Philosophy of Science Association* (pp. 287-299). Academic Press.

Bizer, C., Heath, T., & Berners-Lee, T. (2009). Linked Data - The story so far. *International Journal on Semantic Web and Information Systems*, *5*(3), 1–22. doi:10.4018/jswis.2009081901

Blaumer, A. (2013). *The LOD cloud is dead, long live the trusted LOD cloud.* Retrieved April 15, 2014 from http://blog.semantic-web.at/2013/06/07/the-lod-cloud-is-dead-long-live-the-trusted-lod-cloud/

Block, R. E., & Leroux, P. (1985). Generalized dual coalgebras of algebras, with applications to cofree coalgebras. *Journal of Pure and Applied Algebra*, *36*(1), 15–21. doi:10.1016/0022-4049(85)90060-X

Blue Ribbon Task Force on Sustainable Digital Preservation and Access. (2010). *Sustainable Economics for a Digital Planet: Ensuring Long-Term Access to Digital Information.* Final Report of the Blue Ribbon Task Force on Sustainable Digital Preservation and Access.

Bogardi, J. J., Dudgeon, D., Lawford, R., Flinkerbusch, E., Meyn, A., & Pahl-Wostl, C. et al. (2012). Water security for a planet under pressure: Interconnected challenges of a changing world call for sustainable solutions. *Current Opinion in Environmental Sustainability*, *4*(1), 35–43. doi:10.1016/j.cosust.2011.12.002

Boice, R. (1992). *The new faculty member.* San Francisco: Jossey-Bass.

Boldrini, E., Craglia, M., Mazzetti, P., & Nativi, S. (2015). The Brokering Approach for enabling collaborative scientific research. In *Collaborative Knowledge in Scientific Research Networks.* IGI Global Press.

Bolotnikova, E. S., Gavrilova, T. A., & Gorovoy, V. A. (2011). To a Method of Evaluating Ontologies. *Journal of Computer and Systems Sciences International*, *50*(3), 448–461. doi:10.1134/S1064230711010072

Bommer, J. J., Scherbaum, F., Bungum, H., Cotton, F., Sabetta, F., & Abrahamson, N. A. (2005). On the use of logic trees for ground-motion prediction equations in seismic-hazard analysis. *Bulletin of the Seismological Society of America*, *95*(2), 377–389. doi:10.1785/0120040073

Bonaiuti, G. (2006). E-learning 2.0, Il futuro dell'apprendimento in rete, tra formale e informale. *I quaderni di Form@re, n.6.*

Bonano, E. J., & Apostolakis, G. E. (1991). Theoretical foundation and practical issues for using expert judgements in uncertainty analysis of high-level radioactive waste disposal. *Radioactive Waste Management and the Nuclear Fuel Cycle*, *16*(2), 137–159.

Bond, C. E., Gibbs, A., Shipton, Z. K., & Jones, S. (2007). What do you think this is? 'Conceptual uncertainty' in geoscience interpretation. *GSA Today*, *17*(11), 4–10. doi:10.1130/GSAT01711A.1

Bond, C. E., Lunn, R. J., Shipton, Z. K., & Lunn, A. D. (2012). What makes an expert effective at interpreting seismic images? *Geology*, *40*(1), 75–78. doi:10.1130/G32375.1

Bonner, B. L. (2000). The Effects of Extroversion on Influence in Ambiguous Group Tasks. *Small Group Research*, *31*(2), 225–244. doi:10.1177/104649640003100205

Bonner, B. L. (2004). Expertise in Group Problem Solving: Recognition, Social Combination, and Performance. *Group Dynamics*, *8*(4), 277–290. doi:10.1037/1089-2699.8.4.277

Boose, E., Ellison, A., Osterweil, L., Clarke, L., Podorozhny, R., & Hadley, J. et al. (2007). Ensuring reliable datasets for environmental models and forecasts. *Ecological Informatics*, *2*(3), 237–247. doi:10.1016/j.ecoinf.2007.07.006

Borgatti, S. P. (2002). *NetDraw Software for Network Visualization*. Analytic Technologies.

Borgatti, S. P., Everett, M. G., & Freeman, L. C. (2002). *UCINET for Windows: Software for social network analysis. Harvard Analytic Technologies.*

Borgman, C. L. (2012). The conundrum of sharing research data. *Journal of the American Society for Information Science and Technology*, *63*(6), 1059–1078. doi:10.1002/asi.22634

Borisoff, D. (1998). Strategies for effective mentoring and for being effectively mentored: A focus on research institutions. *Journal of the Association for Communication Administration*, *27*, 84–96.

Börner, K., Contractor, N., Falk-Krzesinski, H. J., Fiore, S. M., Hall, K. L., & Keyton, J. et al. (2010). A Multi-Level Systems Perspective for the Science of Team Science. *Science Translational Medicine*, *2*(49), 49cm24. doi:10.1126/scitranslmed.3001399 PMID:20844283

Bos, N., Zimmerman, A., Olson, J., Yew, J., Yerkie, J., Dahl, E., & Olson, G. (2007). From shared databases to communities of practice: A taxonomy of collaboratories. *Journal of Computer-Mediated Communication*, *12*(2), 652–672. doi:10.1111/j.1083-6101.2007.00343.x

Botts, M., Percivall, G., Reed, C., & Davidson, J. (2008). *OGC® sensor web enablement: Overview and high level architecture.*Lecture Notes in Computer Science, 4540, 175–190. doi:10.1007/978-3-540–79996-2_10

Bowman, A., Crowther, C., Pybus, J., & Kirkham, R. (2008). *Virtual Research Environment for the Study of Documents and Manuscripts*. Oxford eResearch Conference 2008. Retrieved from http://bvreh.humanities.ox.ac.uk/files/Oxford-eResearch2008-VRE-SDM.pdf

Boyack, K. W., Klavans, R., & Borner, K. (2005). Mapping the backbone of science. *Scientometrics*, *64*(3), 351–374. doi:10.1007/s11192-005-0255-6

Bozeman, B., & Corley, E. (2004). Scientists' collaboration strategies: Implications for scientific and technical human capital. *Research Policy*, *33*(4), 599–616. doi:10.1016/j.respol.2004.01.008

Bracken, L. J., Wainwright, J., Ali, G. A., Tetzlaff, D., Smith, M. W., Reaney, S. M., & Roy, A. G. (2013). Concepts of hydrological connectivity: Research approaches, pathways and future agendas. *Earth-Science Reviews*, *119*(0), 17–34. doi:10.1016/j.earscirev.2013.02.001

Brochhausen, M. M., Spear, A. D., Cocos, C. C., Weiler, G. G., Marti'n, L. L., Anguita, A. A., & Tsiknakis, M. M. (2011). The ACGT Master Ontology and its applications — Towards an ontology-driven cancer research and management system. *Journal of Biomedical Informatics*, *44*(1), 8–25. doi:10.1016/j.jbi.2010.04.008 PMID:20438862

Bruner, J. S., & Postman, L. (1949). On the perception of incongruity: A paradigm. *Journal of Personality*, *18*, 206-223. Retrieved from http://www.jstor.org/stable/192762

Bruner, J. S., & Postman, L. (1949). On the perception of incongruity: A paradigm. *Journal of Personality*, *18*(2), 206–223. doi:10.1111/j.1467-6494.1949.tb01241.x

Budgen, D., Rigby, M., Brereton, P., & Turner, M. (2007). A data integration broker for healthcare systems. *Computer*, *40*(4), 34–41. doi:10.1109/MC.2007.112

Buesseler, K. (2001). Ocean biogeochemistry and the global carbon cycle; an introduction to the US Joint Global Ocean Flux Study. *Oceanography (Washington, D.C.)*, *14*(4), 5. doi:10.5670/oceanog.2001.01

Bull, G., Thompson, A., Searson, M., Garofalo, J., Park, J., Young, C., & Lee, J. (2008). Connecting Informal and Formal Learning Experience in tha Age of Participatory Media. *Contemporary Issues in Technology & Teacher Education*, *8*(2), 100–107.

Bunakov, V. (2013). Core semantic model for generic research activity. In *Proceedings of 15th All-Russian Conference "Digital Libraries: Advanced Methods and Technologies, Digital Collections" (RCDL 2013)*. Retrieved from http://purl.org/net/epubs/work/10938269

Bunakov, V., & Matthews, B. (2013). Data curation framework for facilities science. In *Proceedings of2nd International Conference on Data Technologies and Applications (DATA 2013)*. doi:10.5220/0004593302110216

Bunakov, V., Jones, C., & Matthews, B. (2015). Towards the interoperable data environment for Facilities Science. In P. Diviacco, P. Fox, C. Pshenichny, & A. Leadbetter (Eds.), *Collaborative Knowledge in Scientific Research Networks*. Hershey, PA: IGI Global.

Burke, R. J., & McKeen, C. A. (1997). Benefits of mentoring relationships among managerial and professional women: A cautionary tale. *Journal of Vocational Behavior*, *51*(1), 43–57. doi:10.1006/jvbe.1997.1595

Butterfield, M. L., Pearlman, J. S., & Vickroy, S. C. (2008). A system-of-systems engineering GEOSS: Architectural approach. *Systems Journal, IEEE*, *2*(3), 321–332. doi:10.1109/JSYST.2008.925973

Buytaert, W., Baez, S., Bustamante, M., & Dewulf, A. (2012). Web-Based Environmental Simulation: Bridging the Gap between Scientific Modeling and Decision-Making. *Environmental Science & Technology*, *46*(4), 1971–1976. doi:10.1021/es2031278 PMID:22260091

Buzan, T. (2005). *Mind Map handbook*. London: Thorsons.

Byrne, M. W., & Keefe, M. R. (2002). Building Research Competence in Nursing Through Mentoring. *Journal of Nursing Scholarship*, *34*(4), 391–396. doi:10.1111/j.1547-5069.2002.00391.x PMID:12501744

Cabrera, A., & Cabreta, E. F. (2002). Knowledge-sharing Dilemmas. *Organization Studies*, *23*(5), 687–710. doi:10.1177/0170840602235001

Cabrera, E. F., & Cabreta, A. (2005). Fostering knowledge sharing through people management practices. *International Journal of Human Resource Management*, *16*(5), 720–735. doi:10.1080/09585190500083020

Callender, C., & Cohen, J. (2005). There Is No Special Problem About Scientific Representation. *Theoria*, *55*, 67–81.

Callender, C., & Cohen, J. (2006). There is No Special Problem About Scientific Representation. *Theoria*, *21*(1), 67–85.

Calvani, A., Bonaiuti, G., & Fini, A. (2008). Lifelong Learning: quale ruolo può svolgere l'e-learning 2.0? *Journal of e-Learning and Knowledge Society, 4,* 47-56.

Cameron, A., & Askew, N. (Eds.). (2011). *EUSeaMap - Preparatory Action for development and assessment of a European broad-scale seabed habitat map final report*. Retrieved January 27, 2014, from http://jncc.gov.uk/euseamap

Campbell, J. L., Rustad, L. E., Porter, J. H., Taylor, J. R., Dereszynski, E. W., & Shanley, J. B. et al. (2013). Quantity is Nothing without Quality: Automated QA/QC for Streaming Environmental Sensor Data. *Bioscience*, *63*(7), 574–585. doi:10.1525/bio.2013.63.7.10

Cantoni, L., Botturi, L., & Succi, C. (2007). eLearning, Capire, progettare, comunicare. Milano, Italy: Franco Angeli.

Carbonell, B. (2009). The Syntax of Objects and the Representation of History. *History and Theory*, *47*(2), 122–137. doi:10.1111/j.1468-2303.2009.00501.x

Carbotte, S., Arko, R., Ferrini, V., Stocks, K., Miller, S., Maffei, A., Chandler, C., Smith, S., & Bourassa, M. (2013). Rolling Deck to Repository (R2R): Working with international partners to improve data discovery and access. *Bollettino di Geofisica Teorica ed Applicata*, *54*(Supplement 1), 38-39.

Carnap, R. (1947). Meaning and Necessity. Chicago, IL: The University of Chicago Press.

Carnap, R. (1942). *Introduction to semantics*. Cambridge, MA: Harvard University Press.

Carnap, R. (1947). *Meaning and Necessity: A Study in Semantics and Modal Logic*. University of Chicago Press.

Carusi, A., & Torsten, R. (2010). *Virtual Research Environment collaborative landscape study*. Bristol: JISC.

Castelli, D., Manghi, P., & Thanos, C. (2013). A vision towards Scientific Communication Infrastructures. *International Journal on Digital Libraries*, *13*(3-4), 155–169. doi:10.1007/s00799-013-0106-7

Castronova, A. M., Goodall, J. L., & Ercan, M. B. (2013). Integrated modeling within a Hydrologic Information System: An OpenMI based approach. *Environmental Modelling & Software*, *39*(0), 263–273. doi:10.1016/j.envsoft.2012.02.011

Cawyer, C. S., Simons, C., & Davis, S. (2002). Mentoring to facilitate socialization: The case of the new faculty member. *International Journal of Qualitative Studies in Education*, *15*(2), 225–242. doi:10.1080/09518390110111938

Chadwick, A., Arts, R., Bernstone, C., May, F., Thibeau, S., & Zweigel, P. (2008). *Best Practice for the Storage of CO2 in Saline Aquifers. In Observations and guidelines from the SACS and CO2STORE projects*. Amersham: British Geological Survey by Halstan & Co. Ltd.

Chakravartty, A. (1995). Truth and Representation in Science: Two Inspirations from Art. In R. Frigg & M. Hunter (Eds.), *Beyond Mimesis and Convention: Representation in Art and Science*, (pp. 33-50). Berlin: Springer.

Chandler, C., Groman, R., Allison, D., Wiebe, P., & Glover, D. (2013). One approach to ensuring long-term access to ocean research data. *Bollettino di Geofisica teorica ed applicata, 54* (suppl. 1), 213-5.

Chandrasekaran, B., Josephson, J. R., & Benjamins, V. R. (1999). What are ontologies, and why do we need them? *Intelligent Systems and their Applications, 14*(1), 20-26.

Chao, G. T., Walz, P. M., & Gardner, P. D. (1992). Formal and informalmentorships: Acomparison on mentoring functions and contrast with nonmentored counterparts. *Personnel Psychology*, *45*(3), 619–636. doi:10.1111/j.1744-6570.1992.tb00863.x

Chave, A. D., Ampe, T., Arrott, M., Graybeal, J., James, M., & Meisinger, M., … Schofield O. (2011). Cyberinfrastructure for the US Ocean Observatories Initiative. Underwater Technology (UT), *2011 IEEE Symposium 2011 Workshop on Scientific Use of Submarine Cables and Related Technologies (SSC)*. doi:10.1109/UT.2011.5774109

Chen, X., Mark, B., Patrick, S., Robertson, M., Jon, K., & Wang, Z. (2013). Bibliometrical study method on tidal freshwater marshes loss and research orientation. *Acta Ecologica Sinica*, *33*(4), 237–240. doi:10.1016/j.chnaes.2013.05.001

Cheshire, C., & Cook, K. (2004). The emergence of trust networks under uncertainty - implications for internet interactions. *Analyse & Kritik*, *26*, 220–240.

Chilingarian, I., Cayatte, V., Revaz, Y., Dodonov, S., Durand, D., & Durret, F. et al. (2009). A Population of Compact Elliptical Galaxies Detected with the Virtual Observatory. *Science*, *326*(5958), 1379–1382. doi:10.1126/science.1175930 PMID:19797624

Chong, S. A. (2009). Mentoring: Are We Doing It Right? *Annals of the Academy of Medicine, Singapore*, *38*(7), 643–646. PMID:19652858

Chua, A. (2001). Relationship between the types of knowledge shared and types of communications channels used. *Journal of Knowledge Management Practice*. Retrieved from http://www.tlainc.com/articl26.htm

Churchland, P. M. (1979). *Scientific realism and the Plasticity of Mind*. Cambridge, UK: Cambridge University Press. doi:10.1017/CBO9780511625435

Cinque, M. (2010). *Agire creativo – Teoria, formazione e prassi dell'innovazione personale*. Milano, Italy: Franco Angeli.

Clark, D., Krumm, J. M., & Bieleski, J. (1992). Broker: A system integration approach. In *Proceedings of the Second International Conference on System Integration (ICSI'92)* (pp. 162-170). IEEE. doi:10.1109/ICSI.1992.217273

Claro, D. P., de Oliveira Claro, P. B., & Hagelaar, G. (2006). Coordinating collaborative joint efforts with suppliers: The effects of trust, transaction specific investments and information network in the Dutch flower industry. *Supply Chain Management*, *11*(3), 216–224. doi:10.1108/13598540610662112

Cluet, S., Delobel, C., Siméon, J., & Smaga, K. (1998). Your mediators need data conversion! *SIGMOD Record*, *27*(2), 177–188. doi:10.1145/276305.276321

Cogburn, D. L. (2003). HCI in the so-called developing world: What's in it for everyone. *Interaction*, *10*(2), 80–87. doi:10.1145/637848.637866

Cohen, S. G., & Bailey, D. R. (1997). What makes teams work: Group effectiveness research from the shop floor to the executive suite. *Journal of Management*, *23*(3), 239–290. doi:10.1177/014920639702300303

Coles, S. J., & Gale, P. A. (2012). Changing and Challenging Times for Service Crystallography. *Chemical Science*, *3*(3), 683–689. doi:10.1039/c2sc00955b

Collins, K. B., & Ison, R. L. (2010). Trusting emergence: Some experiences of learning about integrated catchment science with the Environment Agency of England and Wales. *Water Resources Management*, *24*(4), 669–688. doi:10.1007/s11269-009-9464-8

Conlon, T. (1997). Towards Diversity: Advancing Knowledge-based Modelling with Knowledge Acquisition. In *Proceedings of 8th International PEG Conference* (pp. 379-386). Sozopol.

Conner, L. G., Ames, D. P., & Gill, R. A. (2013). HydroServer Lite as an open source solution for archiving and sharing environmental data for independent university labs. *Ecological Informatics*, *18*, 171–177. doi:10.1016/j.ecoinf.2013.08.006

Corleto, A., & Tomassetti, V. (2015). Beyond Learning: E-learning Models for Scientific Research. In P. Diviacco, P. Fox, C. Pshenichny, & A. Leadbetter (Eds.), *Collaborative Knowledge in Scientific Research Networks*. Hershey, PA: IGI Global.

Cortner, H., & Moote, M. (1994). Trends and issues in land and water resources management: Setting the agenda for change. *Environmental Management*, *18*(2), 167–173. doi:10.1007/BF02393759

Costello, M. J. (2009). Motivating online publication of data. *Bioscience*, *59*(5), 418–427. doi:10.1525/bio.2009.59.5.9

Coward, R., & Ellis, J. (1977). *Language and Materialism: Developments in Semiology and the Theory of the Subject*. London: Routledge & Kegan Paul.

Cox, M. D. (1997). Long-term patterns in a mentoring program for junior faculty: Recommendations for practice. In D. DcZure (Ed.), To improve the academy (pp. 225–268). Stillwater, OK: New Forum Press and the Professional and Organizational Development Network in Higher Education.

Cox, S. (2012). *Best practice in formalizing a SKOS vocabulary*. Solid Earth and Environment GRID, Commonwealth Scientific and Industrial Organisation (CSIRO). Retrieved from https://www.seegrid.csiro.au/wiki/Siss/VocabularyFormalizationInSKOS

Cox, S. J. D., Simons, B. A., & Walker, G. (2011). Adoption of earth science informatics standards: Key factors. *Geophysical Research Abstracts*, *13*, 5407.

Cross, R., Parker, A., Prusak, L., & Borgatti, S. (2001). Knowing what we know: Supporting knowledge creation and sharing in social networks. *Organizational Dynamics*, *30*(2), 100–120. doi:10.1016/S0090-2616(01)00046-8

Currall, S. C., & Judge, T. A. (1995). Measuring trust between organizational boundary role persons. *Organizational Behavior and Human Decision Processes*, *64*(2), 151–170. doi:10.1006/obhd.1995.1097

Curtis, A., & Wood, R. (2004b). Geological Society of London Special Publication: Geological Prior Information: Informing Science and Engineering (vol. 239). Academic Press.

Curtis, A. (2012). The science of subjectivity. *Geology*, *40*(1), 95–96. doi:10.1130/focus012012.1

Curtis, A., & Wood, R. (2004a). *Optimal elicitation of probabilistic information from experts*. doi:10.1144/GSL.SP.2004.239.01.09

Dalkir, K. (2011). *Knowledge management in theory and practice* (2nd ed.). MIT.

Dame, R. F. (2005). Oyster reefs as complex ecological systems. In R. F. Dame & S. Olenin (Eds.), *The comparative roles of suspension-feeders in ecosystems* (pp. 331–343). Dordrecht, The Netherlands: Springer. doi:10.1007/1-4020-3030-4_19

Darwin, A., & Palmer, E. (2009). Mentoring circles in higher education. *Higher Education Research & Development, 28*(2), 125–136. doi:10.1080/07294360902725017

DataCite Metadata Working Group. (2011). *DataCite Metadata Schema for the Publication and Citation of Research Data*. V2.2. Retrieved from http://schema.datacite.org/meta/kernel-2.2/doc/DataCite-MetadataKernel_v2.2.pdf

Davenport, T. H., & Prusak, L. (1998). *Working Knowledge*. Boston, MA: Harvard Business School Press.

Davenport, T. H., & Prusak, L. (2000). *Working Knowledge: How Organizations Manage What They Know*. Boston, MA: Harvard Business School Press.

Davies, J., van Harmelen, F., & Fensel, D. (Eds.). (2002). *Towards the Semantic Web: Ontology-Driven Knowledge Management*. New York, NY: John Wiley & Sons, Inc. doi:10.1002/0470858060

Davis, J. H. (1982). Social interaction as a combinatorial process in group decision. In H. Brandstatter, J. H. Davis, & G. Stocker-Kreichgauer (Eds.), *Group decision making*. London: Academic Press.

Davis, J. H. (1992). Some compelling intuitions about group consensus decisions and internal aggregation phenomena: Selected examples, 1950–1990. *Organizational Behavior and Human Decision Processes. Special Issue: Group Decision Making, 52*, 3–38.

Davis, J. H. (1996b). Group decision making and quantitative judgements: A consensus model. In E. Witte & J. H. Davis (Eds.), *Understanding group behavior: Consensual action by small groups* (Vol. 1). Hillsdale, NJ: Erlbaum.

Davis, J. H., Au, W., Hulbert, T. L., Chen, X. P., & Zarnoth, P. (1997). The effects of group size and procedural influence on consensual judgements of quantity: The example of damage awards and mock civil juries. *Journal of Personality and Social Psychology, 73*(4), 703–718. doi:10.1037/0022-3514.73.4.703

Davis, J. H., Hulbert, L., & Au, W. T. (1996a). Procedural influence on group decision making: The case of straw polls—Observation and simulation. In R. Y. Hirokawa & M. S. Poole (Eds.), *Communication and group decision making*. Beverly Hills, CA: Sage Publications.

De Langhe, R. (2012). The Kuhnian paradigm. *Topoi, 32*(1), 65–73. doi:10.1007/s11245-013-9153-3

De Longueville, B. (2010). Community-based geoportals: The next generation? Concepts and methods for the geospatial Web 2.0. *Computers, Environment and Urban Systems, 34*(4), 299–308. doi:10.1016/j.compenvurbsys.2010.04.004

De Man, W. H. E. (2006). Understanding SDI; complexity and institutionalization. *International Journal of Geographical Information Science, 20*(3), 329–343. doi:10.1080/13658810500399688

de Saussure, F. (1983). *Course in General Linguistics* (R. Harris, Trans.). London: Duckworth. (Original work published 1916)

De Vries, J. (Ed.). (2005). *More than the sum of its parts: 10 years of the Leadership Development for Women Programme at UWA*. Perth, Australia: The University of Western Australia.

Degraer, S., Verfaillie, E., Willems, W., Adriaens, E., Vincx, M., & Van Lancker, V. (2008). Habitat suitability modelling as a mapping tool for macrobenthic communities: An example from the Belgian part of the North Sea. *Continental Shelf Research, 28*(3), 369–379. doi:10.1016/j.csr.2007.09.001

Del Moral Pérez, M. E., Cernea, D. A., & Villalustre Martinez, L. (2011). Scenari connettivisti e progetto di Learning Object adattabili alla diversità cognitiva. *TD Tecnologie Didattiche, 19*(2), 102–111.

del Rio, J., Toma, D. M., O'Reilly, T. C., Broring, A., Dana, D. R., & Bache, F. et al. (2013). Standards-based plug & work for instruments in ocean observing systems. *IEEE Journal of Oceanic Engineering, 99*, 1–14. doi:10.1109/JOE.2013.2273277

DeRose, S., Maler, E., Orchard, D., & Walsh, N. (2010). *XML linking language (XLink) version 1.1*. Retrieved 1/28/2014 from http://www.w3.org/TR/xlink11/#concepts

Det Norske Veritas. (2010). *CO2QUALSTORE Guideline for Selection and Qualification of Sites and Projects for Geological Storage of CO2*. Author.

DeTienne, K. B., Dyer, G., Hoopes, C., & Harris, S. (2004). Toward a model of effective knowledge management and directions for future research: Culture, leadership, and CKOs. *Journal of Leadership & Organizational Studies*, *10*(4), 26–43. doi:10.1177/107179190401000403

Detsky, A. S., & Barelocher, M. O. (2007). Academic Mentoring – How to Give It and How to Get It. *Journal of the American Medical Association*, *297*(19), 2134–2136. doi:10.1001/jama.297.19.2134 PMID:17507350

Dewulf, A., Francois, G., Pahl-Wostl, C., & Taillieu, T. (2007). A framing approach to cross-disciplinary research collaboration: Experiences from a large-scale research project on adaptive water management. *Ecology and Society*, *12*(2).

Diaz, L., Granell, C., Gould, M., & Huerta, J. (2011). Managing user-generated information in geospatial cyber-infrastructures. *Future Generation Computer Systems-the International Journal of Grid Computing-Theory Methods and Applications*, *27*(3), 304–314. doi:10.1016/j.future.2010.09.002

Dicheva, D., & Aroyo, L. (2004). Concepts and Ontologies in Web-based Educational Systems. *Int. J. Cont. Engineering Education and Life-long Learning*, *14*(3), 187–190.

Dicheva, D., & Dichev, C. (2005). Authoring Educational Topic Maps: Can We Make It Easier? In *Proceedings of 5th IEEE International Conference on Advanced Learning Technologies, ICALT 2005*, (pp.216-219). Kaohsiung, Taiwan: IEEE. doi:10.1109/ICALT.2005.73

Dicheva, D., & Dichev, C. (2007). Helping Courseware Authors to Build Ontologies: the Case of TM4L. In *Proceedings of 13th Int. Conference on Artificial Intelligence in Education* (pp. 77-84). Academic Press.

DiCicco-Bloom, B., & Crabtree, B. F. (2006). The qualitative research interview. *Medical Education*, *40*(4), 314–321. doi:10.1111/j.1365-2929.2006.02418.x PMID:16573666

Dietrich, F., & List, C. (2007). Judgment aggregation under constraints. In T. Boylan & R. Gekker (Eds.), *Economics, Rational Choice and Normative Philosophy*. London: Routledge.

Diviacco, P. (2012), Addressing Conflicting Cognitive Models in Collaborative E-Research: A Case Study in Exploration Geophysics in Collaborative and Distributed E-Research. In A.A. Juan, T. Daradoumis, M. Roca, S.E. Grasman, & J. Faulin (Eds.), Collaborative and Distributed E-Research: Innovations in Technologies, Strategies and Applications (pp. 247-275). Hershey, PA: IGI Global.

Diviacco, P. (2015). Reconciling Knowledge and Collaborative e-Research. In Collaborative Knowledge in Scientific Research Networks. Hershey, PA: IGI Global.

Diviacco, P., & Wardell, N. (2003). Reprocessing and dissemination of Antarctic Seismic Data. *Terra Antarctica Reports 2003, 9*, 133-136.

Diviacco, P., Busato, A. (2013). The Geo-Seas Seismic data viewer: a tool to facilitate control of data access. *Bollettino di Geofisica Teorica ed Applicata, 54* (3), 257-270.

Diviacco, P., Lowry, R., & Schaap, D. (2011). Marine seismic metadata for an integrated European scale data infrastructure: the FP7 Geo-Seas project. *Bollettino di Geofisica Teorica ed Applicata.* Retrieved 01/2013 from http://www.odip.org/documents/odip/downloads/23/bgta0051_diviacco.pdf

Diviacco, P. (2005). An open source, web based, simple solution for seismic data dissemination and collaborative research. *Computers & Geosciences*, *31*(5), 599–605. doi:10.1016/j.cageo.2004.11.008

Diviacco, P. (2012). Addressing Conflicting Cognitive Models in Collaborative E-Research: A Case Study in Exploration Geophysics. In *Collaborative and Distributed E-Research: Innovations in Technologies, Strategies and Applications*. Hershey, PA: IGI Global Press. doi:10.4018/978-1-4666-0125-3.ch012

Diviacco, P. (2015). Reconciling Knowledge and Collaborative e-Research. In *Collaborative Knowledge in Scientific Research Networks*. IGI Global Press.

Diviacco, P., & Busato, A. (2013, September). The Geo-Seas seismic data viewer: A tool to facilitate control of data access. *BGTA*, *54*(3), 257–270. doi:10.4430/bgta0078

Diviacco, P., & Busato, A. (2015). Maps, graphs and annotations as Boundary Objects in Knowledge Networks, Distributed Cognition and Collaborative e-Research. In P. Diviacco, P. Fox, C. Pshenichny, & A. Leadbetter (Eds.), *Collaborative Knowledge in Scientific Research Networks*. Hershey, PA: IGI Global.

Diviacco, P., Glaves, H., & Shaap, D. (2011). *Marine Seismic Metadata for an Integrated European Scale Data Infrastructure: The Fp7 Geo-Seas Project*. BGTA. Doi:10.4430/bgta0051

Diviacco, P., & Pshenichny, C. A. (2010). Concept-referenced spaces in computer-supported collaborative work. In *Geophysical Research Abstracts* (Vol. 12). Vienna, Austria: EGU.

Diviacco, P., & Pshenichny, C. A. (2011). A case study on the use of event bushes as a formal rep- resentation for computer supported collaborative work in the geosciences. In *Geophysical Research Abstracts* (Vol. 12). Vienna, Austria: EGU.

Diviacco, P., Pshenichny, C., Carniel, R., Khrabrykh, Z., Shterkhun, V., & Mouromtsev, D. et al. (Submitted to). Organization of a geophysical information space by using an event-bush-based collaborative tool. *Earth Science Informatics*.

Dolan, M., Thorsnes, T., Leth, J., Alhamdani, Z., Guinan, J., & Van Lancker, V. (2013). Standardisation and harmonisation in seabed habitat mapping: role and added value of geological data and information. Part B: Terrain characterisation. *Geo-Seas report 10.5B*. Retrieved on April 7, 2013, from http://www.geo-seas.eu/content/content.asp?menu=0040028_000000

Douglas, C. A. (1997). *Formal mentoring programs in organizations: An annotated bibliography*. Greensboro, NC: Center for Creative Leadership.

Dreher, G. F., & Ash, R. A. (1990). A comparative study of mentoring among men and women in managerial, professional, and technical positions. *The Journal of Applied Psychology*, *75*(5), 539–546. doi:10.1037/0021-9010.75.5.539

Duerr, R. E., Downs, R. R., Tilmes, C., Barkstrom, B., Lenhardt, W. C., & Glassy, J. et al. (2011). On the utility of identification schemes for digital earth science data: An assessment and recommendations. *Earth Science Informatics*, *4*(3), 139–160. doi:10.1007/s12145-011-0083-6

Duffy, T., & Cunningham, D. (1996). Constructivism: Implications for the design and delivery of instruction. In Handbook of research for educational telecommunications and technology (pp. 170-198). Academic Press.

Duffy, T. M., & Jonassen, D. H. (1992). *Constructivist and the technology of instruction: A conversation*. Hillsdale, NJ: Lawrence Erlbaum Associates.

Duhem, P., & Wiener, P. P. (1954). La Théorie Physique: Son Objet et sa Structure [The Aim and Structure of Physical Theory]. Princeton University Press.

Dumbill, E. (2013). Making sense of big data. *Big Data*, *1*(1), 1–2. doi:10.1089/big.2012.1503

Dumontier, M., & Wild, D. (2012). Linked Data in Drug Discovery. *IEEE Internet Computing*, *16*(6), 68–71. doi:10.1109/MIC.2012.122

Dyck, B., Starke, F. A., Mischke, G. A., & Mauws, M. (2005). Learning to Build a Car: An Empirical Investigation of Organizational Learning. *Journal of Management Studies*, *42*(2), 387–416. doi:10.1111/j.1467-6486.2005.00501.x

Dyer, J. H., & Nobeoka, K. (2000). Creating and managing a high-performance knowledge-sharing network: The Toyota case. *Strategic Management Journal*, *21*(3), 345–367. doi:10.1002/(SICI)1097-0266(200003)21:3<345::AID-SMJ96>3.0.CO;2-N

Eco, U. (1976). *A Theory of Semiotics*. London: Macmillan.

Eco, U. (1979). *A Theory of Semiotics*. Indiana University Press.

Edwards, P. N. (2010). *A Vast Machine: Computer Models, Climate Data, and Politics of Global Warming*. MIT Press.

Edwards, P. N., Jackson, S. J., Chalmers, M. K., Bowker, G. C., Borgman, C. L., & Ribes, D. et al. (2013). *Knowledge Infrastructures: Intellectual Frameworks and Research Challenges*. Ann Arbour, MI: Deep Blue.

Edwards, P. N., Mayernik, M. S., Batcheller, A. L., Bowker, G. C., & Borgman, C. L. (2011). Science friction: Data, metadata, and collaboration. *Social Studies of Science*, *41*(5), 667–690. doi:10.1177/0306312711413314 PMID:22164720

Ehrich, L. C., Hansford, B., & Tennent, L. (2004). Formal mentoring programs in education and other professions: A review of the literature. *Educational Administration Quarterly*, *40*(4), 518–540. doi:10.1177/0013161X04267118

Eldredge, N., & Gould, S. J. (1972). Punctuated equilibria: an alternative to phyletic gradualism. In *Models in paleobiology* (pp. 82–115). San Francisco: Freeman Cooper.

Eliot, T. S. (1935). The rock. In *The complete Poems and Plays 1909-1950*. San Diego, CA: Harcourt Brace & Company.

Eliseev, F. I. (2003). *S Kornilovskim konnim*. Moscow: Astrel Press. Retrieved 18 March 2014 from http://www.dk1868.ru/history/s_korn_konn1.htm (in Russian).

Engelhardt, W., & Zimmerman, J. (1982). Theory of Earth Science. New York: Cambridge.

Engelhardt, W., & Zimmerman, J. (1982). *Theory of earth science*. Cambridge, UK: Cambridge University Press.

Ensher, E. A., Thomas, C., & Murphy, S. E. (2001). Comparison of traditional, step-ahead, and peer mentoring on protégés' support, satisfaction, and perceptions of career success: A social exchange perspective. *Journal of Business and Psychology*, *15*(3), 419–438. doi:10.1023/A:1007870600459

Eppler, M. J., & Bresciani, S. (2013). Visualization in management: From communication to collaboration: A response to Zhang. *Journal of Visual Languages and Computing*, *24*(2), 146–149. doi:10.1016/j.jvlc.2012.11.003

Eppler, M. J., & Burkhard, R. A. (2008). Knowledge Visualization. In M. Jennex (Ed.), *Knowledge Management: Concepts, Methodologies, Tools, and Applications* (pp. 781–793). Hershey, PA: Information Science Reference.

Esposito, G., & Mantese, G. (2003). *E-learning: una guida operativa*. Milano, Italy: Franco Angeli.

Essential Science Indicators. (n.d.). Retrieved September 28, 2013, from http://www.accesowok.fecyt.es/

European Committee for Standardization (CEN). (2014). *European Guide to good Practice in Knowledge Management* (CWA 14924 report). Retrieved January, 2014, from http://www.cen.eu/cen/Sectors/Sectors/ISSS/CWAdownload/Pages/Knowledge%20Management.aspx

European Union. (2007, April 25). Directive 2007/2/EC of the European Parliament and of the Council of 14 March 2007 establishing an Infrastructure for Spatial Information in the European Community (INSPIRE). *Official Journal of the European Union, 50*.

Fagenson, E. A. (1989). The mentor advantage: Perceived career/job experiences of protégés versus non- protégés. *Journal of Organizational Behavior*, *10*(4), 309–320. doi:10.1002/job.4030100403

Falconer, L. (2006). Organizational learning, tacit information, and e-learning: A review. *The Learning Organization*, *13*(2), 140–151. doi:10.1108/09696470610645476

Falkenmark, M. (2001). The greatest water problem: The inability to link environmental security, water security and food security. *International Journal of Water Resources Development*, *17*(4), 539–554. doi:10.1080/07900620120094073

Farjoun, M. (2010). Beyond Dualism: Stability and Change As a Duality. *Academy of Management Review*, *35*(2), 202–225. doi:10.5465/AMR.2010.48463331

Farmer, W. (2008). The seven virtues of simple type theory. *Journal of Applied Logic*, *6*(3), 267–286. doi:10.1016/j.jal.2007.11.001

Feigenbaum, E. A. (1984). Knowledge engineering: the applied side of artificial intelligence. In *Proceedings of a Symposium on Computer Culture: The Scientific, Intellectual, and Social Impact of the Computer.* New York, NY: New York Academy of Sciences. doi:10.1111/j.1749-6632.1984.tb16513.x

Feyerablend, P. (1975). *Against Method*. London: New Left Books.

Fillenbaum, S. (1959). Some stylistic aspects of categorizing behavior. *Journal of Personality*, *27*(2), 187–195. doi:10.1111/j.1467-6494.1959.tb01828.x PMID:13665498

Fini, A. (2008). E-Learning 2.0: natura e sviluppo di una comunità. Esame di un caso. *Journal of e-Learning and Knowledge Society, 4*, 61 – 69.

Flanagin, A. J. (2002). The elusive benefits of the technological support of knowledge management. *Management Communication Quarterly, 16*(2), 242–248. doi:10.1177/089331802237237

Flannery, D., Matthews, B., Griffin, T., Bicarregui, J., Gleaves, M. L., ... Kleese, K. (2009). ICAT: Integrating Data Infrastructure for Facilities Based Science. In *Proceedings of e-Science: Fifth IEEE International Conference on e-Science*. IEEE.

Flemming, B. W. (2000). A revised textural classification of gravel-free muddy sediments on the basis of ternary diagrams. *Continental Shelf Research, 20*(10-11), 1125–1137. doi:10.1016/S0278-4343(00)00015-7

Fodor, J. (1983). *The Modularity of Mind*. Cambridge, MA: MIT Press.

Fogarty, M., & Powell, T. (2002). An overview of the U.S. GLOBEC Program. *Oceanography (Washington, D.C.), 15*(2), 4–12. doi:10.5670/oceanog.2002.17

Folk, R. L. (1954). The distinction between grain size and mineral composition in sedimentary-rock nomenclature. *The Journal of Geology, 62*(4), 344–359. doi:10.1086/626171

Fonesca, F., Davis, C., & Camara, G. (2003). Bridging ontologies and conceptual schemas in geographic information integration. *GeoInformatica, 7*(4), 355–378. doi:10.1023/A:1025573406389

Ford, J.R., Pharaoh, T.C., Hulbert, A.G., Voncent, C.J. & Cooper, A.H. (2009). CASSEM Work Package One – British East Midlands and Lincolnshire Geological modelling. *British Geological Survey Commissioned Report*, CR/08/152. Available on request.

Foucault, M. (2002). Archaeology of Knowledge. London: Routledge.

Fox, P. (2013a). *How environmental informatics is preparing us for the era of Big Data*. Paper presented at 2013 Fall Meeting, San Francisco, CA.

Fox, P. (2013b). *Progress in open-world, integrative, collaborative science data platforms*. Paper presented at 2013 Fall Meeting. San Francisco, CA.

Fox, P., McGuinness, D., Cinquini, L., West, P., Garcia, J., Benedict, J., & Middleton, D. (2009). Ontology-supported scientific data frameworks: The Virtual Solar-Terrestrial Observatory experience. *Computers & Geosciences, 35*(4), 724–738. doi:10.1016/j.cageo.2007.12.019

Frame, I., Austen, K. F., Calleja, M., Dove, M. T., White, T. O. H., & Wilson, D. J. (2009). New tools to support collaboration and virtual organizations. *Philosophical Transactions of the Royal Society A: Mathematical, Physical and Engineering Sciences, 367*(1890), 1051-1056.

Fraser, M. (2004). *Supporting Virtual Research Environments: How can we help?* Retrieved June 27, 2011 from http://users.ox.ac.uk/~mikef/rts/vre/img2.html

Fredericks, J. (2015). Persistence of knowledge across layered architectures. In P. Diviacco, P. Fox, A. Leadbetter, & C. Pshenichny (Eds.), *Collaborative Knowledge in Scientific Research Networks*. Hershey, PA: IGI Global Press.

Freeman, L. C. (1979). Centrality in Social Networks Conceptual Clarification. *Social Networks, 1*(3), 215–239. doi:10.1016/0378-8733(78)90021-7

Freer, J. E., Clark, M. P., Odoni, N. A., Coxon, G., McMillan, H. K., Souvigent, M., et al. (2012). *Understanding the main uncertainties in hydrological impact ensembles of regional climate models predictions for large catchments in the UK*. Paper presented at the American Geophysical Union 45th Annual Fall Meeting. San Francisco, CA.

Gabrielli, G. (2006). *Conoscenza, apprendimento, cambiamento. La gestione dei programmi di knowledge e change management*. Milano, Italy: Franco Angeli.

Gaeta, M., Orciuoli, F., & Ritrovato, P. (2009). Advanced ontology management system for personalised e-Learning. *Knowledge-Based Systems*, *22*(4), 292–301. doi:10.1016/j.knosys.2009.01.006

Gangemi, A., Catennaci, C., Ciaramita, M., & Lehman, J. (2006). Modelling ontology evaluation and validation. In Y. Sure & J. Domingue (Eds.), *The Semantic Web: Research and Applications, LNCS 4011* (pp. 140–154). Berlin: Springer-Verlag.

Gangné, M. (2009). A model of knowledge-sharing motivation. *Human Resource Management*, *48*(4), 571–589. doi:10.1002/hrm.20298

Gardiner, M. (1999). *Making a difference: Flinders University mentoring scheme for early career women researchers*. Adelaide, Australia: Flinders University Staff Development and Training Unit.

Gattuso, J. P., Dawson, N. A., Duarte, C. M., & Middelburg, J. J. (2005). Patterns of publication effort in coastal biogeochemistry: A bibliometric survey (1971 to 2003). *Marine Ecology Progress Series*, *294*, 9–22. doi:10.3354/meps294009

Gauker, C. (1995). *Thinking Out Loud*. Princeton, NJ: Princeton University Press.

Gavrilova, T. (2011). Knowledge Engineering: Structural View. In *New Trends in Software Methodologies, Tools, and Techniques Frontiers:Proceedings of the Tenth SoMet_11* (pp. 219 – 227). IOS Press.

Gavrilova, T. A., Ravodin, R. A., Bolotnikova, E. S., & Kotko, A. V. (2012). Collaborative development of Dermatovenerology Ontology. In *Proceedings of Int. conf. Med-e-Tel 2012: Global Telemedicine and eHealth Updates: Knowledge Resources*, (vol. 5, pp. 39-43). ISfTeH, G.D. of Luxembourg.

Gavrilova, T., Bolotnikova, E., & Gorovoy, V. (2012). New Ergonomic Metrics for Educational Ontology Design and Evaluation. In *New Trends in Software Methodologies, Tools, and Techniques Frontiers» in Artificial Intelligence and Applications series:Proceedings of the Tenth SoMet_12* (vol. 246, pp. 361-378). IOS Press.

Gavrilova, T., Bolotnikova, E., Leshcheva, I., Blagov, E., & Yanson, A. (2013). Measuring Psychological Impact on Group Ontology Design and Development: An Empirical Approach. In P. Klinov & D. Mouromtsev (Eds.), Knowledge Engineering and the Semantic Web (pp. 29-43). Heidelberg, Germany: Springer.

Gavrilova, T. A., & Voinov, A. (1996). Visualized Conceptual Structuring for Heterogeneous Knowledge Acquisition. In *Proceedings of International Conference on Educational Multimedia and Hypermedia, EDMEDIA'96* (pp. 258-264). Boston: MIT.

Gavrilova, T., Leshcheva, I., & Bolotnikova, E. (2012). Using visual conceptual models in teaching. In *Proceedings of the 8th International Conference on Education ICE-2012* (pp. 191–197). Research and Training Institute of East Aegean.

Geng, Q., Townley, C., Huang, K., & Zhang, J. (2005). Comparative knowledge management: A pilot study of Chinese and American universities. *Journal of the American Society for Information Science and Technology*, *56*(10), 1031–1044. doi:10.1002/asi.20194

Gentzen, G. (1969). Investigations into logical deduction. In M., Szabo (Ed.), The Collected Papers of Gerhard Gentzen (pp. 68-128). Amsterdam, The Netherlands: North-Holland (original work published 1934).

GEO. (2011). *Report of the Architecture and Data Committee (ADC), Document 19, GEO-VIII Plenary – 16-17 November 2011*. Retrieved from http://www.earthobservations.org/documents/geo-viii/19_Report%20of%20the%20Architecture%20and%20Data%20Committee%20(ADC).pdf

Geraci, A., Katiki, F., McMonegal, L., Meyer, B., Lane, J., Wilson, P., & Springsteel, F. (1991). *IEEE standard computer dictionary: Compilation of IEEE standard computer glossaries*. IEEE.

Giere, R. (2002). Distributed Cognition in Epistemic Cultures. *Philosophy of Science*, *69*(4), 637–644. doi:10.1086/344627

Giere, R. (2002, December). Discussion Note: Distributed Cognition in Epistemic Cultures. *Philosophy of Science*, 69.

Gilbert, M., & Cordey-Hayes, M. (1996). Understanding the process of knowledge transfer to achieve successful technological innovation. *Technovation*, *16*(6), 301–312. doi:10.1016/0166-4972(96)00012-0

Gill, A. (2009). Knowledge management initiatives at a small university. *International Journal of Educational Management*, *23*(7), 604–616. doi:10.1108/09513540910990834

Girault, C., & Pradat-Peyre, J.-F. (2009). High level Petri Nets. In M. Diaz (Ed.), *Petri Nets: Fundamental models, verification and applications*. Hoboken, NJ: Wiley.

Gläser, J. (2001). *Producing Communities as a Theoretical Challenge*. Paper presented at the TASA 2001 Conference. Sydney, Australia.

Gloria, M. J. K., McGuinness, D. L., Luciano, J. S., & Zhang, Q. (2013). Exploration in web science: instruments for web observatories. In *Proceedings of the 22nd international conference on World Wide Web companion*. Academic Press.

Glover, D., Chandler, C., Doney, S., Buesseler, K., Heimerdinger, G., Bishop, J., & Flierl, G. (2006). The US JGOFS data management experience. *Deep-sea Research. Part II, Topical Studies in Oceanography*, *53*(5–7), 793–802. doi:10.1016/j.dsr2.2006.01.018

Godfray, H. C. J., Beddington, J. R., Crute, I. R., Haddad, L., Lawrence, D., & Muir, J. F. et al. (2010). Food security: The challenge of feeding 9 billion people. *Science*, *327*(5967), 812–818. doi:10.1126/science.1185383 PMID:20110467

Godfrey-Smith, P. (2003). *Theory and reality*. Chicago: Chicago University Press. doi:10.7208/chicago/9780226300610.001.0001

Gogina, M., Glockzin, M., & Zettler, M. L. (2010). Distribution of benthic macrofaunal communities in the western Baltic Sea with regard to near-bottom environmental parameters. 2. Modelling and prediction. *Journal of Marine Systems*, *80*(1-2), 57–70. doi:10.1016/j.jmarsys.2009.10.001

Golder, S., & Huberman, B. A. (2005). *The structure of collaborative tagging systems*. HP Labs Technical Report. Retrieved august 2011 from http://www.hpl.hp.com/research/idl/pa- paers/tags

Gomez, B., Pearthree, G., & Cutcher-Gershenfeld, J. (2013). Cyber-infrastructure and synergistic opportunities across the geosciences. *Eos, Transactions, American Geophysical Union*, *94*(49), 473. doi:10.1002/2013EO490005

Gómez-Pérez, A., Fernández-López, M., & Corcho, O. (2004). *Ontological Engineering with examples from the areas of Knowledge Management, e-Commerce and the Semantic Web*. London: Springer-Verlag.

Goodman, N. (1976). *Language of Art: An Approach to a Theory of Symbols*. Indianapolis, IN: Hackett.

Goovaerts, P. (1997). *Geostatistics for natural resources evaluation*. New York: Oxford University Press.

Gordeyeva, I. (2010). *Enterprise 2.0: theoretical foundations of social media tools influence on knowledge sharing practices in organizations* (Master's Thesis). Retrieved from http://essay.utwente.nl/59921/1/MA_thesis_I_Gordeyeva.pdf

Gordon, B., & Grace, P. (2012). Emergent middleware: Tackling the interoperability problem. *Internet Computing*, *16*(1), 78–82. doi:10.1109/MIC.2012.7

Gorela, K., & Biloslavo, R. (2015). Relationship between Senior and Junior Researcher: Challenges & Opportunities for Knowledge Creating and Sharing. In P. Diviacco, P. Fox, C. Pshenichny, & A. Leadbetter (Eds.), *Collaborative Knowledge in Scientific Research Networks*. Hershey, PA: IGI Global.

Grassberger, P. (1995). The Bak-Sneppen model for punctuated equilibrium. *Physics Letters. [Part A]*, *200*(11), 277–282. doi:10.1016/0375-9601(95)00179-7

Gray, B. (2008). Enhancing transdisciplinary research through collaborative leadership. *American Journal of Preventive Medicine*, *35*(2), S124–S132. doi:10.1016/j.amepre.2008.03.037 PMID:18619392

Graybeal, J. (2009). *Choosing and Implementing Established Controlled Vocabularies*. Marine Metadata Interoperability Project. Retrieved April 16, 2014 from http://marinemetadata.org/guides/vocabs/cvchooseimplement

Graybeal, J., Isenor, A. W., & Rueda, C. (2011). Semantic mediation of vocabularies for ocean observing systems. *Computers & Geosciences*, *40*, 120–131. doi:10.1016/j.cageo.2011.08.002

Greene, S., Johnes, P. J., Freer, J., Bloomfield, J. P., Reaney, S. M., Macleod, C. J. A., & Odoni, N. A. (2012). *The Environmental Virtual Observatory (EVO): can cloud-based modelling provide new understanding of nutrient cycling processes from catchment to national scale?* Paper presented at the 17th International Nitrogen Workshop: Innovations for Sustainable use of Nitrogen Resources. Wexford, Ireland.

Greene, S., Johnes, P. J., Bloomfield, J. P., Reaney, S. M., Lawley, R., & Freer, J. et al. (Manuscript submitted for publication). Improving national capability in nutrient flux modelling using a high resolution geospatial framework. *Environmental Modelling & Software*.

Grice, H. P. (1989). *Studies in the way of words*. Cambridge, MA: Harvard University press.

Grosslight, L., Unger, C., Jay, E., & Smith, C. L. (1991). Understanding models and their use in science: Conceptions of middle and high school students and experts. *Journal of Research in Science Teaching*, *28*(9), 799–822. doi:10.1002/tea.3660280907

Groth, P., & Moreau, L. (2013). *An overview of the PROV family of documents*. World Wide Web Consortium. Retrieved from http://www.w3.org/TR/prov-overview/

Group on Earth Observations. (2005). *Global Earth Observation System of Systems GEOSS*. Noordwijk: ESA Publications Division.

Group on Earth Observations. (2013). *Report on Progress*. Geneva: Author.

Group on Earth Observations. (2014). *About GEO*. Retrieved from Group On Earth Observations: https://www.earthobservations.org/about_geo.shtml

Gruber, T. (1993). A translation approach to portable ontology specifications. *Knowledge Acquisition*, *5*(2), 199–220. doi:10.1006/knac.1993.1008

Gruber, T. (2008). Collective knowledge systems: Where the Social Web meets the Semantic Web. *Web Semantics: Science, Services, and Agents on the World Wide Web*, *6*(1), 4–13. doi:10.1016/j.websem.2007.11.011

Guarino, N., & Giaretta, P. (1998). Ontologies and Knowledge Bases: Towards a Terminological Clarification. In Towards Very Large Knowledge Bases: Knowledge Building & Knowledge Sharing (pp. 25-32). IOS Press.

Gupta, B., Iyer, L. S., & Aronson, J. E. (2000). Knowledge management: Practices and challenges. *Industrial Management & Data Systems*, *100*(1), 17–21. doi:10.1108/02635570010273018

Gurney, R., Emmett, B., McDonald, A., Blair, G., Buytaert, W., Freer, J., et al. (2011). *The environmental virtual observatory: a new vision for catchment science*. Paper presented at the American Geophysical Union, Fall Meeting. New York, NY.

Haberl, H., Winiwarter, V., Andersson, K., Ayres, R. U., Boone, C., & Castillo, A. et al. (2006). From LTER to LTSER: Conceptualizing the socioeconomic dimension of long-term socioecological research. *Ecology and Society*, *11*(2), 13.

Hackman, J. R., & Morris, C. G. (1975). In L. Berkowitz (Ed.), *Group tasks, group interaction process, and group performance effectiveness: A review and proposed integration*. In Advances in experimental social psychology (Vol. 8, pp. 45–99). New York: Academic Press.

Haldin-Herrgard, T. (2000). Difficulties in diffusion of tacit knowledge in organizations. *Journal of Intellectual Capital*, *1*(4), 357–365. doi:10.1108/14691930010359252

Hall, W., De Roure, D., & Shadbolt, N. (2009). The evolution of the Web and implications for eResearch. *Philosophical Transactions of the Royal Society a-Mathematical Physical and Engineering Sciences*, *367*(1890), 991-1001.

Halpin, H., Hayes, P., McCusker, J., McGuinness, D., & Thompson, H. (2010). When owl:sameAs isn't the same: an analysis of identity in Linked Data. *Lecture Notes in Computer Science*, *6496*, 305–320. doi:10.1007/978-3-642-17746-0_20

Hampton, S. E., Strasser, C. A., Tewksbury, J. J., Gram, W. K., Budden, A. E., & Batcheller, A. L. et al. (2013). Big data and the future of ecology. *Frontiers in Ecology and the Environment*, *11*(3), 156–162. doi:10.1890/120103

Hankin, S., Blower, J., Carval, T., Casey, K., Donlon, C., & Lauret, O. ... Harlan, J. (2010). NetCDF-CF-OPeNDAP: Standards for ocean data interoperability and object lessons for community data standards processes. In *Proceedings of OceanObs'09: Sustained Ocean Observations and Information for Society* (Vol. 2). Venice, Italy: ESA Publication WPP-306. doi:10.5270/OceanObs09.cwp.41

Hanson, N. R. (1958) Patterns of Discovery. Cambridge, UK: Cambridge University Press.

Hanson, P. C. (2007). A grassroots approach to sensor and science networks. *Frontiers in Ecology and the Environment*, *5*(7), 343–343. doi:10.1890/1540-9295(2007)5[343:AGATSA]2.0.CO;2

Haraway, D. (1988). Situated Knowledges: The Science Question in Feminism and the Privilege of Partial Perspective. *Feminist Studies*, *14*(3), 575–599. doi:10.2307/3178066

Haring, M. J. (1999). The case for a conceptual base for minority mentoring programs. *Peabody Journal of Education*, *74*(2), 5–14. doi:10.1207/s15327930pje7402_2

Harmon, S. G., Guo, Q., Hanson, P., Hofmann, A., Jones, B., Knoblock, C., et al. (2011). *ISI Technical Report Number ISI-TR-674 October, 2011*. Information Sciences Institute, University of Southern California.

Harris, P. T. (2012a). Surrogacy. In P. T. Harris & E. Baker (Eds.), *Seafloor geomorphology as benthic habitat: GeoHAB atlas of seafloor geomorphic features and benthic habitats* (pp. 93–108). London: Elsevier Insights. doi:10.1016/B978-0-12-385140-6.00005-0

Harris, P. T. (2012b). Biogeography, benthic ecology, and habitat classification schemes. In P. T. Harris & E. Baker (Eds.), *Seafloor geomorphology as benthic habitat: GeoHAB atlas of seafloor geomorphic features and benthic habitats* (pp. 61–92). London: Elsevier Insights. doi:10.1016/B978-0-12-385140-6.00004-9

Harris, R. J. (2009). Improving tacit knowledge transfer within SMEs through e-collaboration. *Journal of European Industrial Training*, *33*(3), 215–231. doi:10.1108/03090590910950587

Hayes, J., & Allinson, C. W. (1998). Cognitive style and the theory and practice of individual and collective learning in organizations. *Human Relations*, *51*(7), 847–871. doi:10.1177/001872679805100701

Head, K. H. (1980). *Manual of soil laboratory testing*. London: Pentech Press.

Healy, C. C., & Welchert, A. J. (1990). Mentoring Relations: A Definition to Advance Research and Practice. *Educational Researcher*, *19*(9), 17–21. doi:10.3102/0013189X019009017

Hendler, J., & Berners-Lee, T. (2010). From the Semantic Web to social machines: A research challenge for AI on the World Wide Web. *Artificial Intelligence*, *174*(2), 156–161. doi:10.1016/j.artint.2009.11.010

Hendriks, P. (1999). Why share knowledge? The influence of ICT on the motivation for knowledge sharing. *Knowledge and Process Management*, *6*(2), 91–100. doi:10.1002/(SICI)1099-1441(199906)6:2<91::AID-KPM54>3.0.CO;2-M

Hevner, A. R. (2007). The three cycle view of design science research. *Scandinavian Journal of Information Systems*, *19*(2), 87.

Hey, T., Tansley, S., & Tolle, K. (2009). *The Fourth Paradigm: Data-Intensive Scientific Discovery*. Microsoft Research.

Higgins, M. C. (2000). The more the merrier? Multiple developmental relationships and work satisfaction. *Journal of Management Development*, *19*(4), 277–296. doi:10.1108/02621710010322634

Higgins, M. C. (2001). Reconceptualizing mentoring at work: A developmental network perspective. *Academy of Management Rewiew*, *26*(2), 254–288.

Hilbert, D., & Bernays, P. (1934). *Crundlagen der Mathematik (Fundamentals of Mathematics)* (Vol. 1). Heidelberg, Germany: Springer-Verlag.

Hildrum, J. M. (2009). Sharing Tacit Knowledge Online: A Case Study of e-Learning in Cisco's Network of System Integrator Partner Firms. *Industry and Innovation*, *16*(2), 197–218. doi:10.1080/13662710902764360

Hinsz, V. B. (1990). Cognitive and consensus processes in group recognition memory performance. *Journal of Personality and Social Psychology*, *59*(4), 705–718. doi:10.1037/0022-3514.59.4.705

Hislop, D. (2002). Mission impossible? Communicating and sharing knowledge via information technology. *Journal of Information Technology*, *17*(3), 165–177. doi:10.1080/02683960210161230

Hislop, D. (2005). *Knowledge Management in Organizations*. Oxford: Oxford University Press.

Hobbie, J., Carpenter, S., Grimm, N., Gosz, J., & Seastedt, T. (2003). The US Long Term Ecological Research program. *Bioscience*, *53*(1), 21–32. doi:10.1641/0006-3568(2003)053[0021:TULTER]2.0.CO;2

Hoff, H., Iceland, C., Kuylenstierna, J., & te Velde, D. W. (2012). Managing the Water-Land-Energy Nexus for Sustainable Development. *UN Chronicle*, *49*(1&2).

Hollander, J. D., Paul, H., & Peters, R. (2011). Introduction: The Metaphor of Historical Distance. *History and Theory*, *50*(4), 1–10. doi:10.1111/j.1468-2303.2011.00599.x

Holmes, D. R., Hodgson, P. K., Simari, R. D., & Nishimura, R. A. (2010). Mentoring making the transition from mentee to mentor. *Circulation*, *121*(2), 336–340. doi:10.1161/CIRCULATIONAHA.108.798321 PMID:20083720

Holmes, J., & Clark, R. (2008). Enhancing the use of science in environmental policy-making and regulation. *Environmental Science & Policy*, *11*(8), 702–711. doi:10.1016/j.envsci.2008.08.004

Hooff, B. V. D., & Weenen, F. D. L. V. (2004). Committed to share: Commitment and CMC use as antecedents of knowledge sharing. *Knowledge and Process Management*, *11*(1), 13–24. doi:10.1002/kpm.187

Horsburgh, J. S., & Reeder, S. L. (2014). Data visualization and analysis within a Hydrologic Information System: Integrating with the R statistical computing environment. *Environmental Modelling & Software*, *52*(0), 51–61. doi:10.1016/j.envsoft.2013.10.016

Horsburgh, J. S., Tarboton, D. G., Maidment, D. R., & Zaslavsky, I. (2008). A relational model for environmental and water resources data. *Water Resources Research*, *44*(5), n/a. doi:10.1029/2007WR006392

Horsburgh, J. S., Tarboton, D. G., Maidment, D. R., & Zaslavsky, I. (2011). Components of an environmental observatory information system. *Computers & Geosciences*, *37*(2), 207–218. doi:10.1016/j.cageo.2010.07.003

Horsburgh, J. S., Tarboton, D. G., Piasecki, M., Maidment, D. R., Zaslavsky, I., Valentine, D., & Whitenack, T. (2009). An integrated system for publishing environmental observations data. *Environmental Modelling & Software*, *24*(8), 879–888. doi:10.1016/j.envsoft.2009.01.002

Hsia, T. L., Lin, L. M., Wu, J. H., & Tsai, H. T. (2006). A Framework for Designing Nursing Knowledge Management Systems. *Interdisciplinary Journal of Information. Knowledge and Management*, *1*, 14–21.

Hsu, M.-H., Ju, T., Yen, C.-H., & Chang, C.-M. (2007). Knowledge sharing behaviour in virtual communities: The relationship between trust, self-efficacy, and outcome expectations. *International Journal of Human-Computer Studies*, *65*(2), 153–169. doi:10.1016/j.ijhcs.2006.09.003

Hughes, R. I. G. (1997). Models and representation. *Philosophy of Science Supplement:Proceedings of the 1996 Biennial Meeting*, *64*, 325-336.

Humble, J., & Farley, D. (2010). *Continuous Delivery: Reliable Software Releases Through Build, Test, and Deployment Automation*. Addison Wesley.

Hussein, H. K., & El-Sattar, A. (2012). State of the Art in Interactive Storytelling Technology: An Approach Based on Petri Nets. In P. Pawlewski (Ed.), *Petri Nets - Manufacturing and Computer Science*. Rijeka, Croatia: InTech.

Hutchinson, G. E. (1957). Concluding remarks. *Cold Spring Harbor Symposia on Quantitative Biology*, *22*(0), 415–427. doi:10.1101/SQB.1957.022.01.039

I2S2 Partners. (2011). *I2S2 Idealised Scientific Research Activity Lifecycle Model*. Retrieved from http://www.ukoln.ac.uk/projects/I2S2/documents/I2S2-ResearchActivityLifecycleModel-110407.pdf

IBM. (1993). *IBM Dictionary of Computing*. New York: McGraw-Hill.

Inkpen, A. C., & Currall, S. C. (2004). The coevolution of trust, control, and learning in joint ventures. *Organization Science*, *15*(5), 586–599. doi:10.1287/orsc.1040.0079

Inkpen, A., & Dikur, I. (1998). Knowledge management processes and international joint ventures. *Organization Science*, *9*(4), 454–468. doi:10.1287/orsc.9.4.454

IPCC. (2012). *Cambio climático 2007: Informe de síntesis*. IPCC.

Ipe, M. (2003). Knowledge sharing in organizations: A conceptual framework. *Human Resource Development Review*, *3*(4), 337–359. doi:10.1177/1534484303257985

Isaac, A., & Summers, E. (2009). *SKOS Simple Knowledge Organization System Primer*. Retrieved 1/28/2014 from http://www.w3.org/TR/2009/NOTE-skos-primer-20090818

Iseler, R., & Bizer, C. (2013). Active learning of expressive linkage rules using genetic programming. *Journal of Web Semantics*, *23*, 2–15. doi:10.1016/j.websem.2013.06.001

ISIS. (2011). *ISIS Data Management Policy*. Retrieved from http://www.isis.stfc.ac.uk/user-office/data-policy11204.html

ISO IEC 15909-2. (2011). *Systems and software engineering: High-level petri nets: Part 2: Transfer format*. Retrieved August 10, 2014 from http://www.iso.org/iso/catalogue_detail.htm?csnumber=43538

ISO. (2003). *ISO 19115:2003 Geographic information -- Metadata*. Geneve: ISO.

ISO. (2005). *ISO 19119:2005 Geographic information -- Services*. Geneve: ISO.

Jackson, N. (2003). Introduction to brokering in higher education. In *Engaging and changing higher education through brokerage* (pp. 3–20). Ashgate.

Jackson, V. A., Palepu, A., Szalacha, L., Caswell, C., Carr, P. L., & Inui, T. (2003). Having the right chemistry: A qualitative study of mentoring in academic medicine. *Academic Medicine*, *78*(3), 328–334. doi:10.1097/00001888-200303000-00020 PMID:12634219

James, J. W. C., Pearce, B., Coggan, R. A., Arnott, S. H. L., Clark, R., Plim, J. F., et al. (2010). The south coast regional environmental characterisation. *British Geological Survey Open Report OR/09/51*.

Jensen, K. (1981). Coloured Petri nets and the invariant-method. *Theoretical Computer Science*, *14*(3), 317–336. doi:10.1016/0304-3975(81)90049-9

Jewitt, G. (2002). Can integrated water resources management sustain the provision of ecosystem goods and services? *Physics and Chemistry of the Earth*, *27*(11-22), 887-895.

Johannessen, J. A., Olaisen, J., & Olsen, B. (2001). Mismanagement of tacit knowledge: The importance of tacit knowledge, the danger of information technology, and what to do about it. *International Journal of Information Management*, *21*(1), 3–20. doi:10.1016/S0268-4012(00)00047-5

Johnsrud, L. K. (1990). Mentoring for administrative staff: Relationships that help and relationships that hinder. In K. M. Moore & S. B. Twombley (Eds.), *Administrative careers and the marketplace* (pp. 57–66). San Francisco: Jossey Bass.

Jonassen, D. H. (1999). Designing constructivist learning environments. In C. M. Reigeluth (Ed.), *Instructional design theories and models* (2nd ed.), (Vol. 2, pp. 215-239). Mahwah, NJ: Lawrence Erlbaum.

Jordan, P., Arnscheidt, A., McGrogan, H., & McCormick, S. (2007). Characterising phosphorus transfers in rural catchments using a continuous bank-side analyser. *Hydrology and Earth System Sciences*, *11*(1), 372–381. doi:10.5194/hess-11-372-2007

Jordan, P., Haygarth, P., Shortle, G., & Harris, R. (2012). Catchment science and policy for agriculture and water quality. *Environmental Science & Policy*, *24*, 1–3. doi:10.1016/j.envsci.2012.09.009

Jordà, S., Kaltenbrunner, M., Geiger, G., & Alonso, M. (2006). *The reacTable: a tangible tabletop musical instrument and collaborative workbench*. ACM SIGGRAPH. doi:10.1145/1179849.1179963

Juan, A., Daradoumis, T., Roca, M., Grasman, S. E., & Faulin, J. (2012). *Collaborative and distributed E-Research: Innovations in Technologies, Strategies and Applications* (pp. 1–419). Hershey, PA: IGI Global. doi:10.4018/978-1-46660-125-3

Kadlec, J., & Ames, D. (2011). Design and Development of Web Services for Accessing Free Hydrological Data from the Czech Republic. In J. Hřebíček, G. Schimak, & R. Denzer (Eds.), *Environmental Software Systems. Frameworks of eEnvironment* (Vol. 359, pp. 581–588). Springer Berlin Heidelberg. doi:10.1007/978-3-642-22285-6_63

Kagan, J. (1966). Reflection-impulsivity: The generality and dynamics of conceptual tempo. *Journal of Abnormal Psychology*, *71*(1), 17–24. doi:10.1037/h0022886 PMID:5902550

Kahneman, D., Slovic, P., & Tversky, A. (Eds.). (1982). *Judgement under Uncertainty: Heuristics and Biases*. Cambridge, UK: Cambridge University Press. doi:10.1017/CBO9780511809477

Kant, I. (1781). *The Critique of Pure Reason*. Retrieved from http://www.gutenberg.org/ebooks/4280

Kaplan, B., & Maxwell, J. A. (2005). Qualitative research methods for evaluating computer information systems. In J. G. Anderson & C. E. Aydin (Eds.), *Evaluating the Organizational Impact of Healthcare Information Systems* (2nd ed., pp. 30–55). New York: Springer. doi:10.1007/0-387-30329-4_2

Kaprow, A. (2003). *Essays on the Blurring of Art and Life*. University of California Press.

Katz, J. S., & Martin, B. R. (1997). What is research collaboration? *Research Policy*, *26*(1), 1–18. doi:10.1016/S0048-7333(96)00917-1

Kerr, N. L. (1992). Issue importance and group decision making. In S. Worchel, W. Wood, & J. A. Simpson (Eds.), *Group process and productivity* (pp. 68–88). Newbury Park, CA: Sage.

Keyser, D. J., Lakoski, J. M., Lara-Cinisomo, S., Schultz, D. J., Williams, V. L., Zellers, D. F., & Pincus, H. A. (2008). Advancing institutional efforts to support research mentorship: A conceptual framework and self-assessment tool. *Academic Medicine*, *83*(3), 217–225. doi:10.1097/ACM.0b13e318163700a PMID:18316865

Khan, O. J., & Jones, N. B. (2011). Harnessing tacit knowledge for competitive advantage: An internal social network approach. *Journal for International Business and Entrepreneurship Development*, *5*, 232–248. doi:10.1504/JIBED.2011.038031

Kholodnaya, M. (2004). *Kognitivnye stili. O prirode individual'nogo uma* (2nd ed.). Saint Petersburg: Piter. (in Russian)

Kim, P., Ferrin, D., Cooper, C., & Dirks, K. (2003). Removing the shadow of suspicion: The effects of apology versus denial for repairing competence-versus integrity-based trust violations. *The Journal of Applied Psychology*, *89*(1), 104–118. doi:10.1037/0021-9010.89.1.104 PMID:14769123

Kinchin, I. M., De-Leij, F. A. A. M., & Hay, D. B. (2005). The evolution of a collaborative concept mapping activity for undergraduate microbiology. *Journal of Further and Higher Education*, *29*(1), 1–14. doi:10.1080/03098770500037655

Kirchler, E., & Davis, J. H. (1986). The influence of member status differences and task type on group consensus and member position change. *Journal of Personality and Social Psychology*, *51*(1), 83–91. doi:10.1037/0022-3514.51.1.83

Kitcher, P. (1978). *The nature of mathematical knowledge*. Oxford, UK: Oxford University Press.

Kitzinger, J. (1995). Introducing Focus Groups. BMJ. *British Medical Journal*, *311*(29), 299–302. doi:10.1136/bmj.311.7000.299 PMID:7633241

Kleene, S. C. (1952). *Introduction to Mathematics*. New York: D. Van Nostrand Co., Inc.

Klein, K. L. (2011). *From History to Theory*. Berkeley, CA: University of California Press.

Knight, C., Gašević, D., & Richards, G. (2006). An Ontology-Based Framework for Bridging Learning Design and Learning Content. *Journal of Educational Technology & Society*, *9*(1), 23–37.

Knobel, C. (2007). *Understanding infrastructure: Dynamics, tensions, and design*. Academic Press.

Knorr Cetina, K. (1999). *Epistemic Cultures: How the Science make knowledge*. Cambridge, MA: Harvard University Press.

Koberg, C. S., Boss, R. W., Chappell, D., & Ringer, R. C. (1994). Correlates and consequences of protégé mentoring in a large hospital. *Group & Organization Management, 19*(2), 219–239. doi:10.1177/1059601194192007

Konert, M., & Vandenberghe, J. (1997). Comparison of laser grain size analysis with pipette and sieve analysis: A solution for the underestimation of the clay fraction. *Sedimentology, 44*(3), 523–535. doi:10.1046/j.1365-3091.1997.d01-38.x

Kotlarsky, J., & Oshri, I. (2005). Social ties, knowledge sharing and successful collaboration in globally distributed system development projects. *European Journal of Information Systems, 14*(1), 37–48. doi:10.1057/palgrave.ejis.3000520

Kowalczyk, S., & Shankar, K. (2011). Data sharing in the sciences. *Annual Review of Information Science & Technology, 45*(1), 247–294. doi:10.1002/aris.2011.1440450113

KPMG. (2000). *Knowledge management research report 2000*. KPMG Consulting. Retrieved April 5, 2011, from http://www.providersedge.com/docs/km_articles/KPMG_KM_Research_Report_2000.pdf

Kragt, M. E., Robson, B. J., & Macleod, C. J. A. (2013). Modellers' roles in structuring integrative research projects. *Environmental Modelling & Software, 39*(0), 322–330. doi:10.1016/j.envsoft.2012.06.015

Kramer, D. M., & Wells, R. P. (2005). Achieving Buy-In Building Networks to Facilitate Knowledge Transfer. *Science Communication, 26*(4), 428–444. doi:10.1177/1075547005275427

Kram, K. E. (1985). *Mentoring at work: Developmental relationships in organizational life*. Glenview, IL: Scott Foresman.

Kram, K. E., & Hall, D. T. (1996). Mentoring in a context of diversity and turbolence. In E. E. Kossek & S. Lobel (Eds.), *Managing diversity: Guman resource strategies for transforming the workplace* (pp. 108–136). Cambridge, MA: Blackwell.

Krasnov, P.N. (1922). *Na vnutrennem fronte* (vol. 1). Berlin, Germany: Archive of Russian Revolution (in Russian).

Kraus, K. (1909). *Sprüche und Widersprüche, Aphorismen*. Frankfurt, Germany: Bibliothek Suhrkamp.

Kubicek, H., & Cimander, R. (2009). Three dimensions of organizational interoperability. *European Journal of ePractice, 6*.

Kuhn, T. (1962). *The Structure of Scientific Revolutions*. Chicago: University of Chicago Press.

Kuhn, T. (1977). *The essential tension*. Chicago: Chicago University Press.

Kuhn, T. (2000). *The road since structure: Philosophical essays 1970-1993*. Chicago: Chicago University Press.

Kuhn, T. S. (1962). *The structure of scientific revolutions*. Chicago, IL: University of Chicago Press.

Kuklick, B. (1969). The Mind of the Historian. *History and Theory, 8*(3), 313–331. doi:10.2307/2504139

Kumpati, M. (1985). *US Patent 4774661*. Washington, DC: US Patent Office.

Kunkel, R., Sorg, J., Eckardt, R., Kolditz, O., Rink, K., & Vereecken, H. (2013). TEODOOR: A distributed geodata infrastructure for terrestrial observation data. *Environmental Earth Sciences, 69*(2), 507–521. doi:10.1007/s12665-013-2370-7

Kutepov, A. P. (1934). *General Kutepov*. Sbornik statei. Retrieved 18 March 2014 from http://www.dk1868.ru/history/Kutep2.htm (in Russian).

Kuznetsov, S. O., Obiedkov, S., & Roth, C. (2007) Reducing the Representation Complexity of Lattice-Based Taxonomies. In PrissU.PolovinaS.HillR. (Eds.), *Proc. 15th International Conference on Conceptual Structures (ICCS 2007) (LNAI), (vol. 4604*, pp. 241-254). Springer. doi:10.1007/978-3-540-73681-3_18

Lai, I. L. A. (2005). Knowledge management for Chinese medicines: A conceptual model. *Information Management & Computer Security, 133*(3), 244–255. doi:10.1108/09685220510602059

Lakatos, I. (1970). Falsification and the methodology of scientific research programmes. In Criticism and the Growth of Knowledge. Cambridge, UK: Cambridge University Press. doi:10.1017/CBO9781139171434.009

Lam, A. (2000). Tacit knowledge, organizational learning and societal institutions: An integrated framework. *Organization Studies*, *21*(3), 487–513. doi:10.1177/0170840600213001

Langan, S. J., Wade, A. J., Smart, R., Edwards, A. C., Soulsby, C., & Billett, M. F. et al. (1997). The prediction and management of water quality in a relatively unpolluted major Scottish catchment: Current issues and experimental approaches. *The Science of the Total Environment*, *194*, 419–435. doi:10.1016/S0048-9697(96)05380-6

Lark, R. M., Dove, D., Green, S. L., Richardson, A. E., Stewart, H., & Stevenson, A. (2012). Spatial prediction of seabed sediment texture classes by cokriging from a legacy database of point observations. *Sedimentary Geology*, *281*, 35–49. doi:10.1016/j.sedgeo.2012.07.009

Lassi, M., & Sonnenwald, D. H. (2010). *Identifying factors that may impact the adoption and use of a social science collaboratory: A synthesis of previous research*. Academic Press.

Lassoued, Y., Leadbetter, A., Clements, O., Grant, M. (2011). *NETMAR semantic framework specification – Version 1*. NETMAR: Open Service Network for Marine Environmental Data. European Commission within the Seventh Framework Programme (2007-2013) Theme ICT 2009.6.4 Project 249024. Deliverable 4.3.1.

Latour, B., & Woolgar, S. (1979). *Laboratory life: The construction of scientific facts*. Princeton, NJ: Princeton University Press.

Laughlin, P. R. (1988). Collective induction: Group performance, social combination processes, and mutual majority and minority influence. *Journal of Personality and Social Psychology*, *54*(2), 254–267. doi:10.1037/0022-3514.54.2.254

Laughlin, P. R. (1992). Influence and Performance in Simultaneous Collective and Individual Induction. *Organizational Behavior and Human Decision Processes*, *51*(3), 447–470. doi:10.1016/0749-5978(92)90021-X

Lave, J., & Wenger, E. (1991). *Situated Learning: Legitimate Peripheral Participation*. Cambridge, UK: Cambridge University Press. doi:10.1017/CBO9780511815355

Lawrence, B., Jones, C., Matthews, B., Pepler, S., & Callaghan, S. (2011). Citation and peer review of data: Moving towards formal data publication. *Int. Journal of Digital Curation*, *6*(2), 4–37. doi:10.2218/ijdc.v6i2.205

Lawson, C., & Lorenz, E. (1999). Collective learning, tacit knowledge and regional innovative capacity. *Regional Studies*, *33*(4), 305–317. doi:10.1080/713693555

Leadbetter, A. (2015). Examining trust in collaborative research networks. In P. Diviacco, P. Fox, C. Pshenichny, & A. Leadbetter (Eds.), *Collaborative Knowledge in Scientific Research Networks*. Hershey, PA: IGI Global.

Leadbetter, A., Arko, R., Chandler, C., Shepherd, A., & Lowry, R. (2013a). Linked Data: An oceanographic perspective. *Journal of Ocean Technology*, *8*(3), 7–12.

Leadbetter, A., Arko, R., Chandler, C., Shepherd, A., & Lowry, R. (2015). Loose integration of local information to generate collaborative marine science knowledge. In *Collaborative Knowledge in Scientific Research Networks*. IGI Global Press.

Leadbetter, A., Lowry, R., & Clements, O. (2013b). Putting meaning into NETMAR – the open service network for marine environmental data. *International Journal of Digital Earth*. doi:10.1080/17538947.2013.781243

Leadbetter, A., Raymond, L., Chandler, C., Pikula, L., Pissiersens, P., & Urban, E. (2013). *Ocean Data Publication Cookbook*. Paris: UNESCO.

Lee, B., Kwon, O., & Kim, H. (2011). Identification of dependency patterns in research collaboration environments through cluster analysis. *Journal of Information Science*, *37*(1), 67–85. doi:10.1177/0165551510392147

Lehrer, K., & Wagner, C. (1981). *Rational Consensus in Science and Society: A Philosophical and Mathematical Study*. Dordrecht: Reidel. doi:10.1007/978-94-009-8520-9

Leiserowitz, A., Maibach, E., Roser-Renouf, C., Smith, N., & Dawson, E. (2013). Climategate, public opinion, and the loss of trust. *The American Behavioral Scientist*, *57*(6), 818–837. doi:10.1177/0002764212458272

Levin, S. A., Grenfell, B., Hastings, A., & Perelson, A. S. (1997). Mathematical and computational challenges in population biology and ecosystems science. *Science*, *275*(5298), 334–343. doi:10.1126/science.275.5298.334 PMID:8994023

Lewontin, R. (1978). Adaptation. *Scientific American*, *239*(3), 155–169. doi:10.1038/scientificamerican0978-212 PMID:705323

Leydesdorff, L. (2004). Clusters and maps of science journals based on bi-connected graphs in Journal Citation Reports. *The Journal of Documentation*, *60*(4), 371–427. doi:10.1108/00220410410548144

Leydesdorff, L., & Rafols, I. (2009). A Global Map of Science Based on the ISI Subject Categories. *Journal of the American Society for Information Science and Technology*, *60*(2), 348–362. doi:10.1002/asi.20967

Li, J., Wang, M.-H., & Ho, Y.-S. (2011). Trends in research on global climate change: A Science Citation Index Expanded-based analysis. *Global and Planetary Change*, *77*(1), 13–20. doi:10.1016/j.gloplacha.2011.02.005

Likens, G. E., & Bormann, F. H. (1974). Linkages between terrestrial and aquatic ecosystems. *Bioscience*, *24*(8), 447–456. doi:10.2307/1296852

Lin, C., Kao, M., & Chang, K. (2010). Is more similar, better? Interacting effort of the cognitive-style congruency and tacitness of knowledge on knowledge transfer in the mentor-protègè dyad. *Asian Journal of Social Psychology*, *13*(4), 286–292. doi:10.1111/j.1467-839X.2010.01325.x

Lindmark, S. (2009). *Web 2.0: Where Does Europe Stand. JRC Scientific and Technical Reports, Institute for Prospectice Technological Studies (IPTS), Joint Research Center (JRC)*. Seville: European Commission.

List, C. (2005). Group knowledge and group rationality: A judgement aggregation perspective. *Episteme: A Journal of Social Epistemology, 2*(1), 25-38

List, C. (2005). The Probability of Inconsistencies in Complex Collective Decisions. *Social Choice and Welfare*, *24*(1), 3–32. http://personal.lse.ac.uk/list/PDF-files/inconsistencies.pdf doi:10.1007/s00355-003-0253-7

List, C., & Pettit, P. (2002). Aggregating Sets of Judgments: An Impossibility Result. *Economics and Philosophy*, *18*, 89–110.

Liu, H.-C., Lin, Q.-L., Mao, L.-X. L.-X., & Zhang, Z.-Y. (2013). Dynamic Adaptive Fuzzy Petri Nets for Knowledge Representation and Reasoning. *Systems, Man, and Cybernetics: Systems. IEEE Transactions*, *43*(6), 1399–1410.

Li, X., & Lara-Rosano, F. (2000). Adaptive fuzzy petri nets for dynamic knowledge representation and inference. *Expert Systems with Applications*, *19*(3), 235–241. doi:10.1016/S0957-4174(00)00036-1

LOD. (2014). Retrieved from http://linkeddata.org

Long, D. (2006). BGS detailed explanation of seabed sediment modified folk classification. In *MESH guide to marine habitat mapping*. Retrieved December 19, 2013, from http://www.searchmesh.net/default.aspx?page=1752

Long, J. C., Cunningham, F. C., & Braithwaite, J. (2013). Bridges, brokers and boundary spanners in collaborative networks: A systematic review. *BMC Health Services Research*, *13*(1), 158. doi:10.1186/1472-6963-13-158 PMID:23631517

Lopez-Ferrer, M. (2015). International funding and collaboration in Sea Level Rise research. In *Collaborative Knowledge in Scientific Research Networks*. IGI Global Press.

Lovett, G. M., Likens, G. E., Buso, D. C., Driscoll, C. T., & Bailey, S. W. (2005). The biogeochemistry of chlorine at Hubbard Brook, New Hampshire, USA. *Biogeochemistry*, *72*(2), 191–232. doi:10.1007/s10533-004-0357-x

Lowe, T. D., & Lorenzoni, I. (2007). Danger is all around: Eliciting expert perceptions for managing climate change through a mental models approach. *Global Environmental Change*, *17*(1), 131–146. doi:10.1016/j.gloenvcha.2006.05.001

Lumpkin, A. (2011, October). A model for mentoring university faculty. *The Educational Forum*, *75*(4), 357–368. doi:10.1080/00131725.2011.602466

Lundvall, B. A., & Johnson, B. (1994). The learning economy. *Journal of Industry Studies*, *1*(2), 22–42. doi:10.1080/13662719400000002

Lyall, C., Bruce, A., Tait, J., & Meagher, L. (2011). *Interdisciplinary Research Journeys: Practical Strategies for Capturing Creativity*. London: Bloomsbury. doi:10.5040/9781849661782

Lycan, W. (1996). *Consciousness and Experience*. Cambridge, MA: MIT Press.

MacKinnon, D., & Tetzlaff, D. (2009). Conceptualising scale in regional studies and catchment science–towards an integrated characterisation of spatial units. *Geography Compass*, *3*(3), 976–996. doi:10.1111/j.1749-8198.2009.00231.x

Macleod, C. J. A. (2010). *What can we learn from systems based approaches: from systems biology to earth systems science?* Paper presented at the 2010 International Congress on Environmental Modelling and Software, Modelling for Environment's Sake, Fifth Biennial Meeting. Ottawa, Canada.

Macleod, C. J. A., Blackstock, K. L., & Haygarth, P. M. (2008). Mechanisms to improve integrative research at the science-policy interface for sustainable catchment management. *Ecology and Society*, *13*(2).

Macleod, C. J. A., & Ferrier, R. C. (2011). Temperate grasslands in catchment systems: the role of scale, connectivity and thresholds in the provision and regulation of water quality and quantity. In G. Lemaire, J. Hodgson, & A. Chabbi (Eds.), *Grassland productivity and ecosystem services* (pp. 229–238). CABI. doi:10.1079/9781845938093.0229

Macleod, C. J. A., & Haygarth, P. M. (2010). Integrating water and agricultural management under climate change. *The Science of the Total Environment*, *408*(23), 5619–5622. doi:10.1016/j.scitotenv.2010.02.041 PMID:20304465

Macleod, C. J. A., Scholefield, D., & Haygarth, P. M. (2007). Integration for sustainable catchment management. *The Science of the Total Environment*, *373*(2-3), 591–602. doi:10.1016/j.scitotenv.2006.12.029 PMID:17239423

Maljers, D., Verfaillie, E., Gunnink, J., Van Heteren, S., & Van Lancker, V. (2006). Interpolating grain-size maps: transnational initiative between Belgium, Netherlands and UK. *MESH Technical Group Meeting*. Retrieved January 9, 2014, from http://www.searchmesh.net/default.aspx?page=1936

Manola, F., & Miller, E. (2004). *RDF Primer*. World Wide Web Consortium. Retrieved from http://www.w3.org/TR/rdf-primer/

Marcos, E., Vara, J. M., & de Castro, M. (2012). Author order: What science can learn from the arts. *Communications of the ACM*, *55*(9), 39–41. doi:10.1145/2330667.2330680

Margulis, L., Corner, J., & Hawthorne, B. (Eds.). (2007). *Ian McHarg: Conversations with students: Dwelling with nature*. Princeton Architectural Press.

Martí, J., Aspinall, W. P., Sobradelo, R., Felpeto, A., Geyer, A., & Ortiz, R. et al. (2008). A long-term volcanic hazard event tree for Teide-Pico Viejo stratovolcanoes (Tenerife, Canary Islands). *Journal of Volcanology and Geothermal Research*, *178*(3), 543–552. doi:10.1016/j.jvolgeores.2008.09.023

Martin Cantos, A. (2011). Journal and country rankings in instrumentation and ocean engineering. *Martech*. Retrieved from http://upcommons.upc.edu/e-prints/handle/2117/13401

Martin, J. S., & Marion, R. (2005). Higher education leadership roles in knowledge processing. *The Learning Organization*, *12*(5), 140–151. doi:10.1108/09696470510583520

Martin, P. (2002). Knowledge representation in CGLF, CGIF, KIF, Frame-CG and Formalized-English. In *Proceedings of ICCS'02, 10th International Conference on Conceptual Structures* (pp. 77-91). Borovets, Bulgaria: Springer. doi:10.1007/3-540-45483-7_7

Marwick, A. D. (2001). Knowledge management technology. *IBM Systems Journal*, *40*(4), 814–830. doi:10.1147/sj.404.0814

Matthews, B., Bunakov, V., & Jones, C. (2013). *Investigations as research objects within facilities science*. Paper presented at the 1st Workshop on Linking and Contextualizing Publications and Datasets. Malta.

Matthews, B., Kourousias, G., Yang, E., & Griffin, T. (2012). Model of the data continuum in Photon and Neutron Facilities. *PaNdata ODI, Deliverable D6.1*. Retrieved from http://pan-data.eu/sites/pan-data.eu/files/PaNdataODI-D6.1.pdf

Matthews, B., Sufi, S., Flannery, D., Lerusse, L., Griffin, T., Gleaves, M., & Kleese, K. (2009). *Using a Core Scientific Metadata Model in Large-Scale Facilities.* Paper presented at the 5th International Digital Curation Conference. London, UK.

Maul, P., Savage, D., Benbow, S., Walke, R. R., & Bruin, R. (2005). Development of a FEP database for the geological storage of carbon dioxide. Greenhouse Gas Control Technologies, 1, 701-709.

Mauz, I., Peltola, T., Granjou, C., Van Bommel, S., & Buijs, A. (2012). How scientific visions matter: Insights from three long-term socio-ecological research (LTSER) platforms under construction in Europe. *Environmental Science & Policy*, *19*, 90–99. doi:10.1016/j.envsci.2012.02.005

Mayer, R., Davis, J., & Schoorman, F. D. (1995). An integrative model of organizational trust. *Academy of Management Review*, *20*(3), 709–734. doi:10.5465/AMR.1995.9508080335

Mayfield, M. (2010). Tacit knowledge sharing: Techniques for putting a powerful tool in practice. *Development and Learning in Organizations*, *24*(1), 24–26. doi:10.1108/14777281011010497

Mayor, E., & Cordey-Hayes, M. (2000). Knowledge translation: A new perspective on knowledge transfer and foresight. *Foresight*, *2*(4), 411–423. doi:10.1108/14636680010802762

Mayo-Wilson, C., Zollman, K., & Danks, D. (2013). Wisdom of the crowds vs. groupthink: Learning in groups and isolation. *International Journal of Game Theory*, *42*(3), 695–723. doi:10.1007/s00182-012-0329-7

Mazella, G. (2015). Knowledge building and computer tools. In P. Diviacco, P. Fox, C. Pshenichny, & A. Leadbetter (Eds.), *Collaborative Knowledge in Scientific Research Networks*. Hershey, PA: IGI Global.

Mazzetti, P., Nativi, S., & Caron, J. (2009). RESTful implementation of geospatial services for Earth and Space Science applications. *International Journal of Digital Earth*, *2*(S1), 40–61. doi:10.1080/17538940902866153

McCormick, T., & Heaven, R. (2013). *Geoscientific vocabularies and Linked Data at the British Geological Survey – progress and pragmatism.* Paper presented at 2013 Fall Meeting, American Geophysical Union. San Francisco, CA.

McGonigle, D. F., Harris, R. C., McCamphill, C., Kirk, S., Dils, R., Macdonald, J., & Bailey, S. (2012). Towards a more strategic approach to research to support catchment-based policy approaches to mitigate agricultural water pollution: A UK case-study. *Environmental Science & Policy*, *24*, 4–14. doi:10.1016/j.envsci.2012.07.016

McGrath, J. E. (1984). *Groups: Interaction and performance*. Englewood Cliffs, NJ: Prentice Hall.

McGregor, D., & Cutcher-Gershenfeld, J. (2005). *The human side of enterprise*. McGraw-Hill (first published in 1960).

McGuinness, D. L. (2012). *Community science- the next frontier*. Paper presented at the AGU. San Francisco, CA.

McGuinness, D. (2003). Ontologies come of age. In D. Fensel, J. Hendler, H. Lieberman, & W. Wahlster (Eds.), *Spinning the Semantic Web: bringing the World Wide Web to its full potential* (pp. 171–196). Cambridge, MA: MIT.

Mejlgaard, N., Bloch, C., Degn, L., Ravn, T., & Nielsen, M. W. (2012). *Monitoring Policy and Research Activities on Science in Society in Europe (MASIS) Final synthesis report*. Bruxelles, Belgium: European Commission.

Merlin, P. M. (1974). *A Study of the Recoverability of Computing Systems*. (PhD Thesis). University of California, Irvine, CA.

Merton, R. K. (1973). *The Normative Structure of Science. In The Sociology of Science*. Chicago: The University of Chicago Press.

MESH Consortium. (2007). MESH guide to marine habitat mapping. *MESH, Mapping European Seabed Habitats*. Retrieved August 24, 2013, from http://www.searchmesh.net/default.aspx?page=1616

Mesner-Andolšek, D. (2011). Knowledge hoarding or sharing. In V. Dermol (Ed.), *Knowledge as business opportunity: Proceedings of the Management, Knowledge and Learning International Conference 2011*, (pp. 167-176). Celje, Slovenia: International School for Social and Business Studies.

Mesot, J. (2012). A need to rethink the business model of user labs? *Neutron News*, *23*(4), 2–3. doi:10.1080/10448632.2012.725324

Meyer, E. (2009). Moving from small science to big science: Social and organizational impediments to large scale data sharing. *e-Research. Transformation in Scholarly Practice*, *3*(2), 147–159.

Meyer, M. (2010). The rise of the knowledge broker. *Science Communication*, *32*(1), 118–127. doi:10.1177/1075547009359797

Meyers, H. W., & Smith, S. (1999). Coming home-mentoring new teachers: A school-university partnershipmto support the development of teachers from diverse ethnic backgrounds. *Peabody Journal of Education*, *74*(2), 75–89. doi:10.1207/s15327930pje7402_7

Michener, W., Porter, J., Servilla, M., & Vanderbilt, K. (2011). Long term ecological research and information management. *Ecological Informatics*, *6*(1), 13–24. doi:10.1016/j.ecoinf.2010.11.005

Miles, A., & Bechhofer, S. (2009). *SKOS simple knowledge organization system reference*. World Wide Web Consortium. Retrieved from http://www.w3.org/TR/2009/REC-skos-reference-20090818/

Milne, A. E., Macleod, C. J. A., Haygarth, P. M., Hawkins, J. M. B., & Lark, R. M. (2009). The wavelet packet transform: A technique for investigating temporal variation of river water solutes. *Journal of Hydrology (Amsterdam)*, *379*(1-2), 1–19. doi:10.1016/j.jhydrol.2009.09.038

Mintzberg, H. (1983). *Structure in Fives: Designing Effective Organizations*. Englewood Cliffs, NJ: Prentice-Hall.

Mizoguchi, R., & Bourdeau, J. (2000). Using Ontological Engineering to Overcome Common AI-ED Problems. *International Journal of Artificial Intelligence in Education*, *11*(2), 107–121.

Moller, K., & Svahn, S. (2004). Crossing east-west boundaries: Knowledge sharing in intercultural business networks. *Industrial Marketing Management*, *33*(3), 219–228. doi:10.1016/j.indmarman.2003.10.011

Montanari, A., Young, G., Savenije, H. H. G., Hughes, D., Wagener, T., & Ren, L. L. et al. (2013). "Panta Rhei—Everything Flows": Change in hydrology and society—The IAHS Scientific Decade 2013–2022. *Hydrological Sciences Journal*, 1–20.

Mooney, H. (2011). Citing data sources in the social sciences: Do authors do it? *Learned Publishing*, *24*(2), 99–108. doi:10.1087/20110204

Morgan, M. G., Pitelka, L. F., & Shevliakova, E. (2001). Elicitation of expert judgements of climate change impacts on forest ecosystem. *Climatic Change*, *49*(3), 279–307. doi:10.1023/A:1010651300697

Morgan, R. M., & Hunt, S. D. (1994). The commitmenttrust theory of relationship marketing. *Journal of Marketing*, *58*(3), 20–38. doi:10.2307/1252308

Morrison, E. W. (1993). Newcomer information seeking: Exploring types, modes, sources and outcomes. *Academy of Management Journal*, *36*(1), 557–589. doi:10.2307/256592

Mouromtsev, D. I., Pshenichny, C. A., & Yakovlev, A. V. (2014). Semantic and structural delineation of market scenarios by the event bush method. *Decision Support Systems and Electronic Commerce*, *57*, 34–41. doi:10.1016/j.dss.2013.07.008

Moya-Anegon, F., Vargas-Quesada, B., Herrero-Solana, V., Chinchilla-Rodriguez, Z., Corera-Alvarez, E., & Munoz-Fernandez, F. J. (2004). A new technique for building maps of large scientific domains based on the cocitation of classes and categories. *Scientometrics*, *61*(1), 129–145. doi:10.1023/B:SCIE.0000037368.31217.34

Murdock, J., Buckner, C., & Allen, C. (2010). Evaluating Dynamic Ontologies. In A. Fred, J.L.G. Dietz, K. Liu, & J. Filipe (Eds.), *Knowledge Discovery, Knowledge Engineering and Knowledge Management. Second International Joint Conference* (pp. 258-275). Valencia, Spain: Springer.

Murphy, G., & Salomone, S. (2013). Using social media to facilitate knowledge trabsfer in complex engineering environments: A primer for educators. *European Journal of Engineering Education*, *38*(1), 1–15. doi:10.1080/03043797.2012.742871

Murray, M. (1991). *Beyond the myths and magic of mentoring: How to facilitate an effective mentoring program*. San Francisco: Jossey-Bass.

Murray, S. R., & Peyrefitte, J. (2007). Knowledge Type and Communication Media Choice in the Knowledge Transfer Process. *Journal of Managerial Issues*, *19*(1), 111–133.

Muste, M. V., Bennett, D. A., Secchi, S., Schnoor, J. L., Kusiak, A., & Arnold, N. J. et al. (2013). End-to-End Cyberinfrastructure for Decision-Making Support in Watershed Management. *Journal of Water Resources Planning and Management*, *139*(5), 565–573. doi:10.1061/(ASCE)WR.1943-5452.0000289

Muste, M., Kim, D., Arnold, N., Whiteaker, T., Just, C., & Kruger, A. (2010). Digital catchment inception using community project components. *Proceedings of the ICE-Water Management*, *163*(1), 13–25. doi:10.1680/wama.2010.163.1.13

National Research Council. (2003). *Government data centers: meeting increasing demands*. Washington, DC: The National Academies Press. Retrieved 4/10/2014 from http://www.nap.edu/catalog/10664.html

National Research Council. (2012). *The future of scientific knowledge discovery in open networked environments: Summary of a workshop*. Washington, DC: The National Academies Press. Retrieved 4/10/2014 from http://www.nap.edu/catalog/18258.html

Nativi, S., & Bigagli, L. (2009). Discovery, mediation, and access services for Earth Observation data. *IEEE Journal of Selected Topics in Applied Earth Observations and Remote Sensing*, 233-240.

Nativi, S., Khalsa, S., Domenico, B., Craglia, M., Pearlman, J., Mazzetti, P., & Rew, R. (2011). *The brokering approach for earth science cyberinfrastructure*. NSF EarthCube White Paper.

Nativi, S., Caron, J., Domenico, B., & Bigagli, L. (2008). 09 01). Unidata's Common Data Model mapping to the ISO 19123 Data Model. *Earth Science Informatics*, *1*(2), 59–78. doi:10.1007/s12145-008-0011-6

Nativi, S., Craglia, M., & Pearlman, J. (2012). The brokering approach for multidisciplinary interoperability: A position paper. *International Journal of Spatial Data Infrastructures Research*, *7*, 1–15.

Nativi, S., Craglia, M., & Pearlman, J. L. (2013). Earth science infrastructures interoperability: The brokering approach. *IEEE Journal of Selected Topics in Applied Earth Observations and Remote Sensing*, *6*(3), 1118–1129. doi:10.1109/JSTARS.2013.2243113

Nativi, S., Mazzetti, P., Craglia, M., & Pirrone, N. (2013). The GEOSS solution for enabling data interoperability and integrative research. *Environmental Science and Pollution Research International*, 1–16. PMID:24243262

Nature. (2010). Editorial: Closing the climategate. *Nature, 468*, 345.

Navigli, R., & Velardi, P. (2008). From Glossaries to ontologies: extracting semantic structure from textual definitions. In P. Buitelaar & P. Cimiano (Eds.), *Ontology learning and population: bridging the gap between text and knowledge* (pp. 71–87). Amsterdam: IOS Press.

Neches, R., Fikes, R. E., Finin, T., Gruber, T., Patil, R., Senator, T., & Swartout, W. R. (1991). Enabling Technology for Knowledge Sharing. *AI Magazine*, *12*(3), 36–56.

Neon Project. (n.d.). Retrieved from http://www.neon-project.org/

Neri, A., Aspinall, W. P., Cioni, R., Bertagnini, A., Baxter, P. J., & Zuccaro, G. et al. (2008). Developing an Event Tree for probabilistic hazard and risk assessment at Vesuvius. *Journal of Volcanology and Geothermal Research*, *178*(3), 397–415. doi:10.1016/j.jvolgeores.2008.05.014

Newell, S., & Swan, J. (2000). Trust and inter-organizational networking. *Human Relations*, *53*(10), 287–328.

Newman, M. (2005). Power laws, Pareto distributions and Zipf's law. *Contemporary Physics*, *46*(5), 323–351. doi:10.1080/00107510500052444

Nielsen, M. A. (2012). *Reinventing Discovery: The New Era of Networked Science*. Princeton, NJ: Princeton University Press.

NOAA, National Coastal Data Development Center. (2013). *Introduction to Geospatial Metadata – ISO 191** Metadata*. Retrieved 4/28/2014 from ftp://ftp.ncddc.noaa.gov/pub/Metadata/Online_ISO_Training/Intro_to_ISO/presentations/2_ISO_101.pptx

Nonaka, I. (1994). A Dynamic theory of Organizational Knowledge Creation. *Organization Science*, *5*(1), 14–37. doi:10.1287/orsc.5.1.14

Nonaka, I., & von Krogh, G. (2009). Tacit Knowledge and Knowledge Conversion: Controversy and Advancement in Organizational Knowledge Creation Theory. *Organization Science*, *20*(3), 635–652. doi:10.1287/orsc.1080.0412

Novak, J. D., & Cañas, A. J. (2006). The Theory Underlying Concept Maps and How to Construct Them. *Technical Report IHMC CmapTools 2006-01*. Florida Institute for Human and Machine Cognition. Retrieved online December 12, 2006, from http://cmap.ihmc.us/Publications/ResearchPapers/TheoryUnderlyingConceptMaps.pdf

Novak, J. D. (1998). *Learning, Creating, and Using Knowledge: Concept Maps as Facilitative Tools in Schools and Corporations*. Mahwah, NJ: Lawrence Erlbaum & Assoc.

Noy, N., Chugh, A., Liu, W., & Musen, M. (2006). A Framework for Ontology Evolution in Collaborative Environments. In CruzI.DeckerS.AllemangD.PreistC.SchwabeD.MikaP.UscholdM.AroyoL. (Eds.), *Proceedings of the 5th International semantic Web Conference* (pp. 544-558). Athens, GA: Springer. doi:10.1007/11926078_39

NRC Ocean Studies Board. (2000). *50 years of ocean discovery: National Science Foundation 1950-2000*. Washington, DC: The National Academies Press.

O'Donnell, A. M., Dansereau, D. F., & Hall, R. H. (2002). Knowledge Maps as Scaffolds for Cognitive Processing. *Educational Psychology Review*, *14*(1), 71–86. doi:10.1023/A:1013132527007

O'Hagan, A., Buck, C. E., Daneshkhah, A., Eiser, J. R., Garthwaite, P. H., & Jenkinson, D. J. et al. (2006). *Uncertain Judgements: Eliciting Expert's Probabilities*. Wiley.

OAIS. (2012). Reference Model for an Open Archival Information System. *CCSDS 650.0-M-2 (Magenta Book) Issue 2, June 2012*. Retrieved from http://public.ccsds.org/publications/archive/650x0m2.pdf

Ohtsubo, Y., Masuchi, A., & Nakanishi, D. (2002). Scheme Model in a Group Polarization Context. Majority Influence Process in Group judgement: Test of the Social judgement. *Group Processes & Intergroup Relations*, *5*, 249–261. doi:10.1177/1368430202005003005

Olimpo, G. (2010). Società della conoscenza, educazione, tecnologia. *TD Tecnologie Didattiche*, *50*, 4–16.

Oliver, R. (2001). Developing e-learningenvironments that support knowledge construction in higher education. In Working for Excellence in the E-conomy (pp. 407-16). Churchlands: We-B Centre.

Olson, G. M., Zimmerman, A., & Bos, N. (2008). *Scientific Collaboration on the Internet*. Cambridge, MA: The MIT Press. doi:10.7551/mitpress/9780262151207.001.0001

Olson, J., Ellisman, M., James, M., Grethe, J., & Puetz, M. (2008). Biomedical informatics research network (BIRN). In G. Olson, A. Zimmerman, & N. Bos (Eds.), *Scientific Collaboration on the Internet*. Cambridge, MA: MIT Press. doi:10.7551/mitpress/9780262151207.003.0013

Olson, R. K., & Connelly, L. M. (1995). Mentoring through predoctoral fellowships to enhance research productivity. *Journal of Professional Nursing*, *11*(5), 270–275. doi:10.1016/S8755-7223(05)80007-9 PMID:7593970

Oltramari, A., & Ferrario, R. (2009). *Formal Ontologies Meet Industry*. Amsterdam: IOS Press.

Omran, E., Bregt, A., & Crompvoets, J. W. (2007). Spatial data sharing: A cross-cultural conceptual model. *Research and Theory in Advancing Spatial Data Infrastructure Concepts*, 79-97.

Online Etimology Dictionary. (n.d.). *Etimology for broker*. Retrieved from Online Etimology Dictionary: http://www.etymonline.com/index.php?allowed_in_frame=0&search=brocour&searchmode=none

OPSI. (1988). *Data Protection Act 1998*. London: Office of Public Sector Information. Retrieved from http://www.legislation.gov.uk/ukpga/1998/29/contents

Osborne, T. (2004). On mediators: Intellectuals and ideas trade in the knowledge society. *Economy and Society*, *33*(4), 430–447. doi:10.1080/0308514042000285224

Osimo, D. (2008). *Web 2.0 in government: Why and how (Scientific and Technical Reports, Institute for Prospectice Technological Studies [IPTS], Joint Research Center [JRC])*. Seville: European Commission.

Ostroff, C., & Kozlowski, S. W. (1993). The role of mentoring in the information gathering process of newcomers during early organizational socialization. *Journal of Vocational Behavior*, *42*(2), 170–183. doi:10.1006/jvbe.1993.1012

Outram, F., Lloyd, C., Jonczyk, J., Benskin, C. M., Grant, F., & Dorling, S. et al. (2013). High-resolution monitoring of catchment nutrient response to the end of the 2011-2012 drought in England, captured by the demonstration test catchments. *Hydrology and Earth System Sciences Discussions*, *10*(12), 15119–15165. doi:10.5194/hessd-10-15119-2013

Oxford University Press. (2014). *Definition of broker in Oxford dictionary (British & World English)*. (M. Proffitt, Ed.). Retrieved from Oxford Dictionaries: http://www.oxforddictionaries.com/us/definition/english/broker

Paczuski, M., Maslov, S., & Bak, P. (1996). Avalanche dynamics in evolution, growth, and depinning models. *Physical Review E: Statistical Physics, Plasmas, Fluids, and Related Interdisciplinary Topics*, *53*(1), 414–443. doi:10.1103/PhysRevE.53.414 PMID:9964272

Pahl-Wostl, C. (2007). Transitions towards adaptive management of water facing climate and global change. *Water Resources Management*, *21*(1), 49–62. doi:10.1007/s11269-006-9040-4

Pahl-Wostl, C., Lebel, L., Knieper, C., & Nikitina, E. (2012). From applying panaceas to mastering complexity: Toward adaptive water governance in river basins. *Environmental Science & Policy*, *23*(0), 24–34. doi:10.1016/j.envsci.2012.07.014

Panahi, S., Watson, J., & Partridge, H. (2012). Social media and tacit knowledge sharing: developing a conceptual model. World Academy of Science, Engineering and Technology, (64), 1095-1102.

PaNData collaboration. (2013). *Counting Users of the European Photon and Neutron Facilities 2012 Survey*. Retrieved from http://pan-data.eu/Users2012-Results

Panteli, N., & Sockalingam, S. (2005). Trust and conflict within virtual inter-organizational alliances: A framework for facilitating knowledge sharing. *Decision Support Systems*, *39*(4), 599–617. doi:10.1016/j.dss.2004.03.003

Papazoglou, M. P., Traverso, P., Dustar, S., & Leymann, F. (2007). Service-oriented computing: State of the art and research challenges. *Computer*, *40*(11), 38–45. doi:10.1109/MC.2007.400

Parsons, M. A. (2011). Making data useful for modelers to understand complex Earth systems. *Earth Science Informatics*, *4*(4), 197–223. doi:10.1007/s12145-011-0089-0

Parsons, M., & Fox, P. (2013). Is data publication the right metaphor? *Data Science Journal*, *12*(0), 32–46. doi:10.2481/dsj.WDS-042

Patel, V. L., Arocha, J. F., & Zhang, J. (2004). Thinking and reasoning in medicine. In *Cambridge handbook of thinking and reasoning*. Cambridge, UK: Cambridge University Press.

Pawlowski, S. D., Robey, D., & Raven, A. (2000). Supporting shared information systems: boundary objects, communities, and brokering. In *Proceedings of the twenty first international conference on Information systems*. Association for Information Systems.

Peirce, C. S. (1958). Collected Writings (8 Vols.). (C. Hartshorne, P. Weiss, & A. W. Burks, Eds.). Cambridge, MA: Harvard University Press.

Peirce, C. S. (1958). *The Collected Works of Charles Sanders Peirce*. Harvard University Press.

Pennington, D. D. (2008). Cross-Disciplinary Collaboration and Learning. *Ecology and Society*, *13*(2).

Pennington, D. D. (2011). Bridging the Disciplinary Divide: Co-Creating Research Ideas in eScience Teams. *Computer Supported Cooperative Work-the Journal of Collaborative Computing*, *20*(3), 165–196. doi:10.1007/s10606-011-9134-2

Pera, M. (1994). Scienza e Retorica. University of Chicago Press.

Perry, M., & Herring, J. (Eds.). (2012). *GeoSPARQL - A geographic query language for RDF data*. Open Geospatial Consortium. Retrieved April 16, 2014 from http://www.opengeospatial.org/standards/geosparql

Peterson, E. R., Rayner, S. G., & Armstrong, S. J. (2009). Researching the psychology of cognitive style and learning style: Is there really a future? *Learning and Individual Differences*, *19*(4), 518–523. doi:10.1016/j.lindif.2009.06.003

Petliura, S. V. (1918). Potreba viiskovoi literature. *Knigar*, *7*, 375–376.

Petri, C. A. (1962). *Kommunication mit Automaten*. (PhD Dissertation). University of Bonn, Bonn, Germany.

Petri, C. A. (1973). Concepts of Net Theory. In *Mathematical Foundations of Computer Science: Proceedings of Symposium and Summer School*. Bratislava, Slovakia: Mathematical Institute of the Slovak Academy of Sciences.

Petri, C. A. (1977). General Net Theory. In B. Shaw (Ed.), *Computing System Design:Proceedings of the Joint IBM University of Newcastle upon Tyne Seminar*. Newcastle Upon Tyne, UK: University of Newcastle upon Tyne.

Pettigrew, T. F. (1958). The measurement and correlates of category width as a cognitive variable. *Journal of Personality*, *26*(4), 532–544. doi:10.1111/j.1467-6494.1958.tb02350.x

Pfister, R. A., & Eppler, M. J. (2012). The benefits of sketching for knowledge management. *Journal of Knowledge Management*, *16*(2), 372–382. doi:10.1108/13673271211218924

Phillips, L. D. (1999). Group elicitation of probability distributions: are many heads better than one? In J. Shantaeu, B. Mellors, & D. Schum (Eds.), *Decision Science and Technology: Reflections on the contributions of Ward Edwards* (pp. 313–330). Norwell, MA: Kluwer Academic Publishers. doi:10.1007/978-1-4615-5089-1_17

Pierce, G. (1998). Developing new university faculty through mentoring. *The Journal of Humanistic Education and Development*, *37*(1), 27–38. doi:10.1002/j.2164-4683.1998.tb00403.x

Piwowar, H. A., Day, R. S., & Fridsma, D. B. (2007). Sharing detailed research data is associated with increased citation rate. *PLoS ONE*, *2*(3), e308. doi:10.1371/journal.pone.0000308 PMID:17375194

Plattener, S. (1998). A Most Ingenious Paradox: The Market for Contemporary Fine Art. *American Anthropologist*, *100*(2), 482–493. doi:10.1525/aa.1998.100.2.482

Polanyi, M. (1966). The Tacit Dimension. New York: Anchor Day Books.

Polanyi, M. (1958). *Personal Knowledge Towards a Post-critical Philosophy*. London: Routledge and Kegan Paul Ltd.

Polanyi, M. (1966). *The Tacit Dimension*. London: Routledge and Kegan Paul Ltd.

Polanyi, M. (1983). *The tacit dimension*. Gloucester, MA: Peter Smith Publisher. (Original work published 1966)

Pololi, L., & Knight, S. (2005). Mentoring faculty in academic medicine: A new paradigm? *Journal of General Internal Medicine*, *20*(9), 866–870. doi:10.1111/j.1525-1497.2005.05007.x PMID:16117759

Polson, D., Curtis, A., Vivalda, C., & Saunier, S. (2009). Process to Track the Evolving Perception of Risk During CO2 Storage Projects. In *Proceedings of Offshore Europe 2009*. SPE 124703.

Polson, D., & Curtis, A. (2010). Dynamics of uncertainty in geological interpretation. *Journal of the Geological Society*, *167*(1), 5–10. doi:10.1144/0016-76492009-055

Polson, D., & Curtis, A. (2015). Assessing individual influence on group decisions in geological carbon capture and storage problems. In P. Diviacco, P. Fox, C. Pshenichny, & A. Leadbetter (Eds.), *Collaborative Knowledge in Scientific Research Networks*. Hershey, PA: IGI Global.

Polson, D., Curtis, A., & Vivalda, C. (2012). The evolving perception of risk during reservoir evaluation projects for geological storage of CO2. *International Journal of Greenhouse Gas Control*, *9*, 10–23. doi:10.1016/j.ijggc.2012.02.010

Popova, O. V. (2011). *Political analysis and prediction*. St. Petersburg, Russian Federation: Aspect Press. (in Russian)

Popper, K. (1959). The logic of scientific discovery. New York: Basic Books.

Popper, K. (2004). Conjectures and refutations: the growth of scientific knowledge. (Reprinted. ed.). London: Routledge.

Popper, K. (1934). *The logic of scientific discovery*. New York, NY: Basic Books.

Popper, K. (1963). *Conjectures and refutations*. London, UK: Routledge.

Porter, J. H., Hanson, P. C., & Lin, C.-C. (2012). Staying afloat in the sensor data deluge. *Trends in Ecology & Evolution*, *27*(2), 121–129. doi:10.1016/j.tree.2011.11.009 PMID:22206661

Post, A. L., Wassenberg, T. J., & Passlow, V. (2006). Physical surrogates for macrofaunal distributions and abundance in a tropical gulf. *Marine & Freshwater Research*, *57*(5), 469. doi:10.1071/MF05182

Protégé. (n.d.). Retrieved from http://protege.stanford.edu

Pshenichny, C. A. (2013). Adjustment of the Event Bush Method to Chemical and Related Technological Tasks. In D. Mouromtsev, C. Pchenichniy, & D. Ignatov (Eds.), *Modeling States, Events, Processes and Scenarios* (pp. 74-81). Retrieved 18 March 2014 from http://iccs2013.hbcse.tifr.res.in/workshops/copy2_of_text.pdf

Pshenichny, C. A., & Kanzheleva, O. M. (2011). Theoretical foundations of the event bush method. In K. Sinha, L. Gundersen, J. Jackson, & D. Arctur (Eds.), Societal Challenges and Geoinformatics (pp. 139-165). Boulder, CO: GSA Special Paper 482. doi:10.1130/2011.2482(12)

Pshenichny, C. A., & Khrabrykh, Z. V. (2002). Knowledge Base of Formation of Subaerial Eruption Unit. In S. Leroy & I. Stuart (Eds.), *Environmental Catastrophes and Recovery in the Holocene (Abstracts)*. London: Brunel University. Retrieved 18 March 2014 from: http://atlas-conferences.com/cgi-bin/abstract/caiq-22

Pshenichny, C. A., & Mouromtsev, D. I. (2013). Representation of the Event Bush Approach in Terms of Directed Hypergraphs. In H.D. Pfeiffer, D.I. Ignatov, J. Poelmans, & N. Gadiraju (Eds.), Conceptual Structures for STEM Research and Education (pp. 287–298). Mumbai, India: Springer. doi:10.1007/978-3-642-35786-2_22

Pshenichny, C. A., Carniel, R., & Akimova, V. L. (2005). Decreasing the uncertainty of BBN technique by means of complex formal approach tovolcanological information treatment. *Geophysical Research Abstracts*, *7*, EGU05-A-01016.

Pshenichny, C. A., Carniel, R., & Diviacco, P. (2013). *Engineering of Dynamic Knowledge in Exact Sciences: First Results of Application of the Event Bush Method in Physics. ICCS 2013 - MSEPS 2013 – Modeling States*. Events, Processes and Scenarios.

Pshenichny, C. A., & Mouromtsev, D. I. (2015). Grammar of Dynamic Knowledge for Collaborative Knowledge Engineering and Representation. In P. Diviacco, P. Fox, C. Pshenichny, & A. Leadbetter (Eds.), *Collaborative Knowledge in Scientific Research Networks*. Hershey, PA: IGI Global.

Pshenichny, C. A., Nikolenko, S. I., Carniel, R., Sobissevitch, A. L., Vaganov, P. A., & Khrabrykh, Z. V. et al. 2008). The event bush as a potential complex methodology of conceptual modelling in the geosciences. In Sanchez-MarreJ. BejarJ.ComasJ.RizzoliA.GuarisoG. (Eds.), *International Congress on Environmental Modelling and Software*, (vol. 2, pp. 900-912). Barcelona, Spain: International Environmental Modelling and Software Society.

Pshenichny, C. A., Nikolenko, S. I., Carniel, R., Vaganov, P. A., Khrabrykh, Z. V., & Moukhachov, V. P. et al. (2009). The Event Bush as a Semantic-based Numerical Approach to Natural Hazard Assessment (Exemplified by Volcanology). *Computers & Geosciences*, *35*(5), 1017–1034. doi:10.1016/j.cageo.2008.01.009

Pudovkin, A. (1993). Citation Relationships Among Marine Biology Journals and Those in Related Fields. *Marine Ecology Progress Series*, *100*(1-2), 207–209. doi:10.3354/meps100207

Putnam, H. (1975). Mind, Language and Reality. In Philosophical Papers, (vol. 2). Cambridge, UK: Cambridge University Press. doi:10.1017/CBO9780511625251

Quin, L. (2013). *Extensible markup language (XML)*. World Wide Web Consortium. Retrieved from http://www.w3.org/XML/

Quinn, P., Hewett, C., Muste, M., & Popescu, I. (2010). Towards new types of water-centric collaboration. *Proceedings of the ICE-Water Management*, *163*(1), 39–51. doi:10.1680/wama.2010.163.1.39

Ragins, B. R., & Cotton, J. (1991). Easier said than done: Gender differences in perceived barriers to gaining a mentor. *Academy of Management Journal*, *34*(4), 939–951. doi:10.2307/256398

Ragins, B., & Cotton, J. L. (1999). Mentor functions and outcomes: A comparison of men and women in formal and informal mentoring relationships. *The Journal of Applied Psychology*, *84*(4), 529–550. doi:10.1037/0021-9010.84.4.529 PMID:10504893

Raisanen, T., & Oinas-Kukkonen, H. (2008). A system architecture for the 7C knowledge environment. *Information Modelling and Knowledge Bases XIX*, *166*, 217 Retrieved from http://www.epd.aau.dk/uploads/media/ejc_7c_information_system_architecture_final.pdf

Ramanan, R. A., Phillips, R. S., Davis, R. B., Silen, W., & Reede, J. Y. (2002). Mentoring in medicine: Keys to satisfaction. *The American Journal of Medicine*, *112*(4), 336–341. doi:10.1016/S0002-9343(02)01032-X PMID:11893387

Ramani, S., Gruppen, L., & Kachur, E. K. (2006). Twelve tips for developing effective mentors. *Medical Teacher*, *28*(5), 404–408. doi:10.1080/01421590600825326 PMID:16973451

Ramchandani, C. (1973). *Analysis of Asynchronous Concurrent Systems by Timed Petri Nets*. (PhD Thesis). Massachusetts Institute of Technology, Cambridge, MA.

RDF. (2014). Retrieved from http://www.w3.org/RDF/

Redman, C. L., Grove, J. M., & Kuby, L. H. (2004). Integrating social science into the long-term ecological research (LTER) network: Social dimensions of ecological change and ecological dimensions of social change. *Ecosystems (New York, N.Y.)*, *7*(2), 161–171. doi:10.1007/s10021-003-0215-z

Reeves, T. C. (2003). Storm clouds on the digital education horizon. *Journal of Computing in Higher Education*, *15*(1), 3–26. doi:10.1007/BF02940850

Reichman, O. J., Jones, M. B., & Schildhauer, M. P. (2011). Challenges and Opportunities of Open Data in Ecology. *Science*, *331*(6018), 703–705. doi:10.1126/science.1197962 PMID:21311007

Reill, P. H. (1986). Narration and Structure in Late Eighteenth-Century Historical Thought. *History and Theory*, *25*(3), 286–298. doi:10.2307/2505192

Reiss, H., & Kröncke, I. (2005). Seasonal variability of infaunal community structures in three areas of the North Sea under different environmental conditions. *Estuarine, Coastal and Shelf Science*, *65*(1-2), 253–274. doi:10.1016/j.ecss.2005.06.008

Renosh, P. R., Schmitt, F. G., Loisel, H., Sentchev, A., & Mériaux, X. (2014). High frequency variability of particle size distribution and its dependency on turbulence over the sea bottom during re-suspension processes. *Continental Shelf Research*, *77*, 51–60. doi:10.1016/j.csr.2014.01.024

Renzl, B. (2008). Trust in management and knowledge sharing: The mediating effects of fear and knowledge documentation. *Omega*, *26*(2), 206–220. doi:10.1016/j.omega.2006.06.005

Ribarić, S., Pavešić, N., & Zadrija, V. (2009). Intersection Search for a Fuzzy Petri Net-Based Knowledge Representation Scheme. Knowledge-Based and Intelligent Information and Engineering Systems (LNCS), (vol. 5711, pp. 1-10). Berlin: Springer.

Ribeiro, R., & Collins, H. (2007). The bread-making machine: Tacit knowledge and two types of action. *Organization Studies*, *28*(9), 1417–1433. doi:10.1177/0170840607082228

Rinner, C., Kessler, C., & Andrulis, S. (2008). The use of Web 2.0 concepts to support deliberation in spatial decision-making. *Computers, Environment and Urban Systems*, *32*(5), 386–395. doi:10.1016/j.compenvurbsys.2008.08.004

Roberts, J. (2000). From Know-how to Show-how? Questioning the Role of Information and Communication Technologies in Knowledge Transfer. *Technology Analysis and Strategic Management*, *12*(4), 429–443. doi:10.1080/713698499

Rodríguez, J. G., & Uriarte, A. (2009). Laser diffraction and dry-sieving grain size analyses undertaken on fine- and medium-grained sandy marine sediments: A note. *Journal of Coastal Research*, *25*, 257–264. doi:10.2112/08-1012.1

Rogalchuk, V., & Solomin, K. (2015). Perspectives of Use of Petri Nets in Collaborative Research. In P. Diviacco, P. Fox, C. Pshenichny, & A. Leadbetter (Eds.), *Collaborative Knowledge in Scientific Research Networks*. Hershey, PA: IGI Global.

Rollett, H. (2003). *Knowledge Management: Processes and Technologies*. Boston: Kluwer Academic Publishers. doi:10.1007/978-1-4615-0345-3

Romolini, M., Record, S., Garvoille, R., Marusenko, Y., & Stuart Geige, R. (2013). The Next Generation of Scientists: Examining the Experiences of Graduate Students in Network-Level Social-Ecological Science. *Ecology and Society*, *18*(3), art42. doi:10.5751/ES-05606-180342

Rorty, R. (1989). *Contingency, Irony and Solidarity*. Cambridge, UK: Cambridge University Press. doi:10.1017/CBO9780511804397

Rösseler, O., & Seaman, B. (2011). *Neosentience: The benevolence engine*. Intellect Ltd.

Rossi, P. (2009). *Tecnologia e costruzione di mondi*. Roma, Italy: Armando Editore.

Rozenberg, G., & Grzegorz. (1997). *Handbook of Graph Grammars and Computing by Graph Transformations* (vols. 1–3). Singapore: World Scientific Publishing.

Rueda, C., Bermudez, L., & Fredericks, J. (2009). The MMI Ontology Registry and Repository: A portal for marine metadata interoperability. *OCEANS 2009, MTS/IEEE Biloxi – Marine Technology for Our Future: Global and Local Challenges*. Retrieved from http://ieeexplore.ieee.org/stamp/stamp.jsp?tp=&arnumber=5422206&isnumber=5422059

Runge, A. K., Scherbaum, F., Curtis, A., & Roggelsen, C. (2013). An Interactive Tool for the Elicitation of Subjective Probabilities in Probabilistic Seismic-Hazard Analysis. *Bulletin of the Seismological Society of America*, *103*(5), 2862–2874. doi:10.1785/0120130026

Rüsen, J. (1996). Some Theoretical Approaches to Intercultural Comparative Historiography. *History and Theory*, *35*(4), 5–22. doi:10.2307/2505441

Sahay, B. S. (2003). Understanding trust in supply chain relationships. *Industrial Management & Data Systems*, *103*(8), 553–563. doi:10.1108/02635570310497602

Salas, F. R., Boldrini, E., Maidment, D. R., Nativi, S., & Domenico, B. (2012). Crossing the digital divide: An interoperable solution for sharing time series and coverages in Earth sciences. *Natural Hazards and Earth System Sciences*, *12*(10), 3013–3029. doi:10.5194/nhess-12-3013-2012

Sands, R. G., Parson, L. A., & Duane, J. (1991). Faculty Mentoring Faculty in a Public University. *The Journal of Higher Education*, *62*(2), 174–193. doi:10.2307/1982144

Savage, D., Maul, P., Benbow, S., & Walke, R. (2004). *A generic FEP database for the assessment of long-term performance and safety of the geological storage of CO2*. Retrieved from http://www.co2captureandstorage.info/docs/QuintessaReportIEA.pdf

Scandura, T. A. (1992). Menorship and career mobility: An empirical investigation. *Journal of Organizational Behavior*, *13*(2), 169–174. doi:10.1002/job.4030130206

Scardamaglia, M., & Bereiter, C. (2003). Knowledge Building. In *Encyclopedia of Education* (2nd ed., pp. 1370–1373). New York: Macmillan Publishing Company.

Schaap, D., & Lowry, R. (2010). SeaDataNet – Pan-European infrastructure for marine and ocean data management: Unified access to distributed data sets. *International Journal of Digital Earth*, *3*(1), 50–69. doi:10.1080/17538941003660974

Schimel, D., Hargrove, W., Hoffman, F., & MacMahon, J. (2007). NEON: A hierarchically designed national ecological network. *Frontiers in Ecology and the Environment*, *5*(2), 59–59. doi:10.1890/1540-9295(2007)5[59:NAHDNE]2.0.CO;2

Schindler, D. (2009). Lakes as sentinels and integrators for the effects of climate change on watersheds, airsheds, and landscapes. *Limnology and Oceanography*, *54*(6), 2349–2358. doi:10.4319/lo.2009.54.6_part_2.2349

Schlee, J. (1973). U.S. Geological Survey Professional Paper:*Atlantic continental shelf and slope of the United States – sediment texture of the northeastern part (vol. 529L)*. Washington, DC: United States Government Printing Office.

Schmidt, D. C., Stal, M., Rohnert, H., & Buschmann, F. (2013). *Pattern-Oriented Software Architecture, Patterns for Concurrent and Networked Objects* (Vol. 2). John Wiley & Sons.

Schmidt-Kloiber, A., Moe, S. J., Dudley, B., Strackbein, J., & Vogl, R. (2013). The WISER metadatabase: The key to more than 100 ecological datasets from European rivers, lakes and coastal waters. *Hydrobiologia*, *704*(1), 29–38. doi:10.1007/s10750-012-1295-6

Schnotz, W., & Kurschner, C. (2008). External and Internal Representations in the Acquisition and Use of Knowledge: Visualization Effects on Mental Model Construction. *Instructional Science: An International Journal Of The Learning Sciences*, *36*(3), 175–190. doi:10.1007/s11251-007-9029-2

Schreiber, G. (2000). *Knowledge Engineering and Management: The CommonKADS Methodology*. Cambridge, MA: MIT Press.

Searle, J. (1995). *The Construction of Social Reality*. New York: The Free Press.

Sears, J. (1993). Poster Session - a Bibliometric Analysis of the Marine Biotechnology Literature - a 90s Assessment. In *Aquatic Information Resources: Tools of Our Trade:Proceedings of the 18th Annual Iamslic Conference*, (pp. 147–147). Academic Press.

Self, W. (1993). *My Idea of Fun*. London: Bloomsbury Publishing.

Sellars, S., Nguyen, P., Chu, W., Gao, X., Hsu, K.-, & Sorooshian, S. (2013). Computational Earth Science: Big Data Transformed Into Insight. *Eos, Transactions, American Geophysical Union*, *94*(32), 277–278. doi:10.1002/2013EO320001

Semantic Web. (2014). Retrieved from http://www.w3.org/standards/semanticWeb/

Shadbolt, N., & Berners-Lee, T. (2008). Web science emerges. *Scientific American*, *299*(4), 76–81. doi:10.1038/scientificamerican1008-76 PMID:18847088

Shannon, C. E., & Weaver, W. (1949). *A mathematical model of communication*. Chicago, IL: University of Illinois Press.

Shaon, A., Callaghan, S., Lawrence, B., Matthews, B., Osborn, T., Harpham, C., & Woolf, A. (2012). Opening Up Climate Research: A Linked Data Approach to Publishing Data Provenance. *International Journal of Digital Curation*, *7*(1), 163–173. doi:10.2218/ijdc.v7i1.223

Sharratt, M., & Usoro, A. (2003). Understanding knowledge-sharing in online communities of practice. *Electronic Journal of Knowledge Management*, *1*(2), 187–196.

Shedroff, N. (1994). *Information interaction design: A unified field theory of design*. San Francisco: Vivid Studios.

Shepard, F. P. (1954). Nomenclature based on sand-silt-clay ratios. *Journal of Sedimentary Petrology*, *24*, 151–158.

Sherlock, G. (2000). Analysis of large-scale gene expression data. *Current Opinion in Immunology*, *12*(2), 201–205. doi:10.1016/S0952-7915(99)00074-6 PMID:10712947

Sheth, A. P., & Larson, J. A. (1990). Federated database systems for managing distributed, heterogeneous, and autonomous databases. *ACM Computing Surveys*, *22*(3), 183–236. doi:10.1145/96602.96604

Shneiderman, B. (1996). The eyes have it: A task by data type taxonomy for information visualizations. In *Proceedings ofIEEE Symposium on Visual Languages* (pp. 336-343). IEEE Computer Society Press. doi:10.1109/VL.1996.545307

Smith, M., Campbell, D., Polson, D., & Mackay, E. (2010). *CO_2 Aquifer Storage Site Evaluation and Monitoring*. Edinburgh, UK: Heriot Watt University.

Smith, S. G. (2009). Historical Meaningfulness in Shared Action. *History and Theory*, *48*(1), 1–19. doi:10.1111/j.1468-2303.2009.00482.x

Smykla, E., & Zippel, K. (2010). Literature Review: Gender and International Research Collaboration. *Report Prepared with Funding from NSF, OISE, 936970.*

Sneppen, K., Bak, P., Flyvbjerg, H., & Jensen, M. (1995). Evolution as a self-organized critical phenomenon. *Proceedings of the National Academy of Sciences of the United States of America*, *92*(11), 5209–5213. doi:10.1073/pnas.92.11.5209 PMID:7761475

Sniezek, J. A. (1992). Groups under uncertainty: An examination of confidence in group decision making. *Organizational Behavior and Human Decision Processes*, *52*(1), 124–155. doi:10.1016/0749-5978(92)90048-C

Soekijad, M., & Andriessen, E. (2003). Conditions for knowledge sharing in competitive alliances. *European Management Journal*, *21*(5), 578–587. doi:10.1016/S0263-2373(03)00107-5

Solomin, K., & Mandrik, M. (2015). Constructing historical knowledge through graphic boundary objects. In P. Diviacco, P. Fox, C. Pshenichny, & A. Leadbetter (Eds.), *Collaborative Knowledge in Scientific Research Networks*. Hershey, PA: IGI Global.

Song, D. (2009). The tacit knowledge-sharing strategy analysis in the project work. *International Business Research*, *2*(1), 83–85. doi:10.5539/ibr.v2n1p83

Soomai, S. S., MacDonald, B. H., & Wells, P. G. (2013). Communicating environmental information to the stakeholders in coastal and marine policy-making: Case studies from Nova Scotia and the Gulf of Maine/Bay of Fundy region. *Marine Policy*, *40*, 176–186. doi:10.1016/j.marpol.2013.01.017

Sorensen, A. A., Seary, A., & Riopelle, K. (2010). Alzheimer's Disease Research: A COIN Study Using Co-authorship Network Analytics.*1st Collaborative Innovation Networks Conference - Coins2009, 2*(4), 6582–6586.

Sorribas, J., Olive, J., Diviacco, P., & Bermudez, L. (2012). Representing oceanographic vessels by means of sensor web enablement standards. In *Proceedings of the First ACM SIGSPATIAL Workshop on Sensor Web Enablement*. ACM. doi:10.1145/2451716.2451717

Sowa, J. F. (1984). *Conceptual Structures: Information Processing in Mind and Machine*. Reading, MA: Addison-Wesley.

Specia, L., & Motta, E. (2007). Integrating folksonomies with the semantic web. In *Proceedings of the European Semantic Web Conference*. ESWC Press.

Spiridovich, A. I. (1962). *Velikaya voina i fevralskaya revolucia. 1917 – 1922*. Retrieved 18 March 2018 from http://www.rummuseum.ru/lib_s/spirid000.php (in Russian).

Srikanthan, G., & Dalrymple, J. F. (2002). Developing a Holistic Model for Quality in Higher Education. *Quality in Higher Education*, *8*(2), 215–224. doi:10.1080/1353832022000031656

Stafleu, J., Maljers, D., Gunnink, J. L., Menkovic, A., & Busschers, F. S. (2011). 3D modelling of the shallow subsurface of Zeeland, the Netherlands. *Netherlands Journal of Geosciences*, *90*, 293–310.

Stahl, G. (2000). A Model of Collaborative Knowledge-Building. In FishmanB.O'Connor-DivelbissS. (Eds.), *Fourth International Conference of the Learning Sciences* (pp. 70-77). Mahwah, NJ: Erlbaum.

Star, S. L., & Griesemer, J. R. (1989). Institutional ecology, translations and boundary objects: Amateurs and professionals in Berkeley s museum of vertebrate zoology. *Social Studies of Science*, *19*(3), 387–420. doi:10.1177/030631289019003001

Steininger, K., Ruckel, D., Dannerer, E., & Roithmayr, F. (2010). Healthcare knowledge transfer through a web 2.0 portal: An Austrian approach. *International Journal of Healthcare Technology and Management*, *11*(1), 13–30. doi:10.1504/IJHTM.2010.033272

Steinkellner, P. F., & Czerny, E. J. (2010, December). *Educating Managers for a Paradox World – Duality and Paradoxes in Management*. Paper presented at the International Conference on Management Learning 'Management Makes the World Go Around. Vienna, Austria.

Stenmark, D. (2001). Leveraging Tacit Organizational Knowledge. *Journal of Management Information Systems*, *17*(3), 9–24.

Stephens, S. (2012). From Tree to Map: Using Cognitive Learning Theory to Suggest Alternative Ways to Visualize Macroevolution Evolution. *Education and Outreach*, *5*(4), 603–618. doi:10.1007/s12052-012-0457-3

Stevenson, A. (Ed.). (2011). *EMODnet-Geology project final report*. Retrieved on December 30, 2013, from http://www.emodnet-geology.eu

Stevens, T., & Connolly, R. M. (2004). Testing the utility of abiotic surrogates for marine habitat mapping at scales relevant to management. *Biological Conservation*, *119*(3), 351–362. doi:10.1016/j.biocon.2003.12.001

Stewart, B. M., & Krueger, L. E. (1996). An evolutionary concept analysis of mentoring in nursing. *Journal of Professional Nursing*, *12*(5), 311–321. doi:10.1016/S8755-7223(96)80011-1 PMID:8871697

Stojanovic, L. (2004). *Methods and Tools for Ontology Evolution* (Doctoral dissertation, University of Karlsruhe). Retrieved 10 August 2014 from d-nb.info/1001606787/34?origin=publication_detail

Sturrock, J. (1986). *Structuralism*. London: Paladin.

Sveiby, K. E. (1997). *The new organizational wealth: Managing and measuring knowledge-based assets*. San Francisco, CA: Berret-Koehler.

Sveiby, K. E. (2001). A knowledge-based theory of the firm to guide in strategy formulation. *Journal of Intellectual Capital*, *2*(4), 344–358. doi:10.1108/14691930110409651

Sverrison, Á. (2001). Translation networks, knowledge brokers and novelty construction: Pragmatic environmentalism in Sweden. *Acta Sociologica*, *44*(4), 313–327. doi:10.1080/00016990152696402

Swap, W., Leonard, D., Shields, M., & Abrams, L. (2001). Using Mentoring and Storytelling to Transfer Knowledge in the Workplace. *Journal of Management Information Systems*, *18*(1), 95–114.

Syme, S. L. (2008). The science of team science - Assessing the value of transdisciplinary research. *American Journal of Preventive Medicine*, *35*(2), S94–S95. doi:10.1016/j.amepre.2008.05.017 PMID:18619409

Takagi, K., & Lin, H. (2011). Temporal dynamics of soil moisture spatial variability in the Shale Hills Critical Zone Observatory. *Vadose Zone Journal*, *10*(3), 832–842. doi:10.2136/vzj2010.0134

Takeuti, G. (1975). Proof Theory. Studies in Logic and the Foundations of Mathematics (vol. 81).New York: American Elsevier Publishing Co.

Tananbaum, G. (2013). *Implementing an open data policy: A primer for research funders*. Scholarly Publishing and Academic Resources Coalition. Retrieved from http://www.sparc.arl.org/sites/default/files/sparc-open-data-primer-final.pdf

Tansley, S., & Tolle, K. M. (Eds.). (2009). *The fourth paradigm: data-intensive scientific discovery*. Redmond, WA: Microsoft Research.

Taylor, W. A., & Wright, G. H. (2004). Organizational readiness for successful knowledge sharing: Challenges for public secto managers. *Information Resources Management Journal*, *17*(2), 22–37. doi:10.4018/irmj.2004040102

Tennison, J. (2010). *Creating URIs*. Retrieved April 16, 2014 from http://data.gov.uk/resources/uris

Tetzlaff, D., Carey, S., & Soulsby, C. (2013). Catchments in the future North: Interdisciplinary science for sustainable management in the 21st Century. *Hydrological Processes*, *27*(5), 635–639. doi:10.1002/hyp.9707

Tetzlaff, D., Soulsby, C., Bacon, P., Youngson, A., Gibbins, C., & Malcolm, I. (2007). Connectivity between landscapes and riverscapes—a unifying theme in integrating hydrology and ecology in catchment science? *Hydrological Processes*, *21*(10), 1385–1389. doi:10.1002/hyp.6701

Thompson, C. A., Robertson, W. D., & Greenberg, J. (2013). Where Have All the Scientific Data Gone? LIS Perspective on the Data-At-Risk Predicament. *College & Research Libraries*, crl13–crl500.

Thorpe, A. (2009). Environmental eScience. *Philosophical Transactions of the Royal Society A: Mathematical, Physical and Engineering Sciences*, *367*(1890), 801-802.

Thorp, J. H., Thoms, M. C., & Delong, M. D. (2006). The riverine ecosystem synthesis: Biocomplexity in river networks across space and time. *River Research and Applications*, *22*(2), 123–147. doi:10.1002/rra.901

Tippins, M. (2003). Implementing knowledge management in academia: Teaching the teachers. *International Journal of Educational Management*, *17*(7), 339–345. doi:10.1108/09513540310501021

Tolk, A. (2006). What comes after the Semantic Web - PADS Implications for the Dynamic Web. In *Proceedings of the 20th Workshop on Principles of Advanced and Distributed Simulation (PADS'06)*. IEEE. doi:10.1109/PADS.2006.39

Treleaven, L. (2004). A knowledge-sharing approach to organizational change: A critical discourse analysis. In H. Tsoukas & N. Mylonopoulos (Eds.), *Organizations as Knowledge Systems: Knowledge, Learning and Dynamic Capabilities* (pp. 154–180). New York, NY: Palgrave Macmillan.

Trentin, G. (2000). *Dalla formazione a distanza alla comunità di pratica attraverso l'apprendimento in rete.* Retrieved from http://www.tdmagazine.itd.cnr.it/files/pdfarticles/PDF20/Trentin.pdf

Trentin, G. (2001). *Dalla formazione a distanza all'apprendimento in rete.* Milano, Italy: Franco Angeli.

Tress, B., Tress, G., & Fry, G. (2006). Defining concepts and the process of knowledge production in integrative research. In B. Tress, G. Tress, G. Fry, & P. Opdam (Eds.), Landscape research to landscape planning (pp. 13-26). Academic Press. doi:10.1007/978-1-4020-5363-4_2

Tress, G., Tress, B., & Fry, G. (2005). Clarifying integrative research concepts in landscape ecology. *Landscape Ecology*, *20*(4), 479–493. doi:10.1007/s10980-004-3290-4

Trigg, R. H., & Bødker, S. (1994). From implementation to design: tailoring and the emergence of systematization in CSCW. In *Proceedings of the 1994 ACM conference on Computer supported cooperative work* (pp. 45-54). ACM. doi:10.1145/192844.192869

Tsvetkov, V. J. (2010). *Selected essays: Tsarskaya pamiatka, Zaveschanie imperatora, Proschalniy prikaz.* Retrieved 18 March 2014 from http://rusk.ru/st.php?idar=424636 (in Russian).

Turnitsa, C. (2005). *Extending the Levels of Conceptual Interoperability Model. In Proceedings, Summer Computer Simulation Conference 2005.* Philadelphia: Curran Associates, Inc.

Tutte, W. T. (1998). *Graph theory as I have known it.* New York: Oxford University Press.

Udden, J. A. (1914). Mechanical composition of clastic sediments. *Bulletin of the Geological Society of America*, *25*, 655–744.

UNESCO. (1987). GF3: A General Formatting System for geo-referenced data: Volume 2: technical description of the GF3 format and code tables. Paris: Intergovernmental Oceanographic Commission.

UNESCO. (2005). *Towards Knowledge Societies.* Paris, France: UNESCO Publishing.

Urban, E., Leadbetter, A., Moncoiffe, G., Pissiersens, P., Raymond, L., & Pikula, L. (2012). Pilot projects for publishing and citing ocean data. *Eos, Transactions, American Geophysical Union*, *93*(43), 425–426. doi:10.1029/2012EO430001

Vaccari, L., Craglia, M., Fugazza, C., Nativi, S., & Santoro, M. (2012). Integrative research: The EuroGEOSS experience. Academic Press.

van Dam, K., Li, D., Miller, S. D., Cobb, J. W., Green, M. L., & Ruby, C. L. (2011). Challenges in Data Intensive Analysis at Scientific Experimental User Facilities. In Handbook of Data Intensive Computing, (pp. 249-284). Springer. doi:10.1007/978-1-4614-1415-5_10

Van de Ven, A. H. (1976). On the nature, formation, and maintenance of relations among organizations. *Academy of Management Review*, *1*(4), 24–36. doi:10.2307/257722

Van der Meulen, M. J., De Lang, F. D., Maljers, D., Dubelaar, C. W., & Westerhoff, W. E. (2003). *Grondsoorten ent delfstoffen bij naam (Soil types and raw materials by name). Publicatiereeks Grondstoffen, 2003/16*. Delft, The Netherlands: Dienst Weg- en Waterbouwkunde. (in Dutch)

Van Dijk, T. A. G. P., Van Dalfsen, J. A., Van Overmeeren, R. A., Van Lancker, V., Van Heteren, S., & Doornenbal, P. J. (2012). Benthic habitat variations over tidal ridges, North Sea, the Netherlands. In P. T. Harris & E. Baker (Eds.), *Seafloor geomorphology as benthic habitat: GeoHAB atlas of seafloor geomorphic features and benthic habitats* (pp. 241–249). London: Elsevier Insights. doi:10.1016/B978-0-12-385140-6.00013-X

Van Fraassen. (1989). *Laws and Symmetry*. Oxford, UK: Oxford University Press.

Van Lancker, V., & Van Heteren, S. (Eds.). (2013). Standardisation and harmonisation in seabed habitat mapping: role and added value of geological data and information: Part A: Sediment characterisation. *Geo-Seas report 10.5A*. Retrieved on April 7, 2013, from http://www.geo-seas.eu/content/content.asp?menu=0040028_000000

Van Lancker, V., Moerkerke, G., Du Four, I., Verfaillie, E., Rabaut, M., & Degraer, S. (2012). Fine-scale geomorphological mapping of sandbank environments for the prediction of macrobenthic occurrences, Belgian part of the North Sea. In P. T. Harris & E. Baker (Eds.), *Seafloor geomorphology as benthic habitat: GeoHAB atlas of seafloor geomorphic features and benthic habitats* (pp. 251–260). London: Elsevier Insights. doi:10.1016/B978-0-12-385140-6.00014-1

van Renssen, A. (2005). *Gellish: A generic extensible ontological language*. (Doctoral dissertation). Delft University of Technology, Delft, The Netherlands.

Varisco, B. M. (1995). Paradigmi psicologici e pratiche didattiche con il computer. *TD Tecnologie Didattiche*, *7*, 57–68.

Veblen, T. B. (1899). *The Theory of the Leisure Class. An Economic Study of Institutions*. London: Macmillan Publishers.

Velden, T., Haque, A., & Lagoze, C. (2010). A new approach to analyzing patterns of collaboration in co-authorship networks: Mesoscopic analysis and interpretation. *Scientometrics*, *85*(1), 219–242. doi:10.1007/s11192-010-0224-6

Verfaillie, E. (2008). *Development and validation of spatial distribution models of marine habitats, in support of the ecological valuation of the seabed*. (PhD thesis), Ghent University, Ghent, Belgium.

Verfaillie, E., Van Lancker, V., & Van Meirvenne, M. (2006). Multivariate geostatistics for the predictive modelling of the surficial sand distribution in shelf seas. *Continental Shelf Research*, *26*(19), 2454–2468. doi:10.1016/j.csr.2006.07.028

Viator, R. E., & Scandura, T. A. (1991). A study of mentor-protégé relationships in large public accounting firms. *Accounting Horizons*, *5*, 20–30.

Vitolo, C., Elkhatib, Y., Reusser, D., Macleod, C. J. A., & Buytaert, W. (2015). Web technologies for environmental Big Data. *Environmental Modelling & Software, 63,* 185-198.

Vlissides, J., Helm, R., Johnson, R., & Gamma, E. (1995). *Design patterns: Elements of reusable object-oriented software* (Vol. 49). Addison-Wesley.

Walker, A., & Scott, K. (1993). Preparing for leadership in schools: The mentoring contribution. In B. J. Carwell & E. M. Carter (Eds.), *The Return of the Mentor: Strategies for Workplace Learning* (pp. 77–90). London: The Falmer Press.

Wang, W.-M., Peng, X., Zhu, G-N., Hu, J., & Peng, Y-H. (2014). Dynamic representation of fuzzy knowledge based on fuzzy Petri net and genetic-particle swarm optimization. Expert Systems with Applications, 41, 1369–1376.

Warner, A. (2002). A taxonomy primer. Austin, TX: School of Information, University of Texas at Austin. Retrieved from http://www.ischool.utexas.edu/~i385e/readings/Warner-aTaxonomyPrimer.html

Warren, R. (1970). Perceptual restoration of missing speech sounds. *Science*, *167*(3917), 392–393. doi:10.1126/science.167.3917.392 PMID:5409744

Wasburn, M. H., & LaLopa, J. M. (2003). Mentoring faculty for success: Recommendations based on evaluations of a program. *Planning and Changing, 34*(3-4), 250-264.

Wegehenkel, M. (2003). Longterm evaluation of land use changes on catchment water balance--a case study from North-East Germany. *Physics and Chemistry of the Earth Parts A/B/C*, *28*(33-36), 1281–1290. doi:10.1016/j.pce.2003.09.004

Weichselgartner, J., & Kasperson, R. (2010). Barriers in the science-policy-practice interface: Toward a knowledge-action-system in global environmental change research. *Global Environmental Change*, *20*(2), 266–277. doi:10.1016/j.gloenvcha.2009.11.006

Weinberger, D. (2002). *Small pieces loosely joined: a unified theory of the web*. New York, NY: Perseus Books.

Weinberger, D. (2011). *Too big to know: rethinking knowledge now that facts aren't facts, experts are everywhere, and the smartest person in the room is the room*. New York: Basic Books.

Weisberg, M., & Muldoon, R. (2009). Epistemic landscapes and the division of cognitive labor. *Philosophy of Science*, *76*(2), 225–252. doi:10.1086/644786

Wenger, E., McDermott, R., & Snyder, W. M. (2002). Cultivating communities of practice: A guide to managing knowledge. Harvard Business School Press.

Wenger, E. (1998). *Communities of Practice: Learning, Meaning and Identity*. Cambridge, UK: Cambridge University Press. doi:10.1017/CBO9780511803932

Wenger, E. C., & Snyder, W. M. (2000). Communities of practice: The organizational frontier. *Harvard Business Review*, *78*(1), 139–146. PMID:11184968

Wenger, E., McDermott, R., & Snyder, W. (2002). *Cultivating Communities of Practice: A Guide to Managing Knowledge*. Cambridge, MA: Harvard Business School Press.

Wentworth, C. K. (1922). A scale of grade and class terms for clastic sediments. *The Journal of Geology*, *30*(5), 377–392. doi:10.1086/622910

Wertheimer, M. (1938). *Gestalt theory*. Hayes Barton Press.

Whitcomb, M. E. (2004). The future of academic medicine: Caereer development of junior faculty. *Academic Medicine*, *79*(3), 195–196. doi:10.1097/00001888-200403000-00001 PMID:14985190

White, H. (1973). *Metahistory: The Historical Imagination in 19th-century Europe*. Baltimore, MD: Johns Hopkins University.

Whitely, W., Dougherty, T. W., & Dreher, G. F. (1991). Relationship of career mentoring and socioeconomic origin to managers' and professionals' early career progress. *Academy of Management Journal*, *34*(2), 331–351. doi:10.2307/256445

Whitley, G. G., & Oddi, L. F. (1998). Graduate student-faculty collaboration in research and publication. *Western Journal of Nursing*, *20*(5), 572–583. doi:10.1177/019394599802000505 PMID:9775739

Whitley, R. (2000). *The intellectual and social organization of the sciences*. Oxford, UK: Clarendon Press.

Wilbanks, J. (2009). I have seen the paradigm shift, and it is us. In T. Hey, S. Tansley, & K. M. Tolle (Eds.), *The fourth paradigm* (pp. 209–214). Microsoft Research.

Wildenborg, T., Leijnse, T., Kreft, E., Nepveu, M., & Obdam, A. (2004). *Longterm safety assessment of CO2 storage: the scenario approach*. Paper presented at the Seventh international conference on greenhouse gas control technologies. Vancouver, Canada.

Wildenborg, T., Leijnse, A., Kreft, E., Nepveu, M., Obdam, A., & Orlic, B. (2005). Risk assessment methodology for CO2 storage: the scenario approach. In *Carbon dioxide capture for storage in deep geologic formations*. Amsterdam: Elsevier.

Wilkesmann, M., & Wilkesmann, U. (2011). Knowledge transfer as interaction between experts and novices supported by technology. *VINE: The Journal of Information and Knowledge Management Systems*, *41*(2), 96–112. doi:10.1108/03055721111134763

Wilkesmann, U., Fischer, H., & Wilkesmann, M. (2009). Cultural characteristics of knowledge transfer. *Journal of Knowledge Management*, *13*(6), 464–477. doi:10.1108/13673270910997123

Willems, W., Goethals, P., Van den Eynde, D., Van Hoey, G., Van Lancker, V., Verfaillie, E. et al. (2008). Where is the worm? Predictive modelling of the habitat preferences of the tube-building polychaete *Lanice conchilega*. *Ecological Modelling*, *212*(1-2), 74–79. doi:10.1016/j.ecolmodel.2007.10.017

Williamson, C. E., Saros, J. E., & Schindler, D. W. (2009). Sentinels of Change. *Science*, *323*(5916), 887–888. doi:10.1126/science.1169443 PMID:19213905

Willis, Z. (2012). US Integrated Ocean Observing System (IOOS®) – Delivering benefits to science and society. *OCEANS 2012, Yeosu*. Doi: 10.1109/OCEANS-Yeosu.2012.6263613

Wilson, M. (2012). Meeting a scientific facility provider's duty to maximise the value of data. In *DataCite Summer Meeting, Digital Research Data in Practice (DataCite2012)*. Retrieved from http://purl.org/net/epubs/work/62852

Winder, N. (2003). Successes and problems when conducting interdisciplinary or transdisciplinary (= integrative) research. In B. Tress, G. Tress, A. van der Valk & G. Fry (Eds.), Potential and Limitations of Interdisciplinary and Transdisiplinary Landscape Studies (pp. 74-90). Wageningen: Delta Series 2. Alterra Green World Research and Wageningen University.

Wind, J., & Main, J. (1999). *Diving change*. New York: The Free Press.

Witkin, H. A., Oltman, P.K., Raskin, E. & Karp, S.A. (1971). *A manual for the embedded figures tests*. Consulting Psychologists Press.

Witkin, H. A., Moore, C. A., Goodenough, D. R., & Cox, P. W. (1977). Field-dependent and field-independent cognitive styles and their educational implications. *Review of Educational Research*, *47*(1), 1–64. doi:10.3102/00346543047001001

Wocial, L. D. (1995). The role of mentorsin promoting integrity and preventing scientific misconduct in nursing research. *Journal of Professional Nursing*, *11*(5), 276–280. doi:10.1016/S8755-7223(05)80008-0 PMID:7593971

Wolski, M. P., & Richardson, J. P. (2010). Moving researchers across the eresearch chasm. *Ariadne*, (65): 1–10. Retrieved from http://www.ariadne.ac.uk/issue65/wolski-richardson/

Woo, G. (1999). *The mathematics of natural catastrophes*. London, UK: Imperial College Press; doi:10.1142/9781860943867

World Wide Web Consortium (W3C). (n.d.). *Semantic web*. Retrieved 1/28/2014 from http://www.w3.org/standards/semanticweb

Wu, J. G., & Loucks, O. L. (1995). From balance of nature to hierarchical patch dynamics: A paradigm shift in ecology. *The Quarterly Review of Biology*, *70*(4), 439–466. doi:10.1086/419172

Wulf, W. A. (1989). The national collaboratory: A white paper. In T*owards a national collaboratory: Report of an invitational workshop at the Rockefeller University*. Washington, DC: National Science Foundation.

Xiao-Zhong, L. (2003). A Kind of Fuzzy Petri Nets and its Dynamic Characteristics. *Acta Simulata Systematica Sinica,* 2003-S1.

Yang, C. W., Raskin, R., Goodchild, M., & Gahegan, M. (2010). Geospatial Cyberinfrastructure: Past, present and future. *Computers, Environment and Urban Systems*, *34*(4), 264–277. doi:10.1016/j.compenvurbsys.2010.04.001

Yang, E., Matthews, B., & Wilson, M. (2013). Enhancing the Core Scientific Metadata Model to Incorporate Derived Data. *Future Generation Computer Systems, 29*(2), 612–623. doi:10.1016/j.future.2011.08.003

Ye, M., Pohlmann, K. F., & Chapman, J. B. (2008). Expert elicitation of recharge model probabilities for the Death Valley regional flow system. *Journal of Hydrology (Amsterdam), 354*(1-4), 102–115. doi:10.1016/j.jhydrol.2008.03.001

Yeung, D. S., & Tsang, E. C. C. (1994). Fuzzy knowledge representation and reasoning using Petri nets. *Expert Systems withApplications, 7*(2), 281–289. doi:10.1016/0957-4174(94)90044-2

Yi, J. (2006). Externalization of tacit knowledge in online environments. *International Journal on E-Learning, 5*(4), 663–674.

Yoder, L. (1990). Mentoring: A concept analysis. *Nursing Administration Quarterly, 15*(1), 9–19. doi:10.1097/00006216-199001510-00005 PMID:2234538

Yudelson, M., Gavrilova, T., & Brusilovsky, P. (2005). Towards User Modeling Meta Ontology. In L. Ardissono, P. Brna, & A. Mitrovic (Eds.), *UM 2005 (LNAI), (vol. 3538*, pp. 448–452). Berlin: Springer-Verlag.

Zablith, F. (2008). Dynamic ontology evolution. In *Proceedings of the International Semantic Web Conference (ISWC)Doctoral Consortium*. Karlsruhe, Germany: Springer.

Zacharias, S., Bogena, H., Samaniego, L., Mauder, M., Fuß, R., & Pütz, T. et al. (2011). A Network of Terrestrial Environmental Observatories in Germany. *Vadose Zone Journal, 10*(3), 955–973. doi:10.2136/vzj2010.0139

Zack, M. (1998). What knowledge problems can information technology help to solve. In *Proceedings of the Fourth Americas Conferences on Information Systems*, (pp. 644-646). Academic Press.

Zappa F. (1979). *Packard Goose on Joe's Garage* [LP]. Zappa Record, CBS

Zeleny, M. (2005). *Human Systems Management: Integrating Knowledge, Management and Systems*. World Scientific. doi:10.1142/4929

Zhang, W., Qian, W., & Ho, Y.-S. (2009). A bibliometric analysis of research related to ocean circulation. *Scientometrics, 80*(2), 305–316. doi:10.1007/s11192-007-1863-0

Zhu, F., Turner, M., Kotsiopoulos, I., Bennett, K., Russell, M., & Budgena, D. et al. (2004). Dynamic data integration using web services. In *Proceedings of the IEEE International Conference on Web Services (ICWS'04)* (pp. 262-269). IEEE.

Ziegler, C.-N., & Lausen, G. (2005). Propagation models for trust and distrust in social networks. *Information Systems Frontiers, 7*(4/5), 337–358. doi:10.1007/s10796-005-4807-3

Žižek, S. (1991). *Looking Awry. An introduction to Jacques Lacan through Popular Culture*. Cambridge, MA: MIT Press.

Zoschenko, M.M. (1933, July 3). *Kakie u menya byli professii*. Literaturnyi Leningrad (in Russian).

About the Contributors

Paolo Diviacco is a geophysicist at Istituto Nazionale di Oceanografia e di Geofisica Sperimentale (OGS). He holds a PhD in exploration geophysics, and has 20 years of experience in seismic data processing. He maintained always an interest in computer science and programming, developing software in the field of geophysics and Web-based data management. At the same time, he has always been interested in the philosophical and sociological aspects of scientific production in general and in the geo-sciences in particular. He has been active in developing ideas and Web-based systems that could support scientists while collaborating and foster the collaborative attitude among research institutions. In this context, he worked within several international projects, such as, for example, FP7 ODIP, FP7 Geo-Seas, FP7 Emodnet, FP7 Eurofleets, and the Antarctic Seismic Data Library.

Peter Fox is Tetherless World Constellation Chair, Professor of Earth and Environmental Science, Computer Science and Cognitive Science, and Director of the Information Technology and Web Science Program at Rensselaer Polytechnic Institute. Fox has a BSc (hons) and PhD in Applied Mathematics (Physics and Computer Science) from Monash University. His research covers the fields of ocean and environmental informatics, computational and computer science, semantic data frameworks, and solar and solar-terrestrial physics. The results are applied to large-scale distributed data science investigations. Fox is President of the Federation of Earth Science Information Partners (ESIP), chair of the International Union of Geodesy and Geophysics Union Commission on Data and Information, and serves on the editorial boards of many prominent Earth and space science informatics journals. In 2012, Fox was awarded the European Geoscience Union, Ian McHarg/Earth and Space Science Informatics Medal, and ESIP's Martha Maiden Lifetime Achievement award for service to the Earth Sciences Information communities.

Cyril Pshenichny is the Geognosis Project leader and an Assistant Professor at the Department of Computer System Design and Security, ITMO University. His PhD defended in 1998 was on composition and history of ancient volcanic rocks. His research path brought him to the issues of artificial intelligence and knowledge engineering from the geoscience, where he first developed the method of event bush now acquiring much wider application. He has authored and coauthored dozens of research papers, gave lectures and conducted seminars throughout the world. Currently, he is a co-editor of the *Collaborative Knowledge in Scientific Research Networks* monograph to be published by IGI Global.

Adam Leadbetter is a data scientist at the British Oceanographic Data Centre, focusing on the development of controlled vocabularies for the marine science domain. He is educated as an oceanographer, including research into coastal sediment transport, but maintained a parallel interest in computer science throughout that education. In addition to his work in the realms of vocabularies and Linked Data, he has a particular interest in developing data into a citable entity for the purposes of scientific reproducibility. He is also co-chair of the technical committee of the International Coastal Atlas Network, defining the work plan to build an interoperable system of Web atlases to link information systems across jurisdictional boundaries.

* * *

Robert Arko is a senior developer in the Lamont-Doherty Earth Observatory (LDEO) Geoinformatics Center, and technical director of the Rolling Deck to Repository (R2R) Program. His background is in Computer Science, specializing in Digital Curation and Semantic Web technologies. He specializes in designing data, information, and knowledge systems for environmental sensors and samples from oceanographic research vessels, and has deployed on 20 field expeditions over the past 20 years.

Roberto Biloslavo is a Full Professor of Management at the University of Primorska, Faculty of Management. His research work is focused on management, leadership, knowledge management, strategic management, wisdom, and sustainable development. He is a former Vice-Rector for Academic Affairs at the University of Primorska. Besides teaching and researching, he consults to different domestic and international companies about vision and mission statement development, knowledge management and leadership development.

Enrico Boldrini has a Master's degree in Computer Science at the University of Florence and is currently a PhD student at the University of Calabria. He is working at the Earth and Space Science Informatics Laboratory (ESSI-Lab) of the National Research Council of Italy (CNR) since 2007, in the context of several projects, such as FP7 ODIP, FP7 SeaDataNet II, FP7 EarthServer, FP7 EuroGEOSS, FP7 GEOWOW, CNR GIIDA, the Italian Project of Interest NextData, ESA HMA-T, ESA ProdTrees, NSF BCube, GEOSS Common Infrastructure (GCI). He is a member of the Italian Arctic Data Center (IADC) working group, European Geosciences Union (EGU), Open Geospatial Consortium (OGC). His work focuses on enabling inter-disciplinary interoperability in the context of Earth Science. His main activities include development and management of discovery and access broker frameworks (e.g. GI-cat, GI-axe, the GEO-DAB), drafting of metadata standards (e.g. SeaDataNet ISO 19115 profile, NetCDF EO conventions), international collaborations.

Vasily Bunakov is a researcher in the Scientific Computing Department of STFC. He obtained his Master's degree from Moscow Institute of Physics and Technology and worked in the information and IT departments of Russian and European high energy physics laboratories, as well as for the London branch of Deutsche Bank where he was a part of a digital content management programme. Vasily has a versatile IT background in system administration, software development, software testing, service management, and project management. He has contributed to a number of EU projects for managing research data, Public Sector Information, and cloud services. His current research interests are focused on long-term digital preservation, data modelling, and Linked Open Data.

Alessandro Busato is a developer and system administrator at Istituto Nazionale di Oceanografia e di Geofisica Sperimentale (OGS) with 16 years of experience as IT and Web analyst and developer. He has always maintained an interest in e-Science through his previous experiences with NewTech at Trieste Area Science Park, developing Java Web applications, and within several European projects at Trieste Elettra Synchrotron. Currently, he is working within several EU collaborative projects, such as, for example, the FP7 Eurofleet project developing a real time metadata logger, within the GeoSeas project developing high resolution Web viewing and processing facilities for seismic data, and within OGS developing a portal to access legacy geophysical data. Previously, he worked at Insiel Spa developing applications for e-Government. He is skilled in Linux/MacOS system administration, Java, Web technologies and platforms, C, Javascript including kits as JQuery Openlayers, XML, and PostgresQL/Postgis.

Cynthia "Cyndy" Chandler is an Information Systems professional at Woods Hole Oceanographic Institution. She has over 30 years of experience integrating information systems technology with oceanography, and has participated in 18 research cruises. The primary focus of her current research is on improving and enhancing data management activities throughout the full research data life cycle, from proposal through preservation. Recent projects include exploration of Semantic Web technologies aimed at connecting resources to improve discovery, facilitating interoperability, and ultimately, making ocean science research data, especially data acquired during research cruises, more accessible to the research community. She is a Principal Investigator with the Biological and Chemical Oceanography Data Management Office, the Rolling Deck to Repository project, and the OceanLink project, all of which are funded by the US National Science Foundation.

Andrea Corleto, graduated in Literature and Philosophy at University of Rome "La Sapienza," is a journalist and researcher. From 2005, he worked on the development of APAT Website (Italian Agency for Environmental Protection). From 2010, he started his collaboration with the Department of Earth Sciences (University of Rome "La Sapienza") and ENEA (Italian National Agency for New Technologies, Energy, and Sustainable Economic Development) on research and testing of advanced technologies for the dissemination and communication of scientific data on Carbon Capture and Storage technologies. Currently employed in ENEA's Technical Unit for Information Systems Development and ICT, he carries out research activities on e-learning and Web content management. He is a member of SIe-L (Italian Society of e-Learning). His main research interests include educational technologies and online and collaborative learning.

Max Craglia, since 2005, works at the Digital Earth and Reference Data Unit, European Commission – Joint Research Centre. The Unit is responsible for the technical coordination of the INSPIRE Directive, creating an infrastructure for Spatial Information in Europe to support environmental policy. Max is responsible for the development of the INSPIRE specifications for Metadata, and for research on the socio-economic assessment of INSPIRE. He coordinates research projects related to the interoperability of multi-disciplinary e-infrastructures that contribute to the Global Earth Observation System of

Systems. Recent activities have taken Digital Earth as a framework for evolving current e-infrastructures and integrate official data with real time data coming from citizens and sensors. This approach is being applied to develop novel indicators of Quality of Life in urban areas to include both quantitative data from sensors and qualitative information from the public. For a selection of recent publications see http://inspire-forum.jrc.ec.europa.eu/pg/pages/view/29954/publications-articles-in-refereed-journals.

Andrew Curtis is the Professor of Mathematical Geoscience at the University of Edinburgh. His research includes wave theory, geophysical imaging of the Earth's interior, inversion, and experimental design theory, and expert elicitation methods. He has a PhD in Geophysics from the University of Oxford in UK, and before joining the University of Edinburgh, he carried out scientific research at the University of Utrecht in The Netherlands, and at Schlumberger Cambridge Research in the UK.

Rogier De Langhe is a post-doctoral research fellow at the Centre for Logic and Philosophy of Science (Ghent University) in Ghent, Belgium. He is the winner of the 2010 Graduate Student Essay Award of the Philosophy of Science Association. His research focuses on the dynamics of scientific change.

Janet Fredericks is an information systems specialist at the Woods Hole Oceanographic Institution, where she is responsible for the information systems and operations of the Martha's Vineyard Coastal Observatory (MVCO). The MVCO is a cabled observatory with network-enabled access to sensors that have been delivering meteorological and oceanographic data in real-time since 2001 (http://www.whoi.edu/mvco). She is in the Coastal Ocean Fluid Dynamics Lab of the Applied Ocean Physics & Engineering Department. In addition, she is serving on the Data Management and Communications Steering Team for the Interagency Ocean Observation Committee (http://www.iooc.us) and is on the board of advisors of the QARTOD project (Quality Assurance in Real Time Ocean Data) of the Integrated Ocean Observing System (http://www.ioos.noaa.gov).

Tatiana Gavrilova is a Head of Information Technologies in Management Department at Graduate School of Management (GSOM) in St. Petersburg University. She is a Head of St. Petersburg Division of Russian Association for Artificial Intelligence (RAAI). Dr. Gavrilova has more than 150 publications in the field of knowledge engineering and knowledge management; she is the author of 5 books (in Russian). Her research interest deals with modern issues of knowledge management such as knowledge codification and structuring using the cognitive approach. She is a leader of several research projects on ontology development and cognitive computing. In 1998 and 2004-2005, she was a Fulbright Visiting Senior Research Scholar and Lecturer in Penn State University and University of Pittsburgh (USA). From 2000, she is an IT-management consultant for one of St. Petersburg leading consulting companies. Dr. Gavrilova has lecturing experience in English at the Universities and business schools of France (HEC), Finland (Aalto et al.), Italy (University of Milano), Poland (Warsaw and Gdansk), and Estonia (Tartu University). Currently, she is teaching the courses on Knowledge Engineering and Knowledge Management for executive and master level in GSOM.

Katjuša Gorela is a graduate in Political Science-Public Administration at the Faculty of Social Sciences and in Italian Language and Literature at the The Faculty of Arts, University of Ljubljana (Slovenia). She is currently a PhD student at the Faculty of Education, University of Primorska. She focuses her research interests on knowledge, education, and pedagogical practice.

Catherine Jones is an experienced project manager and researcher in the information management field with both computer science and library qualifications. She was responsible for the development and implementation of the STFC ePubs Institutional Repository. Her research interests include policy and context for data preservation linking data and publications and versioning of digital objects in institutional repositories. She is the author of a book titled *Institutional Repositories: Context and Culture in an Open Access Environment*. Her current portfolio includes a discovery portal for data from longitudinal studies, preservation policy creation tool through the SCAPE project, and work on context for research objects. She is currently a Databib editor and also reviews for the journal *VINE*.

Irina Leshcheva is Senior Lecturer in Graduate School of Management (GSOM), Saint-Petersburg University. Originally a mathematician, she has worked in the IT industry as a project leader (Artificial Life Co., Siemens, Tercom). Her research spans knowledge engineering, conceptual modelling, cognitive sciences, and ontology engineering. She teaches the courses on Business Simulation, Web Technologies, and Information Management. She has more than 20 scientific publications.

Mayte Lopez-Ferrer, Researcher at INGENIO, a joint Institute of the Spanish Council for Scientific Research (CSIC), and the Polytechnic University of Valencia (UPV), has a degree in Documentation from the Carlos III University of Madrid and PhD in Documentation from the University of Valencia. Previously, she worked at the López Piñero Institute for the History of Science and Medicine, a joint center of the Spanish Council for Scientific Research (CSIC), and the University of Valencia (UV) dedicated to research around historical and social studies on medicine and science. She has been lecturer of University of Valencia and of the Open University of Catalonia (UOC) in their respective Degrees of Documentation. Her research topics are bibliometrics and social network analysis, mainly applied to scientific activity.

Roy Lowry is the technical director of the British Oceanographic Data Centre with over 30 years of experience in marine data management. He was recently awarded the International Oceanographic Data and Information Exchange Achievement Award for his contributions to file format standardisation and semantic technology developments in the field of marine science.

Kit Macleod is a catchment scientist based at the James Hutton Institute in the UK. His expertise is in the theory and practice of understanding temperate hydrological catchments to improve their management. Kit's interest in the use of integrative research to support decision-making was developed during his first post-doctoral position work as a risk analyst with Simon Pollard at the Environment Agency of England and Wales. Since that formative experience, Kit has developed an international reputation for furthering the theory and practice of catchment systems-based integrative research spanning the natural, social, and computational sciences.

Maria V. Mandrik is a senior researcher and the head of Department of Fund Processing, Acquisition, and Institutional Archives of St. Petersburg Branch of the Archive of the Russian Academy of Sciences. She received her Master's and PhD degrees from the Department of Russian History, Faculty of History, St. Petersburg State University, in 1997 and 2000, respectively. She has worked in the Archive since 2002. Her research interests include various aspects of Russian history and related methodological concerns. She has authored and coauthored over 40 research papers and written introductions to editions of works of Russian historians of the 19th century (M. Longinov, M. Liubavsky, D. Nasonov, and N. Rubinstein).

Giuseppe M. R. Manzella worked in the National Agency for New Technologies, Environment, and Sustainable Development and National Research Council. His working experiences in physical oceanography include modelling, data analysis, and information systems. He has participated in national and international oceanographic programs supported by the Italian Ministry of Environment, Italian Ministry of Research, UNESCO IOC, European Commission, and the European Space Agency. He has been involved also in the definition of goals and objectives of the Topic Centre on Marine and Coastal Environment of the European Environment Agency (1995 - 1997). He has been Chairman of the Italian Oceanographic Commission and, in such role, represented Italy in the UNESCO-IOC initiatives and meetings. He is the co-author/author of 45 peer-reviewed papers.

Brian Matthews is leader of the Research Data Group within the STFC's Scientific Information Group. His main focus is in research and development into data management technology to support the scientific programmes of STFC, including the large-scale research facilities. He has extensive experience in metadata for research data cataloguing, exchange, and integration in digital preservation and in data publication. Further, he has undertaken research work in Semantic Web and Linked Data in distributed systems and in data modelling, and has authored over 150 peer-reviewed publications.

Paolo Mazzetti holds a degree in Electronic Engineering and began working on active microwave remote sensing and Earth Observation data sharing. He taught Telematics and Communication Networks at the University of Florence, Italy, for 10 years. He joined CNR in 2005 as research scientist, and he is currently the scientific coordinator of the "Geospatial information and ENvironmental knowledge Sharing" (GENS) Working Group at the Institute for Atmospheric Pollution Research (CNR-IIA), and deputy coordinator of the Earth and Space Science Informatics Laboratory (ESSI-Lab) of CNR-IIA. His main research interest is in the design of architectures for advanced e-infrastructures supporting multidisciplinary applications. He carries out his research activity in the context of several national, European, and international projects and programmes related to geospatial data sharing initiatives such as GEO/GEOSS, INSPIRE, and Copernicus.

Dmitry Mouromtsev is the Head of Intellectual Systems Lab and an Assistant Professor at the Department of Computer System Design and Security, ITMO University, St. Petersburg, Russia. His research interests focus on information systems, artificial intelligence, enterprise resource planning systems, and knowledge management. He received his PhD from NRU ITMO in 2003. He is the General Chair of International Conference on Knowledge Engineering and Semantic Web (KESW) and an editor of Springer CCIS Volume No. 394. His book, *Intelligent Systems in Management*, was published in 2008 in Russian. Much of his research has studied artificial intelligence, semantic technologies, and their use in applications.

Stefano Nativi is Head of the Florence Division of Institute of Atmospheric Pollution Research (CNR-IIA) of the National Research Council of Italy. He coordinates the activities of the Earth and Space Science Informatics Laboratory (ESSI-lab). His research includes multi-disciplinary interoperability, research infrastructures, and brokering technologies for Earth and Space Sciences. He co-founded the Earth and Space Sciences Informatics (ESSI) Division of the European Geosciences Union (EGU) service for five years as president. He is member of the GEO Infrastructure Implementation Board (IIB) and served as co-chair of the GEO Science & Technology Committee for three years. He is part of the INSPIRE (Infrastructure for Spatial Information in the European Commission) Drafting Teams and co-chair of the OGC Earth Systems Science Domain WG and netCDF-CF Standardization WG. He has taught at the Universities of Padua (Italy), Jena (Germany), and Florence (Italy). He is member of the Open Access WG of Science Europe.

Debbie Polson is Postdoctoral Research Associate at the University of Edinburgh. Her current research is on understanding observed changes in the global hydrological cycle and the role of anthropogenic activity in driving changes. She has PhD in Meteorology from the University of Edinburgh Centre for Ecology and Hydrology, and has previously worked on risk assessment in subsurface carbon capture and storage as part of the CO2 Aquifer Storage Site Evaluation and Monitoring (CASSEM) project.

Vyacheslav Rogalchuk is a PhD student and part-time lecturer at the Saint-Petersburg State Transport University, Russia. He graduated from SPbSTU in 2004 with a diploma in Information Systems Engineering. His current research interests include knowledge engineering, software development and lifecycle management, Semantic Web, and Linked Data technologies. He teaches courses on database programming, data structures, and computation theory.

Adam Shepherd is a software engineer at Woods Hole Oceanographic Institution and co-chair of the Drupal Working Group for the Federation of Earth Science Information Partners. He has been an invited speaker at the American Geophysical Union Fall Meeting, presented at Drupalcon North America, and co-authored an essay on oceanographic Linked Data for the *Journal of Ocean Technology*. Adam holds a BS in Computer Science and a Minor in Business Administration from Northeastern University where he was awarded the 2001 Co-op of the Year for his work at the Woods Hole Oceanographic Institution. With 13 years of experience building Web-enabled software, he concentrates on using science-driven use cases for improving ocean science research capabilities. Currently, his research focus has been on semantically enabled Linked Data applications for vocabulary alignment, data validation, and improved data discovery.

Konstantin Solomin (born April 23, 1989) holds a bachelor of Computer Science from St. Petersburg National Research University of Information Technologies, Mechanics, and Optics (ITMO) and is currently working towards his Master's degree. He is also employed in Web development as a front-end application developer in one of St. Petersburg' software companies. He has coauthored several research papers and conference abstracts. Konstatin's Master's project combines program engineering with Bush Event theory basics for building applications that could visually represent and validate data entered as a group of event nodes. Other scientific interests include Petri nets, graph theory, and application of expert knowledge in various fields of study. Professional interests on the other hand involve building RESTful applications and exploring new Javascript frameworks.

Veronica Tomassetti was born in 1975 in Avezzano (AQ). She is a graduate in Sociology (University of Rome "La Sapienza"), holds Master's Degrees in "E-Learning: Methods, Techniques, and Application" (University Tor Vergata) and in Counseling. Since 2012, she has been a researcher in ENEA (Italian National Agency for New Technologies, Energy, and Sustainable Economic Development), is currently employed in the Technical Unit for Information Systems Development and ICT. Her main research interests include the development and application of Learning Management Systems (LMS) and Content Management Systems (CMS), and the production of educational content by means of authoring software and video communication. She is a member of SIe-L (Italian Society of e-Learning).

Sytze van Heteren (PhD Boston University, USA) is a coastal and marine geologist at the Geological Survey of The Netherlands. For more than 25 years, he has conducted and supervised coastal projects in The Netherlands and in the USA. Key areas of expertise include biogeomorphology of shallow siliciclastic seas, coastal and marine sedimentology, coastal morphodynamics, and applied Quaternary geology. He has represented the Survey in various national and international research consortiums, coordinated the coastal-zone and shallow-marine mapping program, analyzed the impact of storm surges on coastal barriers, and helped to build bridges between marine geology, ecology, and archeology.

Vera Van Lancker (PhD Ghent University, Belgium) is a marine geologist at the Royal Belgian Institute of Natural Sciences, Operational Directorate Natural Environment, and visiting professor at Ghent University. Her research focuses on sediment and morphodynamics, seabed habitats, geological resources, and the impact of human activities (aggregate extraction, dredging and disposal of dredged material, windmill farms, beach nourishment), targeting optimized sustainable exploitation of marine environments. As coordinator of several nationally funded projects and task leader in international projects, she has an extensive track record in seabed and water-column mapping and modelling, and in the integration of multidisciplinary datasets. She has been actively involved in the valorization of geological and geophysical data for a wide user community (e.g., EU-FP7 Geo-Seas, DG MARE EMODnet-Geology), publishing 200+ papers and reports (36 peer-reviewed) on various applied aspects of marine geological research.

Index

E

F

G

H

I

J

K

L

M

N

O

P

R

S

T

U

V

W

www.ingramcontent.com/pod-product-compliance
Lightning Source LLC
LaVergne TN
LVHW060353300926
849162LV00006B/1259

* 9 7 8 1 4 6 6 6 6 5 6 7 5 *